SEINFELD, John H. Air pollution; physical and chemical fundamentals. McGraw-Hill, 1975. 523p il tab 74-4296. 22.50. ISBN 0-07-056042-0. C.I.P.

In this rapidly expanding field of knowledge, in which several minor works have made their recent appearance, this book stands out as providing a truly in-depth treatment of air pollution chemistry, atmospheric transport processes, combustion sources, and control methods. It provides a comprehensive mathematical approach to the basic issues in both science and engineering that underlie the air pollution problem. Arrangement of topics is realistic and well balanced, and the textual style is easy to read. Seinfeld's book is very suitable as a text for upper-division undergraduate or graduate students. Much reference material is included. Definitely recommended for library purchase.

AIR POLLUTION
PHYSICAL AND CHEMICAL
FUNDAMENTALS

McGRAW-HILL
BOOK COMPANY
New York
St. Louis
San Francisco
Düsseldorf
Johannesburg
Kuala Lumpur
London
Mexico
Montreal
New Delhi
Panama
Paris
São Paulo
Singapore
Sydney
Tokyo
Toronto

JOHN H. SEINFELD
California Institute of Technology

Air Pollution

PHYSICAL AND CHEMICAL FUNDAMENTALS

This book was set in Times New Roman.
The editors were B. J. Clark and Matthew Cahill;
the cover was designed by Pencils Portfolio, Inc.;
the production supervisor was Leroy A. Young.
The drawings were done by J & R Services, Inc.
Kingsport Press, Inc., was printer and binder.

Library of Congress Cataloging in Publication Data

Seinfeld, John H
Air pollution: physical and chemical fundamentals.

1. Air-Pollution. I. Title.
TD883.S4 628.5′3 74-4296
ISBN 0-07-056042-0

AIR POLLUTION
PHYSICAL AND CHEMICAL
FUNDAMENTALS

3 4 5 6 7 8 9 0 KPKP 7 9 8 7 6

To my parents

CONTENTS

PREFACE

This book is an outgrowth of my experience with a course on the fundamentals of air pollution for advanced undergraduates and graduate students at the California Institute of Technology. In the early stages of developing material for the course, it became apparent that a single textbook suitable for such a course did not exist. In an effort to provide a cohesive development of the study of air pollution, I prepared a set of lecture notes, from which this book has evolved.

The basic aim of the book is to present in a rigorous, quantitative manner many of the necessary fundamentals required for an analysis of the air pollution problem. There exists a number of books which treat the subject from an essentially descriptive point of view. Although such treatments can be worthwhile for the beginner and the nonscientist, they often do not answer in a satisfactory manner such questions as:

How are air pollutants formed at the source?

In what way do conventional control methods abate pollutant formation and emission?

How does one predict the concentrations of airborne pollutants?

What are the chemical processes responsible for the transformation of pollutants in the atmosphere?

Because of the range of disciplines touched by the overall air pollution problem (such as meteorology, atmospheric chemistry, combustion chemistry, and medicine), it is impossible to cover all aspects of the problem in equal levels of detail in a single textbook of moderate length. In selecting material to cover a rather broad subject, an author invariably weights his choices with his own areas of interest. Thus, I have chosen to stress two subject areas in air pollution: the physical and chemical behavior of air pollutants in the atmosphere and the sources and control methods for pollutants. As a result, the disciplines which are drawn upon most heavily are atmospheric chemistry, fluid mechanics, mass transfer, and combustion. I have chosen not to treat several areas in significant detail; these include effects of air pollution (on human beings, animals, plants, and materials), analytical means for the measurement and detection of air pollutants, and legislative and regulatory measures that have been enacted. In general, then, this book has been written to provide a rigorous and in-depth treatment of the *physical* and *chemical* fundamentals of air pollution.

The level of treatment is appropriate for juniors, seniors, and graduate students in engineering, physics, and chemistry. It is assumed that the reader will have had a basic undergraduate course in transport phenomena (momentum, heat, and mass transfer). This prerequisite may make the book somewhat unsuited for those whose prime interest is air pollution chemistry, although only Chaps. 5 and 6 rely heavily on transport phenomena. Similarly, a reader without an interest in air pollution chemistry may wish to omit certain sections of Chap. 4 on first reading.

The book is organized as shown on page xv. Chapter 1 presents an overview of the entire air pollution problem, including sources, atmospheric behavior, and effects. In addition, we discuss in Chap. 1 the issue of air quality standards versus emissions standards and the economic evaluation of control alternatives. In Chap. 2 the classes of air pollutants are discussed with respect to global sources and sinks. Chapters 3 to 6 are devoted to the physical and chemical behavior of air pollutants in the atmosphere. Finally, the mechanisms of pollutant formation and associated methods of control are the subjects of Chaps. 7 and 8. Appendixes elaborating on certain aspects in the main body of the text are also provided.

This book can be used for a course of one or two quarters (or one semester) in length. For courses in which there is insufficient time available to cover the entire book, the following sections may be omitted: 1.7, 4.6, 4.7, 5.5, 6.4, and 6.5. Material which constitutes detailed derivations or extensions of the general level of treatment appears in smaller print. Those desiring a deeper coverage of a subject than that prevailing in the book as a whole are encouraged to undertake these portions. Finally, the problems at the end of the chapters are intended to provide the reader with the opportunity of applying the material presented in the chapters to situations of practical interest. It is important that these problems be considered an integral part of the text and therefore be attempted.

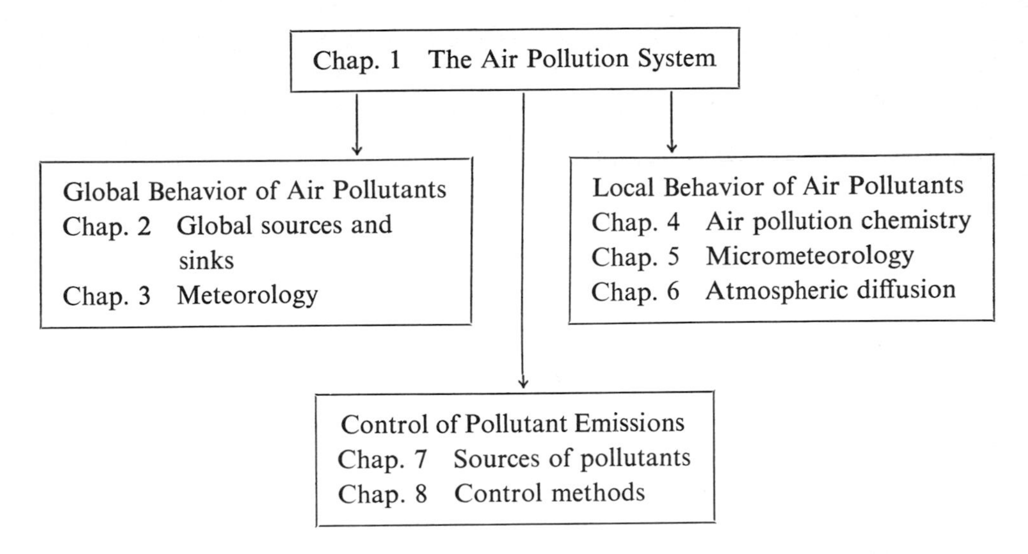

Many people have contributed to this book. Particular acknowledgment is extended to Sheldon K. Friedlander who originally kindled my interest in air pollution and with whom I share the above-mentioned course. I am also greatly indebted to Robert G. Lamb who provided me with many insights on the problem of turbulent diffusion. Special appreciation is given Thomas A. Hecht for his contributions to Chap. 4 and Grant E. Robertson who aided in the preparation of Appendix B. Also, I wish to extend gratitude to Philip M. Roth and James Wei for reading the manuscript and providing valuable comments on its content. To Lenore Kerner I give special acknowledgment for her superb typing. Finally, to my wife Connie goes deep gratitude for her encouragement and understanding.

JOHN H. SEINFELD

AIR POLLUTION
PHYSICAL AND CHEMICAL
FUNDAMENTALS

1

ELEMENTS OF THE AIR POLLUTION PROBLEM

1.1 AIR POLLUTION SYSTEM

Air pollution may be defined as any atmospheric condition in which substances are present at concentrations high enough above their normal ambient levels to produce a measurable effect on man, animals, vegetation, or materials. By "substances" we mean any natural or man-made chemical elements or compounds capable of being airborne. These substances may exist in the atmosphere as gases, liquid drops, or solid particles. As stated, our definition includes any substance, whether noxious or benign; however, in using the term "measurable effect" we will generally restrict our attention to those substances which cause undesirable effects.

The air pollution problem can be simply depicted as a system consisting of three basic components:

$$\underset{\text{Emission sources}}{1} \xrightarrow[\text{Pollutants}]{} \underset{\text{Atmosphere}}{2} \xrightarrow[\text{Mixing and chemical transformation}]{} \underset{\text{Receptors}}{3}$$

The ultimate aim of a study of this system is to provide an answer to the question: What is the optimum way to abate air pollution? It is quite clear that the abatement of air pollution in the large urban centers of the world will require a substantial economic investment as well as, perhaps, changes in patterns of living and energy use. It is unrealistic to speak of no air pollution whatsoever; it is virtually impossible to eliminate entirely all man-made emissions of foreign gases and particles into the atmosphere. It is more sensible to aim toward the reduction of pollutant emissions to a point such that serious adverse effects associated with the presence of pollutants in the air are eliminated. Because of the great expenditure of money that will be required, social and political factors will play a major role in meeting this goal.

Efforts to formulate a coherent strategy for air pollution control have been hampered to a large extent by the inability of those who have studied the various aspects of air pollution to demonstrate clearly the relationship between emission levels and airborne concentrations and between airborne concentrations and the adverse effects (mainly to human health) of air pollutants. In short, the motivation for appropriate action by the political segment of the society will arise only if the necessity for change is clearly evident. The necessity for action must be established by a thorough, scientific analysis of all aspects of the problem.

The study of emission sources, such as the internal combustion engine and the burning of coal and oil in industrial boilers and furnaces, requires a knowledge of the chemistry of combustion as well as of the engineering aspects of the design of such equipment. The understanding of the physical and chemical behavior of pollutants in the atmosphere necessitates a knowledge of meteorology, fluid mechanics, and atmospheric chemistry, as well as of aerosol physics. Finally, the evaluation of the effects of pollutants on people, animals, and plants requires backgrounds in physiology, medicine, and plant pathology. Clearly, the scope of disciplines encompassed by the full air pollution system is too vast for one person to master effectively. Thus, in a one-volume treatment of the fundamentals of air pollution, one must elect to cover some aspects in more depth than others. Although we strive to touch upon each of the elements mentioned above, particular emphasis is given to sources and their control and the atmospheric processes involving air pollutants. It would have been desirable to devote equal space to the third link in the chain, namely the effects of air pollutants; however, considerably less is known about this link than about the other two. In fact, the lack of ability to associate health effects with air pollutant dosages in an unambiguous way has been one of the principal obstacles in gaining public support for air pollution control.

In the broadest sense, air pollution is a global problem since pollutants ultimately become dispersed throughout the entire atmosphere. Customarily, air pollution is thought of as a phenomenon characteristic only of large urban centers and industrialized regions, wherein concentrations often reach values several orders of magnitude greater than ambient background levels. For this reason we will consider primarily air pollution within so-called airsheds, regions of hundreds to tens of thousands of square kilometers which by virtue of meteorology and topography have common air pollution problems.

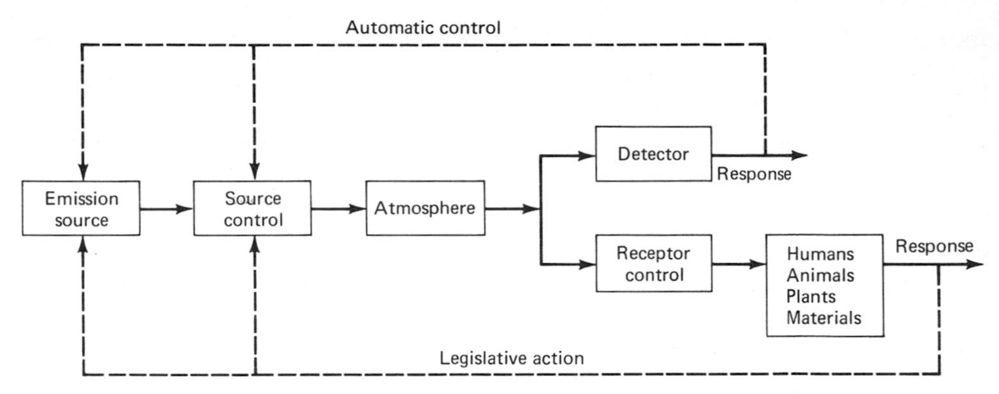

FIGURE 1.1
Air pollution system. Each block represents a given process in the chain of events from the formation of pollutants at the source to the detection of airborne pollutants by receptors. The dashed lines illustrate means by which responses become a basis for the regulation of emission sources and their controls.

Let us consider in slightly more detail the elements of an air pollution problem for a particular airshed. Figure 1.1 summarizes in block-diagram form the components of the air pollution system.

The genesis of air pollution is an *emission source*. Major emission sources are (1) transportation, (2) electric power generation, (3) refuse burning, (4) industrial and domestic fuel burning, and (5) industrial processes. Associated with emission sources are *source controls* which are devices or operating procedures that prevent some of the pollutants produced by the emission source from reaching the atmosphere. Typical source controls include the use of gas-cleaning devices, substitution of a fuel which results in less emissions for one which results in greater emissions, and modifications of the process itself. Pollutants are emitted to the *atmosphere* which acts as a medium for transport, dilution, and physical and chemical transformation. Pollutants may subsequently be detected by instruments or by human beings, animals, plants, or materials. Detection by these various "sensors" is manifested by some response, such as an irritation. Finally, as a result of these responses, emission sources and their controls can be modified either through automatic remote sensing of airborne concentrations or through public pressure and subsequent legislation.

In Fig. 1.1 we have indicated controls at both the sources and the receptors. Three points in the air pollution system are amenable, at least in principle, to control action. First, as we have mentioned, control can be exercised at the source of emission, resulting in lower quantities or a different distribution of effluents reaching the atmosphere or an alteration of the spatial and temporal distribution of emissions. Second, control could be directed to the atmosphere, for example by diverting wind flows or by discharging huge quantities of heat to alter the temperature structure of the atmosphere. Finally, control could be reserved for receptors, for example by extensive use of filtered air conditioning systems or, in the limit, use of gas masks. Of the three, control at the emission source is not only the most feasible but also the most practical. In short, the best way to control air pollution is to prevent contaminants from entering the atmosphere in the first place. Therefore, when we consider air pollution control measures, we will restrict our attention to source controls.

The study of air pollution follows the components shown in Fig. 1.1. The major categories are:

Sources of air pollutants
Control methods
Atmospheric behavior of air pollutants
Effects of air pollutants
Legislative and regulatory measures

In this first chapter we shall discuss briefly each of these five components so as to provide an overview of the air pollution problem. We shall spend somewhat more time in this chapter on those aspects which will not be covered subsequently, namely effects of air pollutants and legislative and regulatory measures.

1.2 AIR POLLUTANTS

The variety of airborne matter is so great that it is difficult to construct tidy classifications. Nevertheless, we begin by placing air pollutants in two general categories:

1 *Primary pollutants:* those emitted directly from sources
2 *Secondary pollutants:* those formed in the atmosphere by chemical interactions among primary pollutants and normal atmospheric constituents

Sampling of effluent streams from various sources provides the types and quantities of primary pollutants emitted from sources, usually in terms of the chemical nature and the physical state (gas, liquid drops, solid particles), and atmospheric measurements serve to identify secondary pollutants.

The composition of dry air at sea level in units of parts per million (ppm) by volume is shown in Table 1.1. There are, of course, several other constituents present at very low background levels, including those species normally classed as air pollutants. Some of these other constituents exhibit significant spatial or temporal variations. For example, background levels of water vapor and ozone are

Water vapor	H_2O	0 to 3% by volume (0 to 30,000 ppm)
Ozone	O_3	0 to 0.07 ppm (ground level)

Depending on the water vapor concentration, the other components of Table 1.1 have lower fractions than shown, although still in the same proportions.

There are two ways in which the concentration of air pollutants is normally expressed. The first, used in Table 1.1, is employed for gaseous pollutants, namely parts per million by volume (volume fraction $\times 10^6$), abbreviated ppm. For example, from Table 1.1 we see that N_2 present at 78.084 percent by volume is 780,840 ppm. By contrast, typical concentrations of gaseous air pollutants are 0.0001 percent by volume, or 1 ppm. To avoid the clumsiness associated with such low percentages, a ppm measure is used. Concentrations are sometimes also expressed in parts per hundred million (pphm) or parts per billion (ppb).

The second common concentration measure is based on the weight of pollutant per volume of air in micrograms per cubic meter, abbreviated $\mu g/m^3$. This measure is generally employed for particulate matter and also sometimes for gases. Conversion between ppm and $\mu g/m^3$ depends on the molecular weight and the volume occupied by a mole of the substance. At standard temperature (25°C) and pressure (1 atm), the relation is

$$\text{conc. in } \mu g/m^3 = \frac{\text{molecular weight}}{0.0245} \times [\text{conc. in ppm}]$$

Those substances usually considered air pollutants can be classified as follows:

1 Sulfur-containing compounds
2 Nitrogen-containing compounds
3 Carbon-containing compounds (excluding carbon monoxide and carbon dioxide)
4 Carbon monoxide and carbon dioxide
5 Halogen compounds
6 Particulate matter
7 Radioactive compounds

Table 1.1 COMPOSITION OF DRY AIR AT SEA LEVEL

Gas	Concentration, ppm
Nitrogen N_2	780,840
Oxygen O_2	209,460
Argon Ar	9340
Carbon dioxide CO_2	315
Neon Ne	18
Helium He	5.2
Methane CH_4	1.0–1.5
Krypton Kr	1.1
Nitrous oxide N_2O	0.5
Hydrogen H_2	0.5
Xenon Xe	0.08

The first point to note about the above list is that classifications are on both a chemical and physical basis, since particulate matter refers to the physical state whereas the other categories refer to the chemical state. This conforms with the standard air pollutant classifications that are used in virtually all publications on the subject. Particulate matter may, in fact, contain sulfur, carbon, and nitrogen compounds, etc. Therefore, we will assume categories 1 to 5 above to refer to gaseous compounds.

Table 1.2 summarizes the classifications of gaseous air pollutants. We shall study the global sources and sinks of these compounds in Chap. 2 and the mechanisms by which they are formed at the source in Chap. 7. As we shall see, in several cases, global natural emissions (but not local emissions in an urban area) of a particular pollutant far exceed man-made (anthropogenic) emissions. This is the case for ammonia (NH_3), the nitrogen oxides (NO and NO_2), and methane (CH_4). Often when referring to the oxides of nitrogen, NO and NO_2, the compact designation NO_X is used. (Sometimes the oxides of sulfur, SO_2 and SO_3, are given a similar designation SO_X.)

Both CO and CO_2 are products of the combustion of carbonaceous fuels, from incomplete and complete combustion, respectively. Actually, CO_2 (together with water, the other product of complete combustion of hydrocarbon fuels) is not normally considered a pollutant. Nevertheless, the global background concentration of CO_2 has been steadily increasing, leading to concern about its possible effect on global meteorology.

Certain halogen compounds such as HF and HCl are produced in metallurgical and other operations. Fluoride compounds are harmful and irritating to human beings, animals, and plants, even when they are present at very low concentrations.

Table 1.2 CLASSIFICATION OF GASEOUS AIR POLLUTANTS

Class	Primary pollutants	Secondary pollutants	Man-made sources
Sulfur-containing compounds	SO_2, H_2S	SO_3, H_2SO_4, MSO_4†	Combustion of sulfur-containing fuels
Nitrogen-containing compounds	NO, NH_3	NO_2, MNO_3†	Combination of N_2 and O_2 during high-temperature combustion
Carbon-containing compounds	C_1–C_5 compounds	Aldehydes, ketones, acids	Combustion of fuels; petroleum refining; solvent use
Oxides of carbon	CO, CO_2	None	Combustion
Halogen compounds	HF, HCl	None	Metallurgical operations

† MSO_4 and MNO_3 denote general sulfate and nitrate compounds, respectively.

However, because the halogen compounds constitute only a small percentage of usual urban air pollutants, we will not consider them further.

By "particulate matter" we refer to any substance, except pure water, that exists as a liquid or solid in the atmosphere under normal conditions and is of microscopic or submicroscopic size but larger than molecular dimensions (about 2 Å). There are several terms used in conjunction with airborne particulate matter (Hidy and Brock, 1970):

1 *Dusts:* solid particles dispersed in a gas by the mechanical disintegration of material, for example clouds that result during the crushing and grinding of rocks and from subjecting powders to blasts of air.

2 *Smokes:* small particles resulting from condensation of supersaturated vapor composed of material of low vapor pressure in relatively high concentrations. The most prominent example is particulate suspensions from combustion processes. Another term, similar to smoke, is *fume*, particles formed by condensation, sublimation, or chemical reaction, of which the predominant part by weight consists of particles with diameters less than 1 micrometer (10^{-6} m). Tobacco smoke and condensed metallic oxides are examples of fume.

3 *Mists:* suspension of liquid droplets formed by condensation of vapor but sometimes by atomization. A mist consists of fairly large particles, exceeding 10 micrometers (μm) in diameter at relatively low concentrations. The size of the particles is the principal property that distinguishes mists from smokes. If the concentration is high enough to obscure visibility, the mist is called a *fog*.

4 *Aerosol:* a cloud of microscopic and submicroscopic particles in air, such as a smoke, fume, mist, or fog.

We will not consider radioactive pollutants because of their specialized nature. The interested reader is referred to Eisenbud (1968).

Secondary pollutants result from chemical reactions in the atmosphere. Several types of reactions may take place in the atmosphere: (1) thermal gas-phase reactions, (2) photochemical gas-phase reactions, and (3) thermal liquid-phase reactions in small liquid drops. Thermal gas-phase reactions are the result of the "collision" of two molecules which possess appropriate energies; they are the customary type of chemical reaction. Photochemical reactions, on the other hand, involve the dissociation or excitation of a molecule upon absorption of radiation by the molecule. Liquid-phase reactions are mostly ionic in nature, possibly catalyzed by substances present in the liquid. Surfaces of liquid drops and solid particles may provide a locus for reactions which might not otherwise take place.

In addition to the secondary sulfur-, carbon-, and nitrogen-containing compounds already mentioned, a very important secondary pollutant is ozone (O_3).

There are virtually no sources of ozone; however, ozone concentrations in certain polluted atmospheres sometimes reach 0.5 ppm.

Although at any time a particular urban atmosphere will undoubtedly contain at least some of each of the primary and secondary pollutants cited above, two relatively distinct types of urban air pollution have emerged. We cite these examples at this point to illustrate the possible roles of both primary and secondary pollutants.

The first and oldest type of air pollution is typified by low temperatures, high concentrations of sulfur compounds (SO_2 and sulfates), water, and particles. The SO_2 and particles result from combustion of coal and high-sulfur-containing fuel oil. Cities with this characteristic type of air pollution are often in cold climates where electric power generation and domestic heating are major sources of emissions. Examples include London, New York, and Chicago. The droplets of sulfuric acid tend to irritate the lungs, and the soot particles from coal combustion cause a layer of grime to form on exposed surfaces. In this case the atmosphere consists mainly of the primary pollutants SO_2 and particles and the secondary pollutant H_2SO_4 formed directly from the SO_2.

The second "type" of air pollution appeared only with the widespread use of gasoline as a motor fuel. Although automobile exhaust was recognized as a potential air pollutant as early as 1915, it was not until about 1945 that the first urban air pollution problem definitely attributable to automobile emissions appeared in Los Angeles. This type of air pollution, once the exclusive property of Los Angeles, now occurs worldwide in metropolitan areas in which there is a heavy use of automobiles. Tokyo, Denver, and Rome are three examples. Although historically this second type of air pollution has been called "smog" (or "photochemical smog"), presumably borrowed from the English condensation of smoke and fog, it is, in fact, neither smoke nor fog but the reactants and products of a complex series of reactions which take place when sunlight irradiates an atmosphere laden with organic gases and oxides of nitrogen. Photochemical smog is characterized by high temperatures, bright sunlight, low humidity, and an eye-burning haze. Although the main primary pollutants in

Table 1.3 COMPARISON OF CONCENTRATION LEVELS BETWEEN CLEAN AND POLLUTED AIR

Component	Clean air	Polluted air
SO_2	0.001–0.01 ppm	0.02–2 ppm
CO_2	310–330 ppm	350–700 ppm
CO	<1 ppm	5–200 ppm
NO_x	0.001–0.01 ppm	0.01–0.5 ppm
Hydrocarbons	1 ppm	1–20 ppm
Particulate matter	10–20 $\mu g/m^3$	70–700 $\mu g/m^3$

photochemical smog are nitric oxide and hydrocarbons, these are rapidly converted to secondary pollutants, ozone, organic nitrates, oxidized hydrocarbons, and so-called photochemical aerosol. It is the secondary pollutants that are responsible for effects such as eye irritation and plant damage.

Table 1.3 shows typical concentration differences for major air pollutants between clean and polluted air.

1.3 ATMOSPHERIC ASPECTS OF AIR POLLUTION

Once pollutants become airborne they are subject to the dispersing action of the atmosphere. Occurring simultaneously with transport by the wind and turbulent mixing are the chemical reactions which transform primary to secondary pollutants. The atmospheric aspects of air pollution can be divided according to:

1 Atmospheric chemistry
2 Meteorology
3 Transport and dispersion of pollutants

Atmospheric chemistry involves the study of the transformation processes affecting airborne pollutants, processes which may take place on time scales of a few seconds to several weeks. The reactions which serve to transform and remove primary pollutants will be discussed in Chaps. 2 and 4.

Meteorology is the study of the dynamics of the atmosphere, particularly pertaining to momentum and energy. Meteorological scales of motion can be categorized as follows:

1 *Macroscale:* phenomena occurring on scales of thousands of kilometers, such as semipermanent high and low pressure areas which reside over the oceans and continents. (The term *synoptic* is commonly used to denote macroscale.)
2 *Mesoscale:* phenomena occurring on scales of hundreds of kilometers, such as land-sea breezes, mountain-valley winds, and migratory high and low pressure fronts.
3 *Microscale:* phenomena occurring on scales of less than 10 km, such as the meandering and dispersion of a chimney plume and the complicated flow regime in the wake of a large building.

Each of these scales of motion plays a role in air pollution, although over different periods of time. For example, micrometeorological effects take place over scales of the order of minutes to hours, whereas mesometeorological phenomena influence transport and dispersal of pollutants over hours to days. Finally, macrometeorological scales of motion have characteristic times of days to weeks. The aspects of macro- and mesoscale meteorology important in air pollution are discussed in Chap. 3.

Prediction of transport and dispersion of pollutants requires knowledge of the effects of winds and turbulence on the motion of particles (or gas molecules) released into the atmosphere. Chapters 5 and 6 are devoted to the fluid-mechanical aspects of pollutant behavior in the atmosphere. Material which is classically considered micrometeorology is included in these two chapters.

With respect to urban air pollution, the region of the atmosphere governing transport and dispersion is the so-called planetary boundary layer, roughly the lowest 1000 m. The planetary boundary layer represents the extent of influence of the earth's surface on wind structure in the atmosphere. Within the planetary boundary layer, winds are influenced by the prevailing high-level flows and the frictional drag of the surface. With respect to air pollution, the key problem associated with the planetary boundary layer is to predict the variation of wind speed and direction with altitude as a function of surface roughness and temperature profile.

The atmospheric *temperature profile* (the variation of temperature with altitude) has an important effect on wind structure and turbulence in the lowest 1000 m. In the troposphere (the 10 to 20 km of the atmosphere closest to the ground) the temperature normally decreases with increasing altitude because of the decrease in pressure with height. The temperature profile against which all others are judged is that observed for a parcel of dry air as it moves upward in a hydrostatically stable atmosphere and expands slowly to lower pressure with no gain or loss of heat. If such a profile exists in the atmosphere, a parcel of air at any height is in neutral equilibrium; that is, it has no tendency either to rise or fall. The atmosphere is, however, very seldom in such delicate equilibrium; the influence of surface heating and large-scale phenomena usually results in a temperature profile different from this reference profile. If the temperature decreases faster with height than the reference profile, air parcels at any height are unstable, that is, if they are displaced either upward or downward, they will continue their movement in the direction in which they were displaced. Such a condition is referred to as *unstable*. On the other hand, if the temperature decreases more slowly with height than the reference profile (or even increases), air parcels are inhibited from either upward or downward motion and the situation is referred to as *stable*. The stability condition of the atmosphere plays an important role in determining the rate of dispersal of pollutants.

The phenomenon of direct interest in predicting the dispersion of air pollutants is *turbulent diffusion*. Actually, turbulent diffusion is something of a misnomer. The phrase refers to the observed spreading of a cloud of marked particles in a turbulent fluid at a rate many orders of magnitude greater than that from molecular diffusion alone. The spreading is really not due to a "diffusion" phenomenon such as results from molecular collisions but rather is a result of the rapid, irregular motion of macroscopic lumps of fluid (called *eddies*) in turbulence. Thus, the scales of length in turbulent diffusion are much greater than in molecular diffusion, with the contribution of the latter to the dispersion of pollutants in turbulence being virtually negligible. The level of turbulence in the planetary boundary layer increases with increased wind speed, surface roughness, and instability. Turbulence, therefore, arises from both mechanical forces (shear, surface friction) and thermal forces (buoyancy).

A key objective of air pollution meteorology is the prediction of the dispersion of pollutants. Such dispersion depends on the following:

Physical nature of the pollutant (gas, particle)
Wind speed and direction
Atmospheric stability
Level of turbulence
Emission conditions (exit velocity, temperature, etc.)
Source configuration (smokestack, highway, etc.)

The role of each of these factors in the atmospheric behavior of pollutants will subsequently be considered in detail.

1.4 EFFECTS OF AIR POLLUTION

Substantial evidence has accumulated that air pollution affects the health of human beings and animals, damages vegetation, soils and deteriorates materials, affects climate, reduces visibility and solar radiation, impairs production processes, contributes to safety hazards, and generally interferes with the enjoyment of life and property. Although some of these effects are specific and measurable, such as damages to vegetation and material and reduced visibility, most are difficult to measure, such as health effects on human beings and animals and interference with comfortable living. As a result, there has been disagreement over the quantitative effects of air pollution. Each of the effects listed above has been the subject of considerable attention, and a number of comprehensive reviews of the effects of air pollution have been written (for example, Stern, 1968, vol. I, chaps. 11–15). In this section we shall present a brief summary of some of the most important established effects of air pollution. We shall provide references which the interested reader may consult for more thorough coverage. We begin with the effects of air pollution on meteorology.

1.4.1 Effects of Air Pollution on Atmospheric Properties

Air pollutants affect atmospheric properties in the following ways:

1 Visibility reduction
2 Fog formation and precipitation
3 Solar radiation reduction
4 Temperature and wind distribution alteration

These effects are primarily associated with the urban atmosphere. In addition, there is much current interest in possible effects of air pollutants, mainly carbon dioxide and particles, on the atmosphere as a whole. We defer treatment of these until Chap.

3, since it is necessary to discuss the atmospheric energy balance before these effects can be assessed.

Perhaps the most noticeable effect of air pollution on the properties of the atmosphere is the reduction in visibility which frequently accompanies polluted air. Visibility limitation is not only aesthetically unpleasing but also may lead to safety hazards. The *prevailing visibility* is defined as the greatest distance in a given direction at which it is just possible to see and identify (1) a prominent dark object in the daytime and (2) an unfocused, moderately intense light source at night, which is attained or surpassed around at least half the horizon circle but not necessarily in continuous sectors. Visibility is reduced by two effects which gas molecules and particles have on visible radiation: absorption and scattering of light. Absorption of certain wavelengths of light by gas molecules and particles is sometimes responsible for atmospheric colorations. However, light scattering is the more important phenomenon responsible for impairment of visibility.

Light scattering refers to the deflection of the direction of travel of light by airborne material. Visibility is reduced when there is significant scattering because particles in the atmosphere between the observer and the object scatter light from the sun and other parts of the sky through the line of sight of the observer. This light decreases the contrast between the object and the background sky, thereby reducing visibility.

Large particles (of size greater than a few micrometers) scatter light by three processes: (1) reflection from the particle surface, (2) diffraction around the edges of the particle, and (3) refraction upon passing through the interior of the particle. Most of the light scattered by large particles is not significantly altered in direction from its original path (the forward direction). Particles most effective in scattering light are those with sizes comparable to the wavelength of visible light (0.4 to 0.7 μm). A much greater fraction of the incident light is scattered away from the forward direction by particles in this size range than by the large particles. Particles of very small size (less than 0.1 μm) scatter light equally in the forward and backward directions. In addition, small particles scatter light of short wavelengths more effectively than light of long wavelengths. This effect is, in fact, responsible for the reddish hue of sunsets, since the shorter-wavelength blue component of sunlight is scattered out of the line of sight, leaving the red components to reach the observer.

Visibility Reduction by Light Scattering

Let us examine the effect of atmospheric particulate concentration on visibility reduction. To do so we consider the case in which a black object is being viewed against a white background. We first define the contrast at a distance x from the object, $C(x)$, as the relative difference between the light intensity of the target and the background,

$$C(x) = \frac{I_B(x) - I_o(x)}{I_B(x)}$$

where $I_B(x)$ and $I_o(x)$ are the intensities (measured in joules cm^{-2} sec^{-1}) of the background and the object, respectively. At the object ($x = 0$), $I_o(0) = 0$, since the object is assumed to be black and therefore absorbs all light incident on it. Thus, $C(0) = 1$. Over the distance x between the object and the observer, $I_o(x)$ will be affected by two phenomena: (1) absorption of light by gases and particles and (2) addition of light which is scattered into the line of sight. Over a distance dx the intensity change dI_o is a result of these two effects. The fraction of I_o diminished is assumed to be proportional to dx, since dx is a measure of the number of particles present. The fractional reduction in I_o is written $dI_o = -(b_{abs} + b_{scat})I_o\,dx$, where $b_{abs}\,dx$ is the fraction of I_o lost by absorption and $b_{scat}\,dx$ is the fraction of I_o lost by scattering out of the line of sight. It is common to assume that b_{abs} and b_{scat} are constants (independent of x) which depend on the concentration of particles present. In addition, the intensity I_o can be increased over the distance dx by scattering of light from the background into the line of sight. The increase can be expressed as $b' I_B(x)\,dx$, where b' is a constant. The net change in intensity is given by

$$dI_o(x) = [b' I_B(x) - (b_{abs} + b_{scat})I_o(x)]\,dx$$

By its definition as background intensity, I_B must be independent of x. Thus, along any other line of sight

$$dI_B(x) = 0 = [b' I_B - (b_{abs} + b_{scat})I_B]\,dx$$

We see that $b' = b_{abs} + b_{scat}$. Thus, we find that the contrast $C(x)$ varies according to

$$dC(x) = -(b_{abs} + b_{scat})C(x)\,dx$$

and therefore that the contrast decreases exponentially with distance from the object,

$$C(x) = e^{-(b_{abs} + b_{scat})x}$$

The coefficient b_{abs} accounts for absorption of light by both gas molecules and particles, and b_{scat} accounts for scattering of light by both gas molecules and particles. We can decompose b_{abs} and b_{scat} into the gas and particulate components as follows:

$$b_{abs} = b_{abs-part} + b_{abs-gas}$$
$$b_{scat} = b_{ps} + b_{Ray}$$

where b_{ps} is the particulate scattering coefficient, and b_{Ray} is the gas-molecule scattering coefficient, called the *Rayleigh scattering* coefficient, Experimentally it has been found that the ratio b_{scat}/b_{Ray} varies from 1.3 to 2.5. A ratio of unity indicates the cleanest possible air ($b_{ps} = 0$). Thus, the higher this ratio, the greater is the contribution of particulate scattering to total light scattering.

Scattering by particulate matter of sizes comparable to the wavelength of visible light (called *Mie scattering*) is mostly responsible for visibility reduction in the atmosphere. The average wavelength of light is about 0.52 μm. Thus, particles in the range 0.1 to 1 μm in radius are the most effective, per unit mass, in reducing visibility. The scattering coefficient b_{scat} has been found to be more or less directly dependent on the atmospheric aerosol concentration in this size range, as long as the relative humidity is sufficiently low that fog formation does not take place (below about 70 percent relative humidity).[1]

The visual range corresponding to a certain contrast C can be computed from

$$x = \frac{-\ln C}{b_{scat}}$$

For the average human eye the threshold limit of C is about 0.02, so that the prevailing visibility (in units of length) can be roughly estimated as $3.9\, b_{scat}^{-1}$.

[1] The detailed dependence of b_{scat} on the composition of the atmosphere is, in reality, quite complex. Those wishing to delve more deeply into this subject should consult Middleton (1952).

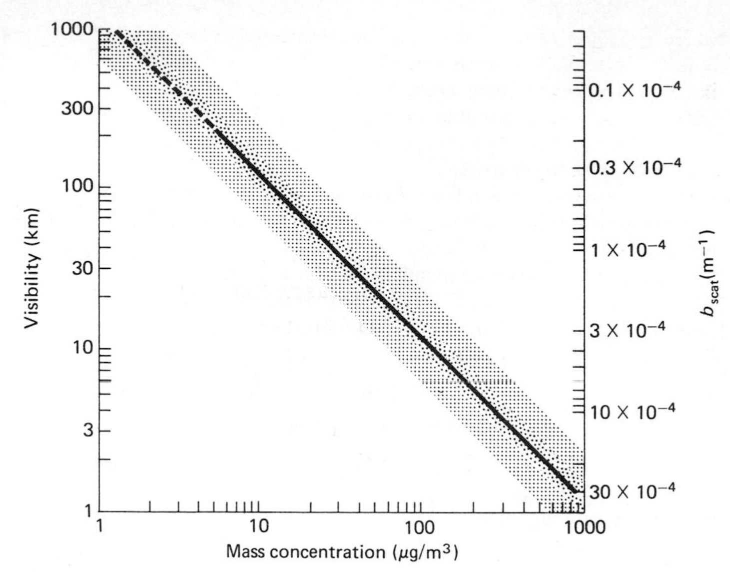

FIGURE 1.2
Visibility as a function of mass concentration of aerosol. Approximately 90 percent of the cases can be expected to fall inside the shaded band. The dashed portion represents the region where Rayleigh scattering by air is important. (*Source: Charlson*, 1968b.)

On the basis of field measurements Charlson (1967, 1968a, 1968b), correlated total aerosol mass concentration m and b_{scat} by

$$\frac{m}{b_{scat}} = 0.31 \text{ g/m}^2$$

This value represents the mode of a distribution with about 90 percent of the measurements between 0.15 and 0.6 g/m^2. This ratio can be viewed as the amount of material per square meter needed to diminish the intensity of light by scattering to a value $1/e$ of the initial intensity. Figure 1.2 shows the regression line obtained by Charlson, together with 90 percent confidence limits. Meteorological range in kilometers is plotted as a function of mass concentration. The scattering coefficient is given as a third ordinate.

In addition to reducing visibility, air pollution affects urban climates with respect to increased fog formation and reduced solar radiation.[1] The frequency of fog formation has been observed to be higher in cities than in the country in spite of the fact that air temperatures tend to be higher and relative humidities tend to be lower in

[1] In discussion of urban climates one must be careful to distinguish between the natural modification of climate due to the concentration of buildings and that directly attributable to air pollution. Climate modifications from air pollution are a result of the "dome" of pollutants existing over a city.

cities as opposed to the country. The explanation for this observation lies in the mechanism of fog formation. With high concentrations of SO_2, the formation of sulfuric acid by oxidation of SO_2 on the surface of particles in a humid environment leads to the formation of small droplets of fog which would not otherwise have formed. For example, in determining the correlation between SO_2 concentrations and fog formation in the German cities of Gelsenkirchen and Hamburg, Georgii (1969) found that 80 percent of all cases of high SO_2 concentrations occurred with visibilities below 5 km. In addition to fog formation, increased precipitation has been linked to areas with high particulate concentrations (Changnon, 1968).

Scattering and absorption of both solar and infrared radiation, as well as emission of radiation, occur within the polluted layer. The net effect of these radiative processes during the night is a marked cooling of the polluted layer. During the day the net effect of pollutants on the temperature of the urban dome depends on the relative magnitude of heating caused by absorption of solar radiation and cooling due to emission of infrared radiation (e.g., see Atwater, 1971). At the ground a reduction in direct and scattered solar radiation is expected because of the blanket of particles over an urban area. The particles are most effective in reducing radiation when the angle of the sun is low, that is, when the path length through the layer is the greatest. Studies in London, for example, have shown that the average duration of bright sunshine in central London (in hours per day) is discernibly less than in the surrounding countryside (Georgii, 1969). In general, the decrease in direct solar radiation due to a polluted layer amounts to 10 to 20 percent.

Additional discussion of the effect of air pollution on urban climates can be found in Lowry (1967), Georgii (1969), Peterson (1969), and Landsberg (1970).

1.4.2 Effects of Air Pollution on Materials

Air pollutants can affect materials by soiling or chemical deterioration. High smoke and particulate levels are associated with soiling of clothing and structures[1], and acid or alkaline particles, especially those containing sulfur, corrode materials, such as paint, masonry, electrical contacts, and textiles. Ozone is particularly effective in deteriorating rubber (Jaffe, 1967). Residents of Los Angeles, where high ozone levels are routinely experienced, must replace automobile tires and windshield wiper blades more frequently than residents in cities where high ozone concentrations are not as common. For a review of the effects of air pollution on materials the reader should consult Yocum and McCaldin (1968).

[1] For example, in a study of the costs of air pollution (primarily due to soiling) in Steubenville and Uniontown, Ohio, Michelson and Tourin (1966) found that the increased cleaning and maintenance costs which they correlated with particulate concentrations in Steubenville were about $84 per person per year more than the same costs in the similar but less polluted city of Uniontown.

1.4.3 Effects of Air Pollution on Vegetation

Pollutants which are known phytotoxicants (substances harmful to vegetation) are sulfur dioxide, peroxyacetyl nitrate (an oxidation product in photochemical smog), and ethylene. Of somewhat lesser severity are chlorine, hydrogen chloride, ammonia, and mercury. In general, the gaseous pollutants enter the plant with air through the stomata in the course of the normal respiration of the plant. Once in the leaf of the plant, pollutants destroy chlorophyll and disrupt photosynthesis. Damage can range from a reduction in growth rate to complete death of the plant. Symptoms of damage are usually manifested in the leaf, and the particular symptoms often provide the evidence for the responsible pollutant. Table 1.4 summarizes the symptoms characteristic of plant damage by several pollutants. For additional information on the effects of air pollution on vegetation, the reader may consult Brandt and Heck (1968), Hindawi (1970), and Jackobson and Hill (1970).

1.4.4 Effects of Air Pollution on Human Health

We now come to the most controversial and probably the most important effect of air pollution, that on human health. First we consider the mechanisms by which pollutants can affect the human body. We then discuss the type of evidence available on the effects of long-term exposure to pollutant levels characteristic of urban areas.

Table 1.4 SUMMARY OF SYMPTOMS AND INJURY THRESHOLDS FOR AIR POLLUTION DAMAGE TO VEGETATION

Pollutant	Symptom	ppm	Sustained exposure time
Ozone (O_3)	Fleck, bleaching, bleached spotting, growth suppression. Tips of conifer needles become brown and necrotic.	0.03	4 hr
SO_2	Bleached spots, bleached areas between veins, chlorosis, growth suppression, reduction in yield.	0.03	8 hr
Peroxyacetyl nitrate (PAN)	Glazing, silvering or bronzing on lower surface of leaves.	0.01	6 hr
HF	Tip and margin burn, chlorosis, dwarfing leaf abscission, lower yield.	0.0001	5 weeks
Cl_2	Bleaching between veins, tip and leaf abscission.	0.10	2 hr
Ethylene (C_2H_4)	Withering, leaf abnormalities, flower dropping, and failure of flower to open.	0.05	6 hr

SOURCE: Hindawi (1970).

Pollutants enter the body through the respiratory system, which can be divided into the upper respiratory system, consisting of the nasal cavity and the trachea, and the lower respiratory system, consisting of the bronchial tubes and the lungs. At the entrance to the lungs, the trachea divides into two bronchial trees which consist of a series of branches of successively smaller diameter. The entire bronchial tree consists of over 20 generations of bifurcations, ending in bronchioles of diameters of about 0.05 cm. At the end of the bronchioles are large collections of tiny sacs called alveoli. It is across the alveolar membranes that oxygen diffuses from the air in the sacs to the pulmonary capillaries and carbon dioxide diffuses in the opposite direction. Although an individual alveolus has a diameter of only about 0.02 cm, there are several hundred million alveoli in the entire lung, providing a total surface area for gas transport of roughly 50 m^2.

It is customary to represent the difference in pressure between the ambient air and air in the lung as the product of the volumetric rate of airflow into the upper respiratory tract times a parameter called the *airway resistance*. If, for a given pressure difference, the volumetric airflow into the lung is lower than for a normal lung, i.e. a larger airway resistance, then a constriction of the air passages has occurred. Certain pollutants affect the airway resistance of the lung.

The respiratory system has several levels of defense against invasion by foreign material. Large particles are filtered from the airstream by hairs in the nasal passage and are trapped by the mucus layer lining the nasal cavity and the trachea. These large particles are unable to negotiate the sharp bends in the nasal passage, and, because of their inertia, impinge on the wall of the cavity as the air rushes down toward the lung. In addition, particles may also be scavenged by fine hairlike cilia which line the walls of the entire respiratory system. These cilia continually move mucus and trapped material to the throat where they are removed by swallowing. Most particles of sizes exceeding 5 μm are effectively removed in the upper respiratory system.

Particles of radii less than a few micrometers generally pass through the upper respiratory system, escaping entrapment. Some of the larger of these particles (about 1 μm in size) are deposited on the bronchial walls immediately behind bifurcations in the bronchial tree. The mechanism for this deposition is believed to be inertial impaction which results from the swirling air motions caused by the bifurcation. Very small particles (radii < 0.1 μm) are strongly influenced by brownian motion (rapid, irregular movement due to collisions of the particle with air molecules). As a result, these particles have a high probability of striking the bronchial walls somewhere in the bronchial tree. There is effectively a "window" in the size range 0.1 to 1 μm, where the particles are too large to be influenced by brownian motion but are too small to be trapped in the upper portion of the lung. Particles in this size range are able to penetrate deep into the lung. Figure 1.3 summarizes the fate of particles in the respiratory system.

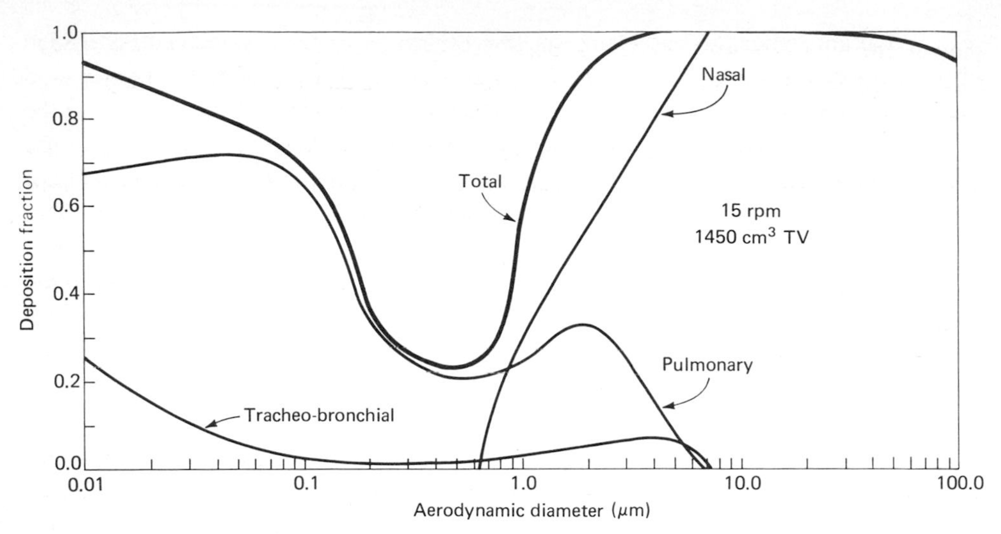

FIGURE 1.3
Deposition of monodisperse aerosols of various diameters in the respiratory tract of man (assuming a respiratory rate of 15 respirations per minute and a tidal volume of 1450 cm^3). (*Source: Task Group on Lung Dynamics*, 1966. *With the permission of Microfirm International Marketing Corporation, exclusive copyright licensee of Pergamon Press journal back files.*)

For gases, the solubility governs what proportion is absorbed in the upper airway and what proportion reaches the terminal air sacs of the lungs. For example, SO_2 is quite soluble and, consequently, is absorbed early in the airway, leading to airway resistance (swelling) and stimulated mucus secretion. On the other hand, CO, NO_2, and O_3 are relatively insoluble and are able to penetrate deep into the lung to the air sacs.[1] Nitrogen dioxide and ozone cause pulmonary edema (swelling) which inhibits gas transfer to the blood. Carbon monoxide is transported from the air sacs to the blood and combines with hemoglobin as oxygen does.

It is important to note that more than one pollutant may induce the same effect. For example, sulfur dioxide and formaldehyde both produce irritation and increased airway resistance in the upper respiratory tract, and both CO and NO_2 interfere with oxygen transport by hemoglobin. Several pollutants usually are present at the same time, and, as a result, observed effects may actually be attributable to the combined action of more than one pollutant. A good example of this is the case of SO_2 and particulate matter. Health effects become far more serious when both are present

[1] Sulfur dioxide is highly soluble in water compared with other gaseous pollutants. For example, the solubility of SO_2 at 20°C is 11.3 g/100 ml, as compared with 0.004, 0.006, 0.003, and 0.169 g/100 ml for O_2, NO, CO, and CO_2, respectively.

than if either occurs separately. A possible explanation for this effect is that SO_2 becomes absorbed on the surface of very small particles and is carried by the particles deep into the lung.

Table 1.5 summarizes some observed relations between pollutant levels and physiological responses. In many cases, we can account for the responses by considering the mechanisms of irritation of individual pollutants.

1.4.4.1 Carbon monoxide The effects of carbon monoxide exposure are reflected in the oxygen-carrying capacity of the blood. In normal functioning, hemoglobin molecules in the red blood cells carry oxygen, which is exchanged for carbon dioxide in the capillaries connecting arteries and veins. Carbon monoxide is relatively insoluble and easily reaches the alveoli along with oxygen. The carbon monoxide diffuses through the alveolar walls and competes with oxygen for one of the four iron sites in the hemoglobin molecule. The affinity of the iron site for CO is about 210 times greater than for O_2, so that this competition is extremely effective. When a hemoglobin molecule acquires a CO molecule it is called *carboxyhemoglobin* (abbreviated COHb). The presence of carboxyhemoglobin decreases the overall capacity of the blood to carry oxygen to the cells. In addition, the presence of CO on one of the iron sites of a hemoglobin molecule not only removes that site as a potential carrier of an O_2 molecule but also causes the other iron sites of the molecule to hold more tightly onto the O_2 molecules they are carrying. Symptoms of CO poisoning depend on the amount of hemoglobin combined with CO. The effects of CO exposure are summarized in Fig. 1.4. Each curve shows the increase of COHb with exposure time for different inhaled concentrations of CO. An equilibrium is reached in each case in about 8 hr. The formation of COHb is a reversible process, with a half-life for dissociation after exposure of about 2 to 4 hr for low concentrations.

The major source of CO in urban areas is automobile exhaust. Levels typical of urban areas range from 5 to 100 ppm. It appears that the most serious danger associated with CO is the exposure of drivers on heavily congested highways to CO levels of the order of 100 ppm. It has been found experimentally that relatively low COHb levels can affect the ability to estimate time intervals, can delay reaction times, and reduce visual sensitivity in the dark. The supposition that CO leads to increased incidences of traffic accidents, by virtue of effects such as these, is indeed a compelling one. For additional discussion of the effect of CO on human health, the reader is referred to Goldsmith and Landaw (1968).

1.4.4.2 Oxides of sulfur Sulfur dioxide is highly soluble and consequently is absorbed in the moist passages of the upper respiratory system. Exposure to SO_2 levels of the order of 1 ppm leads to constriction of the airways in the respiratory tract.

Table 1.5 OBSERVED RELATIONS BETWEEN POLLUTANT LEVELS AND HEALTH EFFECTS

Pollutant	Concentration level producing adverse health effects	Adverse health effects	Reference
Particulate matter and sulfur oxides	*1* 80–100 $\mu g/m^3$ particulates (annual geometric mean)	*1* Increased death rates for persons over 50 years of age	Winkelstein (1967)
	2 130 $\mu g/m^3$ (0.046 ppm) of SO_2 (annual mean) accompanied by particulate concentrations of 130 $\mu g/m^3$	*2* Increased frequency and severity of respiratory diseases in schoolchildren	Douglas and Waller (1966)
	3 190 $\mu g/m^3$ (0.068 ppm) of SO_2 (annual mean) accompanied by particulate concentrations of about 177 $\mu g/m^3$	*3* Increased frequency and severity of respiratory diseases in schoolchildren	Lunn et al. (1967)
	4 105–265 $\mu g/m^3$ (0.037–0.092 ppm) of SO_2 (annual mean) accompanied by particulate concentrations of 185 $\mu g/m^3$	*4* Increased frequency of respiratory symptoms and lung disease	Petrilli et al. (1966)
	5 140–260 $\mu g/m^3$ (0.05–0.09 ppm) of SO_2 (24-hr average)	*5* Increased illness rate of older persons with severe bronchitis	Carnow et al. (1969)
	6 300–500 $\mu g/m^3$ (0.11–0.19 ppm) of SO_2 (24-hr mean) with low particulate levels	*6* Increased hospital admissions for respiratory disease and absenteeism from work of older persons	Brasser et al. (1967)
	7 300 $\mu g/m^3$ particulates for 24 hr accompanied by SO_2 concentrations of 630 $\mu g/m^3$ (0.22 ppm)	*7* Chronic bronchitis patients suffering acute worsening of symptoms	Lawther (1958)

Table 1.5 (*continued*)

Pollutant	Concentration level producing adverse health effects	Adverse health effects	Reference
Carbon monoxide (CO)	*1* 58 mg/m^3 (50 ppm) for 90 min (similar effects) upon exposure to 10 to 17 mg/m^3 (10–15 ppm) for 8 or more hr	*1* Impaired time-interval discrimination	Beard and Wertheim (1967)
	2 Effects upon equivalent exposure to 35 mg/m^3 (30 ppm) for 8 or more hr	*2* Impaired performance in psychomotor tests	Schulte (1963)
	3 Effects upon equivalent exposure to 35 mg/m^3 (30 ppm) for 8 or more hr	*3* Increase in visual threshold	McFarland et al. (1944)
Photochemical oxidants	*1* In excess of 130 $\mu g/m^3$ (0.07 ppm)	*1* Impairment of performance by student athletes	Wayne et al. (1967)
(O_3 and peroxy-organic nitrates)	*2* 490 $\mu g/m^3$ (0.25 ppm) maximum daily value. (This value would be expected to be associated with a maximum hourly average concentration as low as 300 $\mu g/m^3$ (0.15 ppm))	*2* Aggravation of asthma attacks	Schoettlin and Landau (1961)
	3 200 $\mu g/m^3$ (0.1 ppm) maximum daily value	*3* Eye irritation	Heuss and Glasson (1968)

SOURCE: Ross (1972, pp. 32,33).

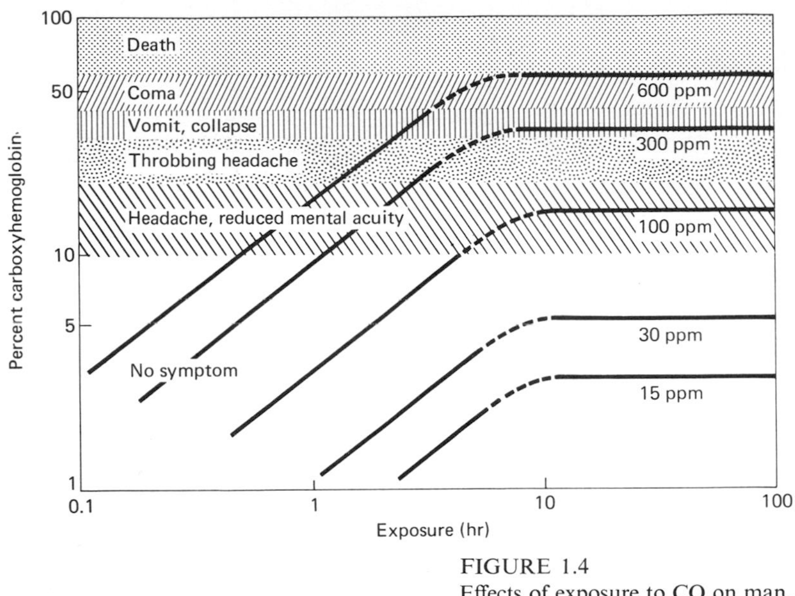

FIGURE 1.4
Effects of exposure to CO on man.

As we have already noted, high SO_2 levels are often associated with high particulate concentrations. The fact that a three- to fourfold increase in the irritant response to SO_2 is observed in the presence of particulate matter is presumably attributable to the ability of the aerosol particles to transport SO_2 deep into the lung. Figure 1.5 summarizes responses to various dosages of SO_2.

1.4.4.3 Oxides of nitrogen There is no available evidence supporting the proposition that nitric oxide (NO) is a health hazard at levels found in urban air. Nitrogen dioxide (NO_2), on the other hand, is transformed in the lungs to nitrosamines, some of which may be carcinogenic. In addition, NO_2 may be transferred to the blood to form a compound called methemoglobin. Nitrogen dioxide is known to irritate the alveoli, leading to symptoms resembling emphysema upon long-term exposure to concentrations of the order of 1 ppm (Project Clean Air, 1970).

1.4.4.4 Photochemical oxidants As we shall see later, the term *photochemical oxidants* refers to the secondary pollutants formed in photochemical smog from reactions involving hydrocarbons and oxides of nitrogen. The principal ingredient in this category is ozone, with smaller amounts of oxygen-containing hydrocarbon compounds. The effect of ozone on pulmonary function is still not thoroughly

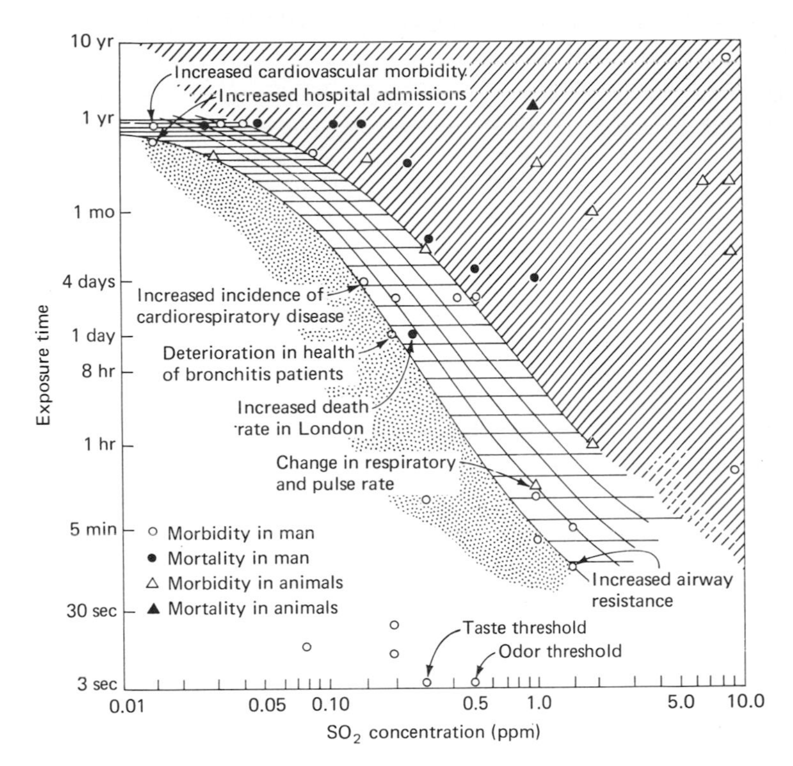

FIGURE 1.5
Health effects due to various exposures to SO_2. Shaded area represents the range of exposures where excess deaths have been reported. Grid area represents the range in which increased morbidity has been reported. Speckled area represents the range of exposures where health effects are suspected. (*Source: Williamson*, 1973.)

understood. In general, ozone at levels of about 1 ppm produces a narrowing of the airways deep in the lung, resulting in increased airway resistance. The effects of long-term exposure to ozone at levels typical of urban air (about 0.1 to 0.2 ppm) have not been established. Experiments with animals have exhibited irreversible changes in pulmonary function after long-term exposure to levels of 1 ppm. A topic of current speculation is that exposure to low levels of ozone accelerates the aging of lung tissue by the oxidation of certain compounds in proteins.

A widespread effect of photochemical smog is eye irritation. The precise mechanism by which certain compounds irritate eyes is not known; in fact, it appears that the compounds responsible for eye irritation in smog may not all have been identified. Those that are known to be irritants and that have been detected in photochemical smog are formaldehyde (HCHO), acrolein (CH_2CHCHO), and members of the family of peroxyacyl nitrates, two of which are

$$CH_3\overset{\displaystyle O}{\overset{\|}{C}}OONO_2 \qquad \text{peroxyacetyl nitrate (PAN)}$$

$$C_6H_5\overset{\displaystyle O}{\overset{\|}{C}}OONO_2 \qquad \text{peroxybenzoyl nitrate (PBzN)}$$

1.4.4.5 Lead Correlation between blood lead levels and exposure to lead in ambient air was exhibited by Goldsmith and Hexter (1967). The mechanisms of lead poisoning are complex. In short, lead inhibits several steps in the formation of hemoglobin. Depending on the mode of entry into the body, up to 60 percent of the total lead ingested can be permanently retained by the body.

Over the years, concentrations of air pollutants in a particular area have, on occasion, reached excessively high levels for periods of several hours to several days. The result has been a number of so-called air pollution episodes, the most serious of which are listed in Table 1.6. During an episode, the person most likely to be seriously injured is one who is either elderly or is in questionable health, perhaps already suffering from a respiratory disease.

Table 1.6 AIR POLLUTION EPISODES

Location	Date	Pollutants	Symptoms and effects
Meuse Valley, Belgium	Dec. 1–5, 1930	SO_2 (9.6–38.4 ppm)	63 excess deaths, chest pain, cough, eye and nasal irritation, all ages affected
Donora, Pa.	Oct. 26–31, 1948	SO_2, particles (0.5–2 ppm)	20 excess deaths, chest pain, cough, eye and nasal irritation, older people mainly affected
Poza Rica, Mexico	Nov. 24, 1950	H_2S	22 excess deaths, 320 hospitalized, all ages affected
London	Dec. 5–9, 1952	SO_2, particles	4000 excess deaths
New York	Nov. 24–30, 1966	SO_2, particles	168 excess deaths

SOURCE: Goldsmith (1968).

Although the results of an occasional air pollution episode are spectacular, of greater concern are the chronic effects on a population living in polluted air. One of the most difficult tasks has been to obtain a quantitative link betwen long-term exposure to air pollution and health effects. To do this, a measured index of pollution (concentrations averaged over a certain period of time), an effect, and a demonstrated relation between the two are needed. The objective, then, is to determine the levels of morbidity (sickness) and mortality for specific diseases that are ascribable to air pollution. The problem in doing this is to separate the effect of air pollution on health from those of personal habits such as smoking and exercise, diet, living conditions, occupational exposures, and hereditary factors. Because of the great number of factors that play a role in determining a person's health, isolation of the effect of one variable, like atmospheric purity, requires data for large populations that presumably differ only in their exposure to air pollutants. The data normally available are the rates of morbidity (based on hospital admissions, lost workdays due to sickness, etc.) and mortality for a particular group, usually defined geographically. Unfortuntely, many of the above factors, such as smoking and diet, are unknown, so that in the analysis it must be assumed (no doubt incorrectly in many cases) that the factors are roughly the same for all large groups or that they vary randomly with respect to the level of air pollution.

Diseases of the respiratory system are generally correlated with air pollution. There are two types of reaction to air pollutants by the respiratory system. The first is acute reaction, such as irritative bronchitis, and the second is chronic reaction, such as chronic bronchitis and pulmonary emphysema.

Bronchitis refers to a condition of inflammation of the bronchial tree. The inflammation is accompanied by increased mucus production and a cough. Airway resistance is increased because of the presence of the thickened mucus layer. Acute bronchitis is generally a short-lasting disease, caused by a virus or foreign material in the lung. Chronic bronchitis, on the other hand, is a sustained inflammation of the bronchial system, leading to an increase in the volume of mucoid bronchial secretion sufficient to cause expectoration. It is frequently accompanied by a cough and shortness of breath. The persistent inflammation leads to swelling of the terminal bronchi and increased airway resistance.

Emphysema is a condition in which the alveoli in the lung become uneven and overdistended due to destruction of the alveolar walls. The disease is accompanied by shortness of breath, particularly following exercise. The destruction of alveoli is progressive, resulting in an increased blood flow necessary to accomplish oxygen transfer and to a decreased ability to eliminate foreign bodies which reach the alveolar region. Emphysema has no known cure and is one of the fastest growing causes of death in the United States.

An extensive review of the literature through 1969 pertaining to population studies on the chronic effects of air pollution is presented by Lave and Seskin (1970). Some of the results of their statistical analysis are summarized in Table 1.7. In the studies considered, morbidity and mortality rates were compared in different geographical areas, among people within an occupational group, and among children. Various methods were used, ranging from individual medical examinations and interviews to tabulations of existing data. To discredit the results, Lave and Seskin point out, one would have to argue that the relationships found by all the investigators are spurious because the level of air pollution is correlated with a third factor, which is the "real" cause of ill health. No such third factor could be found or demonstrated. The available evidence analyzed by Lave and Seskin therefore suggests that (1) air pollution plays a pathogenic (disease-causing) role in chronic bronchitis, emphysema, and asthma and (2) air pollution increases morbidity and/or mortality in patients with preexisting respiratory disease.

Table 1.7 RELATION BETWEEN MORTALITY OR MORBIDITY AND AIR POLLUTION INDICES AS DETERMINED BY LAVE AND SESKIN (1970)

Disease	Air pollution index	Relation
Bronchitis	Sulfur concentration; total concentration of solids	Strong relation between bronchitis mortality and several indices of air pollution. Bronchitis mortality could be reduced by from 25 to 50 percent, depending on the location and pollution index, by reducing pollution to the lowest level currently prevailing in the region.
Lung cancer	Urban versus rural areas	With adjustments made for age and smoking history, incidence of lung cancer is about 1.5 times as great in urban as in rural areas.
Nonrespiratory tract cancer	Suspended particles; smoke density	Stomach cancer significantly related to a particulate deposit index.
Cardiovascular disease	Urban versus rural areas	A substantial abatement of air pollution would lead to 10 to 15 percent reduction in the mortality and morbidity rates for heart disease.
Total respiratory disease	Sulfates; particles	Strong relation between incidences of emphysema and bronchitis and air pollution. Also relations for pneumonia and influenza cited.
Total mortality rate	Sulfates; particles	A 10 percent decrease in the minimum concentration of particles would decrease the total death rate by 0.5 percent. A 10 percent decrease in the minimum concentration of sulfates would decrease the total death rate by 0.4 percent.

1.5 AIR QUALITY STANDARDS VERSUS EMISSIONS STANDARDS

The underlying motivation of a study of air pollution is to implement its control and abatement. More specifically, the objective is to control pollutant sources so that ambient pollutant concentrations are reduced to levels considered safe from the standpoint of undesirable effects. To quantify the objectives it is desirable to know the amount of damage (to all aspects of the environment) caused by each pollutant as a function of exposure to various levels of that pollutant[1]. As we have seen in the previous section, such information is usually difficult to obtain.

The legislative basis for air pollution abatement in the United States is the 1963 Clean Air Act and its 1970 amendments. The 1970 amendments provide for the establishment of two kinds of national ambient air quality standards. *Primary* ambient air quality standards are those requisite to protect public health with an adequate margin of safety. *Secondary* ambient air quality standards specify a level of pollutant concentrations requisite to protect the public welfare from any known or anticipated adverse effects associated with the presence of such air pollutants in the air. Secondary standards are based on damage to crops, vegetation, wildlife, visibility, climate, and on adverse effects to the economy. Thus, an air quality standard is a level to which a pollutant concentration should be reduced to avoid undesirable effects. Air quality standards are not based on technological or economic acceptability; they are dictated solely by the effects of air pollution, not the causes.

Table 1.8 presents the national primary and secondary ambient air quality standards for photochemical oxidants, carbon monoxide, nitrogen dioxide, sulfur dioxide, suspended particulate matter, and hydrocarbons. In one case, the California standards, where different from the Federal standards, are also listed in Table 1.8.

The Clean Air Act requires each state to adopt a plan that provides for the implementation, maintenance, and enforcement of the national air quality standards. It is, of course, emission reductions that will abate air pollution. Thus, the states' plans must contain legally enforceable emission limitations, schedules, and time tables for compliance with such limitations. The control strategy must consist of a combination of measures designed to achieve the total reduction of emissions necessary for the attainment of the air quality standards. The control strategy may include, for example, such measures as emission limitations, emission charges or taxes, closing or relocation of commercial or industrial facilities, periodic inspection and testing of motor vehicle emission control systems, mandatory installation of control devices on motor vehicles, means to reduce motor vehicle traffic, including such measures as

[1] In many cases pollutants cannot be considered independently, since their effects are strongly influenced by the presence of other pollutants. Examples include SO_2 and particles and oxides of nitrogen and hydrocarbons.

Table 1.8 NATIONAL AMBIENT AIR QUALITY STANDARDS

Pollutant	Averaging time	Primary standard[a,b]	Secondary standard[a,b]	Most relevant effect	Comments
Photochemical oxidants (corrected for NO_2)	1 hr[c]	160 $\mu g/m^3$ (0.08 ppm)	Same as primary std.	Eye irritation	This level is below that associated with aggravation of respiratory diseases.
Carbon monoxide	8 hr	10 mg/m^3 (9 ppm)	Same as primary std.	2.5% COHb	This level is below that associated with impairment in time discrimination, visual function, and psychomotor performance.
	1 hr	40 mg/m^3 (35 ppm)		2.5% COHb	
Nitrogen dioxide	Annual average,	100 $\mu g/m^3$ (0.05 ppm)	Same as primary std.	. . .	For 1-hr average at slightly higher dosage, effects are observed in experimental animals. Also produces atmospheric discoloration.
	1 hr[d]	0.25 ppm			
Sulfur dioxide	Annual average,	80 $\mu g/m^3$ (0.03 ppm)	. . .	Respiratory irritation	This level is below that associated with alteration in lung function.
	24 hr	365 $\mu g/m^3$ (0.14 ppm)	. . .		
	3 hr		1300 $\mu g/m^3$ (0.5 ppm)		
	1 hr[d]	0.5 ppm			
Suspended particulate matter	Annual geometric mean,	75 $\mu g/m^3$	60 $\mu g/m^3$	Respiratory disease	See footnote *e*
	24 hr	260 $\mu g/m^3$	150 $\mu g/m^3$		
Hydrocarbons (corrected for methane)	3 hr (6–9 A.M.)[f]	160 $\mu g/m^3$ (0.24 ppm)	Same as primary std.	Eye irritation	Based on oxidant formation.

[a] National standards, other than those based on annual averages or annual geometric means, are not to be exceeded more than once a year. (Environmental Protection Agency, 1971).

[b] Concentration expressed first in units in which it was promulgated. Equivalent units given in parentheses are based upon a reference temperature of 25°C and a reference pressure of 760 mm of mercury. All measurements of air quality are to be corrected to a reference temperature of 25°C and a reference pressure of 760 mm. of Hg.

[c] Corrected for SO_2 in addition to NO_2.

[d] State of California standard. (State of California Air Resources Board, 1970.)

[e] This standard applies to suspended particulate matter in general. It is not intended to be a standard for toxic particles such as asbestos, lead, or beryllium. Because size distribution influences the effect of particulate matter on health, the standard will be reevaluated as data on health effects related to size distribution become available.

[f] Hydrocarbons alone have not been found to damage health. However, as we have noted, hydrocarbons react with oxides of nitrogen to produce photochemical smog. A Los Angeles study showed that high-school cross-country runners had impaired performances when photochemical oxidants (primarily ozone) ranged between 0.03 and 0.30 ppm for 1 hr prior to the start of a race. An ambient oxidant standard was set at 0.08 for 1 hr. Working backward from this value, based on air monitoring data, an ambient hydrocarbon standard of 0.24 ppm for a 3-hr duration was chosen.

gasoline rationing, parking restrictions, and carpool lanes on freeways, and expansion and promotion of the use of mass transportation facilities.

Each state must also provide a contingency plan to control pollutant sources during periods of air stagnation when air quality is predicted to reach levels which would constitute imminent and substantial endangerment to human health. The contingency plans are to provide for emission reductions based on "warning" and "emergency" episode concentrations. The warning levels indicate that additional control is necessary if meteorological conditions can be expected to remain the same for a period of 12 hours. At the emergency levels, the most stringent controls are necessary to insure that concentrations do not reach levels where imminent and substantial endangerment to the health of any significant portion of the population will occur. Table 1.9 summarizes the national warning, emergency, and significant harm levels.

The Clean Air Act also empowers the Environmental Protection Agency (EPA) to establish standards of performance for new stationary sources and modified existing sources. In 1971 standards of performance were promulgated for limiting emissions of sulfur oxides, nitrogen oxides, particulate matter, and sulfuric acid mist

Table 1.9 WARNING, EMERGENCY, AND SIGNIFICANT HARM LEVELS FOR AMBIENT AIR QUALITY

Pollutant	Warning levels	Emergency levels	Significant harm levels
Photochemical oxidant	800 $\mu g/m^3$, 1-hr av (0.4 ppm)	1200 $\mu g/m^3$, 1-hr av (0.6 ppm)	800 $\mu g/m^3$, 4-hr av (0.4 ppm) 1200 $\mu g/m^3$, 2-hr av (0.6 ppm) 1400 $\mu g/m^3$, 1-hr av (0.7 ppm)
Carbon monoxide	34 mg/m^3, 8-hr av (30 ppm)	46 mg/m^3, 8-hr av (40 ppm)	57.5 mg/m^3, 8-hr av (50 ppm) 86.3 mg/m^3, 4-hr av (75 ppm) 144 mg/m^3, 1-hr av (125 ppm)
Nitrogen dioxide	2260 $\mu g/m^3$, 1-hr av (1.2 ppm) 565 $\mu g/m^3$, 24-hr av (0.3 ppm)	740 $\mu g/m^3$, 24-hr av (0.4 ppm)	3750 $\mu g/m^3$, 1-hr av (2.0 ppm) 938 $\mu g/m^3$, 24-hr av (0.5 ppm)
Sulfur dioxide	1600 $\mu g/m^3$, 24-hr av (0.6 ppm)	2100 $\mu g/m^3$, 24-hr av (0.8 ppm)	2620 $\mu g/m^3$, 24-hr av (1.0 ppm)
Particulate matter	625 $\mu g/m^3$, 24-hr av	875 $\mu g/m^3$, 24-hr av	1000 $\mu g/m^3$, 24-hr av
Sulfur dioxide and particulate matter combined	Product of $\mu g/m^3$ for both is equal to 261×10^3, 24-hr av	Product of $\mu g/m^3$ for both is equal to 393×10^3, 24-hr av	Product of $\mu g/m^3$ for both is equal to 490×10^3, 24-hr av

from: (1) fossil-fuel-fired electric generators of more than 250 million Btu/hr heat input (1 Btu = 1055 joules), (2) incinerators of more than 50 tons/day charging rate, (3) portland cement plants, (4) nitric acid plants, and (5) sulfuric acid plants. Since then, additional standards have been passed for other stationary sources.

Control of motor vehicle emissions began in California in 1960. Federal controls were initiated in 1967 by adopting California's standards for crankcase and exhaust emissions. The 1970 amendments to the Clean Air Act set motor vehicle exhaust emission standards for new cars through the 1976 model year The standards for light-duty vehicles (those of 6000 lb gross weight or less) required a 90 percent reduction in hydrocarbon and carbon monoxide emissions over those of 1970 vehicles and by 1976, a 90 percent reduction in nitrogen oxide emissions from those of 1971 vehicles

We do not wish to present a detailed list of emissions standards at this point, rather we only want to note the nature of emissions standards. While air quality standards are based largely on health effects, the question arises—On what are emissions standards based? In theory, one would like to work backward from air quality standards to determine those emission standards necessary to meet the air quality standards in a particular region. To do this, we need to relate emission rates to air quality, that is, to represent mathematically the "atmosphere" block in Fig. 1.1. Because of the great importance in obtaining this relation, we will devote considerable attention in this book to this problem. Unfortunately, mathematical models of the atmosphere have only recently reached the stage of development where they can be applied with any confidence to the evaluation of emission level–air quality relations. Therefore, almost all emission standard legislation has been based either on educated guesses of the effect that emission reductions have on air quality, or on the current technological and economic feasilibity of achieving a certain level of emission reduction for the process or piece of equipment in question.

Steps Leading to the Exhaust Emissions Standards in the 1970 Amendments to the Clean Air Act

The motor vehicle exhaust emissions standards contained in the 1970 amendments to the Clean Air Act were essentially based on an analysis performed by Barth (1970). The approach adopted by Barth is termed the rollback method of relating emission reductions to improvements in air quality. The rollback approach is based on first calculating the fractional reduction α in ambient concentrations necessary to achieve desired air quality. In applying rollback to propose and evaluate control strategies, α is calculated from the following relationship

$$\alpha = \frac{c_{mb} - c_{\text{des}}}{c_{mb} - c_b}$$

where c_{mb}, c_{des}, and c_b are the maximum concentration of the pollutant of interest observed in the airshed during a specified yearly period (termed the base year), the desired concentration (the ambient air quality standard), and the background concentration, respectively. It is then assumed that the allowable total daily emissions in the airshed are given by (assuming no growth)

$$E_{da} = (1 - \alpha)\, E_{db}$$

where E_{db} is the total daily emissions estimate for the region during the base year. A projected inventory for the future year of interest must be assembled, considering all factors that will affect emissions. Given these daily emissions, sufficient control measures must be proposed which will reduce emissions in the future year to the allowable level E_{da}, calculated above.

Using the rollback approach for the photochemical oxidant, Barth assumed the yearly maximum 1-hr average oxidant concentrations to be linearly proportional to reactive hydrocarbon emissions and independent of NO_X emissions, and maximum NO_2 concentrations to depend on NO_X emissions, independent of hydrocarbon emissions. In the rollback method, the spatial distribution of sources and the extent that each source contributes to the maximum pollutant concentrations are not considered. We see that the choice of the base year may have a substantial influence on the calculated allowable emissions in some future year. For example, a base year which is "good" meteorologically (that is one in which the maximum pollutant concentrations were relatively low due to favorable meteorological conditions) will lead one to the conclusion that allowable emissions in a future year may be greater than those calculated from a base year that is "bad" meteorologically.

To obtain the emission standards for carbon monoxide, hydrocarbons, and nitrogen oxides, Barth employed the rollback approach using a series of safety factors to guarantee a cushion. To begin, he used the most rapid available projection of the growth of the motor vehicle population from 1967 to 1990, a factor of 2.18 (as compared to the Department of Transportation's estimate of 1.64). He then used the highest recorded concentrations in the United States for each of the pollutants (hydrocarbons and NO_X in Los Angeles and CO in Chicago).

Using the primary air quality standard of 9 ppm over an 8-hr period (see Table 1.8), Barth determined that to reduce CO levels in Chicago to this value a 1980 exhaust emission rate of 6.16 g/mi[1] would be required. The primary ambient air quality standard for NO_X is 0.05 ppm annual average (see Table 1.8). To achieve this level in Los Angeles, Barth determined that the 1980 NO_X auto emission standard should be 0.4 g/mi. Finally, on the basis of the hydrocarbon standard of 0.24 ppm over a 3-hr period, Barth obtained an emission standard of 0.41 g/mi.

These figures suggested by Barth were then rounded off by Congress to the 90 percent requirement and moved up to 1975–6 from 1980. In moving up the goal date from 1980 to 1975–6, Congress overlooked the adjustment in the projected growth rate for emissions that was used by Barth. As Congress prepared the 1970 amendments, the CO standard of 6.16 g/mi was toughened to 3.4 g/mi, presumably on the basis that 1970 cars were already being controlled for CO emissions and that this controlled level should be the datum for the 90 percent reduction. Independently, the California Air Resources Board had determined a set of auto emission standards which it calculated would meet public health requirements until the year 2000. These standards were: 17 g/mi for CO, 1.5 g/mi for NO_X and 0.9 g/mi for hydrocarbons.

A number of factors used in formulating the exhaust emission standards have since been revised. The auto population growth factor of 2.18 is now seen to be too high. A value closer to 1.64, that originally suggested by the Department of Transportation, appears reasonable. Second, the primary ambient air quality standard for NO_2 (0.05 ppm annual average) has since been seriously questioned. The ambient air quality standard for NO_X was based on the so-called Chattanooga school children study. Researchers divided a sector around a Chattanooga, Tenn. dynamite plant that was emitting NO_X and studied the health of school children living various distances from the plant. It was found that the children showed an increased tendency toward respiratory disease at 24-hr exposures to 0.062 and 0.109 ppm of NO_X. The ambient NO_X standard was chosen from these data as 0.05 ppm. Now, the Chattanooga school children study has been discredited by reason of faulty data. The school children may have actually been getting sick from acid mists, not NO_X.

Finally, it is now recognized that the rollback approach provides a poor estimate of the effect of hydrocarbon and NO_X emission reductions on oxidant air quality. Laboratory data show that

[1] 1 mi = 1.609 km. This standard expresses the allowable mass of carbon monoxide discharged by a motor vehicle upon traveling 1 mi. The way in which this quantity is determined for a particular vehicle is discussed in Chap. 8.

the maximum concentration of oxidant depends significantly on the initial hydrocarbon-to-NO_x concentration ratio as well as on the absolute levels of these species.

There are basically two critical points in the air quality standard–emission standard problem. The first, as we have already noted, is the difficulty in obtaining accurate and reliable health data. The second is the problem of relating emission levels to air quality. The following section is devoted to a general discussion of relating emission level to air quality.

1.6 EMISSION LEVEL–AIR QUALITY MODELS

An ambient air quality model is a means whereby pollutant emissions can be related to atmospheric pollutant concentrations. Such a model provides, therefore, a link between emission changes from source control measures and the changes in airborne pollutant concentration levels that can be expected to result. The model will involve considerations of emission patterns, meteorology, chemical transformations, and removal processes.

Many difficult questions and complex issues arise in planning for the abatement and control of air pollution in an urban area. Certain key aspects of abatement planning are best addressed through the use of an ambient air quality model; in some cases there are no alternative means of examining the critical issues. Included among those topics for which an ambient air quality model may be particularly useful as an analytical tool are:

Establishment of emission control legislation
Evaluation of proposed emission control techniques and strategies
Planning of locations of future sources of air contaminants
Planning for the control of air pollution episodes
Assessment of responsibility for existing levels of air pollution

1.6.1 Types of Models

Ambient air quality models can be divided, broadly speaking, into two types: physical and mathematical.

Physical models are intended to simulate the atmospheric processes affecting pollutants by means of a small-scale representation of the actual air pollution problem. A physical model sometimes employed to study the dispersion of pollutants consists of a small-scale replica of the urban area or a portion thereof in a wind tunnel. The problems associated with properly duplicating the actual atmospheric scales of turbulent motion make physical models of this variety of very limited usefulness. The second type of physical models are those intended to simulate the chemical processes occurring in the atmosphere. These models, commonly referred to as *smog chambers*, have been used extensively for 20 years as a means of isolating and studying atmospheric chemical processes. A smog chamber ordinarily consists of a well-stirred laboratory reactor in which primary pollutants are introduced at concentrations typical of the atmosphere and in which chemical reactions subsequently are allowed to take place under atmospheric conditions of temperature, pressure, and radiation. A great deal of important information on atmospheric chemical reactions has been derived from data obtained in smog chambers. However, because changing emission patterns and atmospheric transport and diffusion cannot be simulated in a smog chamber, this type of physical model is inadequate as a general-purpose ambient air quality model. Therefore, while useful for isolating certain elements of atmospheric behavior and invaluable for studying certain critical details, physical models cannot serve the needs of ambient air quality models capable of relating emissions to air quality under a variety of meteorological and source emission conditions over an urban area. Recourse must be had to mathematical models, which can broadly be classified under two types: (1) models based on statistical analysis of past air monitoring data, and (2) models based on the fundamental description of atmospheric transport and chemical processes.

The first class of mathematical ambient air quality models are those based on statistical analysis of past air monitoring data. All large urban airsheds in the United States contain a number of air monitoring stations operated under the auspices of local or state air pollution control authorities. At most stations 15-min to 1-hr average pollutant concentration levels are reported. For example, in California, the contaminants normally measured are total oxidants, carbon monoxide, nitric oxide, nitrogen dioxide, total hydrocarbons, sulfur dioxide, suspended particulate matter, and lead. A great deal of information is potentially available from these enormous data bases.

There are a variety of ways in which these data can be analyzed to yield models relating emissions to air quality. The linear rollback method is perhaps the simplest one based directly on air monitoring data. It assumes that present air quality, as measured by the highest concentration experienced during a particular time period, can be linearly scaled down in direct proportion to emission intensity until background air quality is achieved at zero emissions. In dealing with both primary and secondary pollutants, one can propose types of statistical-empirical models based on correlations of early morning primary pollutant concentrations and secondary pollutant concentrations reached later in the day. Finally, by statistical analysis of past air monitoring data, one can obtain probabilities that certain atmospheric concentration levels will be exceeded as a function of source emission levels. Models of the statistical-empirical type are generally based on several years of monitoring data. Thus, meteorological and chemical inputs do not generally enter explicitly into these models, but, of course, do influence the calculated probabilities implicitly.

In principle, the emission level–air quality problem could be resolved with the use of a reliable, fully validated mathematical model based on fundamental description of atmospheric transport and chemical processes. Every urban locale has its own topographic, climatological, and meteorological conditions. The spatial distribution of contaminant emissions, as well as the average composition of these emissions, also varies from city to city. The concentration of secondary pollutants, formed in the atmosphere through the chemical reaction of primary (or emitted) pollutants, is influenced by local meteorology, emissions patterns, intensity of solar radiation, and many other variables.

Mathematical models that attempt to simulate the complex atmospheric processes involved in urban air pollution are based in general on the equations of mass conservation for individual pollutant species. Models based on the equations of conservation of mass, or continuity, of course, cannot predict variations in the wind velocity field (the momentum equations are needed) or the temperature field (the energy equations are needed). Wind and temperature information thus must be input as data. What such models can do, however, is relate in one equation (or set of equations for more than one species) the effects of all the dynamic processes that influence the mass balance on a parcel of air. These include the transport, turbulent diffusion, and reaction of all pollutant species of interest. Also treated in such models are the introduction or removal of species.

A model based on the equations of conservation of mass requires, as a part of its formulation or as data input, information falling into the following categories: emissions, meteorology, and atmospheric chemistry and removal processes. Models may describe the behavior of reactive species, or they may be limited in application to inert species. Furthermore, models may be formulated under the assumption of steady-state behavior, or they may be descriptors of time-varying behavior. Temporal and spatial resolution of models may vary widely. Models may be based on a fixed grid, or they may be formulated so as to trace the variations in concentration in an "air parcel" moving with the average wind field.

An emission inventory for an airshed has the objective of predicting the mass flux of pollutants into the atmosphere as a function of time and location in the airshed. For the purposes of such an inventory, sources can be divided into two classifications: (1) stationary sources (factories, homes, power plants), and (2) moving sources (motor vehicles, aircraft). The inventory would include the location of all significant stationary sources, and of all highways, by geographic coordinates. For each stationary source, information on the quantity of each pollutant emitted as a function of level of activity (such as the quantity of fuel burned or the quantity of raw materials processed) and, if necessary, the variation of level of activity with time of the day, is required. For mobile sources, the volumes of vehicular traffic (and, if possible, speeds) on all the roadways of the region at different times is necessary. Then, given a model for motor vehicle emissions, say, in terms of g pollutant/km traveled, the total pollutant flux from mobile sources can be computed. While the inventory would describe a typical day, daily (weekday versus weekend) and seasonal variations would be included. The emission inventory could easily be altered to accept long-term forecasts relating to locations of new sources, new transportation patterns, and modified activity distributions for stationary and moving sources. The ability to alter the inventory readily in accord with forecasts is important in using the overall simulation model for future studies.

Meteorological information needed as input generally includes wind speed, wind direction and temperature as a function of elevation—all in addition as a function of location, and on a time scale commensurate with that inherent in the model. A kinetic mechanism for atmospheric chemical reactions is an essential component of a model that is capable of predicting concentrations of chemically reacting species, as is the knowledge of reaction-rate constants for those reactions that comprise the mechanism.

Once the model is properly formulated, and information concerning emissions, meteorology, and chemistry is appropriately incorporated, the model may be solved to obtained predicted concentrations of individual species as a function of location and/or time. Air quality data are normally required to assess the accuracy of the model, by comparing predictions with observations. Methods of solution of the equations comprising the model may be simple or complex. In the case of steady-state models based on the conservation of mass of a single species, solutions may be read directly from charts or tables. For more complex formulations, numerical integration of the governing equations on a digital computer is usually necessary.

1.6.2 Temporal Resolution of Models

The temporal resolution of an ambient air quality model (that is, the time period over which the predicted concentrations are averaged) may vary from several minutes to one year. For example, a model may predict the 15-min average pollutant concentration as a function of location in the airshed, or it may predict the yearly average concentration as a function of location. The requirements in implementing a model will be strongly governed by its temporal resolution.

Models based on statistical analysis of past air monitoring data are generally derived from several years of measurements at one or more stations in an airshed. The concentration actually correlated with emission levels may range from a several-minute to a yearly average value. Thus, the model may predict, for example, the probability that the 1-hr average concentration at a certain station will exceed a given level if total source emission levels in the region are at a prescribed value. Since data are generally available on time periods as short as from 15 min to 1 hr, the choice of the time period for averaging of the data for the statistical model is dictated solely by the purposes for which the model is to be used. The input data requirements for a statistical model generally consist only of estimates of source emission levels since meteorology and atmospheric transformation and removal processes are implicit in the reported data.

Models based on the fundamental description of atmospheric transport and chemical processes may have temporal resolution ranging from the order of several minutes to a year. In general, the basis of these models is the partial differential equation of continuity for an individual species. Those models that require the solution in time of a differential equation based on the continuity equation can be called *dynamic* models, since they describe the evolution of pollutant concentrations with time at different locations in the airshed. Thus, dynamic models simulate the actual real-time temporal behavior of air pollutants in the atmosphere. These models require as inputs the spatial and temporal distribution of emissions over the region of interest, the spatial and temporal distribution of pertinent meteorological variables, and information on the time rate of change of concentrations at a point resulting from transformation and removal processes.

If certain simplifying assumptions are invoked in dealing with the equation of continuity, such as steady source rates and meteorology, then the equation can be

Table 1.10 REQUIREMENTS OF MATHEMATICAL AMBIENT AIR QUALITY MODELS BASED ON TEMPORAL RESOLUTION

Inputs	Statistical models based on past air monitoring data	Models based on the fundamental description of atmospheric transport and chemical processes	
		Dynamic models	Steady-state models
Source emissions	Total daily mass emissions. No temporal resolution required, other than year-to-year changes.	Emissions prescribed as function of location and time with as fine a degree of detail as desired in the predicted concentrations.	Emissions prescribed as a function of location. No temporal variation necessary since steady-state conditions are assumed to prevail.
Meteorological variables	No inputs required. (These variables are implicit in the air monitoring data.)	Wind speed and direction, atmospheric stability, radiation intensity required as a function of location and time with as fine a degree of temporal detail as desired in the predicted concentrations.	Wind speed and direction and atmospheric stability categorized according to frequency of occurrence over the time period of averaging. No spatial variation required
Atmospheric transformation and removal processes.	No inputs required. (These processes are implicit in the air monitoring data.)	Chemical-reaction rate equations must be specified as well as the rate of removal by deposition and gas-to-particle transfer.	Inapplicable to species that undergo nonlinear transformations in the atmosphere.

integrated over a long time period to yield a *steady-state* model. A steady-state model is then capable of predicting the spatial distribution of airborne pollutant concentrations under conditions of time-invariant meteorology and source emission rates. The steady-state models developed to date have been restricted to inert contaminants or those that decay by first-order chemical reaction. Steady-state models do not exist for pollutants which undergo nonlinear transformations in the atmosphere (such as hydrocarbons and oxides of nitrogen) because of the difficulty in properly representing the steady-state concentration distributions of reactive species.

Table 1.10 summarizes the requirements of mathematical ambient air quality models based on their temporal resolution.

1.6.3 Spatial Resolution of Models

The spatial resolution of an ambient air quality model (that is, the area over which the predicted concentrations are averaged) may vary from several meters to several thousand kilometers.

A model based on statistical analysis of past air monitoring data might predict concentrations at one or more stations (essentially "point" measurements) or the average of the readings at a number of stations, intended to represent the average pollutant concentrations in the region. In a model based on the solution of the species continuity equations, the concentrations appearing in those equations are essentially point concentrations. The partial differential equation(s) comprising the model are normally solved numerically by a process that requires the continuous concentration field to be approximated by a discrete grid of points. The choice of the spatial grid on which the equations are solved is governed by the degree of spatial detail in the emissions inventory and the meteorological variables. For example, if the emissions inventory and the meteorological variables are available with a spatial resolution of 5 km, then the spatial resolution of the predicted concentrations can be no smaller than 5 km. Sometimes it is desired to predict pollutant concentrations in the immediate vicinity of sources, such as a highway. In such a case, the spatial resolution of the concentrations might be as small as a few meters.

Different pollutants pose different needs relative to the spatial scales of modeling. For example, carbon monoxide is essentially a local problem in the vicinity of heavily traveled highways and intersections. Thus, to assess the effectiveness of motor vehicle emission controls or traffic alterations on CO levels requires a model with spatial resolution the order of the width of a city street. On the other hand, photochemical oxidant is a region-wide problem, caused by massive area-wide emission of hydrocarbons and oxides of nitrogen. A model capable of relating emission changes to air quality changes for photochemical oxidant might only require a minimum spatial resolution of several square kilometers.

Table 1.11 summarizes the requirements of mathematical ambient air quality models based on their spatial resolution.

1.6.4 Model Requirements as Related to Ambient Air Quality Standards

The type of ambient air quality model, as well as its spatial and temporal resolution, will be dictated by the purpose for which the model is to be used. Ordinarily a model is used to determine if airborne pollutant concentrations corresponding to prescribed emission conditions will meet ambient air quality standards. Since ambient air quality standards are derived on the basis of health and other effects, the time periods over which the concentrations are averaged (in the air quality standards) vary depending on the particular pollutant and its effect. Thus, models with different spatial and temporal resolution may be called for depending on which pollutant or pollutants are being considered.

Although not included in our discussion, a future component of urban airshed

Table 1.11 REQUIREMENTS OF MATHEMATICAL AMBIENT AIR QUALITY MODELS BASED ON SPATIAL RESOLUTION

Inputs	Statistical models based on past air monitoring data	Models based on the fundamental description of atmospheric transport and chemical processes	
		Dynamic models	Steady-state models
Source emissions	Total daily mass emissions. Spatial resolution is arbitrary since the data can be correlated with any desired emission measure.	Emissions prescribed as a function of location and time with as fine a degree of spatial resolution as desired in the predicted concentrations.	Emissions prescribed as a function of location. Since steady-state models are usually based on representing the sources as point, line, or area sources, only their geographic location is needed.
Meteorological variables	No inputs required. (These variables are implicit in the air monitoring data.)	Wind speed and direction, atmospheric stability, radiation intensity required as a function of location and time with as fine a degree of spatial detail as desired in the predicted concentrations.	No spatial resolution in meteorological variables permitted during steady-state periods.
Atmospheric transformation and removal processes	No inputs required. (These variables are implicit in the air monitoring data.)	Spatial resolution need not explicitly enter expressions for transformation and removal processes.	Inapplicable to species that undergo nonlinear transformations in the atmosphere.

models is a *population dosage model.* This model would permit evaluation of alternative control policies from the epidemiological viewpoint, by estimating the exposure of the population to pollutants. When health effects can be more closely related to concentration levels, air quality standards could be converted into population dosage standards. A population dosage model would contain (1) the distribution of population by residence, (2) the distribution of population by employment, and (3) the distribution of commuting times and distances. The model would then compute the expected exposure of population groups.

1.7 SYSTEMS ANALYSIS OF AIR POLLUTION ABATEMENT

In general, the goals of air pollution abatement are the meeting of a set of air quality standards based on medical and perhaps also aesthetic and economic effects. As we have seen, at the present time there is insufficient knowledge to permit an unambiguous, quantitative link between air pollutant levels and health and other effects. As additional epidemiological and other data become available, air quality standards will become more definite.

Air pollution abatement programs can be divided into two categories:

1 Long-term control
2 Short-term control (episode control)

Long-term control strategies involve a legislated set of measures to be adopted over a multiyear period. Episode (or short-term) control involves shutdown and slowdown procedures which are adopted over periods of several hours to several days under impending adverse meteorological conditions. An example of a short-term strategy is the emergency procedures for fuel substitution by coal-burning power plants in Chicago when SO_2 concentrations reach certain levels (Croke and Booras, 1969).

Figure 1.6 illustrates the elements of a comprehensive regional air pollution control strategy, consisting of both long- and short-term measures. Under each of the two types of measures are listed some of the requirements for setting up the control strategy. The air quality objectives of long- and short-term strategies may be quite different. For long-term control, a typical objective might be to reduce to 20 the expected number of days per year that the maximum hourly average airshed concentration of a certain pollutant exceeds a given value. On the other hand, a goal of short-term control is ordinarily to keep the maximum concentration of a certain pollutant below a given value at all stations in the airshed on that particular day. As we will discuss later, each of these two types of control may require different modeling capabilities.

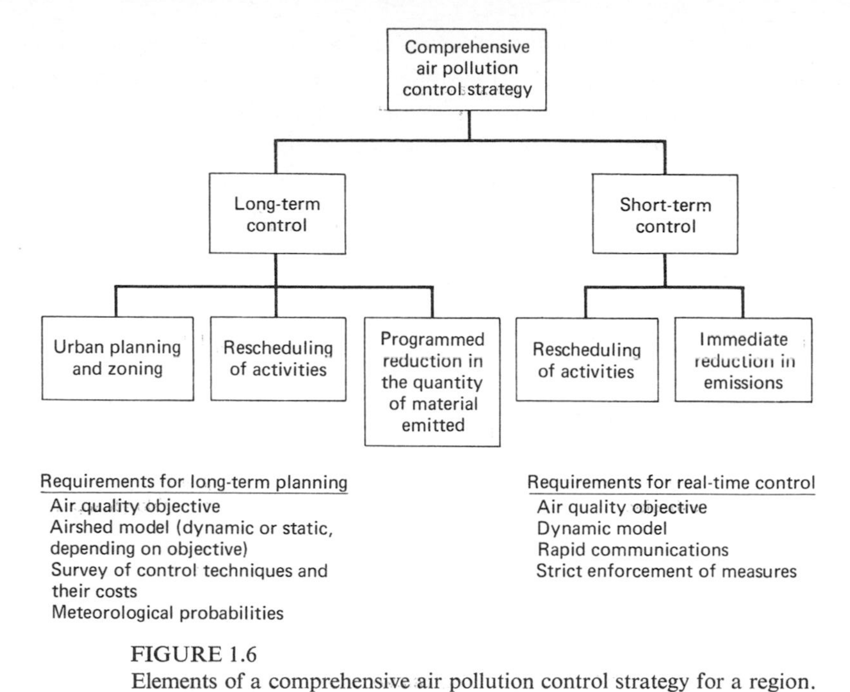

FIGURE 1.6
Elements of a comprehensive air pollution control strategy for a region.

The alternatives for abatement policies depend on whether long- or short-term control measures are being considered. Some examples of long-term air pollution policies are:

Enforcing standards that restrict the pollutant content of combustion exhaust
Requiring used motor vehicles to be outfitted with exhaust control devices
Requiring new motor vehicles to meet certain emissions standards
Prohibiting or encouraging the use of certain fuels in power plants
Establishing zoning regulations for the emission of pollutants
Encouraging the replacement of existing fossil-fuel-burning power plants with nuclear plants
Encouraging the use of electrically powered or natural-gas-powered vehicles for fleets

Short-term controls are of an emergency nature and more stringent than long-term controls that are continuously in effect. Examples of short-term control strategies are:

Prohibiting automobiles with fewer than three passengers from using freeways
Prohibiting the use of certain fuels in some parts of the city
Prohibiting certain activities, such as incineration of refuse

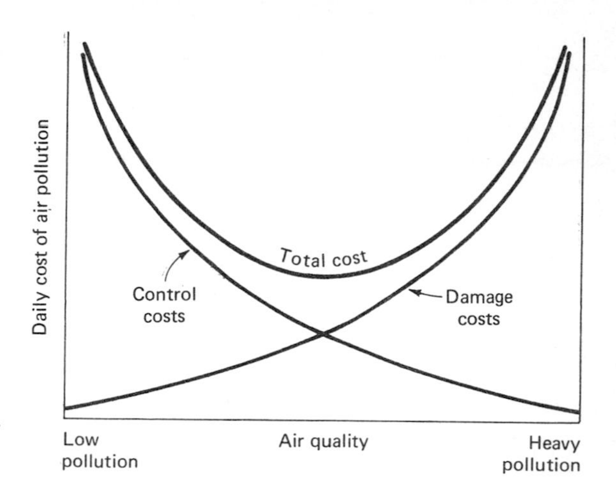

FIGURE 1.7
Total cost of air pollution as a sum of control costs and damage costs.

Let us focus our attention primarily on long-term control of air pollution for a region. It is clear that potentially there are a large number of control policies which could be applied by an air pollution control agency to meet desired air quality goals. The question then is: How do we choose the "best" policy from among all the possibilities? It is reasonable first to establish criteria by which the alternative strategies are to be judged.

Within the field of economics, there is an elaborate hierarchy of techniques called cost/benefit analysis, within which all the consequences of a decision are reduced to a common indicator, invariably dollars. This analysis employs a single measure of merit, namely the total cost, by which all proposed programs can be compared. A logical inclination is to use total cost as the criterion by which to evaluate alternative air pollution abatement policies. The total cost of air pollution control can be divided into a sum of two costs:

1 Damage costs: the costs to the public of living in polluted air, e.g. tangible losses such as crop damage and deteriorated materials and intangible losses such as reduced visibility and eye and nasal irritation
2 Control costs: the costs incurred by polluters (and the public) in order to reduce emissions, e.g. direct costs such as the price of equipment which must be purchased and indirect costs such as induced unemployment as a result of plant shutdown or relocation

We show in Fig. 1.7 the qualitative form of these two costs and their sum as a function of air quality; poor air quality has associated with it high damage costs and low control costs, whereas good air quality is just the reverse. The key problem is: How do we compute these curves as a function of air quality? Consider first the question of quantifying damage costs.

Damage costs to materials and crops, cleaning costs due to soiling, etc., although not easy to determine, can be estimated as a function of pollutant levels (Ridker, 1967). However, there is the problem of translating into monetary value the increased sickness and death resulting from air pollution. One way of looking at this problem is to ask: How much are people willing to spend to lower the incidence of disease, prevent disability, and prolong life? Attempts at answering this question have focused on the amount that is spent on medical care and the value of earnings missed as a result of sickness or death. Lave and Seskin (1970) stated that "while we believe that the value of earnings forgone as a result of morbidity and mortality provides a gross underestimate of the amount society is willing to pay to lessen pain and premature death caused by disease, we have no other way of deriving numerical estimates of the dollar value of air pollution abatement." Their estimates are summarized in Table 1.12. These estimates are so difficult to make that we must conclude that at this time it is not possible to derive a quantitative damage-cost curve such as that shown in Fig. 1.7.

There are actually other reasons why a simple cost/benefit analysis of air pollution control is not feasible. Cost is not the only criterion for judging the consequences of a control measure. Aside from cost, social desirability and political acceptability are also important considerations. For example, a policy relating to zoning for high and low pollution activities would have important social impacts on groups living in the involved areas, and it would be virtually impossible to describe the resulting population movements in dollar terms.

It therefore appears that the most feasible approach to determining air pollution abatement strategies is to treat the air quality standards as constraints not to be violated and to seek the combination of strategies that achieve the required air quality at minimum cost of control. In short, we attempt to determine the minimum cost of achieving a given air quality level through emission controls, i.e. to determine the con-

Table 1.12 ESTIMATED HEALTH COSTS OF AIR POLLUTION IN 1970

Disease	Total annual estimated cost (millions of dollars)	Estimated percentage decrease in disease for a 50% reduction in air pollution	Estimated savings incurred for a 50% reduction in air pollution (millions of dollars)
Respiratory disease	4887	25	1222
Lung cancer	135	25	33
Cardiovascular disease	4680	10	468
Cancer	2600	15	390
Total savings			~ $2100

SOURCE: Lave and Seskin (1970).

trol cost curve in Fig. 1.7, where each point on the curve represents an optimal combination of strategies. We now present a general framework whereby one may determine the least-cost set of control measures to achieve a given air quality standard for an airshed.

Determination of Least-cost Air Pollution Control Strategies

An airshed system can be visualized to consist of a number of components: (1) various pollutant-emitting sources such as motor vehicles, power plants, industries, aircraft, etc.; (2) various chemical species, i.e. pollutants; (3) a multitude of methods for controlling the emissions of the various sources; and (4) meteorological parameters, consisting of wind speeds and directions, stability conditions, radiation intensities, etc., and their frequency of occurrence throughout the year.

A set of control measures applied to the sources will produce a given level of pollutant emissions having a certain spatial and temporal distribution, resulting in some level of air quality. Conversely, for a given degree of air quality there may exist many possible combinations of control measures that can achieve the same air quality. Therefore, the question arises: Which set of measures is, in some sense, the best?

We can envision the analysis of this question as consisting of two steps: (1) the relation between control methods and emission levels, and (2) the relation between emission levels and air quality. The control method–emission level problem consists of determining the least-cost allocation of controls to achieve certain specified mass emission levels from a variety of emitting sources. The emission level–air quality problem consists of the conversion of emission levels into spatial (and, perhaps, temporal) distributions of airborne pollutant concentrations. The former problem, the control method–emission level relation, has received considerable attention, e.g. see Kohn (1969, 1970, 1971a,b,c), Burton and Sanjour (1970), Trijonis (1972), and Seinfeld and Kyan (1971). The latter problem, that relating emission levels to air quality, has already been discussed in Sec. 1.6.

Let us now consider the formulation of the control method–emission level problem for air pollution control, that is, to determine that combination of control measures employed in a certain year that will give mass emissions not greater than prescribed values and do so at least cost. Let $E_1, \ldots, E_N$ represent measures of the mass emissions[1] of N pollutant species (for example, these could be the total daily emissions in the entire airshed in a particular year or the mass emissions as a function of time and location during a day); then we can express the control cost C (say in dollars per day) as $C = C(E_1, \ldots, E_N)$. In order to illustrate the means of minimizing C we take a simple example (Kohn, 1969).

Let us consider a hypothetical airshed with one industry, cement manufacturing. The annual production is 2.5×10^6 barrels of cement, but this production is currently accompanied by 2 kg of particulate matter per barrel lost into the atmosphere. Thus, the uncontrolled particulate emissions are 5×10^6 kg/yr. It has been determined that emissions should not exceed 8×10^5 kg/yr. There are two available control measures, both electrostatic precipitators: Type 1 will reduce emissions to 0.5 kg/bbl and costs \$0.14/bbl; type 2 will reduce emissions to 0.2 kg/bbl but costs \$0.18/bbl. Let

$$X_1 = \text{bbl/yr of cement produced with type 1 units installed}$$
$$X_2 = \text{bbl/yr of cement produced with type 2 units installed}$$

The total cost of control in dollars per year is thus

$$C = 0.14X_1 + 0.18X_2$$

We would like to minimize C by choosing X_1 and X_2. But X_1 and X_2 cannot assume any values; their total must not exceed the total cement production,

$$X_1 + X_2 \leq 2.5 \times 10^6 \tag{1.1}$$

[1] Note that E_i is 0 if i is purely a secondary pollutant.

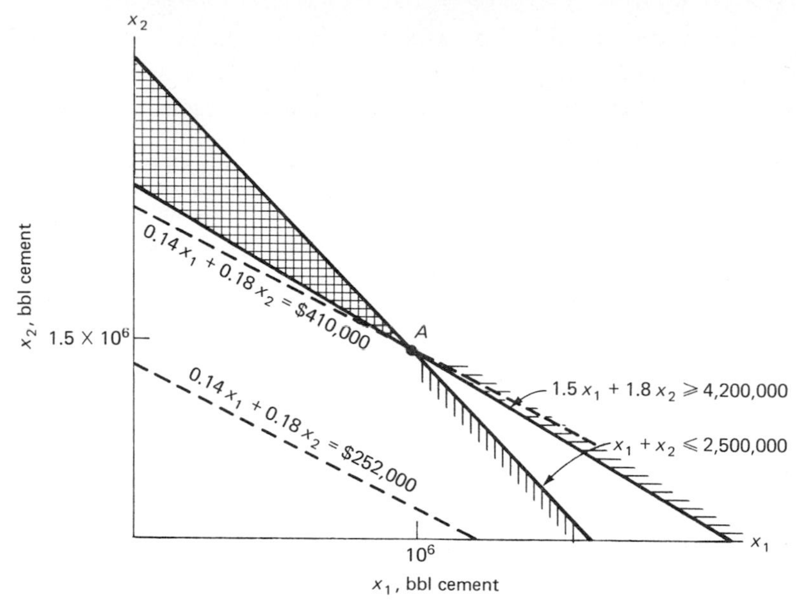

FIGURE 1.8
Least-cost strategy for cement industry example. (*Source: Kohn*, 1969.)

and a reduction of at least 4.2×10^6 kg of particulate matter must be achieved,

$$1.5X_1 + 1.8X_2 \geq 4.2 \times 10^6 \tag{1.2}$$

and both X_1 and X_2 must be nonnegative,

$$X_1, X_2 \geq 0 \tag{1.3}$$

The complete problem is to minimize C subject to Eqs. (1.1) to (1.3). In Fig. 1.8 we have plotted lines of constant C in the X_1X_2 plane. The lines corresponding to (1.1) and (1.2) are also shown. Only X_1, X_2 values in the cross-hatched region are acceptable. Of these, the minimum cost set is $X_1 = 10^6$ and $X_2 = 1.5 \times 10^6$ with $C =$ \$410,000. If we desire to see how C changes with the allowed particulate emissions, we solve this problem repeatedly for many values of the emission reduction (we illustrated the solution for a reduction of 8×10^5 kg of particulate matter per year) and plot the minimum control cost C as a function of the amount of reduction.

The problem that we have described falls within the general framework of *linear programming* problems. Linear programming refers to the problem of minimizing a linear function subject to linear equality or inequality constraints. Its application requires that control costs and reductions remain constant, independent of the level of control. In a more general manner, the problem can be stated formally as follows. Let

$S_i =$ magnitude of emission source i, $i = 1, 2, \ldots, s$
number of 1969 motor vehicles in airshed
number of boilers of a certain size in airshed, etc.

$L_i =$ supply limits of the ith fixed-supply commodity, $i = 1, 2, \ldots, l$
total cubic meters of available natural gas
total barrels of available low-sulfur fuel oil, etc. (only for those commodities in fixed limited supply)

X_j = level of control activity j, $j = 1, 2, \ldots, r$
number of exhaust control devices on 1969 motor vehicles
number of boilers with flue-gas recirculation
number of cubic meters of natural gas burned, etc.
c_j = cost (\$/day) of one unit of control activity j, $j = 1, 2, \ldots, r$
E_i^0, E_i = daily emissions of species i (kg) without control and with control, respectively, $i = 1, 2, \ldots, N$
A_{ij} = number of units of source i controlled by one unit of control activity j, $i = 1, 2, \ldots, s$; $j = 1, 2, \ldots, r$
$A_{ij} = 1$ for one exhaust control device on one 1969 motor vehicle
$A_{ij} = 1$ for one boiler with flue-gas recirculation
D_{ij} = amount of the ith limited-supply input used by one unit of the jth control activity, $i = 1, 2, \ldots, l$; $j = 1, 2, \ldots, r$
$D_{ij} = 1$ for 1 cubic meter of natural gas used as control method j
B_{ij} = reduction in the mass emissions of species i (kg/day) by one unit of the jth control activity, $i = 1, 2, \ldots, N$; $j = 1, 2, \ldots, r$
B_{ij} = kg/day of hydrocarbons reduced by one exhaust control device on one 1969 motor vehicle
B_{ij} = kg/day of NO_X reduced by one boiler with flue-gas recirculation

The general problem is to minimize the total control cost C,

$$C = \sum_{i=1}^{r} c_i X_i \tag{1.4}$$

subject to the following constraints:

1 At least a specified reduction, $E_i^0 - E_i$, in total daily mass emissions of each pollutant be achieved:

$$\sum_{j=1}^{r} B_{ij} X_j \geq E_i^0 - E_i\dagger \qquad i = 1, 2, \ldots, N \tag{1.5}$$

2 The number of sources controlled not exceed the total number of sources in the airshed:

$$\sum_{j=1}^{r} A_{ij} X_j \leq S_i \qquad i = 1, 2, \ldots, s \tag{1.6}$$

3 The amount of fixed-supply inputs not be exceeded:

$$\sum_{j=1}^{r} D_{ij} X_j \leq L_i \qquad i = 1, 2, \ldots, l \tag{1.7}$$

4 The control units be nonnegative:

$$X_i \geq 0 \tag{1.8}$$

As we have noted, this formulation assumes that control activities are homogeneous in their effect on emissions (constant B_{ij}), in their consumption of limited commodities (constant D_{ij}), and in their costs (c_j constant).

† An alternative to this constraint is to ask that a set reduction be achieved exactly, i.e.

$$\sum_{i=1}^{r} B_{ij} X_j = E_i^0 - E_i$$

In many ways this is more convenient than an inequality constraint.

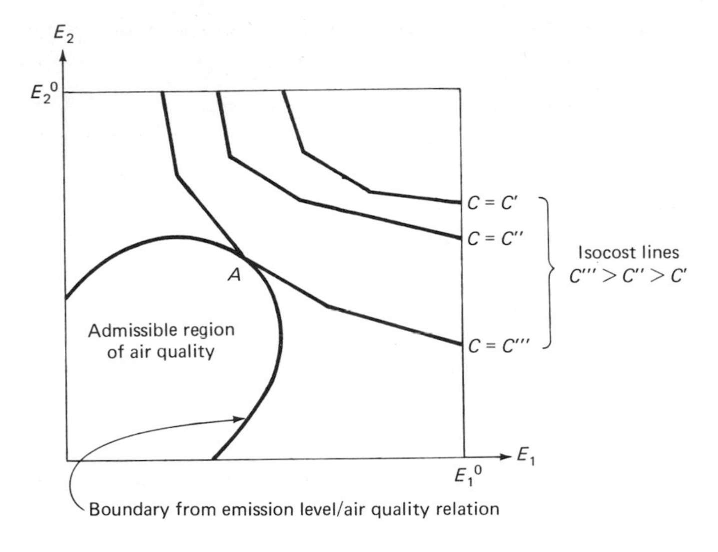

FIGURE 1.9
Single-year minimum-cost strategy for two pollutants, using linear programming.

Figure 1.9 illustrates the form of the results of the linear-programming control cost/emission level problem for two pollutants. The axes are the mass emission levels of the two pollutants, with the uncontrolled levels indicated by $E_1{}^0$ and $E_2{}^0$. We would solve the above linear programming problem for several sets of E_1, E_2 values [using the equality constraint in (1.5)], each time obtaining a value of the minimum cost C necessary to achieve the particular emission levels (or reductions $E_1{}^0 - E_1$ and $E_2{}^0 - E_2$). Lines of constant costs can then be drawn in the E_1E_2 plane. We have shown three such lines in Fig. 1.9, C', C'', and C''', where $C''' > C'' > C'$. As drawn, each node on a line of constant cost represents a solution of a separate linear programming problem. After determining the control cost–emission level relationship, the next step is to determine the emission level–air quality relation. The relation between mass emission levels and air quality will virtually always be nonlinear. We have indicated in Fig. 1.9 a possible region of admissible air quality. The least-cost solution is at point A, where the admissible air quality region just intersects the line of least cost.

We have been discussing primarily the problem of long-term air pollution control. An air pollution simulation system, however, is also a key ingredient in a real-time control system implemented by a municipal air pollution control authority. The objective of a real-time control system is to monitor continuously concentrations at a number of stations (and perhaps also at the stacks of a number of important emission sources) and, with these measurements and weather predictions as a basis, to prescribe actions that must be undertaken by sources to avert dangerously high concentrations if the weather is estimated to be adverse. Figure 1.10 shows in schematic, block-diagram form a complete real-time control system for an airshed, as suggested, for

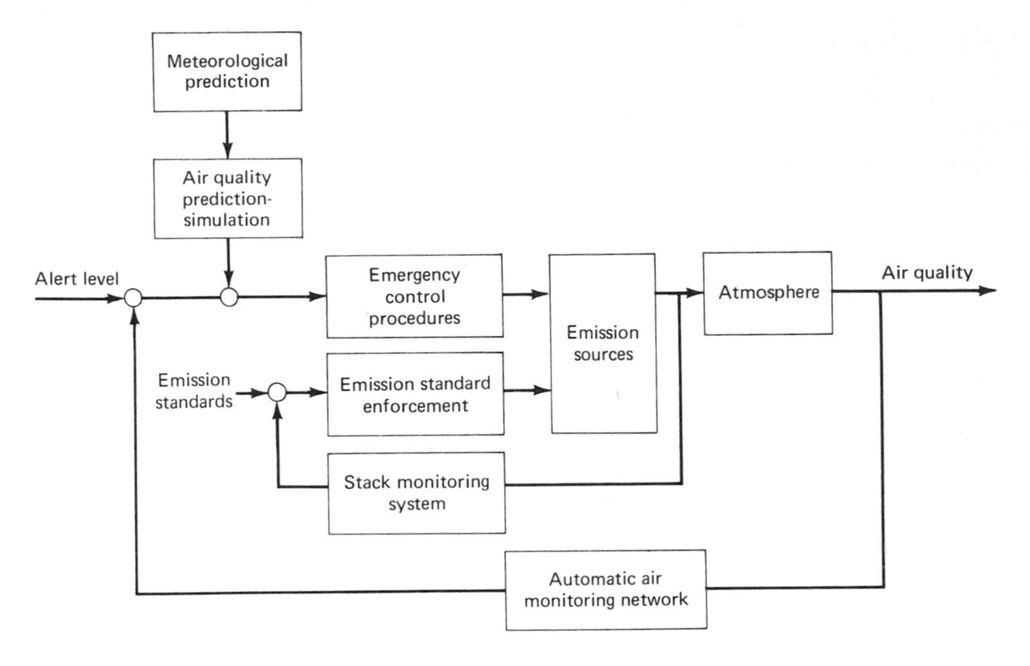

FIGURE 1.10
Elements of a real-time air pollution control system involving automatic regulation of emission sources based on atmospheric measurements.

example, by Savas (1967). Let us examine each of the loops. The innermost loop refers to an automatic stack-monitoring system at major combustion and industrial sources. If the stack emissions exceed the emission standards, the plant would automatically curtail its process to bring stack emissions below the standard. The emission standards would normally be those legislated measures currently in force. The next loop represents a network of automatic monitoring stations which feed their data continuously to a central computer which compares current readings with air quality "danger" values. These values are not necessarily the same as the air quality standards discussed earlier. For example, if the air quality standard for SO_2 is 0.04 ppm for an 8-hr average, the alert level might be 0.5 ppm for a 1-hr average. In such a system we would not rely entirely on measurements to initiate action, since once pollutants reach dangerous levels it is difficult to restore the airshed quickly to safe levels. Thus, we would want to predict the weather to 3–48 hr in advance, say, and use the information from this prediction, combined with the feedback system, in deciding what action, if any, to take. Typical real-time control actions were listed earlier.

1.8 SUMMARY

A thorough analysis of each of the components of the air pollution problem would require several volumes. Clearly, in a textbook of moderate size each component cannot be covered in detail. We have chosen as the objectives of this book to study how pollutants are formed, how they may be controlled, and how they behave in the atmosphere. We have already discussed briefly the effects of air pollution, and these will not be considered further in this book. Because of the rapidly changing status of air pollution control legislation, we will not attempt coverage of this subject. [For a history and compilation of air pollution legislation through 1971 the reader is referred to Ross (1972).] In short, our focus will be on the engineering aspects of sources and controls and on the physics and chemistry of pollutant behavior in the atmosphere.

REFERENCES

Air Quality Criteria for Carbon Monoxide, *U.S. Dept. Health, Education and Welfare Publ. AP*-62, 1969.

Air Quality Criteria for Hydrocarbons, *U.S. Dept. Health, Education and Welfare Publ. AP*-63, 1969.

Air Quality Criteria for Particulate Matter, *U.S. Dept. Health, Education and Welfare Publ. AP*-49, 1969.

Air Quality Criteria for Photochemical Oxidants, *U.S. Dept. Health, Education and Welfare Publ. AP*-63, 1969.

Air Quality Criteria for Sulfur Oxides, *U.S. Dept. Health, Education and Welfare Publ. AP*-50, 1969.

ATWATER, M. A.: The Radiation Budget for Polluted Layers of the Urban Environment, *J. Appl. Meteorol.*, **10**:2, 205 (1971).

BARTH, D.: Federal Motor Vehicle Emission Goals for CO, HC, and NO_x, Based on Desired Air Quality Levels, *J. Air Pollut. Control Assoc.*, **20**:519 (1970).

BEARD, R. R., and G. A. WERTHEIM: Behavioral Impairment Associated with Small Doses of Carbon Monoxide, *Am. J. Public Health*, **57**:2012 (1967).

BRANDT, C. S., and W. W. HECK: Effects of Air Pollutants on Vegetation, in A. C. Stern (ed.), "Air Pollution," vol. I, Academic Press, New York, 1968.

BRASSER, L. G., P. E. JOOSTING, and D. VON ZUILEN: Sulfur Dioxide—To What Level Is It Acceptable? Report G-300, Research Institute for Public Health Engineering, Delft, Netherlands, July, 1967.

BURTON, E. S., and W. SANJOUR: A Simulation Approach to Air Pollution Abatement Program Planning, *Socio-Econ. Plann. Sci.*, **4**:147 (1970).

CARNOW, B. W., M. H. LEPPER, R. B. SHEKELLE, and J. STAMLER: The Chicago Air Pollution Study: SO_2 Levels and Acute Illness in Patients with Chronic Broncopulmonary Disease, *Arch. Environ. Health*, **18**:768 (1969).

CHANGNON, S. A., JR.: The La Porte Weather Anomaly, Fact or Fiction, *Bull. Am. Meteorol. Soc.*, **49**:4 (1968).

CHARLSON, R. J.: Atmospheric Aerosol Research at the University of Washington, *J. Air Pollut. Control Assoc.*, **18**:652 (1968b).

CHARLSON, R. J., N. C. AHLQUIST, and H. HORVATH: On the Generality of Correlation of Atmospheric Aerosol Mass Concentration and Light Scatter, *Atmos. Environ.*, **2**:455 (1968a).

CHARLSON, R. J., H. HORVATH, and R. F. PUESCHEL: The Direct Measurement of Atmospheric Light Scattering Coefficient for Studies of Visibility and Pollution, *Atmos. Environ.*, **1**:469 (1967).

CROKE, E. J., and S. G. BOORAS: The Design of an Air Pollution Incident Control Plan, *Air Pollut. Control Assoc. Paper* 69–99, 62nd APCA Annual Meeting, New York, 1969.

DOUGLAS, J. W. B., and R. E. WALLER: Air Pollution and Respiratory Infection in Children, *Brit. J. Prevt. Social Med.*, **20**:1 (1966).

EISENBUD, M.: Sources of Radioactive Pollution, in A. C. Stern (ed.), "Air Pollution," vol. I, Academic Press, New York, 1968.

Environmental Protection Agency: National Primary and Secondary Ambient Air Quality Standards, *Federal Register*, **36**: 8186 (1971).

GEORGII, H. W.: The Effects of Air Pollution on Urban Climates, *Bull. World Health Organ.*, **40**:624 (1969).

GOLDSMITH, J. R.: Effects of Air Pollution on Human Health, in A. C. Stern (ed.), "Air Pollution," vol. I, Academic Press, New York, 1968.

GOLDSMITH, J. R., and A. C. HEXTER: Respiratory Exposure to Lead: Epidemiological and Experimental Dose-Response Relationships, *Science*, **158**:132 (1967).

GOLDSMITH, J. R., and S. LANDAW: Carbon Monoxide and Human Health, *Science*, **162**:1352 (1968).

HENDRICKSON, E. R.: Air Sampling and Quantity Measurement, in A. C. Stern (ed.), "Air Pollution," vol. II, Academic Press, New York, 1968.

HEUSS, J. M., and W. A. GLASSON: Hydrocarbon Reactivity and Eye Irritation, *Environ. Sci. Technol.*, **2**:1109 (1968).

HIDY, G. M., and J. R. BROCK: "The Dynamics of Aerocolloidal Systems," Pergamon, New York, 1970.

HINDAWI, I. J.: Air Pollution Injury to Vegetation, *U.S. Dept. Health, Education and Welfare Publ. AP*-71, Raleigh, N.C., 1970.

JACKOBSON, J. S., and A. V. HILL (eds.): Recognition of Air Pollution Injury to Vegetation: A Pictorial Atlas, TR-7, Air Pollution Control Association, Agricultural Committee, Pittsburgh, 1970.

JAFFE, L. S.: The Effects of Photochemical Oxidants on Materials, *J. Air Pollut. Control Assoc.*, **17**:6, 375 (1967).

KOHN, R. E.: A Mathematical Programming Model for Air Pollution Control, *School Sci. and Math.*, June 1969, p. 487.

KOHN, R. E.: Linear Programming Model for Air Pollution Control: A Pilot Study of the St. Louis Airshed, *J. Air Pollut. Control Assoc.*, **20**:2, 78 (1970).

KOHN, R. E.: Application of Linear Programming to a Controversy on Air Pollution Control, *Manage. Sci.*, **17**:10, B-609 (1971a).

KOHN, R. E.: Air Quality, the Cost of Capital, and the Multiproduct Production Function *South. Econ. J.*, **38**:2, 156 (1971b).

KOHN, R. E., and D. E. BURLINGAME: Air Quality Control Model Combining Data on Morbidity and Pollution Abatement, *Decis. Sci.*, **2**:300 (1971c).

LANDSBERG, H. E.: Man-made Climatic Change, *Science*, **170**:1265 (1970).

LAVE, L. B., and E. P. SESKIN: Air Pollution and Human Health, *Science*, **169**:723 (1970).

LAWTHER, P. J.: Climate, Air Pollution and Chronic Bronchitis, *Proc. Roy. Soc. Med.*, **51**:262 (1958).

LOWRY, W. P.: The Climate of Cities, *Sci. Am.*, **217**:2, 15 (1967).

LUNN. J. E., J. KNOWELDEN, and A. J. HANDYSIDE: Patterns of Respiratory Illness in Sheffield Infant School Children, *Brit. J. Prev. Social Med.*, **21**:7 (1967).

MACFARLAND, R. E., et al.: The Effects of Carbon Monoxide and Altitude on Visual Thresholds *Aviat. Med.*, **51**:6, 381 (1944).

MICHELSON, I., and B. TOURIN: Comparative Method for Studying Costs of Air Pollution, *Public Health Rept.*, **81**:6, 505 (1966).

MIDDLETON, W. E. K.: "Vision Through the Atmosphere," University of Toronto Press, Toronto, 1952.

PETERSON, J. T.: The Climate of Cities: A Survey of Recent Literature, *U.S. Dept. Health, Education and Welfare Publ. AP*-59, Washington, 1969.

PETRILLI, R. L., G. AGNESE, and S. KANITZ: Epidemiology Studies of Air Pollution Effects in Genoa, Italy, *Arch. Environ. Health*, **12**:733 (1966).

Project Clean Air, Task Force Assessments, "Human Health Effects," vol. 2, University of California, Berkeley, Sept. 1, 1970.

RIDKER, R.: "Economic Costs of Air Pollution: Studies in Measurement," Praeger, New York, 1967.

ROSS, R. D. (ed.): "Air Pollution and Industry," Van Nostrand Reinhold, New York, 1972.

SAVAS, E. S.: Computers in Urban Air Pollution Control Systems, *Socio-Econ. Plann. Sci.*, **1**:157 (1967).

SCHOETTLIN, C. E., and E. LANDAU: Air Pollution and Asthmatic Attacks in the Los Angeles Area, *Public Health Rept.*, **76**:545 (1961).

SCHULTE, J. H.: Effects of Mild Carbon Monoxide Intoxication, *Arch. Environ. Health*, **7**:5, 524 (1963).

SEINFELD, J. H., and C. P. KYAN: Determination of Optimal Air Pollution Control Strategies, *Socio-Econ. Plann. Sci.*, **5**:173 (1971).

State of California Air Resources Board, Recommended Ambient Air Quality Standards Applicable to All Air Basins, Sacramento, September, 1970.

STERN, A. C. (ed.): "Air Pollution," vol. I–III, Academic Press, New York, 1968.

Task Group on Lung Dynamics, Deposition and Retention Models for Internal Dosimetry of the Human Respiratory Tract, *Health Phys.* **12**:173 (1966).

TRIJONIS, J. C.: An Economic Air Pollution Control Model-Application: Photochemical Smog in Los Angeles County in 1975, Ph.D. dissertation, California Institute of Technology, Pasadena, California, 1972.

WAYNE, W. S., P. F. WEHRLE, and R. E. CARROLL: Oxidant Air Pollution and Athletic Performance, *J. Am. Med. Assoc.*, **199**:21, 901 (1967).

WILLIAMSON, S.: "Fundamentals of Air Pollution," Addison-Wesley, Reading, Mass., 1973.

WINKELSTEIN, W.: The Relationship of Air Pollution and Economic Status to Total Mortality and Selected Respiratory System Mortality in Man, *Arch. Environ. Health*, **14**:162 (1967).

YOCUM, J. E., and R. O. MCCALDIN: Effects of Air Pollution on Materials and the Economy, in A. C. Stern (ed.), "Air Pollution," vol. I, Academic Press, New York, 1968.

PROBLEMS

1.1 Develop an equation which allows conversion between $\mu g/m^3$ and ppm for gaseous pollutants at any temperature and pressure.

1.2 A measurement of the number concentration of particles in an urban area yields a value of 10^5 particles/cm^3. If it is assumed that all particles have a density of 1 g/cm^3 and an average radius of 0.05 μm, estimate the prevailing visibility in the area.

1.3 The average quantity of air inhaled by an adult is 500 cm^3 per breath. If 15 breaths per minute are taken and the air contains particulate matter at a concentration of 200 $\mu g/m^3$, what is the total hourly mass of particulate matter that reaches the bronchioles if particles have a density of 1 g/cm^3 and an average radius of 0.05 μm?

1.4 Power plants in a metropolitan basin burn about 60 million barrels (mbbl) of fuel oil or equivalent per year. At present, one-sixth of the fuel burned is fuel oil containing 0.5 percent or more sulfur, leading to SO_2 emissions of 60×10^6 kg/yr. The remainder of the fuel burned is natural gas which, for all practical purposes, results in no SO_2 emissions. From the standpoint of SO_2 abatement, it is advantageous to burn as much natural gas as possible. However, because of shortages in the supply of natural gas, some amount of low-sulfur fuel oil (containing less than 0.5 percent sulfur) must be used instead of natural gas. Sulfur dioxide emissions from this fuel oil are 1.0 kg/bbl. Assume that the cost increase for using natural gas and low-sulfur fuel oil over high-sulfur fuel oil is \$45,000/mbbl and \$30,000/mbbl, respectively. What is the optimal allocation of fuels if SO_2 emissions must be reduced to 6×10^6 kg/yr?

1.5 The state of California ambient air quality standard for NO_2 is 0.25 ppm for a 1-hr exposure. The highest 1-hr-average NO_2 concentration measured during 1970 in California was 0.83 ppm in Los Angeles. According to the rollback concept of relating emissions to air quality, air quality levels change in direct proportion to emission changes, with adjustment for background levels. The background level of NO_2 is approximately 0.005 ppm. Determine the fractional reduction in NO_x emissions below the 1970 level required to meet the California NO_2 standard on the basis of the rollback assumption if NO_2 levels are assumed to be directly related to NO_x emissions. At a new car replacement rate of 10 percent per year, determine the year when the California NO_2 standard can be expected to be attained in Los Angeles if the total vehicle mileage traveled remains constant and the emission levels for years after 1970 are:

1970	4.0 g/mi
1971–5	3.0 g/mi
1976–	0.4 g/mi

2

ORIGIN AND FATE OF AIR POLLUTANTS

The atmosphere is composed primarily of nitrogen, oxygen, and several noble gases the concentrations of which have remained remarkably fixed in time. Also present, however, are a number of gases that occur in relatively small and sometimes highly variable amounts. Water vapor, carbon dioxide, and ozone fall in the latter category as well as do the gases considered to be the common urban air pollutants.

In spite of its apparent unchanging nature, the atmosphere is, in reality, a dynamic system, with its gaseous constituents continually being exchanged with vegetation, the oceans, and biological organisms. The so-called cycles of the atmospheric gases involve a number of physical and chemical processes. Gases are produced by chemical processes within the atmosphere itself, by biological activity, volcanic exhalation, radioactive decay, and man's industrial activities. Gases are removed from the atmosphere by chemical reactions in the atmosphere, by biological activity, by physical processes in the atmosphere (such as particle formation), and by deposition and uptake by the oceans and earth. The average residence time of a gas molecule introduced into the atmosphere can range from hours to millions of years, depending on the species. Most of the species considered air pollutants (in a region in which their concentrations exceed substantially the normal background levels) have natural as well as man-made sources. Therefore, in order to assess the effect man-made pollutant emissions may have on the atmosphere as a whole, it is essential to understand the atmospheric cycles of the pollutant gases, including natural and anthropogenic sources as well as the predominant removal mechanisms.

The object of this chapter is to consider the global cycles of the major classes of air pollutants. We will be concerned at this point only with the general types of processes which result in emissions of the various pollutants, such as combustion and biological decay. The specific mechanisms of pollutant formation in man-made sources is deferred until Chap. 7, so that these processes may be discussed in conjunction with control methods in Chap. 8. Thus, we wish to obtain here a perspective on the air pollution problem from a global point of view. Such a perspective will afford the opportunity to assess how seriously man is affecting the natural cycles of the atmosphere. The principal questions we will attempt to answer for each class of pollutants are:

1 What are the sources (natural and anthropogenic) of each class of pollutants, and how do man's contributions affect the overall cycle of these classes?
2 What are the predominant sinks and removal mechanisms for the classes of air pollutants?

Before we discuss individual pollutants, we present, in some sense, a summary of what is to follow in this chapter. This summary is embodied in Table 2.1, due to Junge (1972). In this table are listed all gases with concentrations higher than the pphm level and some important ones in the ppb range. Aside from N_2, O_2, the noble gases, and H_2O, all these gases have both natural and anthropogenic sources. Although the relative contributions of these two sources vary considerably, in general the natural sources of the pollutant species are the dominant ones (on a global scale).

The constituents in Table 2.1 are arranged according to the nature of their cycles. Of great interest in geochemistry are the physical and chemical processes which determine the concentrations in Table 2.1. The total quantity of a species both in the atmosphere and dissolved in the oceans, say M_g, and that deposited on the earth as sediment, say M_s, must equal that which has been exhaled from the earth's interior over time, M_t. Thus,

$$M_t = M_g + M_s$$

where each of these quantities can be expressed in grams per square centimeter of earth's surface. As Junge notes, if $M_g \gg M_s$, most of the constituent has remained in the atmosphere and, for that reason, it can be called an *accumulative* gas. If, on the other hand, $M_g \ll M_s$, it can be assumed that the fraction remaining in the atmosphere is determined by processes resulting in quasi-steady-state conditions. Junge calls such gases nonaccumulative or *equilibrium* gases. The ratio M_g/M_s for the noble gases and N_2 is greater that 1. From CO_2 on down Table 2.1, the ratio $M_g/M_s \leq 10^{-3}$, clearly establishing these species as equilibrium gases. With the possible exception of oxygen, the distinction between the two general categories in the last column of Table 2.1 seems straightforward.

The processes which govern the abundance of the equilibrium gases vary for different gases. The water composition of the atmosphere is controlled by the variation of the vapor pressure of water with temperature. Ozone is determined, as we shall see in Chap. 3, by chemical processes in the upper atmosphere. Methane, hydrogen, nitrous oxide, and carbon monoxide are controlled by atmospheric chemical reactions as well as by biological processes. It is estimated that only one one-thousandth of all CO_2 released to the earth's surface by exhalation has remained in the atmosphere-ocean system, with most existing as carbonate sediments on the land and in water. The processes actually governing the atmospheric CO_2 content have not been clearly determined, although it is most likely that these represent a combination of chemical equilibrium with the oceans and geochemical cycles. For oxygen, the estimated range of M_g/M_s lies between $\frac{1}{50}$ and $\frac{1}{12}$. Most of the earth's oxygen is

Table 2.1 ATMOSPHERIC GASES ARRANGED ACCORDING TO THEIR CYCLE

Gas	Average mixing ratio, ppm	Residence time†	Cycle†	Status
Ar	9300	....	No cycle	Accumulation during earth's history
Ne	18	....		
Kr	1.1	....		
Xe	0.09	....		
N_2	73×10^4	$\sim 10^6$ yr	Cycle predominantly biological and microbiological	
O_2	21×10^4	$\sim 10^1$ yr		?
CO_2	315	~ 15 yr		Quasi-steady-state or equilibrium (H_2O, CO_2?)
CH_4	1.2–1.5	~ 2 yr		
H_2	0.4	~ 5 yr?		
N_2O	0.25	~ 10 yr		
CO	0.12–0.15	~ 0.1 yr		
H_2S-SO_2	Variable in the ppb range	Days to weeks	Sources predominantly microbiological sinks predominantly physico-chemical	
NH_3				
NO-NO_2				
HC‡				
H_2O	Variable	10 days	Cycle physico-chemical	
O_3	Variable	0.3 yr		
He	5.2	10 yr		

SOURCE: Junge (1972).
† Primarily tropospheric; exchange with oceans is not included.
‡ Various hydrocarbons, natural and anthropogenic, except CH_4.

believed to have been produced by photosynthesis. Oxygen is consumed by oxidation of minerals at the earth's surface. Thus, the oxygen concentration in the atmosphere is probably a result of both accumulation and geochemical cycle.

For the rest of this chapter we focus on those species commonly classed as urban air pollutants. In each case we shall attempt to identify the major sources and sinks, as well as to estimate average tropospheric residence times.

2.1 SULFUR-CONTAINING COMPOUNDS

2.1.1 Origin and Fate of Sulfur-containing Compounds

The major sulfur-containing compounds in the atmosphere are SO_2, SO_3, H_2S, H_2SO_4, and sulfate (SO_4^{2-}) salts. The sources of atmospheric sulfur compounds are combustion of fossil fuels, decomposition and combustion of organic matter, sea spray over the oceans, and volcanoes. The emitted sulfur compounds reside in the atmosphere for some period of time and then are redeposited on the ground or in the oceans. Table 2.2 presents estimates of the amounts of sulfur introduced globally into the atmosphere as of the late 1960s. The estimate of total global sulfur emissions varies considerably, depending on the amount of H_2S released from biological processes in the ocean and on the land. In order to balance the global sulfur cycle, a large source of sulfur is needed. Because of this need, a large oceanic source of H_2S has been postulated. In fact, very little is known about H_2S release from either land- or sea-based sources, and for this reason we present two different estimates in Table 2.2.[1]

Of the approximately 200 Tg of sulfur introduced per year into the atmosphere [based on Robinson and Robbins (1968) estimates of H_2S from natural sources], about 50 to 75 Tg comes from anthropogenic sources. It has been estimated that a 4 percent annual growth rate until 1980 and a 3.5 percent rate thereafter in man-made sulfur emissions will, by the year 2000, have increased anthropogenic sulfur emissions to a level comparable to that of natural emissions.

Table 2.3 shows how the value of 133 Tg of SO_2 emitted from man-made sources in 1965 was arrived at by Robinson and Robbins (1968). By far the most abundant source was the burning of coal, with substantial contributions from petroleum combustion and smelting.

In considering the global impact of air pollutants it is of great importance to identify the fate of each species in the atmosphere. In order to account for the observed global concentrations of pollutants, for example, we must know how long an

[1] It now seems quite certain that H_2S is not emitted from ocean surface waters, because dissolved oxygen would rapidly oxidize it. Suspicion has fallen on dimethyl sulfide, which has been found in some seaweed samples, as the possible oceanic source of sulfur.

Table 2.2 ESTIMATED AMOUNTS OF SULFUR INTRODUCED ANNUALLY INTO THE ATMOSPHERE

Form	Source	Amount, Tg[a]	Amount as sulfur, Tg	Reference
SO_2 and SO_3[b]	Electric power generation	60[c]	30	MIGE (1970)[f]
	Industrial sources	24.5[c]	12.2	MIGE (1970)
	Volcanoes[d]	1.35	0.68	Kellogg et al. (1972)
SO_4^{2-}	Sea spray[e]	118	39	Eriksson (1959, 1960)
H_2S	Biological decay in ocean	185	172	Eriksson (1959, 1960)
		27.2	25.2	Robinson and Robbins (1968)
	Biological decay on land	102	95	Eriksson (1963)
		63.5	60	Robinson and Robbins (1968)
	Industrial sources	2.7	2.5	Robinson and Robbins (1968)

[a] 1 Tg $= 10^{12}$ g.
[b] More than 95 percent of SO_X emissions by man are as SO_2; the remainder as SO_3.
[c] These estimates are based on United States emission factors. Because fuels with somewhat lower sulfur contents than those used worldwide are used in the United States, these estimates may be somewhat low. Kellogg et al. (1972) suggested a value of 90 Tg SO_2 per year from man-made sources. An earlier estimate by Robinson and Robbins (1968) for man-made SO_2 emissions in 1965 was 133 Tg/year.
[d] Sulfur compounds emitted by volcanoes are predominantly SO_2 and H_2S, with smaller amounts of SO_3 and sulfates. Rather rapid conversion of SO_2 and H_2S to SO_3 occurs as the hot eruption clouds mix with atmospheric oxygen.
[e] Approximately 90 percent of the sulfate in sea spray is precipitated back into the oceans, with 10 percent passing over the continents.
[f] "Man's Impact on the Global Environment," 1970.

Table 2.3 GLOBAL ANTHROPOGENIC EMISSIONS OF SO_2 IN 1965 AS ESTIMATED BY ROBINSON AND ROBBINS (1968)

Source	Product consumption or production rate, Tg	Mass of SO_2 produced per mass of product consumption or production rate	SO_2 emissions, Tg
Coal	2810	0.033	92.6
Petroleum combustion			
Gasoline	333	0.0009	0.3
Kerosene	83	0.0024	0.2
Distillate	257	0.0070	1.8
Residual	460	0.04	18.4
Smelting			
Copper	5.85	2	11.7
Lead	2.8	0.5	1.4
Zinc	4.0	0.3	1.2
Petroleum refining	$11{,}317 \times 10^6$ bbl	450 g/bbl	5.1
Total			133.0

emitted species remains in the atmosphere before it is removed or converted to a different species by chemical reaction.

We consider first H_2S, which is added to the atmosphere in uncertain and possibly very large quantities from natural sources. Hydrogen sulfide is oxidized to SO_2 rather rapidly. In fact, of the total molecules of SO_2 in the atmosphere at any given time, as much as 80 percent was originally emitted as H_2S and later converted to SO_2. Hydrogen sulfide can be oxidized by atomic and molecular oxygen and ozone. As we will see in Chaps. 3 and 4, ozone is both a natural constituent of the stratosphere and a constituent of urban atmospheres. The reaction between H_2S and O_3, which appears to be the most important oxidizing reaction for H_2S, is

$$H_2S + O_3 \rightarrow H_2O + SO_2$$

This reaction is quite slow in the gas phase, although it can occur much more rapidly on the surface of airborne particles (Hales et al., 1969). The lifetime of 1 ppb H_2S exposed to 0.05 ppm O_3 in the presence of 15,000 particles/cm^3 is estimated to be 2 hr; in the presence of 200 particles/cm^3 it is estimated to be about 28 hr (Air Quality Criteria for Sulfur Oxides, 1969). Hydrogen sulfide, oxygen, and ozone are soluble in water, and thus the rate of oxidation of hydrogen sulfide in fog or cloud droplets could be very fast. In general, the lifetime of a molecule of H_2S before conversion to SO_2 is of the order of hours.

Next we consider SO_2, the principal fate of which in the atmosphere is oxidation to SO_3. This oxidation proceeds by either of two processes, *catalytic* or *photochemical* each of which will be discussed in detail in Chap. 4. The catalytic process is prevalent during high humidity conditions when SO_2 is readily absorbed by water droplets. If the water droplet contains only SO_2, essentially no reaction occurs. However, the presence of a number of foreign substances, such as metal salts (for example, Fe^{3+}, Mn^{2+} salts) or NH_3, rapidly promotes the reaction of SO_2 and dissolved oxygen in the drop, leading to the formation of sulfate, $SO_4{}^{2-}$. For example, dissolved NH_3 not only leads to formation of ammonium sulfate, $(NH_4)_2SO_4$, but may also increase the solubility of SO_2 in the drop. The photochemical process is associated with daytime, low-humidity conditions. The first step of the process is absorption of light by an SO_2 molecule, leading to a more energetic molecule. This activated SO_2 molecule may then react with O_2, at a rate much faster than do ordinary SO_2 molecules, to give SO_3. If oxides of nitrogen and hydrocarbons are also present, this oxidation process can be significantly accelerated.

The SO_3 formed from SO_2 reacts almost immediately with H_2O to form H_2SO_4. The H_2SO_4 readily combines with water droplets to give an H_2SO_4 solution. If NH_3 is present, as we have just noted, $(NH_4)_2SO_4$ is formed. If the H_2SO_4 contacts sodium chloride (NaCl) particles, then sodium sulfate (Na_2SO_4) and HCl result. In general, then, the ultimate fate of SO_2 is conversion to sulfate salts, the lifetime of the SO_2 before conversion being of the order of days.

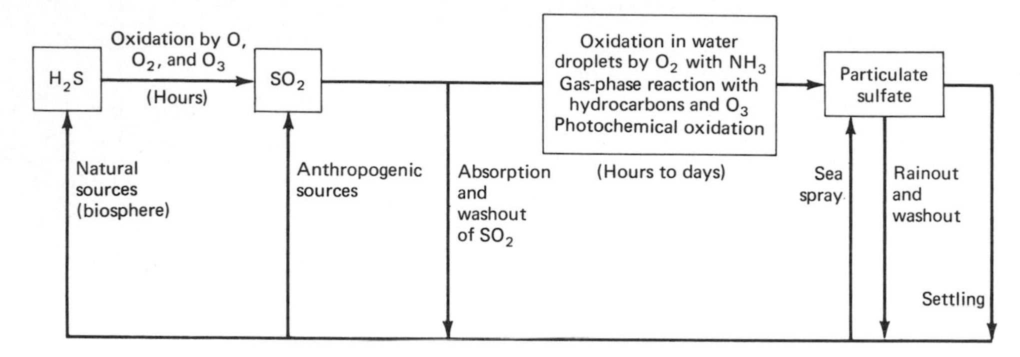

FIGURE 2.1
Fate of atmospheric sulfur compounds. The mean lifetime of each species is indicated under the arrows signifying conversion.

Figure 2.1 summarizes the fate of the atmospheric sulfur compounds together with their estimated mean lifetimes. The ultimate removal of sulfur from the atmosphere occurs by several processes: (1) rainout and washout, (2) diffusion to soil and vegetation, and (3) dry deposition of sulfate particles.

Rainout includes all processes within the clouds that result in removal, whereas washout refers to scavenging by precipitation below the clouds. Beilke and Georgii (1968) estimated the relative contributions of these two processes to sulfate removal in Frankfurt, Germany as 20 and 70 percent, respectively. In Frankfurt, SO_2 concentrations below cloud level are typically those of an urban environment and thus are significantly higher than SO_2 background levels. In clean air, rainout of sulfate aerosol becomes of more relative importance. The amount of sulfate in rainwater also depends on the quantity of sea-salt aerosol in the air.[1] The *excess sulfate* of a rain sample refers to the total sulfate from all sources other than sea salt. The excess sulfate in rainwater can be used as a direct indication of the removal rate of SO_2 and SO_4^{2-} from the atmosphere. Kellogg et al. (1972) estimated the annual scavenging of sulfate by precipitation in the Northern Hemisphere as 240 Tg.

Estimates of the rate of diffusion of SO_2 to soil and vegetation were made by Robinson and Robbins (1968) and Swinbank (1968). In the former study a "deposition velocity" of 1 cm/sec and an ambient SO_2 concentration of 1 $\mu g/m^3$ were assumed, leading to an estimated uptake of about 50 Tg/yr.

The third removal mechanism is deposition of sulfate particles on vegetation and solid surfaces. Assuming that SO_2 and sulfate are removed with reasonably

[1] The sulfate in a sample of rainwater that originated as sea salt may be calculated by

$$\left(SO_4^{2-}\right)_s = \left(\frac{SO_4^{2-}}{Cl^-}\right)_s \times (Cl^-)_{\text{observed}}$$

assuming that no addition or removal of chloride (Cl^-) takes place over land.

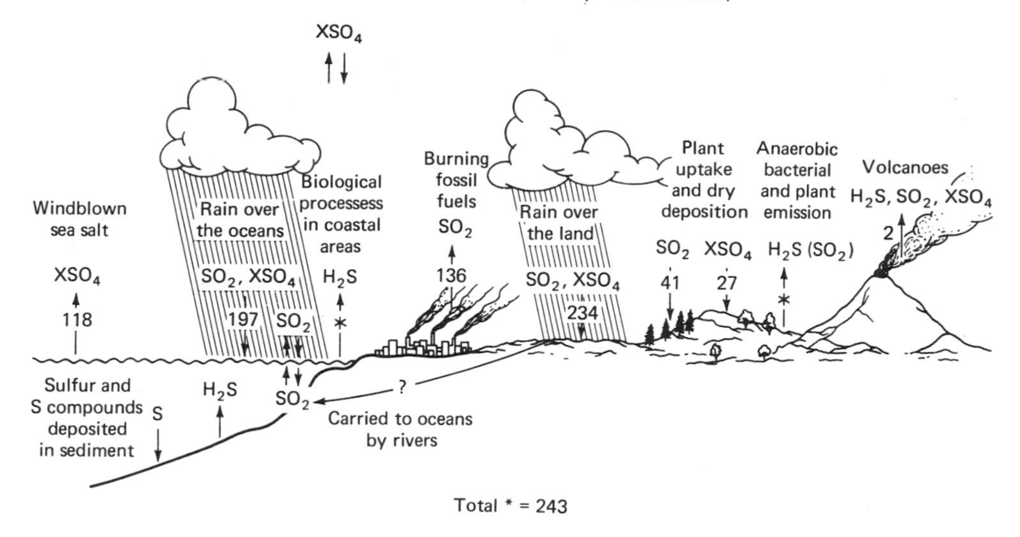

FIGURE 2.2
Sulfur cycle. (*Source: Kellogg et al.*, 1972. *Copyright* 1972 *by the American Association for the Advancement of Science.*)

high efficiency upon deposition, and since the mass concentration of sulfate particles is about equal to that of SO_2 in the troposphere, the mass of annual dry deposition of sulfate over continents can be estimated to be about the same as for diffusion of SO_2 to vegetation (Kellogg et al., 1972).

Kellogg et al. (1972) constructed an overall yearly sulfur cycle, assuming that an approximate balance exists between sources and removal for the entire atmosphere. Figure 2.2 is a schematic of the major sources and sinks. Table 2.4 summarizes the calculations underlying Fig. 2.2.

Having discussed global sources and sinks, we would now like to estimate the mean residence time of sulfur compounds in the atmosphere. In principle, this can be done in two ways. The first is to estimate an upper limit for the residence time of each compound on the basis of the chemical and physical mechanisms of removal. This is essentially what we have already done, and such estimates appear in Fig. 2.1. The second method is based on a source inventory such as that given in Table 2.2 and on measured ambient concentrations. Given the estimated annual global emission rate of a species and given its ambient atmospheric concentration, we can compute the average lifetime of a molecule, assuming the ambient concentration is constant. The more reliable of the two estimation procedures is probably the former, since the latter relies on estimates of the global sources which often are quite uncertain.

Table 2.4 THE ATMOSPHERIC SULFUR BUDGET (UNITS ARE Tg AS SULFATE PER YEAR)

Component	Land	Oceans	Totals
Deposition in the Northern Hemisphere			
Sea-salt SO_4^{2-} in precipitation (118 Tg global total, 10 percent of global total deposited on all land)	7	48	55
Diffusion of SO_2 to surface (theoretically derived flux)	33	0	33
Diffusion of SO_4^{2-} to surface (mass equivalent to SO_2 diffusion)	22	0	22
Excess SO_4^{2-} in precipitation	168	68	236
Totals	230	116	346
Sources of sulfate deposited in the Northern Hemisphere			
Man-made excess SO_4^{2-} (93 percent of worldwide production; 80 percent assumed to be deposited on land)	102	25	127
Sea-salt SO_4^{2-}	7	48	55
Natural excess SO_4^{2-} (total calculated to make a balance; 80 percent assumed to be deposited on land)	131	33	164
Totals	240	106	346
Sources of sulfate deposited in the Southern Hemisphere			
Man-made excess SO_4^{2-} (7 percent of worldwide production; 80 percent deposited on land)	7	2	9
Sea-salt SO_4^{2-} (Tg global total; 10 percent of global total deposited on land)	4	59	63
Natural excess SO_4^{2-} (assumed same source as Northern Hemisphere, proportional to land area; 80 percent deposited on land)	65	16	81
Totals	76	77	153
Deposition in the Southern Hemisphere			
Sea-salt SO_4^{2-} in precipitation	4	59	63
Diffusion of SO_2 to surface (theoretically derived flux, half the concentration in surface air of the Northern Hemisphere, and half the land area)	8	0	8
Diffusion of SO_4^{2-} to surface (mass equivalent to SO_2 diffusion)	5	0	5
Excess SO_4^{2-} in precipitation (total calculated to make a balance; same proportion of precipitation deposition over land as in the Northern Hemisphere)	55	22	77
Totals	72	81	153

SOURCE: Kellogg et al. (1972)

Table 2.4 *(continued)*

NOTES: *1* The deposition of sulfur in the Northern Hemisphere has been estimated from measured concentrations in precipitation and theoretically derived fluxes to the ground, to which is added a proper proportion of the sea-salt sulfate rained out.

2 The total natural excess sulfate produced in the Northern Hemisphere has been calculated to achieve a balance, assuming no transport between hemispheres. The total sulfate and SO_2 deposition in the Southern Hemisphere has been calculated, and from this the natural excess sulfate is also estimated.

Using estimates of the rates of chemical transformation, the lifetimes of atmospheric sulfur compounds can be summarized as follows:

H_2S < 1 day (conversion to SO_2 by O, O_2, O_3)
SO_2 < 3 days (diffusion to vegetation, oxidation to sulfate, washout)
SO_4^{2-} 1 week[1]

Measurements of ambient tropospheric concentrations of sulfur compounds yield:

H_2S 0.002 to 0.02 ppm
SO_2 0.002 to 0.01 ppm
SO_4^{2-} $\sim 2\ \mu g/m^3$

Because the sulfur compounds are removed or chemically altered within hundreds to thousands of kilometers from their sources, their spatial distributions have large variations. For this reason, the measurements of background concentrations have shown wide variation, depending on the part of the world where the measurements were made.

The overall features of the global sulfur cycle are now generally understood. There still exist some gaps in our knowledge, particularly with regard to the biospheric production of gaseous sulfur compounds. In addition, background concentrations of sulfur compounds are only poorly known.

2.1.2 Urban Concentrations of Sulfur Compounds

Let us now consider the short- and long-term concentrations of sulfur compounds reached in urban atmospheres. These concentrations are, of course, the ones of direct concern from an air pollution standpoint. There exist extensive urban monitoring programs for SO_2 and, to a lesser degree, for sulfates in the United States and abroad. As mentioned in Chap. 1, there are a variety of ways to report airborne concentrations, depending on the degree of spatial and temporal averaging employed.

[1] In surface air over land and cloud level. The residence time increases markedly at higher elevations.

For example, one can report yearly average concentration at one location, yearly average concentration over an entire urban area, average maximum daily concentration for a year, etc. A reasonable way to present air pollution data in a manner that conveys as much information as possible is a distribution plot which shows concentrations as a function of the time over which the averaging is carried out or, also, a distribution plot showing the percentage of the time a certain concentration is exceeded. Figures 2.3*a* and 2.3*b* exemplify these two means of data presentation.

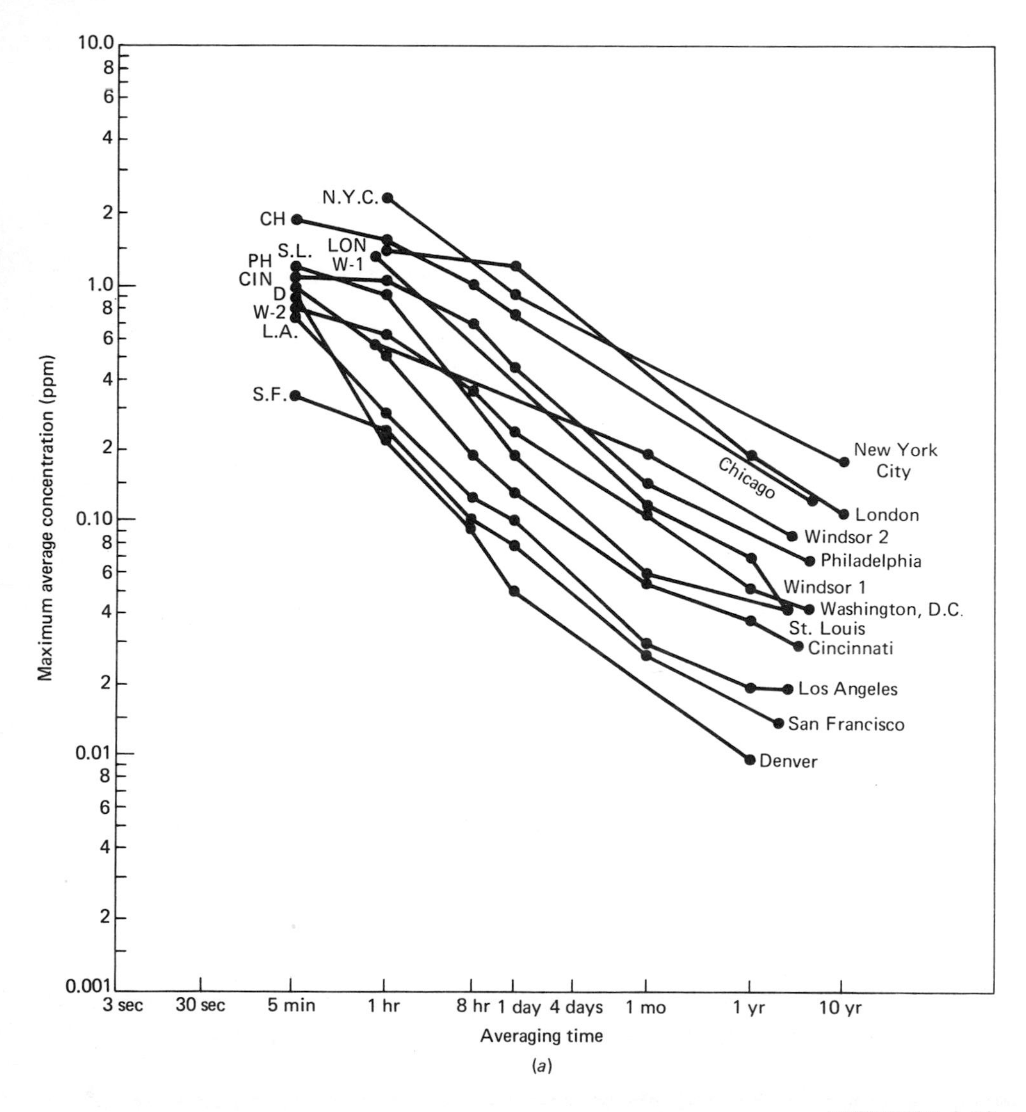

(*a*)

FIGURE 2.3 (*a*)

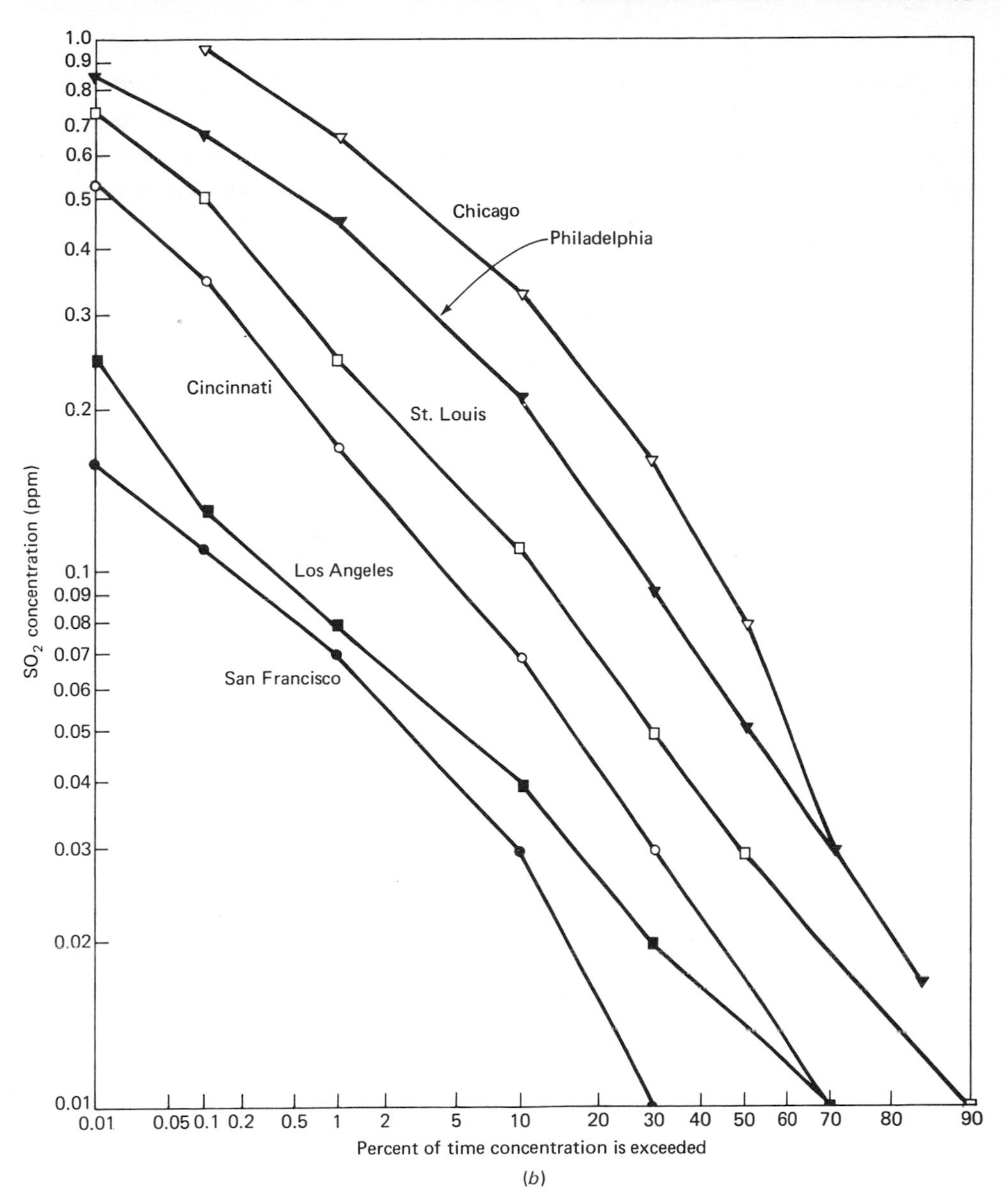

FIGURE 2.3

(*a*) Long-time average SO_2 concentrations (ppm) and associated maximum average concentrations for various time periods for 12 sites in 11 cities. (*Source: Air Quality Criteria for Sulfur Oxides*, 1969.) (*b*) Frequency distribution of SO_2 levels in selected United States cities (1962–1967). (*Source: Air Quality Criteria for Sulfur Oxides*, 1969.)

Figure 2.3*a* shows maximum average SO_2 concentrations for various time periods for 12 sites in 11 cities (Windsor has two sites). For each of the lines shown the last point on the right is the long-term average SO_2 concentration over the interval specified (between 1 and 10 yr). The interval for Denver is 1 yr, and the interval for New York City is 10 yr. As one moves to the left along each line, the other points represent the maximum SO_2 concentration achieved within the interval for the particular averaging time on the abscissa. If, for example, data for one complete year were available in terms of hourly average SO_2 concentrations, there would be one maximum hourly value for that year. If five years of hourly averages were available, five maxima would be obtained (one for each year) and the highest of these would be the point shown corresponding to a 1-hr averaging time. Clearly, as the averaging time decreases, higher concentrations result. From the data shown, the highest 1-hr average concentration is from 10 to 20 times the annual mean.

Figure 2.3*b* shows the frequency distribution of SO_2 levels in several cities for a 1-hr averaging time. For example, in San Francisco 0.03 ppm SO_2 over a 1-hr time period is exceeded only 10 percent of the time, whereas an hourly average SO_2 concentration of 0.03 ppm is exceeded 70 percent of the time in Chicago. Before leaving this section, the reader might wish to compare the concentration levels in Figs. 2.3*a* and 2.3*b* with those in Fig. 1.5.

Reporting of Pollutant Concentration Data

Pollutant concentrations are normally measured at a number of fixed locations in an urban area as a function of time. Although the particular measurement technique determines whether a pollutant concentration may be measured continuously or not, in many cases monitoring data are recorded continuously in time, as illustrated by the record of SO_2 concentrations shown in Fig. 2.4*a*. Continuous data in the form of Fig. 2.4*a* are not easily reported, and so it is common practice to average the data over certain time intervals and report the average concentration over each period. Figure 2.4*a* indicates average concentrations derived from the continuous record. Hourly average concentrations are perhaps the most common way in which urban air pollutant data are reported. As the averaging period is increased, one would expect the maximum average concentration to decrease, as illustrated in Fig. 2.3*a*.

The concentrations recorded at a fixed site as a function of time are a result of the influences of atmospheric turbulence, source emissions, and perhaps also atmospheric chemistry. Because of the random nature of atmospheric turbulence, the concentration of a pollutant at any point in an airshed at any time is a random variable; that is, its value cannot be predicted precisely. Figure 2.4*a* exhibits the essentially random character of the concentration. At any instant the probability that the measured concentration will assume any given value is described by its probability density function (assuming that the function is known). If the random concentration variable is denoted by C, the probability density function $g(c)$ is defined such that the probability that the measured concentration C will lie between c and $c + dc$ is $g(c)\,dc$. Because C changes with time, we must write C as $C(t)$, although for the moment we will neglect the time dependence of C. As we shall see in Chap. 6, it is virtually impossible to determine $g(c)$ except in rather idealized situations. Nevertheless, by plotting the time-average values of $C(t)$ in certain ways, we can draw some general conclusions about the properties of measured concentrations.

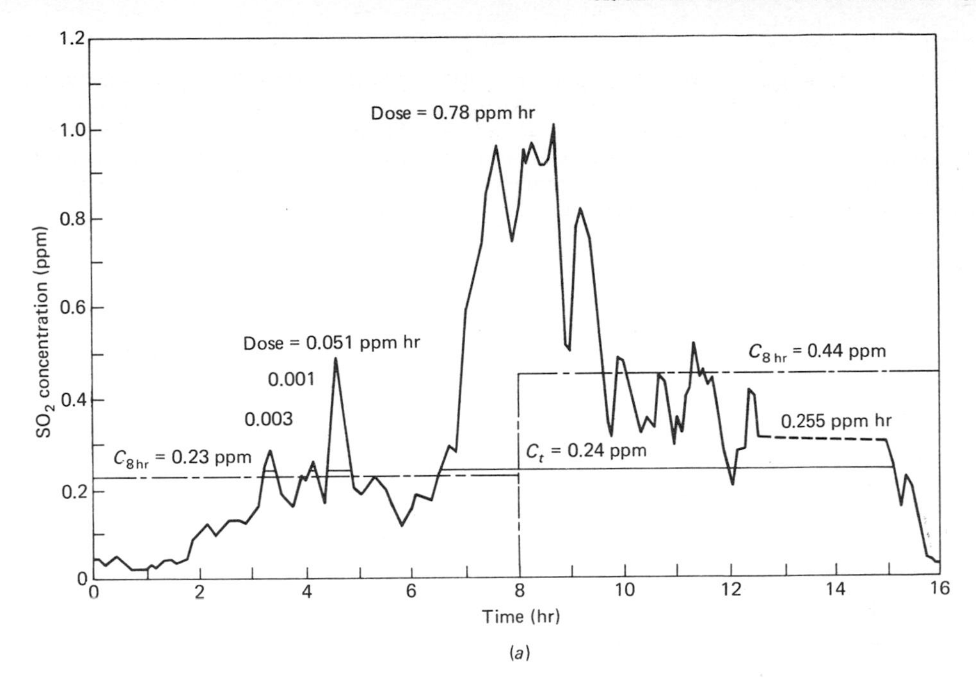

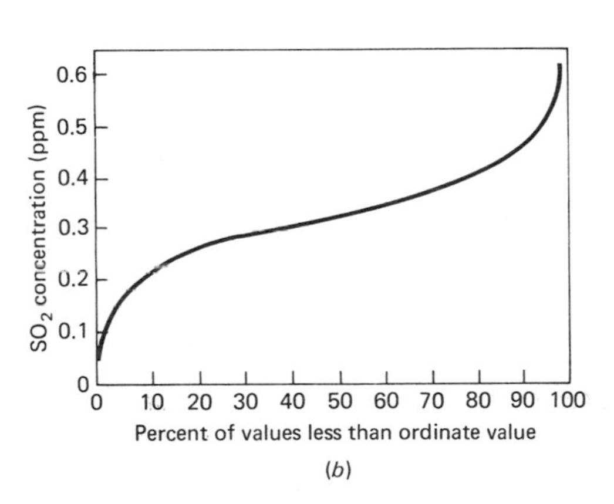

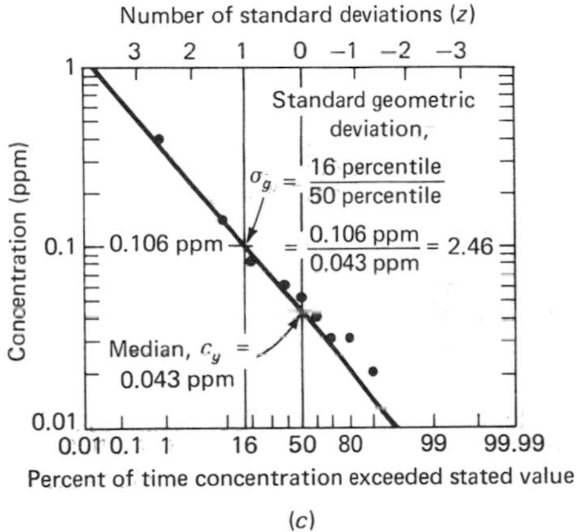

FIGURE 2.4
(*a*) Record of SO_2 concentrations measured at a point in Philadelphia on Jan. 26, 1964. (*Source: Larsen et al.*, 1967.) (*b*) Hypothetical cumulative percentages for SO_2 monitoring data. (*c*) Frequency that various 1-hr-average NO_x concentrations exceeded various values in Washington, D.C. from Dec. 1, 1961 to Dec. 1, 1964. (*Source: Larsen*, 1969.)

The time-average (say hourly) concentrations may be presented in tabular form or in terms of cumulative percentage versus concentration, where the cumulative percentage is the percentage of all the hourly average concentrations over a certain time period (a month, say) having values less than the particular level. Figure 2.4*b* is a typical plot in which both axes are linear scales. Another means of reporting data of this type is a probability graph, in which the concentration is plotted as the ordinate, and the cumulative percentage on a special nonlinear scale is plotted as the abscissa. If the variable (e.g. concentration) being plotted is normally distributed, that is if the probability density function for the variable is given by

$$g(c) = \frac{1}{\sqrt{2\pi}\sigma} e^{-(c-\bar{c})^2/2\sigma^2}$$

where $\bar{c}$ and σ are the mean and standard deviation of C, respectively, then the data points for C will lie on a straight line on a probability graph. Let us examine this property more closely.

In the normal distribution, the integral of $g(c)$ from $-\infty$ to ∞ is unity,

$$\int_{-\infty}^{\infty} g(c)\, dc = 1$$

The probability of observing a value of a normally distributed random variable C equal to or less than some value c is given by

$$\text{Prob}\{C \leq c\} = \frac{1}{\sqrt{2\pi}\sigma} \int_{-\infty}^{c} e^{-(\lambda-\bar{c})^2/2\sigma^2}\, d\lambda$$

It is convenient to express each value of C in terms of the number of standard deviations by which it deviates from the mean $\bar{c}$, that is, $c = \bar{c} + z\sigma$. Expressed in terms of the standard variable z,

$$\text{Prob}\{C \leq \bar{c} + z\sigma\} = \frac{1}{\sqrt{2\pi}} \int_{-\infty}^{z} e^{-\lambda^2/2}\, d\lambda$$

We see that the probability of observing a deviation from the mean z is the same for any normally distributed random variable regardless of the values of $\bar{c}$ and σ.

The probability that $C \leq c$ is just the cumulative percentage divided by 100 for a normally distributed variable. The values of this integral are tabulated for different values of z. If the cumulative percentage of a normally distributed variable is plotted on linear coordinates, an S-shaped curve as in Fig. 2.4*b* is obtained. On a probability graph, however, the abscissa is ruled proportional to the cumulative normal distribution, so that a plot of c versus cumulative percentage will be a straight line if C is a normally distributed random variable. When data are to be plotted in this manner, the data are first tabulated in order of increasing magnitude. Then the total range of the data is divided into equal-size subgroups (say 25) and the cumulative percentage for each subgroup is plotted at the midpoint of that group at a percentage corresponding to that midway between successive groups. The accuracy with which a straight line fits the data on a probability graph is a measure of the closeness of the distribution of the data to a normal distribution. The intersection of the straight line with the 50 percent point is an estimate of the mean $\bar{c}$ of the data. The tabulated values of the integral given above show that 84.13 percent of the total area under the bell-shaped normal curve is included from $-\infty$ to 1σ beyond the mean. Thus, an estimate of the standard deviation can be obtained from the difference between the 84.13 percent point and the 50 percent point.

From plotting measured air pollutant concentrations in this way it was discovered that for a variety of data for different pollutants in different cities the logarithm of the concentration seemed to be normally distributed rather than the concentration itself (Larsen, 1969). A random variable the logarithm of which is normally distributed obeys the *log-normal distribution*

$$G(c) = \frac{1}{\sqrt{2\pi} \ln \sigma_g} e^{-(\ln c - \ln c_g)^2/2 \ln^2 \sigma_g}$$

where c_g and σ_g are the geometric mean and standard deviation, respectively. We shall discuss the properties of the log-normal distribution in more detail later in this chapter. A plot of ln c (or log c) versus cumulative percentage is a straight line for a log-normally distributed variable. Figure 2.4c shows the frequency with which 1-hr-average NO_X concentrations exceeded various values in Washington, D.C., from Dec. 1, 1961 to Dec. 1, 1964. The data are seen to fall on a straight line representing a log-normal distribution. We note that when the logarithm of c is plotted versus cumulative percentage the 50 percent point falls at the value of the geometric mean c_g. The quotient of the ordinate at the 50 and 84.13 percent points (or between the 15.87 and 50 percent points) gives the geometric standard deviation σ_g. From Fig. 2.4c we see that $c_g = 0.043$ ppm and $\sigma_g = 2.46$ for the Washington, D.C., NO_X data.

2.2 CARBON MONOXIDE

2.2.1 Origin and Fate of Carbon Monoxide

With the exception of CO_2,[1] carbon monoxide (CO) is the most abundant air pollutant in the lower atmosphere. Emissions of CO from man-made sources exceed in quantity the mass of anthropogenic emissions of all other air pollutants combined. Previously, CO was thought to be almost exclusively a man-made pollutant, but recent studies have established the existence of natural sources of CO which together far exceed the mass of CO emitted from anthropogenic sources.

Table 2.5 ESTIMATED GLOBAL ANTHROPOGENIC CARBON MONOXIDE SOURCES IN 1970

Source	Emission, Tg
Motor vehicles	222
Other mobile sources	25
Coal combustion	11
Fuel oil combustion	40
Industrial processes	22
Petroleum refining	5
Solid waste disposal	23
Miscellaneous (agricultural burning, etc.)	23
Total	371

SOURCE: Jaffe (1972).

[1] Carbon dioxide is not normally regarded as an air pollutant but rather the normal product of combustion of fossil fuels and organic materials. Nevertheless, the increasing ambient CO_2 level has raised concern about the effects on global meteorology (see Chap. 3).

The major man-made source of CO is the incomplete combustion of fossil fuels. Table 2.5 lists the estimated global anthropogenic sources of CO in 1970.

As mentioned, until recently no important natural sources of CO had been identified. However, the fact that the ever-increasing anthropogenic CO emissions were not leading to increased ambient CO concentrations suggested that perhaps there were huge undiscovered natural sources of CO. Consequently, there has been great current interest in the CO cycle in the atmosphere, particularly in regard to identifying major natural sources.

Weinstock and Niki (1972) have presented a general analysis of the CO balance in nature which we shall now outline. The analysis is based on atmospheric sources and concentrations of both ^{14}CO and CO. The tropospheric CO balance can be expressed in terms of the two material balance equations

$$\frac{d[CO]}{dt} = P_1 + P_2 - k[CO] \tag{2.1}$$

$$\frac{d[^{14}CO]}{dt} = NP_1 + P_3 - k[^{14}CO] \tag{2.2}$$

where [CO] and [^{14}CO] represent the total number of moles of CO and ^{14}CO in the troposphere, respectively, P_1 is the rate of production of CO from "living" carbon of which a mole fraction $N = 1.17 \times 10^{-12}$ is ^{14}CO, P_2 is the rate of production of CO from "dead" carbon (fossil fuels), P_3 is the rate of production of ^{14}CO in the troposphere by cosmic-ray neutrons, and k is an assumed first-order rate constant for removal of both CO and ^{14}CO from the troposphere. Assuming that the tropospheric concentrations of CO and ^{14}CO are unchanging, we can equate the right-hand sides of (2.1) and (2.2) to zero:

$$P_1 + P_2 - k[CO] = 0 \tag{2.3}$$

$$NP_1 + P_3 - k[^{14}CO] = 0 \tag{2.4}$$

One can estimate P_2 at 7×10^{12} moles/yr and P_3 at 290 moles/yr. Based on measurements of ambient CO levels in the troposphere, the total amount of CO in the troposphere is estimated to be 1.7×10^{13} moles. This value is computed in the following way. Junge et al. (1971) gave 0.12 ppm as the average global concentration of CO. Using the average height of the tropopause (the division between the troposphere and stratosphere across which mixing is inhibited, to be discussed in Chap. 3) as 11 km, the troposphere contains 1.4×10^{20} moles. At a concentration of 0.12 ppm, the troposphere contains 1.7×10^{13} moles of CO. The total quantity of ^{14}CO was estimated by similar means to be 45 moles.

The unknown quantities in (2.3) and (2.4) are P_1 and k. Solving for these we get $P_1 = 1.8 \times 10^{14}$ moles/yr $= 5 \times 10^{15}$ g/yr. This source is 25 times greater than the estimated anthropogenic emissions of CO. The calculated value of k is 11 yr^{-1}. The reciprocal of k is the average residence time of CO, 0.09 yr, since CO removal is assumed to be a first-order process. Because of uncertainties in the source estimates, the uncertainty in the P_1 and k values is about 50 percent.

This means of computing an average residence time corresponds to the second method outlined in Subsection 2.1.2. We must use this method because the chemical processes responsible for "living" sources of CO and for destruction of CO are not as well understood as those for the sulfur compounds. Let us now consider what is known about the sources and sinks of CO in the atmosphere.

Suggested removal mechanisms for CO are:

1 Reaction of CO with hydroxyl radicals (OH·) in the troposphere by

$$CO + OH\cdot \rightarrow CO_2 + H\cdot$$

(Levy, 1971; McConnell et al., 1971)
2 Migration to the stratosphere and reaction with OH (Pressman and Warneck, 1970; Junge et al., 1971)
3 Removal of CO by soil (Inman et al., 1971; Ingersoll and Inman, 1972)

Suggested natural sources of CO are:

1 Oxidation of ambient methane (CH_4) by OH,

$$CH_4 + OH\cdot \rightarrow CH_3\cdot + H_2O$$

with subsequent conversion of $CH_3\cdot$ to CO (McConnell et al., 1971; Levy, 1972; Wofsy et al., 1972; Kummler and Baurer, 1972)
2 Oceans (Swinnerton et al., 1970; Wilson et al., 1970; Lamontagne et al., 1971)

Let us discuss briefly each of these possible sinks and sources.

Depending on the concentration of OH in the troposphere, the reaction $CO + OH\cdot \xrightarrow{k'} CO_2 + H$ could be a significant removal mechanism for CO. If this reaction were the main removal mechanism, then the OH concentration required to maintain a steady-state concentration of CO is given by

$$[OH\cdot] = \frac{k}{k'}$$

Taking $k = 11\ yr^{-1} = 3.5 \times 10^{-7}\ sec^{-1}$ and $k' = 1.5 \times 10^{-13}\ cm^3\ molecule^{-1}\ sec^{-1}$, the average OH concentration required to maintain CO in the steady state is 2.3×10^6 molecules/cm^3. On the basis of chemical mechanisms for the lower atmosphere, Levy (1971) and McConnell et al. (1971) estimated the daytime OH concentrations to be 1.2×10^6 molecules/cm^3 and 3×10^6 molecules/cm^3, respectively.

We note that these values apply only in the daytime since a photochemical mechanism is involved in OH production. The good agreement between the two ways of computing the OH concentration confirms the importance of the CO-OH reaction as a sink for CO in the troposphere.

In order to assess the importance of the stratosphere as a sink for CO, we need to know, in addition to estimates of average CO concentration levels, estimates of rates of air transport from the troposphere to the stratosphere and of OH concentration levels in the stratosphere. The effectiveness of the stratosphere as a chemical sink for CO depends on the rate at which CO can be transported into the stratosphere. Air exchange between the troposphere and stratosphere is limited by a layer called the *tropopause*, which acts as a barrier to convective transport. There has been wide interest in determining rates of transfer between the troposphere and stratosphere for questions related to radioactive fallout, stratospheric ozone, and general atmospheric circulation. A number of tracers have been used to estimate rates of exchange, the results of which have led to the following conclusions (Pressman and Warneck, 1970):

1 Mixing in the troposphere, in a given hemisphere, is relatively fast and occurs within a few weeks.

2 The exchange between the Northern and Southern Tropospheres is slower, requiring about 1 yr.

3 The residence time of air in the lower stratosphere, below 20 km, with which we are concerned here, is 1 to 2 yr.

4 Mixing in the stratosphere within a hemisphere is of the order of 1 to 2 yr, and air exchange between the Northern and Southern Stratospheres is about 5 yr.

Pressman and Warneck (1970) and Junge et al. (1971) estimated tropospheric CO residence times based on the stratospheric removal of CO as the predominant CO sink as being 9 and 2.7 yr, respectively. In either case, the stratosphere appears not to be an important sink with respect to a tropospheric residence time of the order of 0.1 yr.

A third possible important sink for CO is uptake by soil. Exploring this possibility, Inman et al. (1971) and Ingersoll and Inman (1972) exposed various soils to test atmospheres. Most of the soils showed a remarkable capability to remove CO from the test atmospheres. The activity of the soils varied from 7.6 to 115 mg CO hr^{-1} m^{-2}, with tropical soils showing the greatest activity and desert soils the least. When the soil was sterilized, CO removal capability was drastically reduced, indicating perhaps that the removal mechanism is the result of biological activity in the soil. On the basis of their data, Inman and Ingersoll were able to estimate the capacity of soil to remove CO from the atmosphere. The average rates of CO uptake in the field study ranged from 12.5 to 59.1 mg CO hr^{-1} m^{-2}. Assuming these values are representative of the average capacities of soils, the capacity of the total surface area of the

United States (7,792,533 km^2) can be estimated to be 1.3×10^3 Tg of CO per year, roughly four times the annual anthropogenic global production of CO. More research on this aspect is needed, for example to ascertain whether soil uptake is peculiar to CO or is a sink mechanism applicable for all gaseous species (Abeles et al., 1971).

We now consider possible natural sources of CO. There is now rather wide agreement that oxidation of methane, initiated by its reaction with the hydroxyl radical, is the predominant natural source of CO. Methane occurs at a concentration of about 1.5 ppm in the lower atmosphere. The sources of methane are natural gas and biological decay in marshes and paddy fields, with a global strength estimated at about 9×10^{13} moles per year. The methane concentration is observed to be globally homogeneous and unchanging in time, which suggests a removal mechanism roughly as great as the source strength and globally homogeneous. The total atmospheric mass of CH_4 is thus about 2.25×10^{14} moles, leading to an estimate of the average residence time of CH_4 of 2.5 yr. The global ambient CO concentration has been determined to be about 0.12 to 0.15 ppm (although there is still some uncertainty as to whether there are significant differences in ambient CO levels between the Northern and Southern Hemispheres). The hydroxyl radical concentration estimated to be necessary for CO removal, namely 10^6 to 3×10^6 molecules/cm^3, can also provide an explanation for the natural source of CO and for the natural sink for CH_4 through the reaction.

$$CH_4 + OH\cdot \rightarrow CH_3\cdot + H_2O$$

which has a rate constant (at 25°C) of 10^{-14} cm^3 molecule^{-1} sec^{-1}. Oxidation of CH_4 by OH initiates a rather complex series of reactions which results in formation of CO (Levy, 1972; Kummler and Baurer, 1972). Wofsy et al. (1972) estimated the total source of CO by this route to be nearly a factor of 10 greater than anthropogenic CO emissions. If methane is, in fact, the natural source of CO, this accounts for the failure to observe any significant increase in CO levels over the past 20 yr in spite of the doubling of combustion sources over this time period. Although Swinnerton et al. (1970) have shown that the oceans are supersaturated with respect to CO, an observation confirmed by Junge et al. (1971), estimates by both of these groups of the amount of CO released by oceans are of the order of 10 percent of the combustion sources of CO.

In summary, consideration of data on the concentration of ^{14}CO in the atmosphere leads to an estimate of approximately a one-month average residence time for CO. Assuming that CO levels are not changing in time, natural sources and sinks of CO ten times greater than anthropogenic emissions are arrived at. If the sink for CO is reaction with hydroxyl radicals, then the concentration of OH necessary to sustain this sink is, at the same time, sufficient to provide the natural source of CO by oxidation of methane.

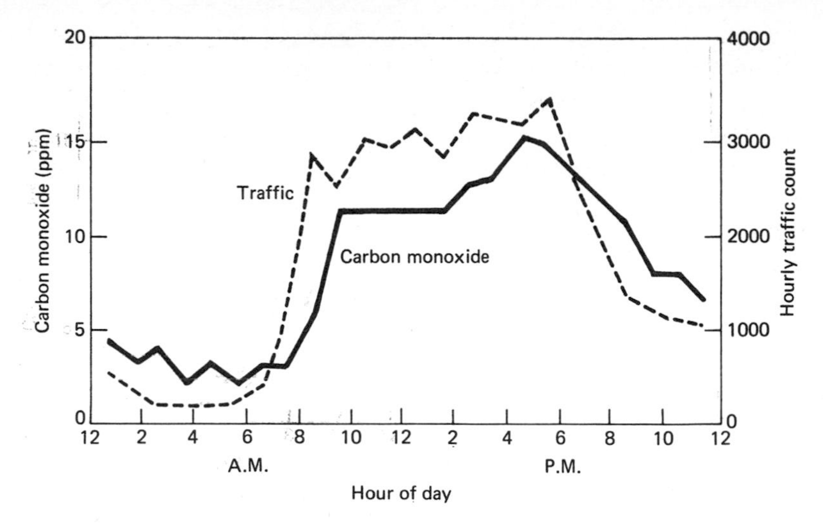

FIGURE 2.5
Hourly average CO concentration and traffic count in midtown Manhattan. (*Source: Johnson et al.*, 1968. *Copyright* 1968 *by the American Association for the Advancement of Science*.)

2.2.2 Urban Concentrations of Carbon Monoxide

Within an urban area combustion far outweighs natural sources of CO. Since CO is produced primarily by motor vehicles, urban CO emissions and concentration levels correlate closely with traffic volume. Figure 2.5 shows the hourly average CO concentration and traffic count in midtown Manhattan. The absence of well-defined rush-hour peaks is probably indicative of the saturation level of traffic throughout the day.

Table 2.6 MAXIMUM AVERAGE CARBON MONOXIDE CONCENTRATIONS FOR 1962 TO 1967 IN FIVE UNITED STATES CITIES

City	8-hr max. av. CO conc., ppm	5-min max. av. CO conc., ppm
Chicago	44	78
Denver	37	73
Los Angeles	32	81
Philadelphia	36	67
Washington	34	49

During conditions of prolonged stagnation, CO concentrations have reached dangerously high levels. Measurements in London from October 1956 to October 1957 yielded readings as high as 360 ppm at street level on a calm day. Listed in Table 2.6 are 8-hr and 5-min maximum average CO concentrations for the period 1962 to 1967 for five American cities. These levels should be compared with those in Fig. 1.4.

2.3 NITROGEN-CONTAINING COMPOUNDS

2.3.1 Origin and Fate of Nitrogen-containing Compounds

The important nitrogen-containing compounds in the atmosphere are N_2O, NO, NO_2, NH_3, and salts of NO_4^-, NO_3^-, and NH_4^+. The first of these, nitrous oxide N_2O, is a colorless gas which is emitted almost totally by natural sources, principally by bacterial action in the soil and by reaction between N_2 and O and O_3 in the upper atmosphere. The gas is employed as an anesthetic and is commonly referred to as "laughing gas." Chemically it is inert at ordinary temperatures and is not considered an air pollutant. The second, nitric oxide (NO), is emitted by both natural and anthropogenic sources. The burning of fuels at high temperatures is the primary anthropogenic source of NO. Nitrogen dioxide (NO_2) is emitted in small quantities along with NO and is also formed in th[illegible]atmosphere by oxidation of NO. Both NO and NO_2 are considered air pollut[illegible]her oxides of nitrogen, such as N_2O_3, N_2O_4, NO_3, and N_2O_5, exist in t[illegible]n very low concentrations and are not of concern as air pollutant[illegible]ily emitted by natural sources, ammonia (NH_3) can be an[illegible] in quantities large enough to produce local concentrat[illegible]entration. Finally, nitrate and ammonium salt[illegible]antities but result from conversion of NO[illegible]

Table 2.7[illegible], and NH_3, and Table 2.8 pr[illegible]nd NO_2 from

[illegible] AND

Spec[illegible]	[illegible]	Anthropogenic sources, Tg
NO and N[illegible]	[illegible]	48
NH_3	[illegible]	3.8

SOURCE: Robinson[illegible]

anthropogenic sources. The combustion of coal, oil, and gasoline accounts for the majority of the man-made emissions. Relative contributions of NO_X from motor vehicles and stationary sources vary from 50-50 in heavily industrialized areas to 90-10 in nonindustrialized areas with heavy traffic.

The most abundant oxide of nitrogen in the lower atmosphere is N_2O. As we noted, N_2O results from biological sources. In addition, N_2O appears to be a major natural source of NO in the upper troposphere and the stratosphere as a result of the reaction

$$N_2O + O \rightarrow 2NO$$

where the atomic oxygen involved is produced by photolytic dissociation of ozone. Nitric oxide and nitrogen dioxide can react with oxygen atoms to be interconverted or they can form NO_3, N_2O_4, and N_2O_5. Subsequent reaction of these unstable species with H_2O, OH, or HO_2 yields nitrous (HNO_2) and nitric (HNO_3) acid. Natural background concentrations of NO_2 are thought to be the result of reaction of NO with ambient ozone. Atmospheric electrical discharges may also produce small amounts of NO_2. The anthropogenic emissions of NO and NO_2 are estimated to exceed the natural production of NO by reaction of N_2O and O.

Atmospheric NH_3 results basically from biological decay at the earth's surface. The fate of NH_3 in the atmosphere is one of the following:

1 Absorption on wet surfaces to form NH_4^+

2 Reaction with acidic material in either gaseous or condensed phases to form NH_4^+

3 Oxidation to NO_3^-

Routes 1 and 2 account for the fate of approximately 75 percent of the NH_3, and route 3 for the remaining 25 percent.

Table 2.8 ESTIMATED GLOBAL ANTHROPOGENIC EMISSIONS OF NITROGEN OXIDES (AS NO_2) IN 1965

Source	Emissions, Tg NO_2
Coal combustion	24.4
Petroleum processes and combustion	20.2
Natural gas combustion	1.9
Miscellaneous (incineration, forest fires, etc.)	1.5
Total	48.0

SOURCE: Robinson and Robbins (1968).

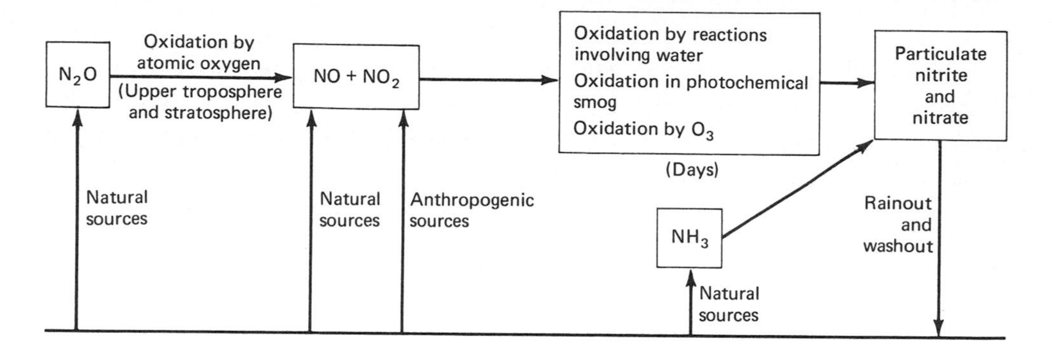

FIGURE 2.6
Nitrogen cycle for the nitrogen-containing compounds in air pollution.

The oxides of nitrogen are usually removed from the atmosphere by particle formation. Ultimately most of the oxides of nitrogen are converted to nitrates, which in turn are removed by rainout, washout, and dry deposition.

Figure 2.6 summarizes the nitrogen cycle for those species important in air pollution. In general, knowledge of the atmospheric nitrogen cycle is quite incomplete, certainly less so than that of the sulfur or CO cycles. Table 2.9 presents estimates of ambient concentrations and average residence times of nitrogen-containing air pollutants.

2.3.2 Urban Concentrations of Nitrogen Oxides

The oxides of nitrogen play an extremely important role in urban air pollution. The chemistry of NO and NO_2 in urban air pollution will occupy much of our attention in Chap. 4.

Table 2.9 AMBIENT CONCENTRATIONS AND ESTIMATED RESIDENCE TIMES OF ATMOSPHERIC NITROGEN COMPOUNDS

Compound	Source	Ambient concentration	Residence time
NO	Combustion	1.0 ppb (as NO_2)	5 days
NO_2	Combustion		
NO_2	Biological action		
NH_3	Biological action	6 ppb	2 weeks
NO_3^-	NO_2 oxidation	0.2 $\mu g/m^3$	
NH_4^+	NH_3 conversion	1.0 $\mu g/m^3$	2–8 days

Maximum hourly average NO_X concentrations of 0.25 ppm or higher occur regularly in Los Angeles and other large cities where concentrations of motor vehicles are high. For example, in 1963–1964, the average number of days per year during which maximum hourly average concentrations of $NO + NO_2$ greater than 0.25, 0.50, and 1 ppm were measured for some California cities were (The Oxides of Nitrogen in Air Pollution, 1967):

Location	0–0.24 ppm	>0.25 ppm	>0.50 ppm	>1.0 ppm
Los Angeles	164	201	89	5
Long Beach	191	174	92	5
Oakland	252	113	32	3
San Diego	289	76	18	1

2.4 HYDROCARBONS

2.4.1 Origin and Fate of Hydrocarbons

Up to now we have dealt with clearly identifiable species, such as SO_2, CO, and NO. Unfortunately, such a vast variety of hydrocarbons is emitted into the atmosphere from both natural and anthropogenic sources that it is not possible to measure all the individual species or to estimate their individual rates of emission. Therefore, when discussing hydrocarbons in air pollution we often cannot be more specific than the quantities of certain *classes* present.

Brief Review of Hydrocarbons Important in Air Pollution

Let us review briefly the classifications of hydrocarbons, particularly those of interest in air pollution. The carbon atom has four valence electrons and can therefore share bonds with from one to four other atoms. The nature of the carbon-carbon bonding in a hydrocarbon molecule basically governs the properties (as well as the nomenclature) of the molecule.

In some sense the simplest hydrocarbon molecules are those in which all the carbon bonds are shared with hydrogen atoms except for a minimum number required for carbon-carbon bonds. Molecules of this type are referred to as *alkanes* or, equivalently, as *paraffins*. The general chemical formula of alkanes is C_nH_{2n+2}. The first four paraffins having a straight chain structure are

CH_4	methane
$CH_3—CH_3$	ethane
$CH_3—CH_2—CH_3$	propane
$CH_3—CH_2—CH_2—CH_3$	*n*-butane

Alkanes need not have a straight chain structure. If a side carbon chain exists, the name of the longest continuous chain of carbon atoms is taken as the base name, which is then modified to include the type of group. Typical examples of substituted alkanes are (the numbering system is indicated below the carbon atoms):

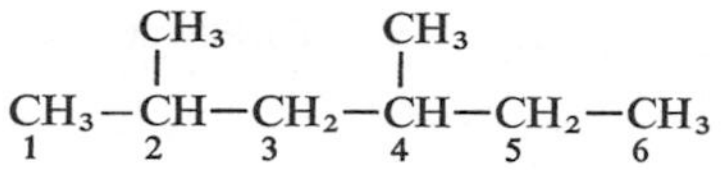

$$\begin{array}{c} \qquad CH_3 \qquad\qquad CH_3 \\ CH_3-\underset{2}{CH}-\underset{3}{CH_2}-\underset{4}{CH}-\underset{5}{CH_2}-\underset{6}{CH_3} \end{array}$$

2,4-Dimethyl hexane

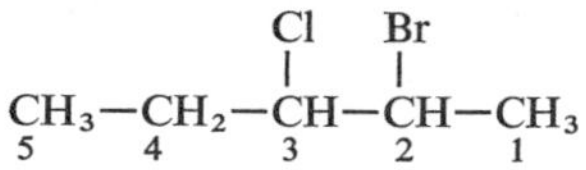

$$\underset{5}{CH_3}-\underset{4}{CH_2}-\underset{3}{\overset{Cl}{|}{CH}}-\underset{2}{\overset{Br}{|}{CH}}-\underset{1}{CH_3}$$

2-Bromo-3-chloropentane

Alkanes may also be arranged in a ring structure, in which case the molecule is referred to as a *cycloalkane*. The name of a cycloalkane is obtained simply by adding the prefix *cyclo-* to the name of the normal alkane having the same number of carbon atoms as in the ring. Cycloalkanes have the general chemical formula C_nH_{2n}. Examples are

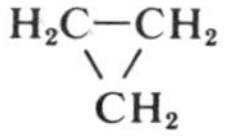

Cyclopropane	Cyclobutane	Cylopentane	Cyclohexane
H_2C-CH_2, CH_2 (3-membered ring)	H_2C-CH_2 / H_2C-CH_2 (4-membered ring)	H_2C-CH_2 / H_2C, CH_2 / $C H_2$ (5-membered ring)	$C H_2$ / H_2C, CH_2 / H_2C, CH_2 / $C H_2$ (6-membered ring)

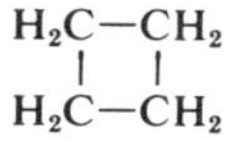

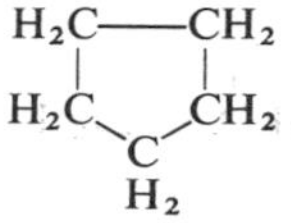

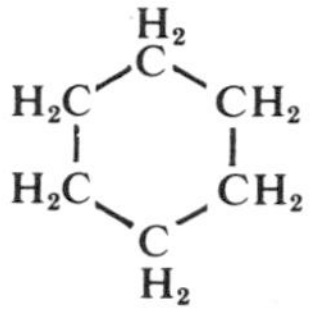

The alkanes and the cycloalkanes generally react by replacement of a hydrogen atom. Once a hydrogen atom is removed from an alkane, the involved carbon atom has an unpaired electron and the molecule becomes a *free radical*, in this case an *alkyl* radical. Examples of alkyl radicals are

$CH_3\cdot$	methyl
$CH_3CH_2\cdot$	ethyl
$CH_3CH_2CH_2\cdot$	*n*-propyl
$CH_3\dot{C}HCH_3$	isopropyl
$CH_3CH_2CH_2CH_2\cdot$	*n*–butyl

Alkyl radicals are often simply designated R·, where R denotes the chemical formula for any member of the alkyl group. The unpaired electron in a free radical makes the species extremely reactive. As we shall see, free radicals play an important role in atmospheric chemistry.

The next class of hydrocarbons of interest in air pollution is the *alkenes*. In this class two neighboring carbon atoms share a pair of electrons, a so-called double bond. Alkenes are also known as *alkylenes* or *olefins*. The location of the carbon atom nearest to the end of the molecule that is the first of the two carbon atoms sharing the double bond is often indicated by the number of the carbon atom. Examples of common alkenes are

$CH_2{=}CH_2$	ethylene
$CH_3CH{=}CH_2$	propylene
$CH_3CH_2CH{=}CH_2$	1-butene
$CH_3CH{=}CHCH_3$	2-butene

Molecules with two double bonds are called *alkadienes*, an example of which is

$$CH_2{=}CH-CH{=}CH_2$$

1,3-Butadiene

Molecules with a single triple bond are known as *alkynes*, the first in the series of which is acetylene, $HC{\equiv}CH$.

Compared with the alkanes, the alkenes are much more reactive. This is so because the double bond of the alkene is more easily attacked than the single carbon-carbon bonds of the alkanes. Also, the alkenes react by addition to the double bond rather than by substitution of hydrogen, as in the case of the alkanes. The alkenes, or, as we shall refer to them, the olefins, are important components of the hydrocarbon mix in the urban atmosphere.

Double-bonded hydrocarbons may also be arranged in a ring structure. This class of molecules, of which the basic unit is benzene,

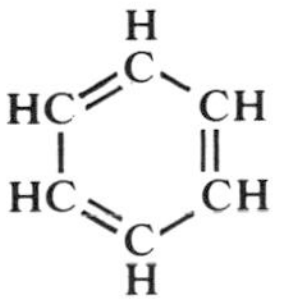

H
HC=C–CH
HC=C–CH
H

is called *aromatics*. Other common aromatic molecules are

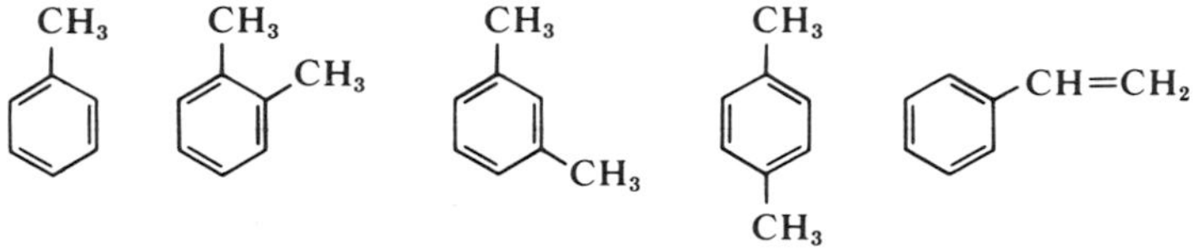

CH_3 CH_3 CH_3 CH_3 CH_3 CH_3 CH_3 $CH{=}CH_2$

Toluene   Orthoxylene   Metaxylene   Paraxylene   Styrene

Aromatics react by substitution and lie intermediate in reactivity between paraffins and olefins.

Hydrocarbons may acquire one or more oxygen atoms. Of the oxygenated hydrocarbons, two classes that are of considerable importance in air pollution are *aldehydes* and *ketones*. In each type of molecule, a carbon atom and an oxygen atom are joined by a double bond. Aldehydes have the general form

$$R{-}\overset{\overset{\displaystyle O}{\|}}{C}{-}H$$

whereas ketones have the structure

$$R{-}\overset{\overset{\displaystyle O}{\|}}{C}{-}R$$

Thus, the distinction lies in whether the carbon atom is bonded to one or two alkyl groups. Examples of aldehydes and ketones are

$H\overset{\overset{O}{\|}}{C}H$	$CH_3\overset{\overset{O}{\|}}{C}H$	$CH_3\overset{\overset{O}{\|}}{C}CH_3$	$CH_3\overset{\overset{O}{\|}}{C}CH_2CH_3$	$CH_2{=}CH\overset{\overset{O}{\|}}{C}H$
Formaldehyde	Acetaldehyde	Acetone	Methylethylketone	Acrolein

Both formaldehyde and acrolein occur in photochemical smog and are rather potent eye irritants.

In this section we will be concerned only with those hydrocarbons which exist in the atmosphere in the gas phase (generally those containing five or fewer carbon atoms). In Sec. 2.5 we will consider particles in the air, many of which consist of relatively nonvolatile hydrocarbons.

Table 2.10 presents estimates of global hydrocarbon emissions, and Table 2.11 shows a breakdown of the 80 Tg/yr of hydrocarbons by source. The combustion of gasoline is the single most important contribution to anthropogenic hydrocarbon emissions.

Table 2.10 ESTIMATED ANNUAL GLOBAL HYDROCARBON EMISSIONS IN 1965

Hydrocarbon	Source	Emissions, Tg
CH_4	Natural (swamps, paddy fields)	1450
Terpenes	Vegetation	154
All	Anthropogenic	80
Total		1684

SOURCE: Robinson and Robbins (1968).

The particular mechanism for removal of hydrocarbons from the atmosphere depends on the individual species involved. Most of the emissions detailed in Table 2.11 occur in urban areas where the conversion of hydrocarbons to other organic compounds takes place rapidly in the presence of oxides of nitrogen. The chemistry of these hydrocarbon removal reactions will be discussed in Chap. 4.

There is a substantial difference between the ambient concentration of methane and other hydrocarbons. Most measurements of ambient CH_4 concentrations yield values between 1.2 and 1.5 ppm. The generally quoted value is 1.5 ppm. Few measurements of low-molecular weight hydrocarbons other than CH_4 have been made

Table 2.11 ESTIMATED ANNUAL GLOBAL ANTHROPOGENIC HYDROCARBON EMISSIONS IN 1965

Source	Emissions, Tg
Coal	
Power	0.18
Industrial	0.63
Domestic and commercial	1.8
Petroleum	
Refineries	5.7
Gasoline	30.8
Evaporation	7.1
Other	0.4
Solvent use	9.1
Incineration	22.6
Wood burning	1.7
Total	80.1

SOURCE: Robinson and Robbins (1968).

in nonurban areas. Consequently, accurate levels of other hydrocarbons have not yet been established. In general, ambient ethane, ethylene, and acetylene concentrations, whenever measured consistently, range two to three orders of magnitude lower that that of methane.

The lack of accurate estimates of mass emissions and ambient concentrations of hydrocarbons other than methane precludes computation of residence times in the atmosphere. In general, lifetimes of higher-molecular-weight hydrocarbons have been estimated as being of the order of days to months, and that of methane as being of the order of years.[1]

With the exception of methane, very little is known about hydrocarbon cycles in the atmosphere. A number of recommendations aimed at improving knowledge of the fate of hydrocarbons in the atmosphere were made at the 1972 Chemist-Meteorologist Workshop. These included:

1 To identify the major organic components present in hydrocarbons from natural sources.

2 To measure reaction rates and establish a reaction scheme for the destruction of naturally produced hydrocarbons in the atmosphere (reactions with trace gases such as O_3, SO_X, NO_X should be included).

3 To determine to what degree naturally produced hydrocarbons act as sources for CO and CH_4.

4 To determine the role that naturally produced hydrocarbons play in the formation of aerosols (gas-to-particle conversion). Determine to what degree this gas-to-particle formation influences the concentration of tropospheric aerosol. Establish the particle formation mechanisms.

2.4.2 Urban Concentrations of Hydrocarbons

The most prevalent hydrocarbon in the atmosphere is methane, at a background level of 1.5 ppm. Since the predominant source of methane is natural and it is virtually inert chemically, it is not generally included when discussing hydrocarbon air pollutants. Thus, the methane concentration is usually subtracted from total hydrocarbon concentrations reported in urban areas.

In a number of studies cited, 56 hydrocarbon species were detected in urban air (Air Quality Criteria for Hydrocarbons, 1970). In fact, the number of hydrocarbons observed is limited only by the sensitivity of the analytical techniques in measuring the

[1] If CH_4 is removed by reaction with OH at a rate of 2.3×10^{15} g/yr, and added at a rate of 1.45×10^{15} g/yr from natural sources, an ambient level of 1.5 ppm leads to a residence time of 1.5 yr.

concentrations. A general characterization of the hydrocarbons in urban air is that of auto exhaust mixed with natural gas and gasoline vapor.

It is important to note that total hydrocarbon concentration is often not an accurate indicator of air pollution potential. Laboratory studies have shown conclusively that rates of reactions of hydrocarbons in the atmosphere differ markedly. Usually the destructive potential of atmospheric hydrocarbons from an air pollution standpoint lies in the products formed by their reactions. As we have mentioned, differences in reactivity of hydrocarbons may be related to their structure. Thus, measurements of total hydrocarbon concentration are often not as revealing as measurements of the concentrations of a number of the more reactive individual species.

2.5 PARTICULATE MATTER

The final class of air pollutants we consider is particulate matter (or aerosols). Particulate matter can be distinguished not only on the basis of chemical composition but also with respect to size. Both of these properties are important in determining the effects of aerosols on atmospheric properties and human health. For example, as we have seen, health effects related to inhaled aerosols depend on (1) the site of deposition in the respiratory tract, which is determined by particle size, and (2) the effect on tissue at the site of deposition, which depends on chemical composition. In addition, the effect of aerosols on light scattering and visibility depends on both particle size and chemical composition.

Figure 2.7 shows a classification of the sizes of atmospheric aerosols.[1] Aerosol sources can be classified as primary and secondary. Primary aerosols are those which are emitted in particulate form directly from sources, such as airborne dust as a result of wind or smoke particles emitted from a smokestack. Secondary aerosols refer to particles produced in the atmosphere, for example, from gas-phase chemical reactions which generate species capable of condensing as a particulate phase. Secondary mechanisms may, in fact, be responsible for the generation of as much aerosol as primary sources. Although primary sources yield particles of all sizes, secondary sources produce mainly very small particles.

Once aerosols are in the atmosphere, their size, number, and chemical composition are changed by several mechanisms until ultimately they are removed by natural processes. Some of the physical and chemical processes which affect the "aging" of atmospheric aerosols are more effective in one regime of particle size than another.

[1] When discussing particle size, we shall always refer to the diameter of the particle as if the particle were spherical. If the particle is, in fact, spherical, the diameter is its true diameter; if not, the diameter is that of a spherical particle which settles at the same velocity as the nonspherical particle.

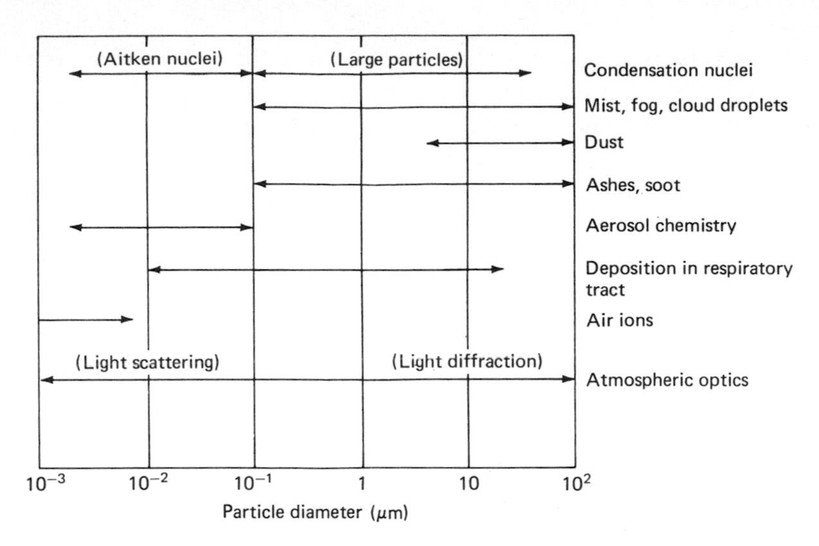

FIGURE 2.7
Sizes of atmospheric particulate matter.

In spite of the specific processes which affect particulate aging, the usual residence time of particles in the lower atmosphere is of the order of a couple of days to a week. Very close to the ground the main mechanism for particle removal is settling and impaction on surfaces, whereas at altitudes above about 100 m, rainout and washout are the predominant removal mechanisms.

In this section we first discuss the origins of atmospheric particulate matter, on the basis of total weight and, to some extent, chemical composition. Then we consider some aspects of urban versus nonurban tropospheric aerosols.

2.5.1 Origin of Particulate Matter

Table 2.12 presents estimates of global sources of particulate matter about 1968. We see that approximately 10 percent of the mass of aerosol emitted on a global basis is of man-made origin.

Let us first consider natural sources, both primary and secondary, of particulate matter. Major primary natural sources include wind-risen dust, sea spray, volcanoes, and forest and brush fires. Secondary natural sources are related to the carbon, sulfur, and nitrogen cycles and to the conversion of gases to particles.

Dust rise by wind is estimated to contribute 200 Tg/yr to the atmosphere. Having the composition of soil, namely silicates and other metals, this dust is usually of a size greater than 0.1 μm in diameter. The total amount of sea salt entering the atmosphere is usually deduced from the contribution of the sea to airborne sulfate.

Eriksson (1959) estimated the annual sulfate production from the sea as 118 Tg (Table 2.2). Since sulfate constitutes 7.7 percent by weight of sea salt, the estimated rate of particulate emission would be 1500 Tg/yr. Junge (1963) suggested that the rate of formation of sea spray is about 600 Tg/yr (assuming that each particle has a diameter of 2 μm). A compromise between these two estimates is 1000 Tg/yr, that given in Table 2.12. Sea-salt particles are generally larger than 0.1 μm in diameter. Volcanoes and forest fires, although sporadic in occurrence, are nonetheless rather important sources of atmospheric particles. If one assumes that 50 percent of the ash from a volcanic eruption remains suspended, on the basis of typical volcanic activity, a yearly emission rate of 4 Tg can be estimated (Hidy and Brock, 1971). Finally, forest and brush fires contribute large numbers of small particles, generally about 0.1 μm in diameter. The estimate of 200 Tg/yr for this source is probably quite conservative, given that the contribution from the United States alone is estimated to be 51 Tg/yr (see Table 2.13).

Table 2.12 ESTIMATED GLOBAL PARTICULATE EMISSIONS IN 1968

Source	Emissions, Tg Natural	Emissions, Tg Anthropogenic
Primary particle production:		
Fly ash from coal	....	36.0
Iron and steel industry emissions	....	9.0
Nonfossil fuels (wood, mill wastes)	....	8.0
Petroleum combustion	....	2.0
Incineration	....	4.0
Agricultural emission	....	10.0
Cement manufacturing	....	7.0
Miscellaneous	....	16.0
Sea salt	1000	
Soil dust	200	
Volcanic particles	4.0	
Forest fires	200	
Subtotal	1404	92.0
Gas-to-particle conversion:		
Sulfate from H_2S	202	
Sulfate from SO_2	....	147
Nitrate from NO_X	430	30
Ammonium from NH_3	269	
Organic aerosol from terpenes, hydrocarbons, etc.	198	27
Subtotal	1099	204
Total	2503	296

Secondary sources of natural aerosol are related to gas-phase chemical reactions which take place in the atmosphere, generating condensable material. Vegetation gives off large amounts of hydrocarbons, particularly compounds of the terpene class. Once in the atmosphere, these hydrocarbons participate in photochemical reactions, producing a vast amount of small particles (diameters less than 0.1 μm). Such particulate clouds are thought to be responsible for the blue hazes observed over forests. The estimate of 198 Tg/yr from this source, given in Table 2.12 and due to Robinson and Robbins (1971), is probably rather conservative; some estimates range as high as 1000 Tg/yr. We have already seen that a component of the sulfur cycle is oxidation of H_2S and SO_2 to produce aerosols. About 60 percent of the H_2S and SO_2 in the atmosphere is eventually converted to sulfate in less than a week's time. The values given in Table 2.12 are those inferred from the gaseous sulfur cycle (Robinson and Robbins, 1971). Finally, the conversion of gaseous nitrogen compounds, principally NH_3, NO, and NO_2, to aerosols is a major source of particulate matter. The estimates of particulate formation from the nitrogen cycle given in Table 2.12 are again due to Robinson and Robbins.

Table 2.13 summarizes the estimated primary sources of particulate matter in the United States in 1968. About 80 percent of the total primary emissions arise from natural sources. Table 2.14 shows a breakdown of the 16.3 Tg/yr of industrial particulate emissions. The values in Table 2.14 were determined from (1) emission factors for an uncontrolled source based on a unit of production, (2) the material processed per year, (3) the average or expected efficiency of control equipment installed on the processes, and (4) the percentage of production capacity equipped with control devices.

Table 2.13 ESTIMATED PARTICULATE EMISSIONS IN THE UNITED STATES IN 1968

Source	Emission, Tg
Natural dusts	57
Forest fires	51
Major stationary industrial sources	16.3
Transportation	1.1
Incineration	0.84
Other sources	1.2
Total	127.4

SOURCE: Vandegrift and Shannon (1971).

The Characterization of an Aerosol Population

A population of particles can be quantitatively characterized by its size and chemical composition distribution (assuming configurational effects such as surface layers and particle densities are constant). Let us assume that there are N available chemical species present of which a particle might consist. Let N_∞ be the total number of particles per unit volume. We can characterize the distribution of chemical species in the particle population by the *composition probability density function* $g(n_1, n_2, \ldots, n_N)$, where $g(n_1, n_2, \ldots, n_N)\, dn_1\, dn_2 \cdots dn_N$ is the fraction of the total number of particles per unit volume containing n_1 to $n_1 + dn_1$ moles of species 1, n_2 to $n_2 + dn_2$ moles of species 2, ..., n_N to $n_N + dn_N$ moles of species N. By its definition

$$\int_0^\infty g(n_1, n_2, \ldots, n_N)\, dn_1\, dn_2 \cdots dn_N = 1$$

Table 2.14 MAJOR INDUSTRIAL SOURCES OF PARTICULATE MATTER IN THE UNITED STATES IN 1968

Source	Efficiency of control, %	Application of control, %	Emission, Tg
Fuel combustion			
Coal			
Electric utility			
Pulverized	92	97	2.45
Stoker	80	87	0.198
Cyclone	91	71	0.165
Industrial boilers			
Pulverized	85	95	0.292
Stoker	85	62	2.02
Cyclone	82	92	0.035
Fuel oil	..	0	0.121
Natural gas and LPG	..	0	0.097
Total from fuel combustion			5.378
Crushed stone, sand, and gravel	80	25	4.17
Grain elevators and other agricultural operations	70	40	1.6
Iron and steel	90–99	35–100	1.22
Cement	94	94	0.84
Forest products	70–95	33–99	0.605
Lime	80–97	25–87	0.52
Clay	80	75	0.425
Primary nonferrous metals	40–98	35–100	0.420
Fertilizer and phosphate rock	80–97	25–100	0.306
Asphalt	97	99	0.198
Ferroalloys	80–99	35–100	0.145
Iron foundries	80	25–33	0.130
Secondary nonferrous metals	90–95	20–95	0.115
Coal cleaning	..	100	0.085
Carbon black	..	..	0.084
Petroleum-catalyst regeneration	..	100	0.041
Acids	95–97	85–90	0.014
Total from other than fuel combustion			10.918
Total major stationary industrial sources			16.3

SOURCE: Vandegrift and Shannon (1971).

If the molar quantities of each of the N species in a particle are known, then, given the molar volumes of each species, the volume and hence the diameter D_p of the particle may be calculated.

Only rarely are enough data available to characterize an aerosol population by its distribution of chemical compositions. Ordinarily, aerosols are described solely by their distribution with respect to size. Considering an aerosol population to consist of a distribution of spherical particles of uniform density but different sizes, we can characterize the size distribution of the population by the size density function $g(D_p)$, where $g(D_p)\,dD_p$ is the fraction of particles having diameters in the range D_p to $D_p + dD_p$. In a dynamic aerosol $g(D_p)$ changes with time and is expressed as $g(D_p, t)$. Most measurements of aerosol properties are concerned with identifying $g(D_p)$.

Instead of characterizing particles by their diameters, it is often convenient to describe them by the logarithm of the diameter, $\ln D_p$. Then the fraction of particles in the size range $\ln D_p$ to $\ln (D_p + dD_p)$ can be expressed as $G(\ln D_p)\,d \ln D_p$, where G is simply the size density function based on $\ln D_p$.

The average surface area per particle of the aerosol population is given by

$$\pi \int_0^\infty D_p^2 g(D_p)\,dD_p$$

and the average mass per particle is

$$\frac{\pi\rho}{6} \int_0^\infty D_p^3 g(D_p)\,dD_p$$

The average particle surface area and mass are directly related to the second and third moments, respectively, of the $g(D_p)$ distribution. We might note that the average particle surface area or mass can also be computed based on G, that is

$$\pi \int_0^\infty D_p^2 G(\ln D_p)\,d \ln D_p$$

$$\frac{\pi\rho}{6} \int_0^\infty D_p^3 G(\ln D_p)\,d \ln D_p$$

In fact, the mean of any function $f(D_p)$ is defined by

$$\overline{f(D_p)} = \int_0^\infty f(D_p) g(D_p)\,dD_p$$

The density functions expressing the distributions of particle surface area and mass (or volume), $g_a(D_p)$ and $g_m(D_p)$, respectively, are related to $g(D_p)$ by

$$g_a(D_p) = \frac{D_p^2 g(D_p)}{\int_0^\infty D_p^2 g(D_p)\,dD_p}$$

$$g_m(D_p) = \frac{D_p^3 g(D_p)}{\int_0^\infty D_p^3 g(D_p)\,dD_p}$$

These distributions are, in general, different in shape from $g(D_p)$.

Let us consider the mean value of the logarithm of D_p,

$$\overline{\ln D_p} = \int_0^\infty \ln D_p g(D_p)\,dD_p$$

In this particular case, the mean value of $\ln D_p$ is also the logarithm of the geometric mean of D_p, D_{pgm}. To see this, we assume a very large number of discrete diameters. The geometric mean of D_p is then defined by

$$D_{pgm} = D_{p_1}^{g_1} D_{p_2}^{g_2} \cdots D_{p_n}^{g_n}$$

where g_i is the fraction of the total particles with diameters D_{p_i}. Taking the logarithm of D_{pgm} and letting $n \to \infty$ yield

$$\ln D_{pgm} = \int_0^\infty \ln D_p g(D_p)\, dD_p \tag{2.5}$$

Aerosol size distributions are often observed to follow a log-normal distribution, that is, one in which the logarithm of the variable is normally distributed. Thus, ln D_p is observed in many cases to satisfy

$$G(\ln D_p) = \frac{1}{(2\pi)^{1/2} \ln \sigma_{gm}} e^{-(\ln D_p - \ln D_{pgm})^2/2 \ln^2 \sigma_{gm}} \tag{2.6}$$

where σ_{gm} is defined by

$$\ln^2 \sigma_{gm} = \int_0^\infty (\ln D_p - \ln D_{pgm})^2 G(\ln D_p)\, d \ln D_p \tag{2.7}$$

and D_{pgm} is given by (2.5). Thus, the distribution is specified by the two parameters: σ_{gm}, the standard geometric deviation, and D_{pgm}, the geometric mean diameter. Because of its importance in aerosol characterization, let us examine some of the properties of the log-normal distribution.

The distribution of the nth moment of $g(D_p)$ or $G(\ln D_p)$ is defined by

$$g_n(D_p) = \frac{D_p{}^n g(D_p)}{\int_0^\infty D_p{}^n g(D_p)\, dD_p} \tag{2.8}$$

or, equivalently, by

$$G_n(\ln D_p) = \frac{D_p{}^n G(\ln D_p)}{\int_0^\infty D_p{}^n G(\ln D_p)\, d \ln D_p} \tag{2.9}$$

Let us consider the quantity $D_p{}^n G(\ln D_p)$, where G is defined by the log-normal distribution (2.6). We need to compute this quantity to determine the distributions of particle surface area ($n = 2$) and mass or volume ($n = 3$) for a population obeying a log-normal distribution. The desired quantity may be expressed as

$$\frac{1}{(2\pi)^{1/2} \ln \sigma_{gm}} e^{n \ln D_p} e^{-(\ln D_p - \ln D_{pgm})^2/2 \ln^2 \sigma_{gm}}$$

Upon combining the two exponentials and expanding the square we obtain

$$\frac{1}{(2\pi)^{1/2} \ln \sigma_{gm}} \exp\left[-\frac{\ln^2 D_p - (2 \ln D_p)(\ln D_{pgm} + n \ln^2 \sigma_{gm}) + \ln^2 D_{pgm}}{2 \ln^2 \sigma_{gm}}\right]$$

By completing the square in the exponential, we get

$$D_p{}^n G(\ln D_p) = \frac{1}{(2\pi)^{1/2} \ln \sigma_{gm}} \exp\left(n \ln D_{pgm} + \frac{1}{2} n^2 \ln^2 \sigma_{gm}\right) \exp\left[-\frac{[\ln D_p - (\ln D_{pgm} + n \ln^2 \sigma_{gm})]^2}{2 \ln^2 \sigma_{gm}}\right] \tag{2.10}$$

A very interesting consequence of (2.10) is that if $G(\ln D_p)$ is log-normal the distribution of the nth moment of G is also log-normal, with a geometric mean diameter of $\ln D_{pgm} + n \ln^2 \sigma_{gm}$ and the same standard geometric deviation as of G, σ_{gm}. Thus, if the number distribution of an aerosol is log-normal, the surface area and mass and volume distributions will also be log-normal.

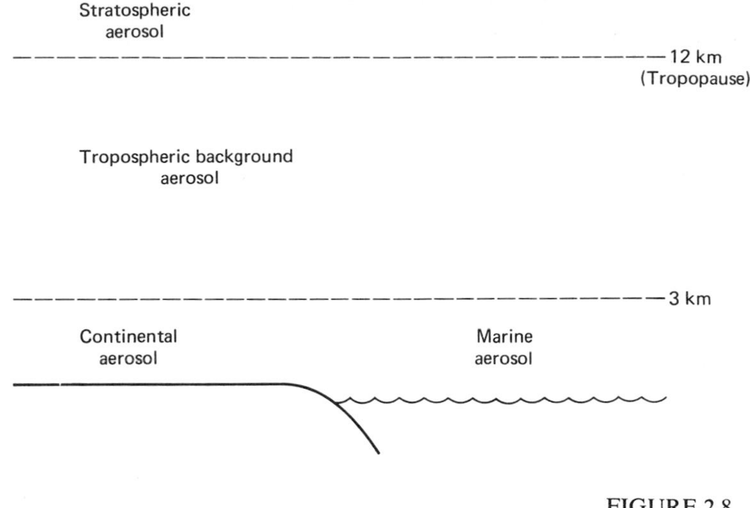

FIGURE 2.8
Global aerosol.

2.5.2 Urban and Background Particulate Matter

Particles in the troposphere can be generally characterized by their sources. The *natural continental aerosol* consists of (1) dust particles raised by the wind, having sizes above about 0.5 μm in diameter; (2) very small paricles ($D_p < 0.4$ μm) resulting from photochemical reactions between ozone and hydrocarbons emitted from vegetation; (3) very small particles from photochemical reactions involving gases such as SO_2, H_2S, NH_3, and O_3; and (4) particles from volcanic eruptions. The *natural marine aerosol* results from the evaporation of sea spray, has essentially the composition of sea salt (NaCl), and is in the size range $D_p > 0.5$ μm. Finally, *man-made aerosols* consist of (1) solid particles formed in combustion processes (smoke) and (2) particles in the range $D_p < 0.5$ μm which result from photochemical reactions involving unburned or partially burned hydrocarbons and oxides of nitrogen.

The number concentration of small particles ($D_p < 0.1$ μm) decreases with increasing altitude, indicating a terrestrial origin. However, the number concentration of particles in the 0.1- to 1-μm range actually goes through a maximum at an altitude of about 18 km. These particles, basically sulfates and nitrates, are most probably formed in the stratosphere by oxidation of gaseous SO_2. The source of SO_2 at this altitude has been speculated to be volcanic eruption ("Man's Impact on the Global Environment," 1970). In spite of these high altitude-particles, the stratospheric aerosol is only a few percent of the total aerosol mass in the atmosphere. The tropospheric background aerosol (that between 3 and 10 km) and continental (including marine) aerosol constitute about 25 and 70 percent by mass, respectively. Figure 2.8 summarizes the global aerosols which we have been discussing.

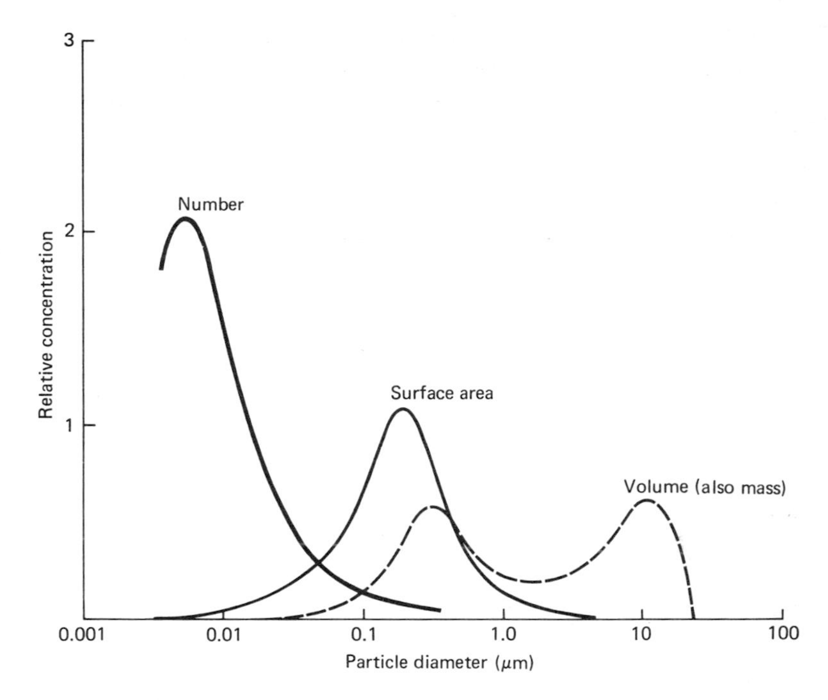

FIGURE 2.9
Typical particulate distributions by number, surface area, and volume.

The removal of particles is primarily a result of rainout and washout. Small particles are scavenged by water droplets in clouds and subsequently rained out. Larger particles can be removed by the same process or by direct washout by falling raindrops. The average lifetime of particles in the troposphere depends, therefore, on the amount of precipitation in the particular area. Typical residence times in the lower troposphere range from several days to a few weeks, whereas those in the upper troposphere may be as long as a month. Stratospheric particles can exist from six months to a year, depending on latitude (shorter residence time over the poles).

In a typical urban atmosphere most of the particles, by number, are smaller than 0.1 μm in diameter. The distribution of particle mass (or volume) over the different size ranges generally exhibits two maxima, one between 0.1- and 1-μm diameter and the other between 1- and 30-μm diameter. These two maxima are a result of two different types of aerosol formation processes. Practically all the fine-particle fraction results from condensation of vapors, whereas the coarse-particle fraction consists of dust, fly ash, and mechanically generated aerosol. Typical number and volume (also mass) distributions are shown in Fig. 2.9. The distribution of particle surface area in an aerosol having the number and volume distributions shown in Fig. 2.9 generally has a single maximum in the range 0.1 to 1 μm in diameter. The number distribution simply expresses the number of particles per volume in the given size range (D_p, $D_p + dD_p$). The volume and mass distributions are proportional to D_p^3, whereas those for surface area are proportional to D_p^2. Thus, the very small particles, although abundant in number, do not contribute much total surface area or mass compared with that of a smaller number of much larger particles. The surface area maximum in the range 0.1 to 1 μm is a result of a balance between a very large number of small particles and a few large particles.

Let us consider some typical observed aerosol size distributions, as shown in Fig. 2.10, in which $g(D_p)$ is plotted as a function of D_p. Note that, in this plot, $g(D_p)$ is *not* normalized, so that

$$N_\infty = \int_0^\infty g(D_p)dD_p$$

where N_∞ is the total number of particles per unit volume.

Measurements of natural continental aerosols have revealed that there is usually only a single peak in the size distribution. Addition of particles in an urban area may lead to a local size distribution with more than one peak, corresponding to the sizes of the emitted particles. As the aerosol ages, the very small particles grow by coagulation and the very large particles settle out by gravity, leading to a smoothing of the size distribution and a return to a single peak.

As seen in Fig. 2.10, both the urban and nonurban spectra are quite broad, indicating the presence of particles from tens of angstroms to tens of micrometers The urban sample is seen to contain many more particles of diameter less than 0.5 μm than any of the nonurban samples. The difference is particularly significant below 0.1 μm in diameter, where a very large number of particles was found in the Pasadena samples. These particles are the result of gas-to-particle conversion processes.

In the correlation of data on continental aerosols, Junge (1963) found that the distribution of large and giant particles often follows the form

$$g(D_p) = aD_p^{-(b+1)} \qquad 0.1\ \mu m < D_p < 10\ \mu m \qquad (2.11)$$

where b has a value close to 3. This so-called Junge distribution is shown in Fig. 2.10. The Pasadena aerosol sample is seen to conform to the Junge distribution throughout the day for sizes above about 0.1-μm diameter. At 0400 hours the size distribution even conformed to the Junge distribution at diameters below 0.1 μm. However, as the aerosol developed throughout the day, the production of very small particles by gas-to-particle conversion increased the number of particles below 0.1 μm diameter to values considerably above that of the European continental aerosol. Regions well removed from the influence of anthropogenic sources may have aerosol size distributions that differ considerably from the Junge distribution, as exemplified by the data

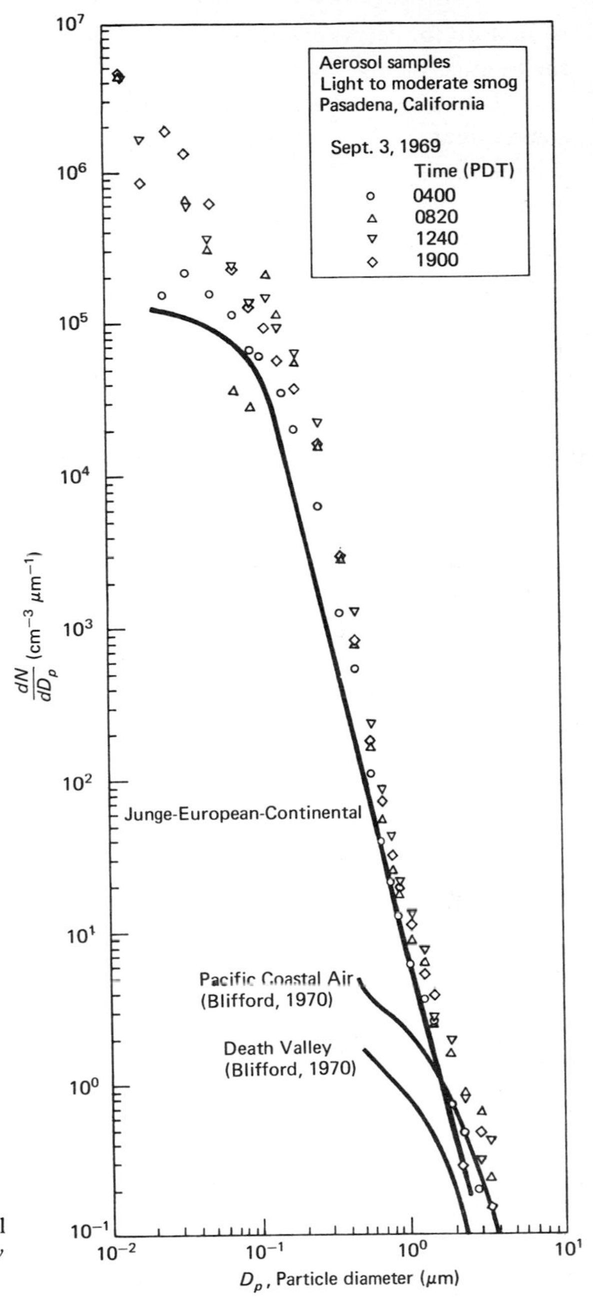

FIGURE 2.10
Comparison of urban and continental aerosol size distributions. (*Source: Hidy and Friedlander*, 1971.)

of Blifford for Pacific coastal air and in Death Valley, California. Clearly in each case the number of large particles is considerably below that of urban air.

Estimates of the relative importance of aerosol sources can be evaluated, to some extent, by measuring the chemical composition of urban and nonurban tropospheric aerosols. The average composition of particles taken from the National Air Sampling Network in the United States is presented in Table 2.15, where average total mass concentrations are shown for four general types of locales ranging from urban to remote. Average particle-mass concentrations range from about 20 $\mu g/m^3$ in clean air to values of 60 to 200 $\mu g/m^3$ in urban areas. (In heavily polluted air, values as high as 2000 $\mu g/m^3$ have been reported.) Number concentrations vary from $10^7/cm^3$ in very clean air to $10^5/cm^3$ in polluted air.

Measured mass fractions of the various chemical constituents can be compared with the estimated fractions of various types of aerosol sources. According to the somewhat low percentages given in Table 2.15, the estimates of natural hydrocarbon emissions, production rates of sulfate and nitrate by reactions in the atmosphere, and sea spray may all be rather high. The basic problem in using measured aerosol compositions is that, unless the measurements are carefully carried out, particles less than 0.5 μm in diameter may not be adequately collected. As we have noted, the secondary aerosol from hydrocarbons from vegetation and from the sulfur and nitrogen cycles exists mainly in sizes below 0.5 μm. Although most of the mass is concentrated in the

Table 2.15 AVERAGE PARTICULATE CONCENTRATIONS FROM THE UNITED STATES NATIONAL AIR SAMPLING NETWORK IN 1966–1967

Suspended Particles, $\mu g/m^3$

Urban	Near-urban	Nonurban intermediate	Remote
102.0	45.0	40.0	21.0

Material	% Total	Range
Benzene soluble organics	5.7	5.1–6.6
Ammonium ion	1.2	0.7–1.7
Nitrate ion	2.5	2.1–3.1
Sulfate ion	14.2	9.9–22.2

SOURCE: Hidy and Brock (1971).

larger particles, as much as 25 percent of the total mass may be below 0.5 μm in diameter.

There have been a number of measurements of particulate concentrations and chemical composition in each of the major cities in the United States. For example, three recently reported analyses of the composition of aerosols in New York City were carried out by Colucci and Begeman (1971), Morrow and Brief (1971), and Cukor et al. (1972); summaries of earlier data for many cities appear in Air Quality Data for Particulate Matter (1969).

In Chap. 4 we shall consider the properties and behavior of aerosols in more detail. At that time we shall examine the nature of the Los Angeles aerosol, as one which exemplifies the effects of both natural and anthropogenic emissions as well as those of the conversion of gaseous reactants to particles.

To conclude we summarize in Table 2.16 sources and dynamic and optical properties of particles by size. Generation mechanisms during combustion are not fully understood; however, it appears that most particles resulting from combustion processes lie in the size range 0.005 to 0.5 μm in diameter. Thus, combustion is listed as a source of particles for sizes up to 1-μm diameter in Table 2.16. Coarse particles, those having diameters greater than 1 μm, generally result from mechanical action. Examples include windborne dust and particles from grinding and spraying. In the latter two operations, particles are usually larger than 10 μm.

Table 2.16 SOURCES AND DYNAMIC AND OPTICAL PROPERTIES OF PARTICLES AS A FUNCTION OF SIZE

Size	Sources	Dynamic properties	Optical properties
<0.1 μm	Condensation, combustion, chemical reactions	Brownian motion, coagulation with other particles	Obey roughly the same laws of light scattering as molecules, little effect on visibility.
0.1–1 μm	Industrial dusts, fly ash from combustion, sea salt	Combination of brownian motion, coagulation, and settling	Particle sizes of the same order as the wavelength of visible radiation; most effective in light scattering and visibility reduction.
>1 μm	Fly ash, erosion, grinding spraying (>10 μm)	Settling; washout by rain, fog, snow; impingement on surfaces	Obey scattering laws of large bodies; not as effective in visibility reduction as smaller particles.

2.6 SUMMARY

The trace gaseous components of the atmosphere can be classified into two groups:

1 CO_2, CH_4, CO, N_2O: Relatively inert gases with lifetimes of the order of months to years. Life cycles of these species are heavily influenced by the rate of emission from natural sources.

2 H_2S, SO_2, NH_3, hydrocarbons, NO: Gases which are rapidly converted to particles, with lifetimes of the order of days to weeks.

Table 2.17 presents a summary of sources, concentrations, and major reactions of atmospheric trace gases.

Table 2.17 SOURCES, CONCENTRATIONS, AND MAJOR REACTIONS OF ATMOSPHERIC TRACE GASES

Gas	Anthropogenic sources	Natural sources	Background concentration	Estimated lifetime	Removal mechanisms
SO_2	Combustion of coal and oil	Volcanoes	0.002–0.01 ppm	3 days	Oxidation to sulfate photochemically or catalytically.
H_2S	Chemical processes	Biological	0.002–0.02 ppm	1 day	Oxidation to SO_2.
CO	Combustion	Oxidation of CH_4, oceans	0.12–0.15 ppm	0.1 yr	Reaction with OH in troposphere and stratosphere. Soil removal.
NO-NO_2	Combustion	Bacterial action in soil	NO: 0.2–2 ppb NO_2: 0.5–4 ppb	5 days	Oxidation to nitrate by photochemical reactions or on aerosol particles.
NH_3	Waste treatment	Biological decay	6–20 ppb	2 weeks	Reaction with SO_2 to form $(NH_4)_2SO_4$. Oxidation to nitrate.
N_2O	None	Biological action in soil	0.25 ppm	4 yr	Photodissociation in stratosphere.
CH_4	Combustion, chemical processes	Swamps, paddy fields	1.5 ppm	1.5 yr	Reaction with OH.

REFERENCES

ABELES, F. B., L. E. CRAKER, L. E. FORRENCE, and G. R. LEATHER: Fate of Air Pollutants: Removal of Ethylene, Sulfur Dioxide, and Nitrogen Dioxide by Soil, *Science*, **173**:914 (1971).

Air Quality Criteria for Hydrocarbons, *U. S. Dept. Health, Education and Welfare Publ. AP–64*, 1970.

Air Quality Criteria for Particulate Matter, *U.S. Dept. Health, Education and Welfare Publ. AP*–49, 1969.

Air Quality Criteria for Sulfur Oxides, *U.S. Dept. Health, Education and Welfare Publ. AP*–50, 1969.

BEILKE, S., and H. GEORGII: Investigation on the Incorporation of Sulfur Dioxide into Fog and Rain Droplets, *Tellus*, **20**:435 (1968).

BLIFFORD, I. H., JR.: Tropospheric Aerosols, *J. Geophys. Res.*, **75**:3099 (1970).

Chemist-Meteorologist Workshop, U.S. Atomic Energy Commission, Washington, D.C., January, 1972.

COLUCCI, J. M., and C. BEGEMAN: Carbon Monoxide in Detroit, New York and Los Angeles Air, *Environ. Sci. Technol.*, **3**:41 (1969).

COLUCCI, J. M., and C. BEGEMAN: Carcinogenic Air Pollutants in Relation to Automotive Traffic in New York, *Environ. Sci. Technol.*, **5**:145 (1971).

CUKOR, P., L. L. CIACCO, E. W. LANNING, and R. L. RUBINO: Some Chemical and Physical Characteristics of Organic Fractions in Airborne Particulate Matter, *Environ. Sci. Technol.*, **6**:633 (1972).

ERIKSSON, E.: The Yearly Circulation of Chloride and Sulfur in Nature; Meteorological, Geochemical, and Pedological Implications. Part I, *Tellus* **11**:375 (1959).

ERIKSSON, E.: The Yearly Circulation of Chloride and Sulfur in Nature; Meteorological, Geochemical, and Pedological Implications. Part II, *Tellus*, **12**:63 (1960).

ERIKSSON, E.: The Yearly Circulation of Sulfur in Nature, *J. Geophys. Res.*, **68**:4001 (1963).

HALES, J. M., J. O. WILKES, and J. L. YORK: The Rate of Reaction Between Dilute H_2S and O_3 in Air, *Atmos. Environ.*, **3**:657 (1969).

HIDY, G. M., and J. R. BROCK: "Dynamics of Aerocolloidal Systems," Pergamon, New York, 1970.

HIDY, G. M., and J. R. BROCK: An Assessment of the Global Sources of Tropospheric Aerosols, in H. M. Englund and W. T. Beery (ed.), "Proceedings of the Second International Clean Air Congress," Academic Press, New York, 1971.

HIDY, G. M., and S. K. FRIEDLANDER: The Nature of the Los Angeles Aerosol, in H. M. Englund and W. T. Beery (ed.), "Proceedings of the Second International Clean Air Congress," Academic Press, New York, 1971.

INGERSOLL, R. B., and R. E. INMAN: Soil as a Sink for Atmospheric Carbon Monoxide, Proceedings of the Conference on Sources, Sinks and Concentrations of Carbon Monoxide and Methane in the Earth's Environment, American Geophysical Union and American Meteorological Society, St. Petersburg, Fla., Aug. 15–17, 1972.

INMAN, R. E., R. B. INGERSOLL, and E. A. LEVY: Soil: A Natural Sink for Carbon Monoxide, *Science*, **172**:1229 (1971).

JAFFE, L. S.: Carbon Monoxide in the Biosphere: Sources, Distribution and Concentrations, Proceedings of the Conference on Sources, Sinks and Concentrations of Carbon Monoxide and Methane in the Earth's Environment, American Geophysical Union and American Meteorological Society, St. Petersburg, Fla, August 15–17, 1972.

JOHNSON, K. L., L. H. DWORETSKY, and A. N. HELLER: Carbon Monoxide and Air Pollution from Automobile Emissions in New York City, *Science*, **160**:67 (1968).

JUNGE, C. E.: "Air Chemistry and Radioactivity," Academic Press, New York, 1963.

JUNGE, C. E.: The Cycle of Atmospheric Gases–Natural and Man Made, *Q. J. Roy. Meteorol. Soc.*, **93**:418, 711 (1972).

JUNGE, C. E., W. SEILER, and P. WARNECK: The Atmospheric ^{12}CO and ^{14}CO Budget, *J. Geophys. Res.*, **76**:2866 (1971).

KELLOGG, W. W., R. D. CADLE, E. R. ALLEN, A. L. LAZRUS, and E. A. MARTELL: The Sulfur Cycle, *Science*, **175**:587 (1972).

KUMMLER, R. H., and T. BAURER: A Temporal Model of Tropospheric Carbon-Hydrogen Chemistry, Proceedings of the Conference on Sources, Sinks and Concentrations of Carbon Monoxide and Methane in the Earth's Environment, American Geophysical Union and American Meteorological Society, St. Petersburg, Fla., Aug. 15–17, 1972.

LAMONTAGNE, R. A., J. W. SWINNERTON, and V. J. LINNENBOM: Nonequilibrium of Carbon Monoxide and Methane at the Air-Sea Interface, *J. Geophys. Res.*, **76**:21, 5117 (1971).

LARSEN, R. I.: A New Mathematical Model of Air Pollutant Concentration Averaging Time and Frequency, *J. Air Pollut. Control Assoc.*, **19**:24 (1969).

LARSEN, R. I., C. E. ZIMMER, D. A. LYNN, and K. G. BLEMEL: Analyzing Air Pollutant Concentration and Dosage Data, *J. Air Pollut. Control Assoc.*, **17**:85 (1967).

LEVY, H. II: Normal Atmosphere: Large Radical and Formaldehyde Concentrations Predicted, *Science*, **173**:141 (1971).

LEVY, H. II: The Tropospheric Budgets for Methane, Carbon Monoxide and Related Species, Proceedings of the Conference on Sources, Sinks and Concentrations of Carbon Monoxide and Methane in the Earth's Environment, American Geophysical Union and American Meteorological Society, St. Petersburg, Fla., Aug. 15–17, 1972.

"Man's Impact on the Global Environment," M.I.T. Press, Cambridge, Mass., 1970.

MCCONNELL, J. C., M. B. MCELROY and S. C. WOFSY: Natural Sources of Atmospheric CO, *Nature*, **233**:187 (1971).

MORROW, N. L., and R. S. BRIEF: Elemental Composition of Suspended Particulate Matter in Metropolitan New York, *Environ. Sci. Technol.*, **5**:786 (1971).

PRESSMAN, J., and P. WARNECK: The Stratosphere as a Chemical Sink for Carbon Monoxide, *J. Atmos. Sci.*, **27**:155 (1970).

ROBINSON, E., and R. C. ROBBINS: Sources, Abundance and Fate of Gaseous Atmospheric Pollutants, Final report of project PR–6755, Stanford Research Institute, Menlo Park, Calif., February, 1968.

ROBINSON, E., and R. C. ROBBINS: Emissions, Concentrations and Fate of Particulate Atmospheric Pollutants, *Am. Petrol. Inst. Publ.* 4076, 1971.

SWINBANK, W. C.: A Comparison Between Predictions of Dimensional Analysis for the Constant-flux Layer and Observations in Unstable Conditions, *Q. J. Roy. Meteorol. Soc.*, **94**:460 (1968).

SWINNERTON, H. S., V. J. LINNENBOM, and R. A. LAMONTAGNE: The Ocean: A Natural Source of Carbon Monoxide, *Science*, **167**:984 (1970).

The Oxides of Nitrogen in Air Pollution, State of California Board of Public Health, Sacramento, Calif., 1967.

VANDEGRIFT, A. E., and L. J. SHANNON: Particulate Pollutant System Study, vol. I; Mass Emissions, vol. II; Fine Particle Emissions, vol. III; Handbook of Emission Properties, Midwest Research Institute Project 3326–B, Washington, 1971.

WEINSTOCK, B., and H. NIKI: Carbon Monoxide Balance in Nature, *Science*, **176**:290 (1972).

WILSON, D. F., J. W. SWINNERTON, and R. A. LAMONTAGNE: Production of Carbon Monoxide and Gaseous Hydrocarbons in Seawater: Relation to Dissolved Organic Carbon, *Science*, **168**:1577 (1970).

WOFSY, S. C., J. C. MCCONNELL, and M. B. MCELROY: Atmospheric CH_4, CO and CO_2, *J. Geophys. Res.*, **77**:4477 (1972).

PROBLEMS

2.1 We have estimated lifetimes of atmospheric species based on the rates of the processes responsible for their chemical destruction. As mentioned, we can obtain estimates of the average residence time of pollutants by another means. Let us assume that the ambient concentrations are unchanging with time. Then, if we have an estimate for the global annual emissions of the particular species from all sources we can estimate the average turnover time for the species. Assume, for the purposes of calculation, that the species of interest are confined to the troposphere. Using the data given in this chapter as well as that given below, estimate the atmospheric residence times of SO_2, CO, and NO + NO_2. Compare your results with those in Table 2.17. Which means of estimating lifetimes do you feel is more reliable?

Total mass of atmosphere $= 5.2 \times 10^{21}$ g
Mass of troposphere $= 4.0 \times 10^{21}$ g
Number of moles in atmosphere $= 1.8 \times 10^{20}$ g moles

2.2 In attempting to project global anthropogenic aerosol emission rates, Hidy and Brock (1971) assumed that both primary and secondary production rates grow exponentially with time. The rate of growth in each source category, in kilograms per year, is given by

$$R_i = a_i + b_i e^{c_i t}$$

Table 1 CONSTANTS FOR PARTICULATE EMISSIONS ESTIMATES

Source i	a_i, kg/yr	b_i, kg/yr	c_i, yr^{-1}
0 Total production of natural aerosols	3.26×10^{12}	0	0
1 Primary emissions	0	1.31×10^{-26}	0.0432
2 Production of secondary aerosol from anthropogenic SO_2 emissions	0	1.45×10^{-10}	0.0242
3 Production of secondary aerosol from anthropogenic NO_X emissions	0	3.54×10^{-13}	0.0266
4 Production of secondary aerosol from anthropogenic hydrocarbon emissions†	1.09×10^{9}	1.18×10^{-31}	0.0468

† All emissions are assumed to remain constant except consumption of fuels by motor vehicles.

where the constants are given in Table 1 for each source class. The total aerosol mass in the troposphere, $m(t)$, is seen to be governed by the balance

$$\frac{dm(t)}{dt} = \sum_{i=1}^{4} R_i - \sum_i S_i + a_0$$

where S_i is the rate of removal of aerosol by mechanism i, and a_0 is the production rate of natural aerosols. Assume that the total rate of removal, $\sum S_i$, is proportional to the total aerosol mass with a characteristic residence time of 3 days. Thus,

$$\sum_i S_i = \frac{m(t)}{\tau}$$

where $\tau = 3$ days. At time $t = 0$, assume that there is no anthropogenic contribution, so that $m(0) = a_0\tau$. Plot the fractional increase in particle loading, $m(t)/m(0)$, for a period of 30 yr. Repeat the calculation for the case in which motor vehicle emissions and SO_2 emissions are reduced by 80 and 50 percent, respectively.

2.3 Show that the mass distribution of particles in an aerosol for which the size distribution is governed by Eq. (2.11) with $b = 3$ is given by (assuming particle density ρ_p independent of diameter D_p)

$$dM = \frac{\pi \rho_p a \ln(10)}{6} d(\log_{10} D_p)$$

What does this result tell you about the mass of particles within a given size range? Specifically, what can be said about the total mass of particles in the range 0.1 to 1 μm diameter as compared with the range 1 to 10 μm? Discuss the implications of this result with respect to light scattering.

2.4 At the terminal settling velocity of a small particle the gravitational force exerted on the particle is balanced by the frictional force retarding the motion. This balance is expressed quantitatively as

$$\frac{\pi}{6} D_p^3 \rho_p g = 3\pi\mu D_p v_s$$

where v_s is the settling velocity. Show that the mean settling velocity of all particles is related to the second moment of the size distribution function $g(D_p)$. Then show that the average mass flux of particles due to settling is related to the fifth moment of $g(D_p)$.

2.5 Assume that the eruption of the volcano Krakatoa in the East Indies in 1883 is estimated to have released approximately 10^{10} kg of particulate matter into the atmosphere. Also assume that this particulate matter was emitted in a monodisperse state with all particles of diameter 0.1 μm and density 1 g/cm^3. You wish to model, in a crude fashion, the concentration dynamics of these particles in the global atmosphere. To do so, assume that the atmosphere can be represented by four well-mixed regions, as shown in Fig. 1. Thus, the concentration of particles in each region is taken to be uniform but variable in time due to transport with the other regions and loss by deposition. The initial injection of particles took place in the Southern Troposphere. After injection, the only mechanism for particle loss is deposition. Assuming that the rate of deposition in the troposphere depends directly on the concentration of particles, write a set of

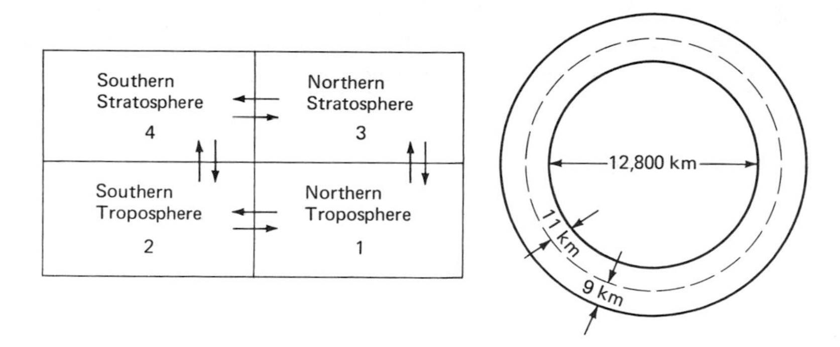

FIGURE 1

Table 2 **DATA FOR PROB. 2.5**

Volumetric flow rates q_{ij} from region i to region j, m^3/m^2-h

j \ i	1	2	3	4
1	....	8.76×10^{-2}	0.165	
2	8.76×10^{-2}	....	......	0.165
3	0.165	.. ..		
4	....	0.165		

Rate constant for particle deposition $= 1.1 \times 10^{-5}\ m^{-2}\ hr^{-1}$

material balance equations for each region. The data in Table 2 will be useful in evaluating the parameters in the equations. The equations can be solved by techniques of linear algebra and matrix theory. Proceed as far as possible toward their solution.

2.6 Many aerosol size distributions are observed to obey the log-normal distribution. Plot the number, surface area, and volume distributions of a log-normally distributed aerosol population. Use $\ln \sigma_{gm} = 1$. Indicate on the figure the geometric mean diameter of the surface area and volume distributions.

2.7 Data on the size distribution of the Pasadena, California, aerosol in September, 1969 were given in Fig. 2.10. In this figure, the ordinate is the nonnormalized number distribution dN/dD_p. It is of interest to determine the volume distribution of this aerosol, particularly the total aerosol volume per unit volume of gas. Replot representative data points from Fig. 2.10 in terms of $dV/d(\log D_p)$, the volume increment per log size interval. Use a linear scale on the ordinate from 0 to 50 and the same abscissa as in Fig. 2.10. What does the area under this curve represent? Finally you wish to compare this volume distribution with that of a log-normal distribution. Choosing the geometric mean diameter of the log-normal distribution to coincide with the value of D_p at the maximum of the experimentally observed volume distribution, determine the value of σ_{gm} which gives the closest fit of the two distributions. Can you explain the deviations of the observed volume distribution from a log-normal distribution?

3

AIR POLLUTION METEOROLOGY

With this chapter we begin a study of the atmospheric aspects of air pollution by discussing some relevant aspects of meteorology. We most often associate air pollution with urban meteorology, the atmospheric conditions over a region of perhaps 100 to 10,000 km^2. An understanding of atmospheric processes on this scale is obviously essential in analyzing urban air pollution. However, more often than not, the local meteorological conditions which lead to elevated pollutant concentrations in an urban area are a result of meteorological processes of a much larger scale. In addition, in tracing the fate of pollutants originating in a large urban area, we require knowledge of the great atmospheric currents that ultimately disperse pollutants over the troposphere. Thus, it is important to understand some elements of synoptic as well as of meso- and micrometeorology in order to have a full picture of the influences of weather and climate on pollutant concentrations.

In this chapter we shall discuss four topics: (1) the atmospheric energy balance, (2) temperature profiles in the lower atmosphere, (3) winds, and (4) air pollution and global climatic change. It is possible that increases in the background concentration of even a minor constituent of the atmosphere may lead to significant changes in atmospheric properties. As we shall see, these changes would most likely be reflected in radiative scattering and absorption processes, and therefore a study of the atmospheric energy balance is essential in understanding potential climatic changes due to air pollutants. Lower atmospheric temperature profiles determine in part the stability of the atmosphere, or, in other words, the degree to which turbulence induced by wind shear, surface roughness, or buoyancy will propagate through the layer. Under strongly stable conditions, disturbances are highly damped and mixing of species is strongly suppressed. It is under such conditions that the worst air pollution episodes have occurred. Finally, the importance of winds to the atmospheric aspects of air pollution is clearly evident. Our discussion of winds in this chapter will be highly qualitative; in Chap. 5 we shall treat air motion in the lower atmosphere from a quantitative standpoint.

3.1 ATMOSPHERIC ENERGY BALANCE

3.1.1 Radiation

Basically all the energy that reaches the earth comes from the sun. The absorption and loss of radiant energy by the earth and the atmosphere are almost totally responsible for the earth's weather, both on a global and local scale. The average temperature on the earth remains fairly constant, indicating that the earth and the atmosphere on the whole lose as much energy by reradiation back into space as is received by radiation from the sun. The accounting for the incoming and outgoing radiant energy constitutes the earth's energy balance. The atmosphere, although it may appear to be transparent to radiation, plays a very important role in the energy balance of the earth. In fact, the atmosphere controls the amount of solar radiation that actually reaches the surface of the earth and, at the same time, controls the amount of outgoing terrestrial radiation that escapes into space. In this section we consider in particular the role of the atmosphere in the earth's energy balance.

The intensity of radiation is measured by the amount of energy transferred per unit area per unit time (typical units are cal cm^{-2} min^{-1}). Although knowledge of the exact nature of all forms of radiation is incomplete, it is known that, with respect to many properties, radiant energy acts as if it were transmitted in the form of waves. The wavelength λ and the frequency v are related by $v = c/\lambda$, where c, the speed of light, is 2.998×10^{10} cm/sec. We will normally characterize radiation by its wavelength, measured in micrometers μm (10^{-4} cm) or angstroms Å (10^{-8} cm).

Radiant energy, arranged in order of its wavelengths, is called the *spectrum* of radiation. The so-called electromagnetic spectrum is shown in Fig. 3.1. The sun radiates over the entire electromagnetic spectrum, although, as we will see, most of the energy is concentrated near the visible portion of the spectrum, the narrow band of wavelengths from 4000 to 7000 Å.

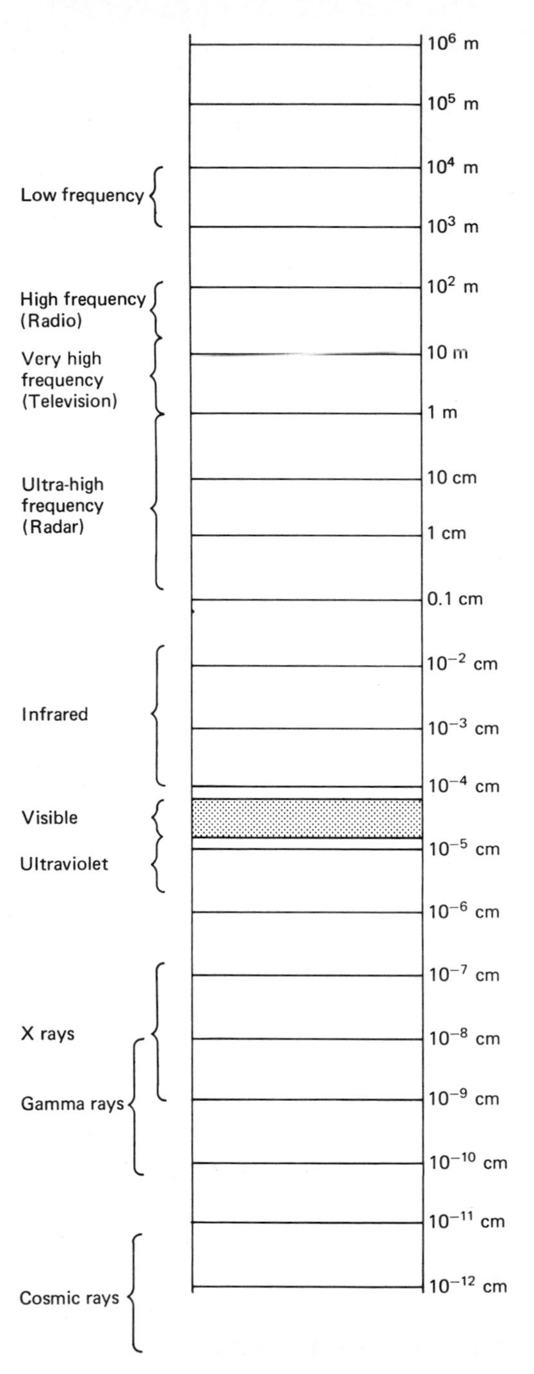

FIGURE 3.1
Electromagnetic spectrum.

Radiation is emitted when an electron drops to a lower level of energy. The difference in energy between the initial and final level, $\Delta\varepsilon$, is related to the frequency of the emitted radiation by Planck's law,

$$\Delta\varepsilon = h\nu = \frac{hc}{\lambda}$$

where $h = 6.63 \times 10^{-34}$ joule-sec. The electromagnetic wave emitted when an electron makes a transition between two energy levels is called a *photon*. When the energy difference $\Delta\varepsilon$ is large, the frequency of the emitted photon is high (very small wavelength) and the radiation is in the x-ray or gamma-ray region. The Planck condition also applies to the absorption of a photon of energy by a molecule. Thus, a molecule can absorb radiant energy only if the wavelength of the radiation corresponds to the difference between two of its energy levels. Since the spacing between energy levels is, in general, different for molecules of different composition and shape, the absorption of radiant energy by molecules of differing structure occurs in different regions of the electromagnetic spectrum.

The amount of energy radiated from a body depends largely on the temperature of the body. It has been demonstrated experimentally that at a given temperature there is a maximum amount of radiant energy that can be emitted per unit time per unit area of a body. This maximum amount of radiation for a certain temperature is called the *blackbody radiation*. A body that radiates for every wavelength the maximum possible intensity of radiation at a certain temperature is called a *blackbody*. This maximum is identical for every blackbody regardless of its constituency. Thus, the intensity of radiation emitted by a blackbody is a function only of the wavelength, absolute temperature, and surface area. The term "blackbody" has no reference to the color of the body. A blackbody can also be characterized by the property that all radiant energy reaching its surface is absorbed.

The energy spectrum of the sun resembles that of a blackbody at 6000°K. (See Fig. 3.2.) The so-called photosphere (the outer layer of the sun) is 400 km thick and varies from 8000°K at the base to about 4000°K at the surface. Thus, it is not in thermodynamic equilibrium as a true blackbody must be, and the sun's spectrum is not exactly a blackbody spectrum. The maximum intensity of incident radiation occurs in the visible spectrum at about 5000 Å. In contrast, Fig. 3.3 shows the emission of radiant energy from a blackbody at 300°K, approximating the earth. The peak in radiation intensity occurs at 100,000 Å in the invisible infrared.

Briefly, we can summarize some of the quantities of importance in the study of radiation:

E = total emissive power, the energy radiated from a surface per unit area per unit time in all directions over all wavelengths

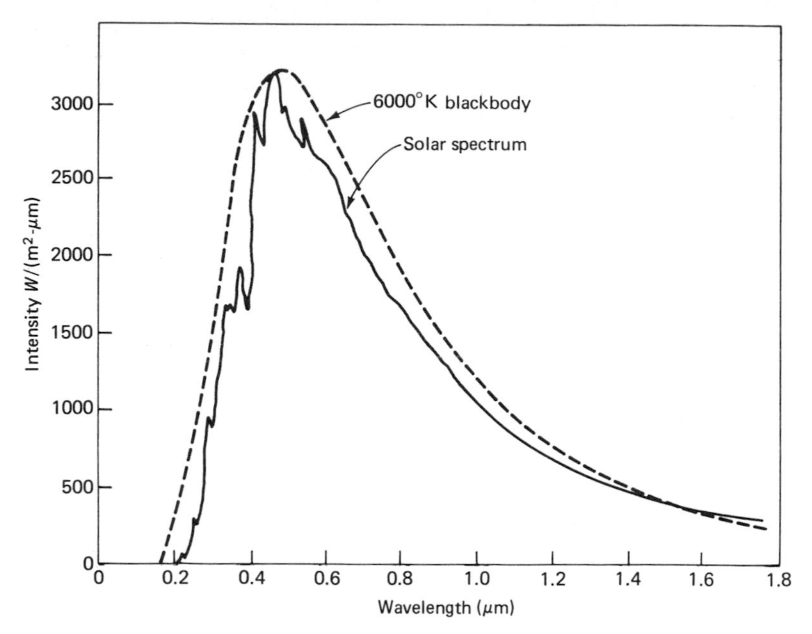

FIGURE 3.2
Solar spectrum and the monochromatic emissivity of a blackbody at 6000°K.

E_λ = monochromatic emissive power, the energy radiated from a surface per unit area per unit time in all directions in the wavelength range λ to $\lambda + d\lambda$

ε = emissivity, ratio of the total emissive power E to that of a blackbody at the same temperature E_B

ε_λ = monochromatic emissivity, ratio of the monochromatic emissive power E_λ to that of a blackbody at the same temperature

ρ = reflectivity, the fraction of the incident radiation that is reflected by a surface (can also define a monochromatic reflectivity ρ_λ)

τ = transmissivity, the fraction of the incident radiation transmitted through a medium per unit thickness along the path of the mean (can also define a monochromatic transmissivity τ_λ).

α = absorptivity, the fraction of the incident radiation that is absorbed by the surface (can also define a monochromatic absorptivity α_λ). Note that $\rho + \tau + \alpha = 1$.

It is important to point out that both absorptivity and reflectivity are properties of the atmosphere. Gases can absorb radiant energy, and particles scatter oncoming radiation back in the direction of is origin. We alluded to these effects in Chap. 1.

In order to understand the spectra of Figs. 3.2 and 3.3, we must consider certain of the laws of radiation. *Kirchhoff's law* relates the emission of radiation of a given

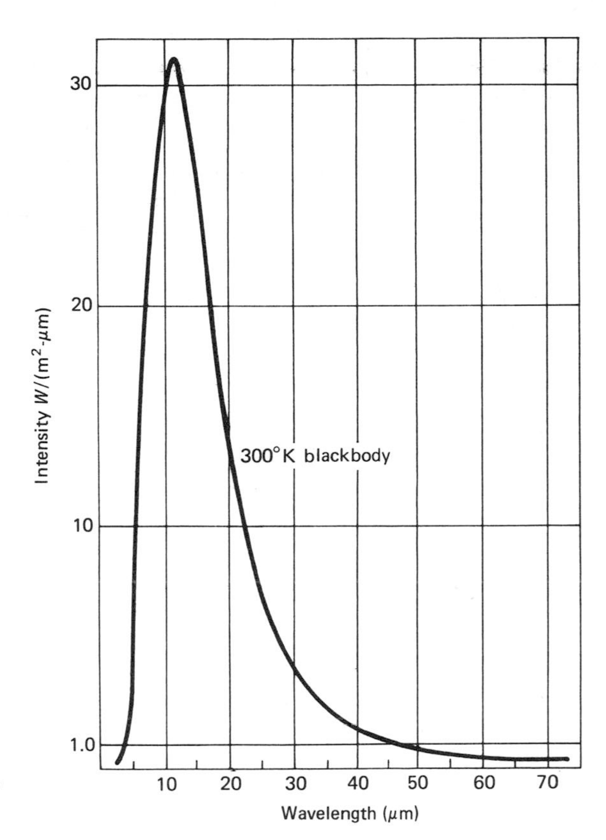

FIGURE 3.3
Monochromatic emissivity of a blackbody at 300°K.

wavelength at a given temperature to the absorption of radiation of the same wavelength. The law states that every body absorbs radiation of exactly those wavelengths which it is capable of emitting at the same temperature, in other words that the absorptivity by a body of radiant energy of a given wavelength at a given temperature is equal to its emissivity in that wavelength at the same temperature:

$$\varepsilon_\lambda = \alpha_\lambda$$

Planck's law relates the monochromatic emissive power of a blackbody with temperature and wavelength:

$$E_{\lambda B} = \frac{2\pi c^2 h \lambda^{-5}}{e^{ch/k\lambda T} - 1}$$

where k is Boltzmann's constant (1.4×10^{-23} joule/°K). The significance of Planck's law can be seen by examining Figs. 3.2 and 3.3. Figure 3.2 shows the emissive power of a blackbody at 6000°K, the approximate average temperature of the outer layer of the sun. Figure 3.3, on the other hand, shows the emissive power of a blackbody at 300°K, close to the temperature of the earth. Both of these curves obey Planck's law. As can be seen from the curves, the higher the temperature, the greater is the emissive power (at all wavelengths). We also see that, as temperature increases, the maximum value of $E_{\lambda B}$ moves to shorter wavelengths. The wavelength at which the maximum amount of radiation is emitted by a blackbody is found by differentiating Planck's formula with respect to λ, setting the result equal to zero, and solving for λ. The result in terms of angstroms is

$$\lambda_{\max} = \frac{2.897 \times 10^7}{T}$$

where T is in degrees Kelvin. Thus, hot bodies not only radiate more energy than cold ones, they do it at shorter wavelengths. We see that the wavelengths for the maxima of solar and terrestrial radiation are 0.48 and about 10 μm, respectively. The sun, with an effective surface temperature of about 6000°K, radiates about 2×10^5 more energy per square meter than the earth at 300°K.[1]

If Planck's law is integrated over all wavelengths, the total emissive power of a blackbody is found to be

$$E_B = \int_0^\infty E_{\lambda B}\, d\lambda = \frac{2\pi^5 k^4 T^4}{15c^2h^3} = \sigma T^4$$

where $\sigma = 5.673 \times 10^{-8}$ watt m^{-2} °K^{-4}, called the Stefan-Boltzmann constant.

The absorption of radiation by gases is one of the most important aspects of both global meteorology and atmospheric chemistry. Molecules can absorb electromagnetic energy by converting it into vibrational, rotational, or electronic energy. A high-energy photon may eject an electron, leaving the molecule ionized with a positive charge. A photon of even higher energy leads to complete dissociation. For example, N_2, O_2, and O_3, under irradiation, dissociate at wavelengths in the ranges

$$N_2 + h\nu \rightarrow N + N \qquad \lambda < 1200\ \text{Å}$$
$$O_2 + h\nu \rightarrow O + O \qquad \lambda < 2400\ \text{Å}$$
$$O_3 + h\nu \rightarrow O + O_2 \qquad 2200\ \text{Å} < \lambda < 2900\ \text{Å}$$

Rotational energy changes require only about 6000 Å, energies too low to produce excitation of electrons. Even lower energies are needed for vibrational changes.

The solar spectrum is radically altered by absorption as the radiation traverses the atmosphere. The most significant absorbing species in the atmosphere are O_2, O_3, water vapor, CO_2, and dust. Figure 3.4 shows the absorption spectrum for the atmosphere as well as the absorption spectra for O_2, O_3, and H_2O. The absorption spectra are quite complex, but they do indicate that only for certain wavelengths may radiation be transmitted through the atmosphere without appreciable loss of energy through absorption. Absorption is so strong in some spectral regions that no solar

[1] Incoming solar intensity is 1.92 cal cm^{-2} min^{-1}, the so-called solar constant.

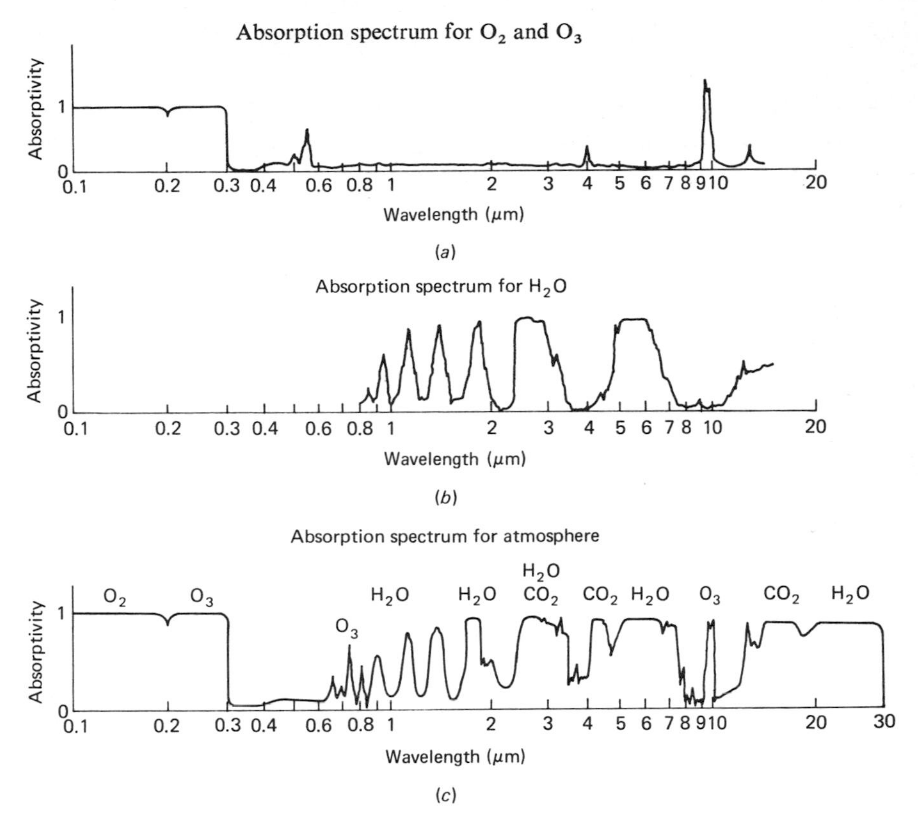

FIGURE 3.4
Absorption spectra for (*a*) molecular oxygen and ozone, (*b*) water, and (*c*) the atmosphere. (*Source: Miller*, 1966.)

energy in those regions reaches the surface of the earth. For example, absorption by O_2 and O_3, followed by dissociation as shown above, is responsible for absorption of practically all the incident radiation with wavelengths shorter than 2900 Å. However, atmospheric absorption is not strong from 3000 to about 8000 Å, forming a "window" in the spectrum. About 40 percent of the solar energy is concentrated in the region of 4000 to 7000 Å. We see that H_2O absorbs in a complicated way, and mostly in the region where the sun's and earth's radiation overlap. Thus, from 3000 to 8000 Å the atmosphere is essentially transparent. From 8000 to 20,000 Å, terrestrial long-wave radiation is moderately absorbed by water vapor in the atmosphere.

Why the molecules in Fig. 3.4 absorb at the particular regions of the spectrum shown can be determined only through elaborate calculations involving the permitted energy levels of the molecules. In general, however, the geometry of the molecule explains why H_2O, CO_2, and O_3 interact strongly with radiation above 4000 Å but

N_2 and O_2 do not. In H_2O, for example, the center of the negative charge is shifted toward the oxygen nucleus and the center of positive charge toward the hydrogen nuclei, leading to a separation between the centers of positive and negative charge, a so-called electric dipole moment. Molecules with dipole moments interact strongly with electromagnetic radiation because the electric field of the wave causes oppositely directed forces and therefore accelerations on electrons and nuclei at one end of the molecule as compared with the other. Similar arguments hold for ozone; however, nitrogen and oxygen are symmetric and thus are not strongly affected by radiation above 4000 Å. The CO_2 molecule is linear but can be easily bent, leading to an induced dipole moment. A transverse vibrational mode exists for CO_2 at 15 μm, just where the earth emits most of its infrared radiation.

It is important to note that the molecules which are responsible for the most pronounced absorption of both solar and terrestrial radiation are the minor constituents of the atmosphere, not nitrogen and oxygen. Thus, ozone in the upper atmosphere effectively absorbs all solar radiation below 2900 Å, whereas water vapor and carbon dioxide absorb much of the long-wave terrestrial radiation. The case of ozone is a particularly interesting one. The background concentration of ozone in the troposphere (the lowest 10 to 15 km of the atmosphere) is about 0.03 ppm. In photochemical smog, such as in Los Angeles, ozone concentrations can reach levels as high as 0.5 ppm. Of course, under these circumstances, ozone is considered a pollutant. In the upper atmosphere, however, ozone concentrations reach 0.2 ppm, solely as a result of chemical processes taking place there. Clearly, ozone in the upper atmosphere is not considered a pollutant; quite the contrary, it is an essential ingredient because of its screening of the short-wavelength ultraviolet radiation.

The presence of ozone concentrations as high as 0.2 ppm in the upper atmosphere can be explained on the basis of both photochemical and thermal reactions involving oxygen and ozone. Ozone is formed in the upper atmosphere as the result of the dissociation of oxygen upon absorption of radiation of wavelengths shorter than 2400 Å, followed by subsequent reaction of atomic and molecular oxygen:

$$O_2 + h\nu \xrightarrow{1} O + O \qquad \lambda < 2400 \text{ Å}$$
$$O + O_2 + M \xrightarrow{2} O_3 + M$$

where M represents a third body (another O_2 perhaps) which absorbs the excess energy of the initial product and allows its stabilization. Thus, reaction 2 should really be written in two steps:

$$O + O_2 \rightleftharpoons O_3^*$$
$$O_3^* + M \rightleftharpoons O_3 + M$$

where O_3^* denotes an excited ozone molecule, which if not stabilized by collision with M will decay to O and O_2.

High-altitude measurements reveal that the atmospheric ozone profile goes through a maximum at a height of about 20 km. At very high altitudes (in the region of 100 km) the formation of ozone by reaction 2 must compete with the recombination of oxygen atoms,

$$O + O + M \xrightarrow{3} O_2 + M$$

because of the relatively high concentration of oxygen atoms at this altitude as a result of reaction 1. Because of this competition, ozone concentrations are small at very high altitudes. As the radiation traverses the upper atmosphere, the short-wavelength components are gradually depleted. Close to the earth, the importance of reaction 2 diminishes because of the elimination of almost all radiation of wavelength shorter than 2400 Å and thus the lack of a source of oxygen atoms by reaction 1.

3.1.2 Energy Balance for Earth and Atmosphere

The processes which determine the temperature of the earth and the atmosphere at any location are quite complicated. These processes include absorption and reflection of radiation by gas molecules and particles in the air as well as absorption, reflection, and emission of radiation by the earth's surface. Nevertheless, we can construct a qualitative overall energy balance for the earth and atmosphere.

As the sun's radiation passes through the upper atmosphere, a small amount (about 3 percent) is absorbed by stratospheric ozone. As the radiation traverses the lower atmosphere, it is further reduced by absorption (primarily by water vapor) and by reflection back to space (mainly by clouds but also by dust particles and air molecules). In total, the average absorption by gases, particles, and clouds amounts to about 20 percent of the incoming solar beam. The average reflectivity (called the *albedo*) of the earth and atmosphere together varies from 30 to 50 percent of the incoming radiation, with clouds being responsible for most of this amount. The average value of the albedo, the incoming radiation that is reflected or scattered back to space without absorption, is usually taken to be about 34 percent. It is important to note that the albedo varies considerably, depending on the surface of the earth. For example, in the polar regions, which are covered by ice and snow, the reflectivity of the surface is very high. On the other hand, in the equatorial regions, which are largely covered with oceans, the reflectivity is low, and most of the energy received is absorbed.

In summary, a qualitative balance on incoming radiation yields:

50%	Intercepted by clouds (25% back to space, 23% to earth, 2% absorbed by clouds)
17%	Absorbed by gases and dust in the atmosphere
12%	Scattered by the air (7% back to space, 5% to earth)
19%	Absorbed by the earth
2%	Reflected by the earth back to space
100%	

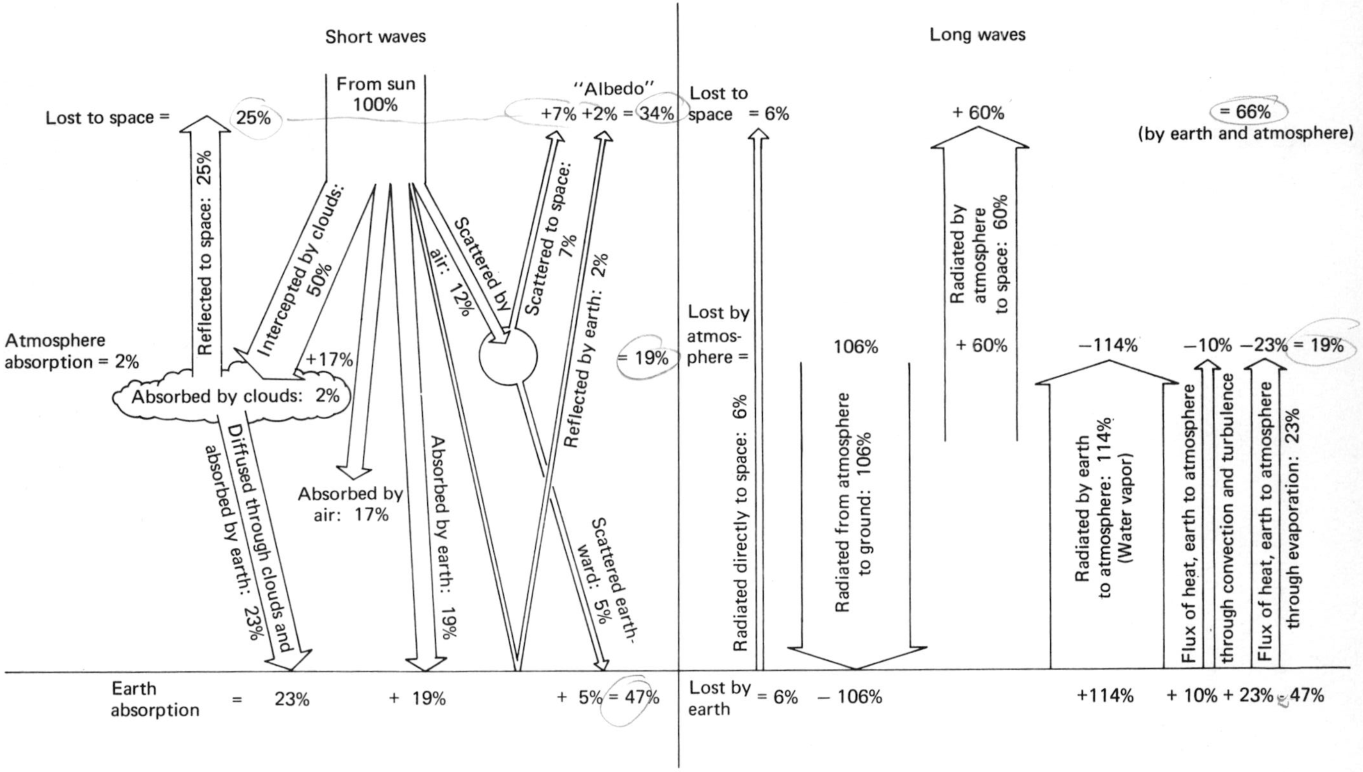

FIGURE 3.5
Earth's energy balance. (*Source: Miller*, 1966.)

Thus, on a basis of 100 units of incoming solar energy, 47 units is absorbed by the earth, 34 units is radiated back to space (25 units of reflection by clouds, 7 units of scattering by the atmosphere, and 2 units of reflection by the earth), and 19 units is absorbed by the atmosphere. This is illustrated on the left-hand side of Fig. 3.5.

The surface of the earth also radiates, but since its temperature is only about 285 to 300°K, the radiation is, as we have seen, in the infrared portion of the spectrum with a maximum intensity at 10^5 Å. The atmosphere almost completely absorbs this long-wave radiation from the earth. As is evident from Fig. 3.4, this radiation is effectively absorbed by CO_2 and H_2O. Some of this radiation is reradiated back to earth and some to space. Thus, whereas the short-wave solar radiation penetrates the atmosphere fairly effectively, the reradiated long-wave energy is kept, by and large, in the atmosphere. The long-wave processes are summarized on the right-hand side of Fig. 3.5.

As a result of both the short- and long-wave radiation processes, the atmosphere continually loses energy. This lost energy is replenished in two ways: (1) conduction of heat from the surface to the atmosphere and (2) evaporation of water from the surface followed by condensation in the atmosphere with its release of latent heat. About two-thirds of the transfer of energy from the earth's surface to the atmosphere occurs by the latter mechanism.

Since the average temperature of the earth remains essentially constant over a long period of time, there is a balance between the incoming solar energy and the outgoing long-wave radiation from the earth. The earth gains and loses energy as follows:

+47 units	Absorbed from sun (short-wave)
−33 units	Lost by evaporation and convection to the atmosphere
−119 units	Long-wave blackbody radiation (113 absorbed by atmosphere, 6 to space)
+105 units	Reradiation back to earth from atmosphere (long-wave)
0 units	Net input of energy

Finally, we can complete the cycle by considering the atmosphere:

+19 units	Short-wave radiation absorbed from sun
+113 units	Long-wave radiation absorbed from earth
−105 units	Long-wave radiation reradiated back to earth
+33 units	Gained by evaporation and convection from earth
−60 units	Long-wave radiation back to space
0 units	Net input of energy

As we have seen, the atmosphere, by absorbing such a high fraction of the long-wave radiation of the earth, acts as an insulation to keep heat near the surface of the earth. This is called the *greenhouse effect.* Water in both vapor and droplet form is the principal agent for this effect. From Fig. 3.4 we see that H_2O is strongly absorbing in both the high- and low-energy portion of the infrared spectrum. Carbon dioxide is next in importance in the greenhouse effect. It is not as important as H_2O because its concentration is lower and because its main infrared absorption is localized in a narrow band near 1.5×10^5 Å.

The fact that the minor constituents of the atmosphere, ozone, carbon dioxide, and water, play such an important role in the energy balance of the earth has led to concern over what the effect of perturbations in the levels of the constituents would be on the radiation balance of the earth. For example, the global background CO_2 concentration has increased from 300 ppm in 1880 to about 330 ppm in 1970. On the basis of the greenhouse effect, we would expect an increase in background CO_2 concentration to lead to an increase in the equilibrium temperature of the earth. We examine this question of climatic changes induced by air pollutants in Sec. 3.4.

3.2 TEMPERATURE IN THE LOWER ATMOSPHERE

The layers of the atmosphere can be classified in a number of ways, such as by temperature, density, and chemical composition. From the standpoint of the dispersion of air pollutants, the most important classification is on the basis of temperature, on which the following layers can be identified (see Fig. 3.6):

1 *Troposphere*: The layer closest to the ground extending to an altitude of 15 km over the equator and 10 km over the poles. Temperature decreases with height at a rate of about 6.5°C/km. Vertical convection keeps the air relatively well mixed.

2 *Stratosphere*: Extends from the tropopause to about 50 km in altitude. Temperature is constant in the lower stratosphere and then increases with altitude owing to the absorption of short-wave radiation by ozone. At the stratopause (the top of the stratosphere) the temperature reaches 270°K. There is little vertical mixing in the stratosphere.

3 *Mesophere*: Extends from 50 to 85 km, over which temperature decreases with altitude until it reaches 175°K, the coldest point in the atmosphere.

4 *Thermosphere*: The uppermost layer. Molecular densities are of the order of 10^{13} molecules/cm^3, as compared with 5×10^{19} at sea level. Intense ultraviolet radiation dissociates N_2 and O_2. Temperatures exceed 1000°K.

With respect to air pollution, our primary interest is in the troposphere to which we shall confine all our attention.

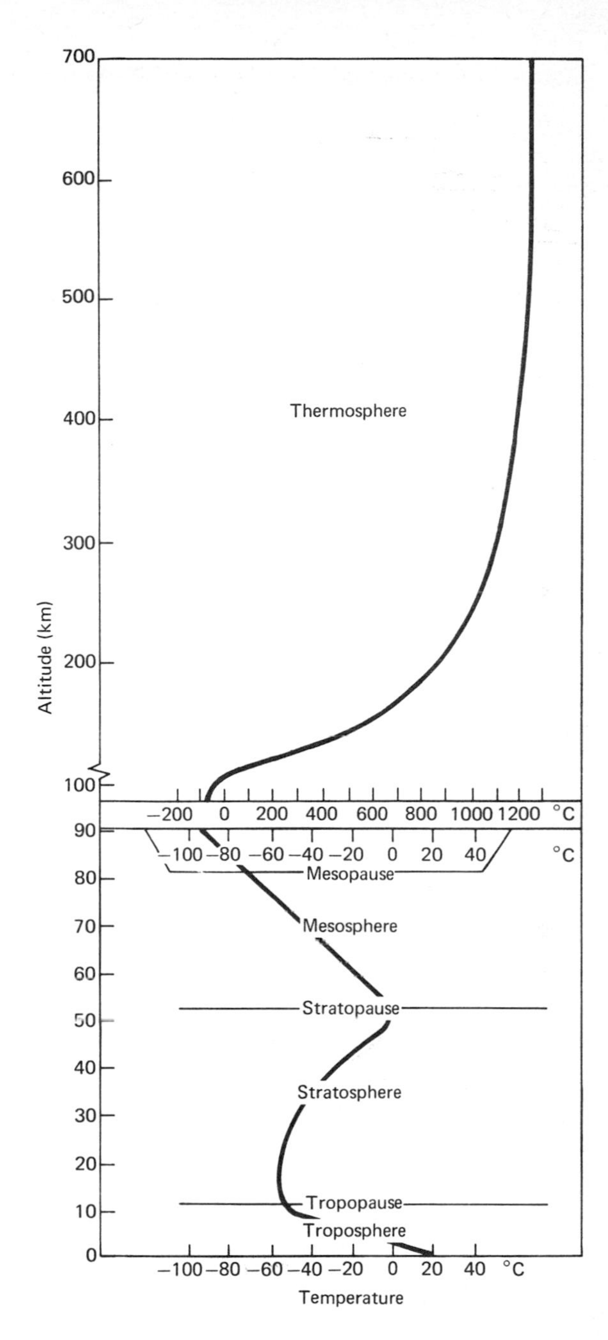

FIGURE 3.6
Atmospheric temperature profile.
(*Source: Miller*, 1966.)

3.2.1 Pressure and Temperature Relationships in the Lower Atmosphere

Air can be considered an ideal gas, so that at any point in the atmosphere

$$p = \frac{\rho RT}{M_a} \tag{3.1}$$

where ρ is the mass density of air (kg/m³), R is the universal gas constant (8.134 joules/°K-mole), and M_a is the molecular weight of air (28.97).

The pressure at any height z is due to the weight of air above. Thus, the change of pressure in the vertical direction obeys the relation

$$\frac{dp(z)}{dz} = -\rho g \tag{3.2}$$

Substituting (3.1) into (3.2), we obtain the general relation between pressure and temperature at any height z:

$$\frac{dp(z)}{dz} = -\frac{gM_a p}{RT} \tag{3.3}$$

If T were constant with height, (3.3) could be integrated directly to yield

$$p(z) = p_0 e^{-gM_a z/RT} \tag{3.4}$$

where p_0 is the pressure at ground level.[1]

As seen in Fig. 3.6, the temperature is not constant in the troposphere but rather decreases with height. Thus, (3.4) is at best a qualitative indication of the general behavior of $p(z)$. Let us see if we can predict the actual temperature profile in the troposphere.

We shall utilize the concept of an *air parcel*, a hypothetical mass of air which may deform as it moves vertically in the atmosphere. The concept of an air parcel is a tenable one as long as the parcel is of such a size that the exchange of air molecules across its boundary is small when compared with the total number of air molecules in the parcel. As such a parcel rises in the atmosphere, it expands to accommodate the lowering pressure; however, it does so in such a way that exchange of heat between the parcel and the surrounding air is negligible. As the parcel expands upon rising, its temperature decreases. The process of vertical mixing in the atmosphere can, for simplicity, be envisioned as one involving a large number of parcels rising and falling. If there is no heat exchange between the parcel and the surrounding air, the parcel and the surrounding air may be at different temperatures (but not different pressures). The relation of the parcel's temperature to that of the air determines whether the parcel will continue rising or falling or whether it will reach a point of equilibrium. Therefore,

[1] The average sea-level pressure is 1.013×10^5 newtons/m² $= 1.013 \times 10^6$ dynes/cm². It is customary to use a pressure unit called the *bar* (10^5 newtons/m²) when dealing with atmospheric pressure. Thus, $p_0 = 1.013$ bars $= 1013$ millibars (mb).

the variation of temperature with altitude in the atmosphere is a key variable in determining the degree to which pollutant-bearing air parcels will mix vertically.

The variation of temperature with height for a rising parcel of dry air which cools adiabatically, that is, with no exchange of heat with its surroundings, is a basic property of the atmosphere. We now will derive the relation for this temperature change, as it will serve as a reference temperature profile against which to compare all actual profiles. To obtain the desired relation we need only the ideal-gas law and the first law of thermodynamics.

The first law of thermodynamics is expressed as

$$dU = dQ - dW$$

where dU is the increase of internal energy of the system, dQ is the heat input to the system across its boundaries, and dW is the energy lost by the system to the surroundings as a result of work done to alter the volume of the system, namely $p\,dV$. The change in internal energy dU is equal to $C_V\,dT$, where C_V is the heat capacity of the system at constant volume.

Our intent is to apply the first law of thermodynamics to an air parcel whose volume is changing as it either ascends or descends in the atmosphere. Ultimately we will combine our result with (3.3), and so it is more convenient to work with pressure and temperature as the variables rather than with pressure and volume. Thus, we convert $p\,dV$ to a form involving p and T. To do this, we express the ideal-gas law as $pV = mRT/M_a$ for a mass m of air. Then,

$$\begin{aligned} d(pV) &= \frac{mR\,dT}{M_a} \\ &= p\,dV + V\,dp \end{aligned}$$

Using this result, together with the adiabatic condition of $dQ = 0$, the first law of thermodynamics reduces to

$$\begin{aligned} C_V\,dT - V\,dp - \frac{mR\,dT}{M_a} \\ = \frac{mRT}{M_a}\frac{dp}{p} - \frac{mR\,dT}{M_a} \end{aligned} \tag{3.5}$$

Rearranging, we obtain

$$\frac{dT}{dp} = \frac{mRT/M_a p}{C_V + mR/M_a} \tag{3.6}$$

Now we have two equations, (3.3) and (3.6), for the relation of T and p with z. Combining these we get

$$\begin{aligned} \frac{dT}{dz} &= -\frac{mg}{C_V + mR/M_a} \\ &= -\frac{g}{\hat{C}_V + R/M_a} \end{aligned}$$

where $\hat{C}_V$ is the heat capacity at constant volume per unit mass of air. We note that $\hat{C}_V + R/M_a = \hat{C}_p$, the heat capacity at constant pressure per unit mass of air. Thus,

$$\frac{dT}{dz} = -\frac{g}{\hat{C}_p} \tag{3.7}$$

which is the rate of temperature change with height for a parcel of dry air rising adiabatically. The right-hand side is a constant for dry air, equal to 1°C/102.39 m or 0.976°C/100 m. This constant is called the *dry adiabatic lapse rate* and is denoted by Γ. Even though we considered only one parcel, (3.7) has general applicability to an atmosphere in which a large number of parcels are rising and falling adiabatically, a point which will be demonstrated in Chap. 5.

If the air contains water vapor, the heat capacity $\hat{C}_p$ must be corrected. If ω is the ratio of the mass of water vapor to the mass of dry air in a given volume of air, the corrected $\hat{C}_p$ is given by

$$\hat{C}'_p = (1 - \omega)\hat{C}_{p_{\text{air}}} + \omega\hat{C}_{p_{\text{water vapor}}} \tag{3.8}$$

Since $\hat{C}_{p_w} > \hat{C}_{p_a}$, $\hat{C}'_p > \hat{C}_{p_a}$. Thus, the rate of decrease of T with z is smaller for a water-bearing atmosphere than for dry air. For example, if 3 percent of the atmospheric pressure is due to water vapor, $-dT/dz = 1°\text{C}/103$ m.

If the parcel contains water vapor, it may cool upon rising until the partial pressure of the water vapor equals the saturation vapor pressure of water. With sufficient nuclei present, condensation may ensue. Then the process is no longer adiabatic. If ΔH is the latent heat of vaporization per gram of water, the release of this heat upon condensation is accounted for in (3.5) by

$$dQ = -\Delta H\, d\omega \tag{3.9}$$

Thus,

$$-\Delta H \frac{d\omega}{dz} = \hat{C}_p \frac{dT}{dz} - V\frac{dp}{dz} \tag{3.10}$$

Using (3.3), we find the lapse rate in a saturated condition to be

$$-\frac{dT}{dz} = +\frac{g}{\hat{C}_p} + \frac{\Delta H}{\hat{C}_p}\frac{d\omega}{dz} \tag{3.11}$$

Since $d\omega/dz$, the rate of change of the ratio of the mass of water vapor to the mass of air, is negative for a rising parcel in which water vapor is condensing, the last term in (3.11) is positive. Thus, the rate of cooling of a rising parcel of moist air is *less* than that for dry air. Since the saturation vapor pressure of water increases very markedly with temperature, the quantity $d\omega/dz$ depends strongly on the temperature. Thus, the wet adiabatic lapse rate is not a constant independent of z. In warm tropical air the wet adiabatic lapse rate is roughly one-third of the dry adiabatic lapse rate, whereas in cold polar regions there is little difference between the two.

3.2.2 Temperature Changes of a Rising (or Falling) Parcel of Air

The relationship between the temperatures and pressures at two heights in an atmosphere with an adiabatic profile is found by integrating (3.6) between any two points. Employing the ideal-gas relation $\hat{C}_p = \hat{C}_V + R/M_a$, and the definition $\gamma = \hat{C}_p/\hat{C}_V$, the result of this integration is

$$\frac{T(z_2)}{T(z_1)} = \left[\frac{p(z_2)}{p(z_1)}\right]^{(\gamma-1)/\gamma} \tag{3.12}$$

For example, if z_1 is taken to be ground level, the temperature θ to which dry air originally in the state T,p would come if brought adiabatically to p_0 is given by

$$\theta = T\left(\frac{p}{p_0}\right)^{-(\gamma-1)/\gamma} \tag{3.13}$$

The temperature θ defined by (3.13) is called the *potential temperature*. We introduce the potential temperature because an actual atmosphere is seldom adiabatic and we want to relate the actual temperature profile to the adiabatic lapse rate. Adiabatic temperature profiles based on potential temperature are vertical on a plot of z versus θ, thereby facilitating such comparisons.

We can further interpret the potential temperature θ as follows: The gradient of θ with z may be expressed in terms of the gradient of absolute temperature T and the adiabatic lapse rate Γ. From (3.13) it is easy to see that

$$\frac{1}{\theta}\frac{d\theta}{dz} = \frac{1}{T}\frac{dT}{dz} - \frac{\gamma-1}{\gamma}\frac{1}{p}\frac{dp}{dz} = \frac{1}{T}\left(\frac{dT}{dz} + \Gamma\right) \tag{3.14}$$

At $z = 0$, $\theta = T$ if p_0 is taken as the surface pressure. Since, in magnitude, θ is quite close to T, (3.14) is often approximated by

$$\frac{d\theta}{dz} \cong \frac{dT}{dz} + \Gamma \tag{3.15}$$

Thus, $d\theta/dz$ is a measure of the departure of the actual temperature profile from adiabatic conditions. Integrating (3.15) with respect to z gives

$$\theta \cong T + \Gamma z \tag{3.16}$$

One might ask: Why does not the atmosphere always have an adiabatic lapse rate as its actual profile? The reason it does not is that other processes such as winds and solar heating of the earth's surface lead to dynamic temperature behavior in the lowest layers of the atmosphere that is seldom adiabatic. These other processes exert a much stronger influence on the prevailing temperature profile than does the adiabatic rising and falling of air parcels.

Let us compute the temperature change with z of an isolated parcel of air (or possibly other gas) as it rises or falls adiabatically through an atmosphere that is not adiabatic. We assume that conduction or convection of heat across the boundary of the parcel will be slow compared with the rate of vertical motion. Thus, an individual parcel is assumed to rise or fall adiabatically, even when the surrounding air is non-adiabatic.

Let T denote the temperature of the air parcel and T' the temperature of the surrounding air. At any height z, the pressure is the same in the parcel as in the atmosphere. The rate of change of T with p in the parcel is given by (3.6), and the rate of change of p with z is given by (3.3). Combining these two relations, we find that

$$\frac{dT}{dz} = -\Gamma \frac{T}{T'} \tag{3.17}$$

Therefore, the rising air will cool at a greater or lesser rate than the adiabatic, depending on whether its temperature is higher or lower than that of the adjacent atmosphere.

If Λ is the actual lapse rate in the atmosphere, then at any height z

$$T'(z) = T_0' - \Lambda z \tag{3.18}$$

Then, from (3.17) and (3.18),

$$\frac{dT}{dz} = -\Gamma \frac{T(z)}{T_0' - \Lambda z} \tag{3.19}$$

Integrating (3.19) with $T(0) = T_0$, the surface temperature of the rising parcel,

$$T(z) = T_0 \left(\frac{T_0' - \Lambda z}{T_0'} \right)^{\Gamma/\Lambda} \tag{3.20}$$

so that, in general,

$$\frac{dT}{dz} = -\Gamma \left(\frac{T_0' - \Lambda z}{T_0'} \right)^{(\Gamma - \Lambda)/\Lambda} \frac{T_0}{T_0'} \tag{3.21}$$

Of course, if $\Lambda = \Gamma$, then

$$\frac{dT}{dz} = -\Gamma \frac{T_0}{T_0'} \tag{3.22}$$

Thus, even if the atmosphere has an adiabatic lapse rate, a parcel of air introduced at the ground at a temperature $T_0 \neq T_0'$ will have a different rate of cooling than the adiabatic.

3.2.3 Atmospheric Stability

The lapse rate in the lower portion of the atmosphere has a great influence on the vertical motion of air. If the lapse rate is adiabatic, a parcel of air displaced vertically is always at equilibrium with its surroundings. Such a condition, in which vertical displacements are not affected by buoyancy forces, is called *neutral* stability. However, because of surface heating and local weather influences, the atmosphere seldom has an adiabatic temperature profile. The atmosphere is either:

1 *Unstable*: Buoyancy forces enhance vertical motion.
2 *Stable*: Buoyancy forces oppose vertical motion.

Let us suppose a warm parcel begins to rise in an atmosphere in which temperature decreases more rapidly with z than the adiabatic rate (its lapse rate exceeds the adiabatic lapse rate). The air parcel cools adiabatically, but the temperature difference between the rising parcel and the surroundings increases with z. If the density of the parcel is ρ and that of the air ρ', the acceleration experienced by the parcel is

$$\text{Acceleration} = g\left(\frac{\rho' - \rho}{\rho}\right)$$

$$= g\left(\frac{T - T'}{T'}\right)$$

Thus, the acceleration increases with z and the parcel continues to rise as long as $T > T'$. We can express the acceleration in terms of the two lapse rates Γ and Λ as follows, if $T_0 \cong T_0'$:

$$\text{Acceleration} = \frac{g(dT/dz - dT'/dz)\, dz}{T'}$$

$$= \frac{g(\Lambda - \Gamma)\, dz}{T'}$$

As long as $\Lambda > \Gamma$, the parcel continues to rise. Similarly, a parcel of air cooler than the surrounding air will continue to descend if its rate of adiabatic heating is less than

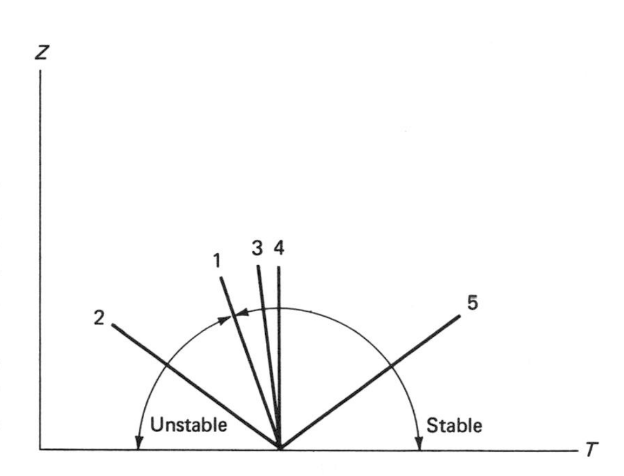

FIGURE 3.7
Temperature profiles in the atmosphere. (1) Adiabatic lapse rate: T decrease with height such that any vertical movement imparted to an air parcel will result in the parcel maintaining the same T or density as the surrounding air. (Neutral stability) 1°C/100 m. (2) Superadiabatic: A rising air parcel will be warmer than its environment so it becomes more buoyant and continues rising. (Unstable.) (3) Subadiabatic: A rising air parcel is cooler than its surroundings so it becomes less buoyant and returns. (Stable.) (4) Isothermal: Temperature constant with height. (Stable.) (5) Inversion: Temperature *increases* with height. (Extremely stable.)

the lapse rate in the atmosphere. Since vertical motion is enhanced by buoyancy, if $\Lambda > \Gamma$ the atmosphere is called *unstable*. Lapse rates Λ for which $\Lambda > \Gamma$ are called *superadiabatic*.

On the other hand, if $\Lambda < \Gamma$, a rising air parcel will cool more rapidly with height than the surroundings and a point will be reached at which the temperature of the parcel equals that of the surroundings. We see that, if $\Lambda < \Gamma$, the acceleration will oppose the motion of a parcel. Thus, any fluctuations in the temperature of an air parcel will cause it to rise or fall, but only for a short distance. When $\Lambda < \Gamma$, the atmosphere is said to be *stable*. Summarizing, the conditions are:

1	$\Lambda = \Gamma$	neutral stability
2	$\Lambda > \Gamma$	unstable (vertical motions enhanced)
3	$\Lambda < \Gamma$	stable (vertical motions suppressed)

These same arguments may be applied to the case of a moist atmosphere. Because of the release of the latent heat of vaporization, a saturated parcel cools on rising at a slower rate than a dry parcel, since

$$\Gamma_{\text{dry}} > \Gamma_{\text{wet}}$$

Thus, a moist atmosphere is inherently *less* stable than a dry atmosphere, and a stable situation with reference to the dry adiabatic lapse rate may actually be unstable for upward displacements of a saturated air parcel.

Figure 3.7 summarizes the types of temperature profiles found in the lower atmosphere, and Fig. 3.8 shows a typical diurnal variation of temperature near the ground. The air mass near the ground is adiabatic only under special circumstances. Adiabatic conditions are reached usually when the sky is heavy with clouds and there is a moderate to high wind. The clouds prevent radiation from reaching the surface and ensure that the temperature of the ground does not differ greatly from the air just above it. The wind serves to mix the air, thereby smoothing out temperature differences. Vertical movement is then a result of mechanical forces, not buoyancy. From an air pollution standpoint, situations in which the temperature increases with height, so-called inversion conditions, are of great importance. Under these conditions the air is very stable, and little mixing of pollutants takes place.

Inversions, as shown in Fig. 3.7, form in one of two ways, through cooling from below or heating from above. Inversions often form, particularly at night, because of radiation cooling at the ground. Horizontal movement of an air mass from above a warm surface (land) to above a cool surface (water) also produces an inversion. (Note that at night the land surface may be cooler than the water.) Such inversions are termed *ground-based* or surface *inversions*. Inversions that are the result of heating from above involve the spreading, sinking, and compression of an air mass as it moves horizontally. As upper layers undergo the greatest elevation change, they experience the greatest degree of compression and thus the greatest increase in temperature. If the temperature increase is sufficient, an inversion will result. The sinking and compression process is termed *subsidence*.

A *frontal inversion* can occur at an interface between two air masses of quite different temperatures, humidity, and pressure. If colder air is advancing, the front is known as a cold front, and vice versa. In each case the warm air overrides the cold air, and in the case of a cold front, the rising warm air leads to condensation and rain following the position of the surface front. An inversion will exist at the interface of both a warm and cold front.

An *advective inversion* is formed when warm air flows over a cold surface or colder air. The inversion can be surface-based, as when warm air flows over cold plains, or elevated, as in the case when a cool sea breeze is overlaid by a warm land breeze.

A *radiational inversion* is illustrated in Fig. 3.8. Such inversions occur frequently when the ground cools at night by radiation. The presence of nocturnal radiational inversions prevents the ventilation of emissions during the night in a city. At night in cities, buildings and streets cool slowly, often resulting in an unstable temperature profile for the first hundred meters or so. But this shallow mixing layer is usually topped by a more stable layer. Figure 3.9 shows a cross section of Cincinnati, Ohio, based on special temperature measurements made near sunrise following a nearly cloudless summer night. The cross section is oriented along the wind direction, roughly from west-southwest on the left to east-northeast on the right. The thin

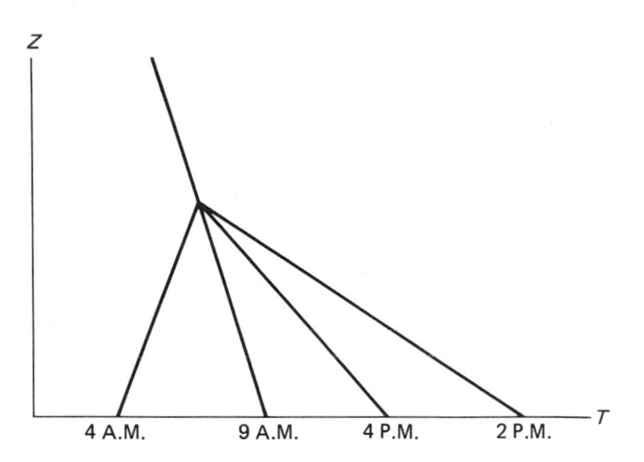

FIGURE 3.8
Typical diurnal variation of temperatures near the ground. 4 A.M.: Radiation from earth to black sky cools ground lower than air producing a ground-based inversion. 9 A.M.: Ground heated rapidly after sunrise. Slightly subadiabatic. 2 P.M.: Continued heating. Superadiabatic. 4 P.M.: Cooling in the afternoon returns the temperature profile to near adiabatic.

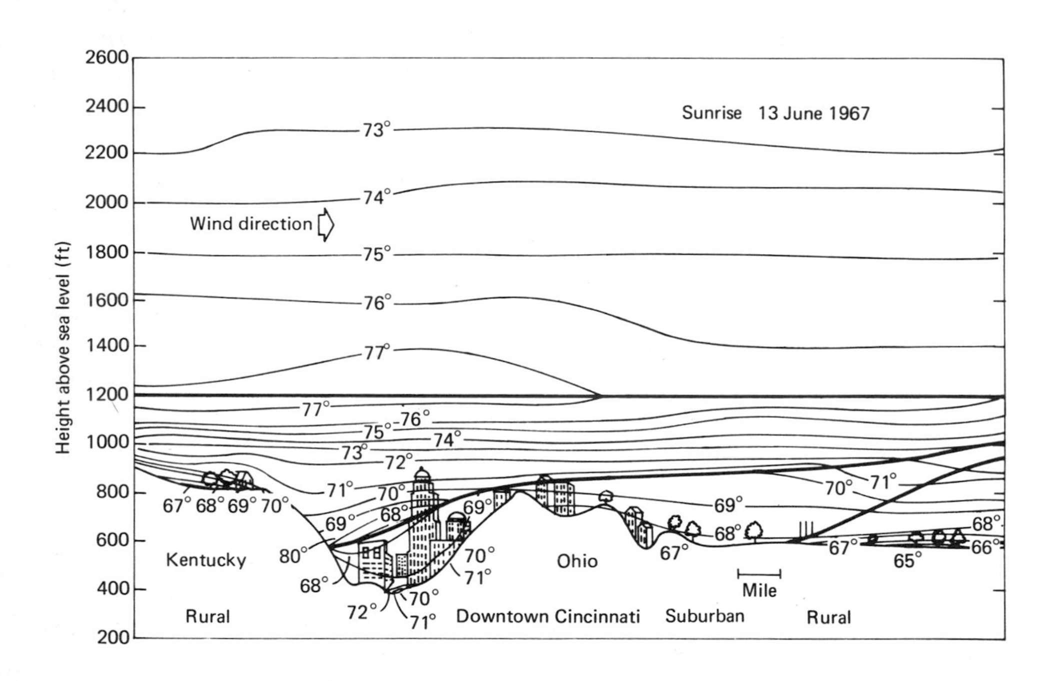

FIGURE 3.9
Cross section of temperature structure through Cincinnati, Ohio, based on measurements near sunrise. Thin lines are isotherms (°F) and heavy lines depict the top or bottom of main temperature inversions. (*Source: Clarke*, 1969.)

lines are isotherms (°F), and the heavy lines denote the base or top of a temperature inversion. On the left side of the figure, in rural Kentucky, the inversion is ground-based, with the temperature increasing from 67°F near the ground to 77°F at the top of the inversion, 400 ft above the surface. Moving down the hill to the urbanized area, the ground-based inversion is replaced by an unstable superadiabatic layer about 200 ft deep. A weak inversion develops again downwind of Cincinnati. We can see what might be called the "urban heat plume." Over downtown Cincinnati, pollutants emitted at ground level would be trapped below the main inversion.

Figure 3.10 shows vertical temperature and SO_2 concentration profiles in New York City at 7:38 A.M. EST on March 9, 1966. We note that a pronounced tempera-

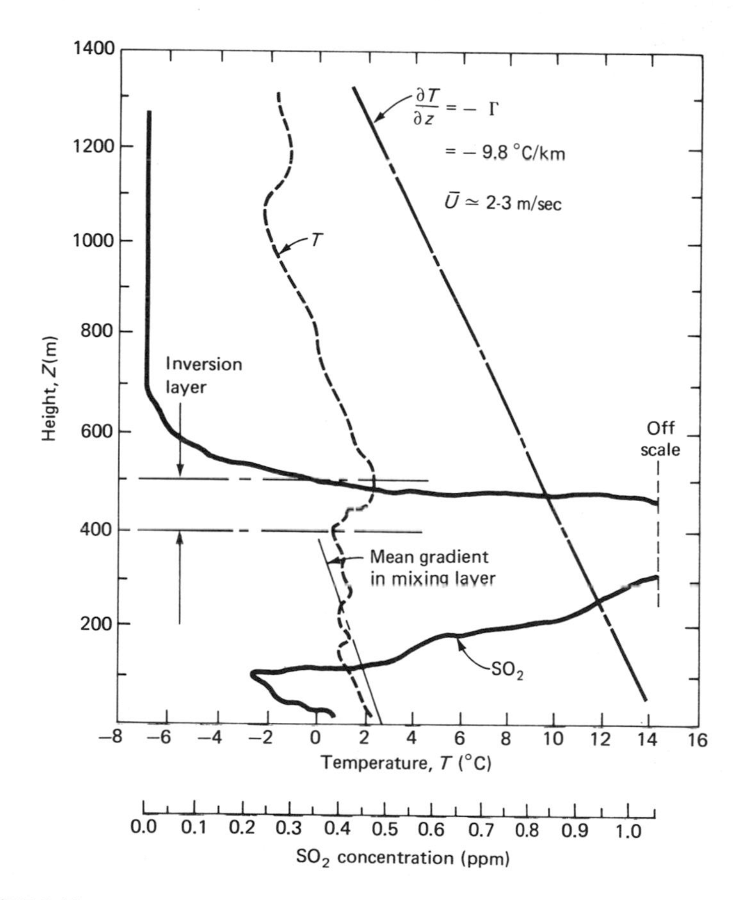

FIGURE 3.10
Vertical temperature and SO_2 concentration profile measurements at Battery Park, New York City at 7:38 A.M. EST on March 9, 1966. (*Source: Davidson,* 1967.)

FIGURE 3.11
View of the Los Angeles City Hall on a clear day. (*Courtesy of Los Angeles County Air Pollution Control District.*)

FIGURE 3.12
View of the Los Angeles City Hall with an inversion layer at 100 m. (*Courtesy of Los Angeles County Air Pollution Control District.*)

FIGURE 3.13
View of the Los Angeles City Hall with an inversion layer at 300 m. (*Courtesy of Los Angeles County Air Pollution Control District.*)

ture inversion beginning at about 400 m inhibits vertical mixing of the SO_2 above the base of the inversion layer. For reference, the dry adiabatic lapse rate is also shown. Figures 3.11 to 3.13 show the Los Angeles City Hall on three different days, corresponding to no smog, a smog layer capped by an inversion of less than 100 m in height, and a smoggy day with an inversion above 300 m.

3.3 WINDS

Even though the total input and output of radiant energy to and from the earth are essentially in balance, they are not in balance at every point on the earth. The amount of energy reaching the earth's surface depends, in part, on the nature of the surface (land versus sea, for example) and the degree of cloudiness, as well as on the latitude of the point. For example, at lower solar angles, in the polar regions the same amount of solar energy as radiated to the tropics must pass through more atmosphere and intercept a larger surface area. The uneven distribution of energy resulting from latitudinal variations in insolation and from differences in absorptivity of the earth's surface leads to the large-scale air motions of the earth. In particular, the tendency

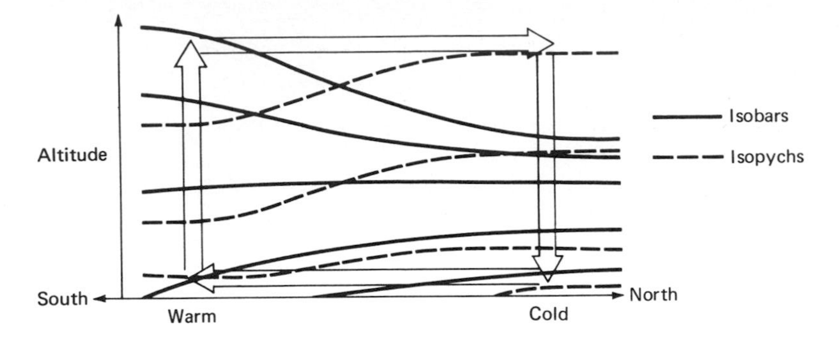

FIGURE 3.14
Thermal circulation in the atmosphere. At the ground the horizontal pressure gradient causes flow from north to south. At the upper level the flow is reversed.

to transport energy from the tropics toward the polar regions, thereby redistributing energy inequalities on the earth, is the overall factor governing the general circulation of the atmosphere.

In order to visualize the nature of the general circulation of the atmosphere, we can think of the atmosphere over either hemisphere as a fluid enclosed within a long, shallow container, heated at one end and cooled at the other. Because the horizontal dimension of the "container" is so much greater than its vertical dimension, the curvature of the earth can be neglected, and the container can be considered to be rectangular. If such a container were constructed in the laboratory and the ends differentially heated as described above, one would observe a circulation of the fluid, consisting of rising motion along the heated wall and descending motion along the cooled wall, flow in the direction of warm to cold at the top of the box, and flow in the direction of cold to warm along the bottom of the box. The situation we have described is a *thermal circulation*, which is illustrated in Fig. 3.14. (Note, that the vertical scale in Fig. 3.14 is greatly exaggerated in comparison with the horizontal scale.) In the atmosphere, then, the tendency is for warm tropical air to rise and cold polar air to sink, with poleward and equatorward flows to complete the circulation.

However, the general circulation of the atmosphere is not as simple as depicted in Fig. 3.14. Another force arises because of the motion of the earth, the Coriolis force. At the earth's surface an object at the equator has a greater tangential velocity than one in the temperate zones. Air moving toward the south, as in Fig. 3.14, cannot acquire an increased eastward (the earth rotates from west to east) tangential velocity as it moves south and, thus, *to an observer on the earth*, appears to acquire a velocity component in the westward direction. Thus, air moving south in the Northern Hemisphere appears to lag behind the earth. To an observer on the earth it appears that the air has been influenced by a force in the westward direction. To an observer

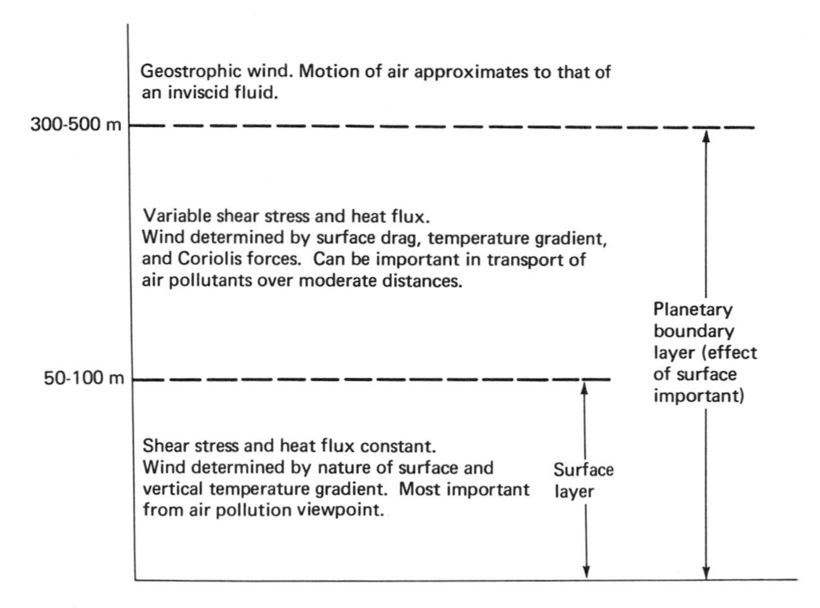

FIGURE 3.15
Regions of the lower atmosphere classified according to airflow.

in space, it would be clear that the air is merely trying to maintain straight-line motion while the earth turns below it. Friction between the wind and the ground diminishes this effect in the lower atmosphere.

From the standpoint of air motion, the atmosphere can be segmented vertically into two layers. Extending from the ground up to about 500 m is the *planetary boundary layer*, the zone in which the effect of the surface is felt and in which the wind speed and direction are governed by horizontal pressure gradients, shear stresses, and Coriolis forces. Above the planetary boundary layer is the *geostrophic layer*, in which only horizontal pressure gradients and Coriolis forces influence the flow. Our main interest in this section is with flow in the geostrophic layer. Chapter 5 is devoted to a detailed treatment of wind and temperature behavior in the planetary boundary layer and, in particular, in the surface layer, the lowest 50 m of the atmosphere. Figure 3.15 shows the regions of the atmosphere, defined on the basis of the type of airflow.

To predict the general pattern of macroscale air circulation on the earth we must consider both the tendency for thermal circulation and the influence of Coriolis forces. Figure 3.16 shows the nature of the general circulation of the atmosphere. At either side of the equator is a thermal circulation, in which warm tropical air rises and cool northern air flows toward the equator. The circulation does not extend all the way

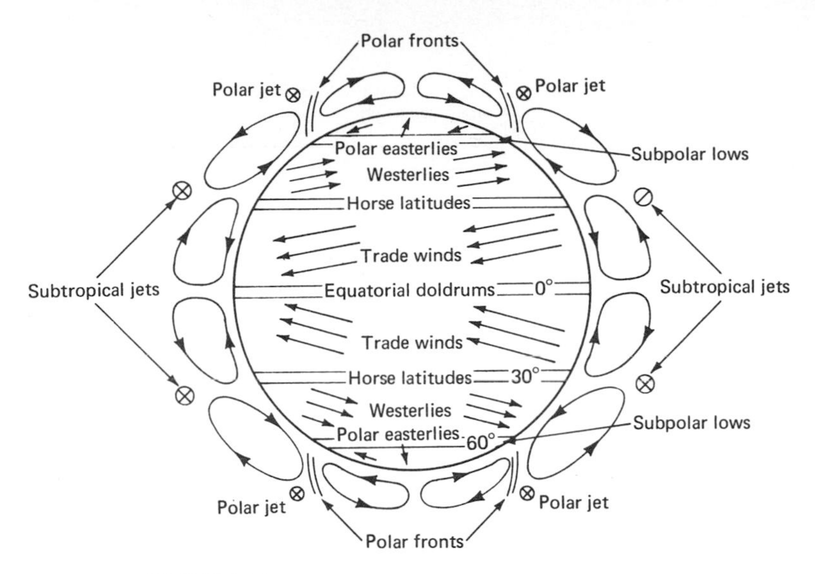

FIGURE 3.16
Schematic representation of the general circulation of the atmosphere.

to the poles because radiative cooling of the upper northward flow causes it to subside (fall) at about 30° N and S latitude. The Coriolis force acting on these cells leads to easterly winds, called the trade winds. The same situation occurs in the polar regions, in which warm air from the temperate zones moves northward in the upper levels, eventually cooling by radiation and subsiding at the poles. The result is the polar easterlies.

In the temperate regions, between 40 and 55° latitude, influences of both tropical and polar regions are felt. The major feature of the temperate regions is large-scale turbulence, which results in the circulation shown in Fig. 3.16. The surface winds in the Northern Hemisphere are westerlies because of the Coriolis force.

At the boundaries between thermal circulation at the equator, 30°, and 55° N and S latitude there are regions of calm. The observed net precipitation near the equator and the polar front is explained by rising moist air which cools. At 30°N and S latitude a strong subsidence of dry air occurs, since the air loses its moisture upon ascension in the equatorial zone. As a result net evaporation of the oceans occurs from 10 to 40° N and S latitude.

Derivation of the Geostrophic Wind Speed

The direction of winds in the geostrophic layer is determined by horizontal pressure gradients and Coriolis forces. As we have discussed, an air parcel moving southward in the Northern Hemisphere as a result of pressure gradients is accelerated toward the west by the Coriolis force. We can actually

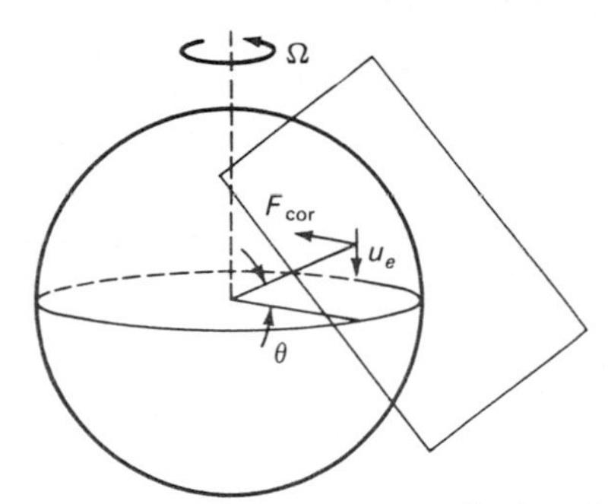

Figure 3.17
Direction of Coriolis force in the Northern Hemisphere.

compute the wind speed and direction at any latitude as a function of the prevailing pressure gradient if we assume that only pressure and Coriolis forces influence the flow.

It can be shown that the acceleration experienced by an object on the surface of the earth (or in the atmosphere) moving with a velocity vector **u** consists of two components, $-\mathbf{\Omega} \times (\mathbf{\Omega} \times \mathbf{r})$ and $-2(\mathbf{\Omega} \times \mathbf{u})$, where $\mathbf{\Omega}$ is the angular rotation vector for the earth and **r** is the radius vector from the center of the earth to the point in question. (For a derivation of these terms see Williamson, 1973.) The first term is simply the centrifugal force, in a direction which acts normal to the earth's surface and is counterbalanced by gravity. The second term, $\mathbf{\Omega} \times \mathbf{u}$, is the Coriolis force. This force arises only when an object, such as an air parcel, is moving, that is, $\mathbf{u} \neq 0$. Even though the Coriolis force is of much smaller magnitude than the centrifugal force, only the Coriolis force has a horizontal component. Since the winds are horizontal in the geostrophic layer, the Coriolis acceleration is given by the horizontal component of the Coriolis term, namely $2u_G\Omega \sin\beta$, where Ω is the rate of rotation of the earth and β is the latitude. The direction of the Coriolis force is perpendicular to the wind velocity, as shown in Fig. 3.17. Wind speed u_G at latitude β lies in the horizontal plane.

In the geostrophic layer it may be assumed that the atmosphere is inviscid (frictionless) and in laminar flow. The equations of continuity and motion for such a fluid are

$$\frac{\partial u}{\partial x} + \frac{\partial v}{\partial y} + \frac{\partial w}{\partial z} = 0 \tag{3.23}$$

and

$$\begin{aligned}
\frac{\partial u}{\partial t} + u\frac{\partial u}{\partial x} + v\frac{\partial u}{\partial y} + w\frac{\partial u}{\partial z} &= -\frac{1}{\rho}\frac{\partial p}{\partial x} + F_x \\
\frac{\partial v}{\partial t} + u\frac{\partial v}{\partial x} + v\frac{\partial v}{\partial y} + w\frac{\partial v}{\partial z} &= -\frac{1}{\rho}\frac{\partial p}{\partial y} + F_y \\
\frac{\partial w}{\partial t} + u\frac{\partial w}{\partial x} + v\frac{\partial w}{\partial y} + w\frac{\partial w}{\partial z} &= -\frac{1}{\rho}\frac{\partial p}{\partial z} + F_z
\end{aligned} \tag{3.24}$$

where u, v, w are the three components of the velocity and F_x, F_y, and F_z are the three components of the external force.

Let the axes be fixed in the earth, with the x axis horizontal and extending to the east, the y axis horizontal and extending to the north, and the z axis normal to the earth's surface. As before, Ω is the angular velocity of rotation of the earth and β the latitude. The components of the Coriolis force in the x, y, and z directions on a particle are the components of $\mathbf{F}_c = -2(\mathbf{\Omega} \times \mathbf{u})$.

$$\begin{aligned}
F_{cx} &= -2\Omega(w\cos\beta - v\sin\beta) \\
F_{cy} &= -2\Omega u\sin\beta \\
F_{cz} &= 2\Omega u\cos\beta
\end{aligned} \tag{3.25}$$

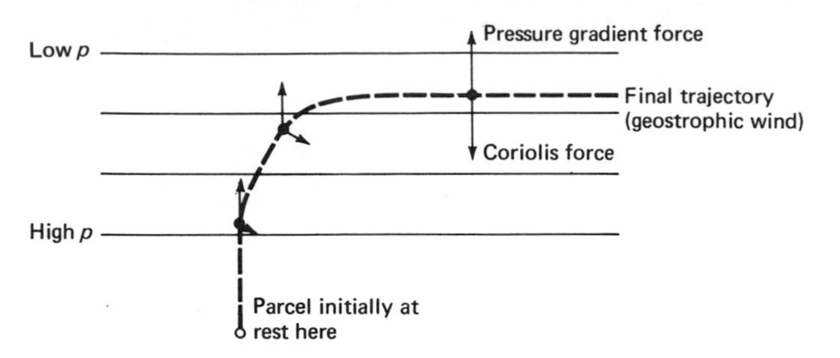

FIGURE 3.18
Approach to geostrophic equilibrium.

At great heights, the vertical velocity component w can usually be neglected relative to the horizontal components u and v. Therefore, substituting (3.25) into (3.24), we obtain, for steady motion,

$$\begin{aligned} u\frac{\partial u}{\partial x} + v\frac{\partial u}{\partial y} &= 2\Omega v \sin\beta - \frac{1}{\rho}\frac{\partial p}{\partial x} \\ u\frac{\partial v}{\partial x} + v\frac{\partial v}{\partial y} &= -2\Omega u \sin\beta - \frac{1}{\rho}\frac{\partial p}{\partial y} \end{aligned} \tag{3.26}$$

We see that the air moves so that a balance is achieved between the pressure gradient and the Coriolis force. Let us consider the situation in which the velocity vector is oriented in the x direction, and so $v = 0$; then

$$u\frac{\partial u}{\partial x} = -\frac{1}{\rho}\frac{\partial p}{\partial x} \tag{3.27}$$

$$0 = -2\Omega u \sin\theta - \frac{1}{\rho}\frac{\partial p}{\partial y} \tag{3.28}$$

We usually denote $2\Omega \sin\beta$ by f, called the Coriolis parameter. From the continuity equation (3.23), we see that $\partial u/\partial x = 0$, since $v = w = 0$. Thus, from (3.27), $\partial p/\partial x = 0$, and the direction of flow is perpendicular to the pressure gradient $\partial p/\partial y$. In addition, from (3.28), we see that the component of the Coriolis force, $-fu$, is exactly balanced by the pressure gradient, $1/\rho\ \partial p/\partial y$. Therefore, the *geostrophic wind speed* u_G is given by

$$u_G = \frac{\partial p/\partial y}{2\rho\Omega \sin\beta} \tag{3.29}$$

The approach to the geostrophic equilibrium for an air parcel starting from rest, accelerated by the pressure gradient and then affected by the Coriolis force, is shown in Fig. 3.18.

The geostrophic balance determines the wind direction at altitudes above about 500 m. In order to describe the air motions at lower levels we must take into account the friction of the earth's surface. The presence of the surface induces a shear in the wind profile, as in a turbulent boundary layer over a flat plate generated in a labora-

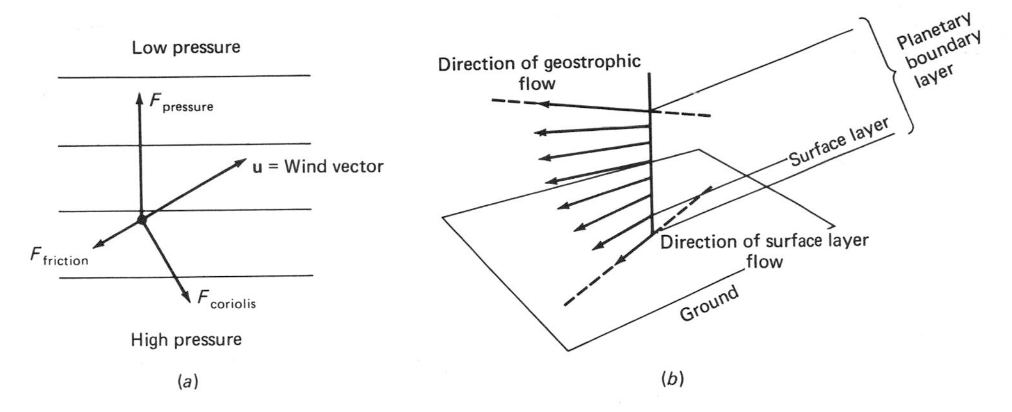

FIGURE 3.19
Variation of wind direction with altitude. (*a*) Balance of forces among pressure gradient, Coriolis force, and friction. (*b*) The Ekman spiral.

tory wind tunnel. In analyzing the geostrophic wind speed we found that for steady flow a balance exists between the pressure force and the Coriolis force. Consequently, steady flow of air at levels near the ground leads to a balance of three forces: pressure force, Coriolis force, and friction force due to the earth's surface. Thus, as shown in Fig. 3.19*a*, the net result of these three forces must be zero for a nonaccelerating air parcel. Since the pressure gradient force F_p must be directed from high to low pressure, and the frictional force F_F must be directed opposite to the velocity u, a balance can be achieved only if the wind is directed at some angle toward the region of low pressure. This angle between the wind direction and the isobars increases as the ground is approached since the frictional force increases. At the ground, over open terrain, the angle of the wind to the isobars is usually between 10 and 20°. Because of the relatively smooth boundary existing over this type of terrain, the wind speed at a 10-m height (the height at which the so-called surface wind is usually measured) is already almost 90 percent of the geostrophic wind speed. Over built-up areas, on the other hand, the speed at a 10-m height may be only 50 percent of the geostrophic wind speed, owing to the mixing induced by the surface roughness. In this case the surface wind may be at an angle of 45° to the isobars.

As a result of these frictional effects, the wind direction commonly turns with height, as shown in Fig. 3.19*b*. The variation of wind direction with altitude is known as the *Ekman spiral*. We defer a derivation of the Ekman spiral until Chap. 5 when we have had the opportunity to discuss turbulent transport of momentum.

The pattern of general circulation shown in Fig. 3.16 does not represent the actual state of atmospheric circulation on a given day. The irregularities of land masses and their surface temperatures tend to disrupt the smooth global circulation patterns we have described. Another influence which tends to break up zonal patterns is the Coriolis force. Air which converges at low levels toward regions of low pressure must also execute a circular motion because of Coriolis forces. The effect of friction at the surface is to direct the winds at low levels in part toward the region of low pressure, producing an inward spiraling motion. This vortex-like motion is given the name *cyclone*. The center of a cyclone is usually a rising column of warm air. Similarly, a low-level diverging flow from a high pressure region will spiral outward. Such a region is called an *anticyclone*. In the Northern Hemisphere the motion of a cyclone is counterclockwise and that of an anticyclone, clockwise. The dimensions of commonly occurring cyclones and anticyclones are from 100 to 1000 km. Most cyclones and anticyclones are born in one part of the world and migrate to another. These are not to be confused with hurricanes or typhoons, which, although they consist of the same type of air motion, are of a smaller scale.

An element of the cyclone-anticyclone phenomenon which has particular importance for air pollution in several parts of the world is the semipermanent subtropical anticyclone, high pressure regions centered over the major oceans. They are called semipermanent because they shift position only slightly in summer and winter. The key feature of the subtropical anticyclone is that the cold subsiding air aloft, which results in the high pressure observed at sea level, is warmed by compression as it descends, often establishing an elevated temperature inversion. The inversion layer generally approaches closer to the ground as the distance from the center of the high pressure increases.

On the eastern side of the subtropical anticyclones the inversion is strengthened by the southerly flow of cool, dry air (recall that in the Northern Hemisphere the rotation in an anticyclone is clockwise). Particularly in coastal areas the low-level air is cooled by contact with the cold ocean, an exchange which tends to strengthen the inversion. Since the air aloft, as well as the southbound low-level flow, is warming, there is little precipitation in these regions. Thus, on the west coasts of continents it is common to find arid, desert-like conditions, such as the deserts of southern California, the Sahara in North Africa, the desert in western Australia, and the coastal plains of South America.

On the other hand, on the western side of the semipermanent anticyclone, inversions are less frequent and the low-level air from the tropics is warm and moist. As it cools on its path to the north, precipitation is heavy. Thus, the eastern coasts of continents in the subtropics are warm and humid, such as the eastern coasts of South America and Africa.

We can now see one of the reasons why Los Angeles is afflicted with air pollution problems. Its location on the west coast of North America in the subtropical region and on the eastern side of the Pacific anticyclone is one in which elevated inversions are

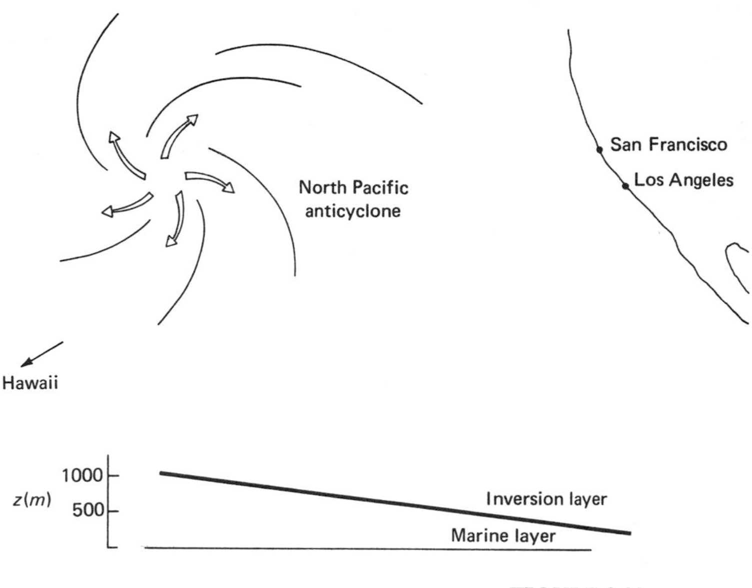

FIGURE 3.20
North Pacific anticyclone.

frequent and strong. The lowest layer of air (the marine layer) is cooled because of its contact with the ocean. Air pollutants are trapped in the marine layer and prevented from vertically exchanging with upper-level air. Such a situation can lead to serious air pollution problems. The base of a subsidence inversion lies typically at an elevation of about 500 m, with the inversion layer extending another 500 to 1000 m upward. Figure 3.20 shows a cross section of the North Pacific anticyclone.

In addition to the semipermanent anticyclones, there are many migratory cyclones and anticyclones in the temperate zones. Formed by confrontations between arctic and tropical air, they have a lifetime of a few weeks and drift with the westerly winds at about 800 km/day. Precipitation is often associated with the rising air over the low pressure center of a cyclone. Cyclones are thus usually accompanied by cloudy skies and inclement weather. On the other hand, anticyclones are characterized by clear skies, light winds, and fair weather. Surprisingly enough, anticyclones lead to air pollution problems, particularly when one temporarily ceases its eastward drift and stagnates for a few days. The classic episode in Donora, Pa. (see Table 1.6) and many in New York occurred under such circumstances. Regions in the United States prone to stagnating anticyclones are the Great Basin between the Rockies and the Sierras, the central basin of California, and the southern Appalachians. In fact, stagnating anticyclones are probably a contributing factor to the haze over the Great Smoky Mountains in eastern Tennessee.

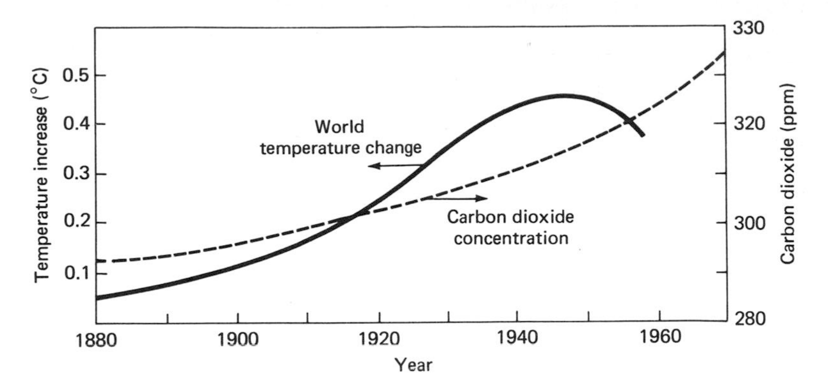

FIGURE 3.21
Trends in mean annual world temperature and CO_2 concentration in the atmosphere. (*Source: Lovelock*, 1971. *With the permission of Microfirm International Marketing Corporation, exclusive copyright licensee of Pergamon Press journal back files.*)

3.4 AIR POLLUTION AND GLOBAL CLIMATIC CHANGE

Over the history of the earth the climate has changed many times. These changes have been attributed to such natural phenomena as variations in the sun's radiation, the earth's orbit, the earth's surface, the composition of the atmosphere, and the circulation of the oceans (Mitchell, 1968; Lamb, 1969). Until about 1900, there was no question that both short- and long-term climatic changes had been due to natural causes. With the advent of large-scale industrialization, man's activities now have the potential of affecting the earth's climate.

Between 1880 and 1940 the mean global temperature rose about 0.4°C. Since 1945, the earth's temperature has fallen about 0.1°C. Whether these changes are attributable to man's activities, natural changes, or a combination of the two is not entirely clear. Even if man has not yet had an effect on global climate, a continuing growth in population and industrial activity certainly has the potential of leading to climatic changes (through the effect of emissions of gases and particles into the atmosphere). Since only small changes in global temperatures can produce enormous effects, such as advance and retreat of the polar ice caps, with associated changes in ocean levels, it is of utmost importance that the potential effects of man on the earth's temperature be carefully assessed.

As we have seen, carbon dioxide, although present only as a trace component, plays a major role in the earth's energy balance (through absorption and emission of infrared radiation). In short, by absorbing terrestrial long-wave radiation and reradiating much of it back to earth, CO_2 prevents cooling of the earth's surface. Thus, an increase in the atmospheric concentration of CO_2 will lead, in theory, to an increase in the surface temperature of the earth.

Figure 3.21 shows the change in ambient CO_2 concentration over the past 70 yr compared with the change in world mean surface temperature. Before we attempt to explain the effect of CO_2 concentration on temperature, we must consider the global CO_2 budget in order to determine the fate of CO_2 introduced into the atmosphere (as we have done for the pollutant species in Chap. 2). In 1950 the ambient CO_2 concentration was about 306 ppm by volume. Since the mass of the atmosphere is 5.14×10^{21} g, there was about 2.39×10^{18} g of CO_2 present. In that year about 0.67×10^{16} g of CO_2 was emitted from the burning of fossil fuel. This represented 0.28 percent of the amount of CO_2 already present (President's Science Advisory Committee, 1965). In 1960 the CO_2 concentration had reached 313 ppm, representing 2.44×10^{18} g of CO_2 in the entire atmosphere. Over the decade 1950 to 1960, it is estimated that 8.24×10^{16} g of CO_2 was added. Based on these data, about 50 percent of the CO_2 produced by combustion of fossil fuels has remained in the atmosphere. The other 50 percent had to go into other reservoirs, the two most likely of which are the oceans and the biosphere[1] (see "Man's Impact on the Global Environment," 1970 for a systematic exploration of the capacity of each of these reservoirs for CO_2). The partitioning of CO_2 between the oceans and the biosphere is too uncertain to permit accurate prediction of the fate of additional CO_2 introduced into the atmosphere. Assuming a 4 percent annual growth rate in fossil fuel combustion and a continuing 50 percent uptake of emitted CO_2 by the oceans and biosphere, it has been estimated that the ambient CO_2 level will be 379 ppm by the year 2000 ("Man's Impact on the Global Environment," 1970).

The decrease in temperature after 1940 indicates that perhaps another effect has overtaken that of CO_2 (assuming that the increase exhibited in Fig. 3.21 was due to CO_2, which, of course, it very well may not have been).[2] In particular, it has been suggested that the global burden of dust and particles is also increasing, with the scattering and absorption effects of these particles more than offsetting the heat-trapping effect of CO_2.

Figure 3.23 compares the amount of carbon burnt, the increase in CO_2, and the increase in turbidity over the period 1860 to 1970. Although the amount of CO_2 has risen at roughly the same rate as the amount of carbon burnt, the level of turbidity

[1] The mass of living things and nonliving organic matter on land and in the oceans.

[2] Lovelock (1971) suggested that global temperature averages can be misleading. He presents Fig. 3.22 which contrasts temperature changes in this century over various parts of the Northern Hemisphere as compared with their values at the beginning of the century. We see that the further north the greater the changes. However, it has been suggested (Lamb, 1969) that current climatic changes are part of a normal cyclic process by which the earth's climate changes with a period of between 100 and 200 yr, attributable to changes in solar output.

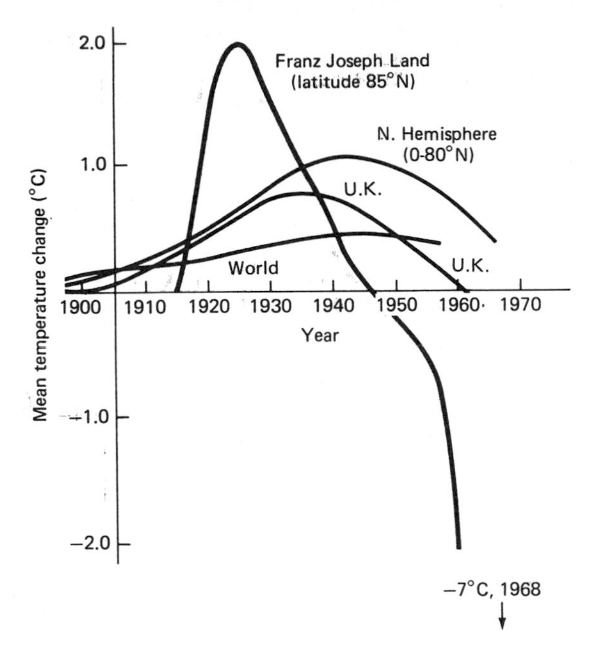

FIGURE 3.22
Trends in mean annual temperature for the world and parts of the Northern Hemisphere. (*Source: Lovelock*, 1971. *With the permission of Microfirm International Marketing Corporation, exclusive copyright licensee of Pergamon Press journal back files.*)

has increased at a much greater rate. Even if combustion is the principal source of particles, the turbidity curve in Fig. 3.23 is not unexpected, since the light obscuration by an aerosol is not a simple linear function of its mass. In the submicrometer size range, turbidity increases with aerosol mass at greater than a linear rate.

We have presented some observational data relating to global temperature changes, CO_2 levels, and turbidity, together with some ideas of the qualitative effect of the latter two variables on the former. The question now arises: Can we predict, based on physical principles, the quantitative effects of CO_2 level and particulate loading on the world's temperature? If so, we can explore the possible effects that increases in each of these variables may have on global temperature. In particular, is it possible that a continued increase in the CO_2 and aerosol content of the atmosphere will eventually cause the global temperature to become unstable, with the earth becoming either very hot or very cold?

We can approach the problem by considering separately the effects of increases in CO_2 and aerosols on global temperature. Such an analysis has been presented by

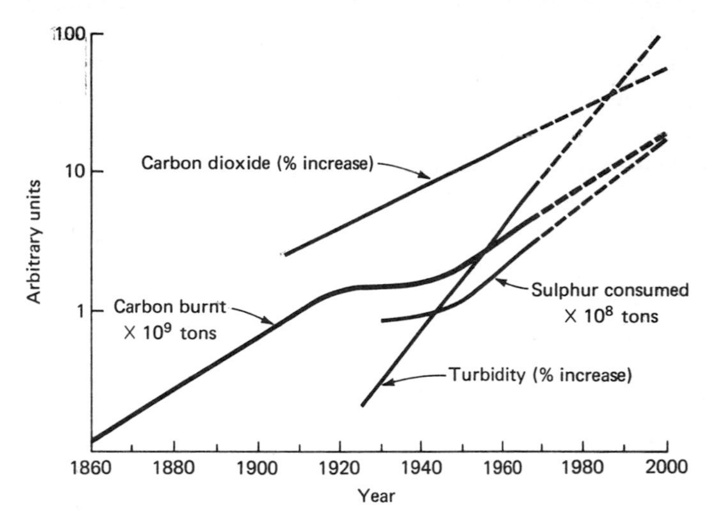

FIGURE 3.23
Trends in combustion of carbon and sulfur from fossil fuels and relation to CO_2 content and atmospheric turbidity. (*Source: Lovelock*, 1971. *With the permission of Microfirm International Marketing Corporation, exclusive copyright licensee of Pergamon Press journal back files.*)

Rasool and Schneider (1971). We consider first CO_2. In their analysis, Rasool and Schneider adopted a model of the atmosphere consisting of vertical distributions of temperature, pressure, water vapor, and cloud cover in order to compute, for different CO_2 levels, the amount of terrestrial long-wave radiation (from 4 to 100 μm) that escapes from the atmosphere. A doubling of CO_2 was found to produce a tropospheric temperature change of 0.8°K. However, as more CO_2 is added to the atmosphere, the rate of temperature increase is proportionately less and less. For a tenfold increase in CO_2, the temperature increase does not exceed 2.5°K. The explanation is that the 15-μm absorption band of CO_2, which is the main source of the absorption, becomes "saturated" and the addition of more CO_2 does not substantially increase the infrared opacity of the atmosphere. Thus, the conclusion is that even an order of magnitude increase in atmospheric CO_2 level should not produce a runaway greenhouse effect on earth.

Now let us consider the effect of particles on global temperature. Atmospheric aerosols in the size range 0.1 to 5 μm in diameter are quite abundant and play an important role in the atmospheric energy balance through scattering and absorption of both solar and terrestrial radiation. (The larger the particle, the greater is its potential for interacting with radiation of longer wavelength.) Scattering of solar radiation by particles in the size range 0.1 to 5 μm is primarily in the forward direction (about 90 percent) with the remainder in the backward direction. Backscattering increases the albedo of the atmosphere, although whether increased backscattering by atmospheric aerosols increases the albedo of the earth and atmosphere together depends on the surface albedo. In addition to scattering, aerosols are capable of absorbing incoming solar radiation. Whereas backscattering reduces the amount of heating both within and below the layer containing the particles, absorption increases the heating within the absorbing layer. Determination of the net effect of these two processes is complicated since the scattering and absorption of both solar and terrestrial radiation depend on aerosol size, number, and composition.[1]

Rasool and Schneider (1971) calculated the magnitude of scattering and absorption of both visible and infrared radiation by atmospheric aerosols. Their analysis, based on the theory of multiple scattering, is quite involved. A critical point in the analysis is the prediction of the rate at which global background opacity will increase with increasing particulate injection. The authors estimate that a fourfold increase in the background global opacity (possible perhaps in the next 50 to 100 yr) would result in a decrease in global temperature by as much as 3.5°K. Such a large decrease sustained over a period of a few years is believed to be sufficient to trigger an ice age.

For additional discussion of the meteorological aspects of global air pollution the reader is referred to Munn and Bolin (1971).

REFERENCES

CLARK, J. F.: Nocturnal Urban Boundary Layer over Cincinnati, Ohio, *Mon. Weather Rev.*, **97**:582 (1969).

DAVIDSON, B.: A Summary of the New York Urban Air Pollution Research Program, *J. Air Pollut. Control Assoc.*, **17**:154 (1967).

LAMB, H. H.: The New Look of Climatology, *Nature*, **223**:120i (1969).

LOVELOCK, J. E.: Air Pollution and Climatic Change, *Atmos. Environ.*, **5**:403 (1971).

"Man's Impact on the Global Environment—Assessment and Recommendations for Action," M.I.T., Cambridge, Mass., 1970.

MCCORMICK, R. A., and J. H. LUDWIG: Climatic Modification by Atmospheric Aerosols, *Science*, **156**:1358 (1967).

MILLER, A.: "Meteorology," Charles E. Merrill Books, Columbus, Ohio, 1966.

MITCHELL, J. J., JR.: Causes of Climatic Change, "Meteorological Monographs," vol. 8, no 30, Lancaster Press, Lancaster, Pa., 1968.

[1] It is important to note that we are interested here primarily in the effects of tropospheric aerosols on surface temperature. Although stratospheric aerosols may warm that layer as a result of absorption, surface air is still likely to be cooled because of the reduction in radiation reaching the earth. For example, after the eruption of Mt. Agung in 1963, a temperature increase of 6 to 7°C occurred in the equatorial stratosphere, but only a slight decrease in global surface temperature took place. ("Man's Impact on the Global Environment," 1970).

MUNN, R. E., and B. BOLIN: Global Air Pollution—Meteorological Aspects: A Survey, *Atmos. Environ.*, **5**:363 (1971).

President's Science Advisory Committee, Restoring the Quality of Our Environment, U.S. Government Printing Office, Washington, 1965.

RASOOL, S. I., and S. H. SCHNEIDER: Atmospheric Carbon Dioxide and Aerosols: Effects of Large Increases on Global Climate, *Science*, **173**:138 (1971).

WILLIAMSON, S. J.: "Fundamentals of Air Pollution," Addison-Wesley, Reading, Mass., 1973.

PROBLEMS

3.1 If the earth-atmosphere system is assumed to be in radiative equilibrium with the sun (that is, it emits as much radiation as it receives), it is possible to estimate a value for the temperature T_e of the earth-atmosphere system. If the planetary albedo is denoted by α and the solar constant by S, show that T_e is given by

$$T_e = \left[\frac{(1-\alpha)S}{4\sigma}\right]^{1/4}$$

Calculate the value of T_e for $\alpha = 0.34$.

The average surface temperature T_s is considerably higher than the earth-atmosphere radiative equilibrium temperature T_e. Why is this so? How would you attempt to compute an estimate for T_s?

3.2 Show that if the atmosphere is isothermal the temperature change of a parcel of air rising adiabatically is

$$T(z) = T_0\, e^{-\Gamma z/T_0'}$$

where T_0 and T_0' are the temperatures of the parcel at the surface and of the air at the surface, respectively.

3.3 A rising parcel of air will come to rest when its temperature T equals that of the surrounding air, T'. Show that the height z where this occurs is given by

$$z = \frac{1}{\Lambda}\left[T_0' - \left(\frac{T_0'^{\Gamma}}{T_0^{\Lambda}}\right)^{1/(\Gamma-\Lambda)}\right]$$

What condition must hold for this result to be valid?

3.4 Show that the condition that the density of the atmosphere does not change with height is

$$\frac{dT}{dz} = -3.42 \times 10^{-2}\ \text{°C/m}$$

3.5 It has been proposed that air pollution in Los Angeles can be abated by drilling large tunnels in the mountains surrounding the basin and pumping the air out into the surrounding deserts. You are to examine the power requirements in displacing the volume of air over the Los Angeles basin. Assume the basin has an area of 4000 km^2 and that the polluted air is confined below an elevated inversion with a mean height of 400 m. The coefficient of friction for air moving over the basin is assumed to be 0.5, and the minimum energy needed to sustain airflow is equal to the energy dissipated by ground friction. Determine the power required to move the air mass 7 km/hr. Compare your result with the capacity of Hoover Dam: 1.25×10^6 kW.

3.6 Elevated inversion layers are a prime factor responsible for the incidence of community air pollution problems. It is interesting to consider the feasibility of eliminating an elevated inversion layer. In principle, this could be done either by cooling all the air from the inversion base upward to a temperature below that at the inversion base or by heating all the air below the top of the inversion to a temperature higher than that at the inversion top.

Show that the energy E required to destroy an elevated inversion by heating from below is given by

$$E = \rho C_p[\Gamma(H_T - H_B) + (T_T - T_B)]\frac{H_T + H_B}{2}$$

where
ρ = average density of air
C_p = heat capacity of air
H_B, H_T = heights of base and top of inversion
T_B, T_T = temperatures of base and top of inversion

Assume that the lapse rate below the base of the inversion is adiabatic and that the rate of temperature increase with height in the inversion is linear.

Estimate the value of E for typical September conditions at Long Beach, California, at 7 A.M. Use

$$H_B = 475 \text{ m} \qquad T_B = 14.1°\text{C}$$
$$H_T = 1055 \text{ m} \qquad T_T = 22.4°\text{C}$$

If the area of the Los Angeles basin is 4000 km^2 and the energy produced by oil burning with 100 percent efficiency is 1.04×10^{12} cal/kg of oil, what is the amount of oil required in order to destroy the inversion over the entire basin?

3.7 The emissions from a single engine to be used on a supersonic transport (SST) flying at an altitude of 20 km at Mach 2.7 on a day with standard weather conditions have been calculated to be ("Man's Impact on the Global Environment," 1970):

Constituent	kg/hr
Ingested air and consumed fuel	
Air	6.25×10^5
Fuel	0.15×10^5
Combustion products	
CO_2	0.468×10^5
H_2O	0.188×10^5
CO	635
NO	635
SO_2	15
Particles	2.2

An estimate of the Federal Aviation Administration is that by 1985 to 1990 500 SST aircraft will be in service, each flying 7 hr a day. Of the 500, 334 would be of United States manufacture and be equipped with four engines, and 166 would be from other countries and would have the equivalent of two engines. You wish to determine the steady-state concentrations of CO_2, H_2O, CO, and NO that would occur after several years of operation, assuming a 2-yr mean residence time for all four products. First, determine the mass emission rates of each of the four pollutants in grams per year as a result of SST operations. Then, assuming that the mass of the stratosphere is 15 percent of the total mass of the atmosphere (5.14×10^{21} g), compute the stratospheric concentrations of the four pollutants in ppm by mass.

Based on these values, what conclusions can you draw about the potential effects of SST emissions on climate? Note that current stratospheric concentrations of CO_2 and H_2O are

CO_2 445 ppm by mass
H_2O 3.0 ppm by mass

What effect are SO_2 and particles likely to have on global climate?

It is important to note that this problem has not considered the *chemical* effect of these emissions on the stratosphere. This is quite another matter, and has been the subject of an intensive multi-year study conducted by the U.S. Department of Transportation. See, for example, "Environmental Impact of Stratospheric Flight," Climatic Impact Committee, National Academy of Sciences, Washington, D.C., 1975.

4

AIR POLLUTION CHEMISTRY

Often the most irritating and deleterious air pollutants are not those emitted directly by sources but those formed in the atmosphere by chemical reactions among the emitted species. Examples cited previously include sulfuric acid and ozone. In an analysis of the effectiveness of means of controlling air pollution it is therefore essential that the chemical process taking place in the atmosphere be understood.

As we noted in Chap. 1, urban smogs have traditionally been divided chemically into two categories: reducing and oxidizing. The former category refers to conditions of high sulfur oxide and particle concentrations, high humidities, and cool temperatures, such as occur in the large metropolitan areas of the northern United States and western Europe in which the burning of coal and fuel oil is the predominant source of primary pollutants. The oxidizing type of air pollution, on the other hand, is characterized by high concentrations of ozone and nitrogen oxides, low humidities, and warm temperatures. This type of air pollution has been given the name *photochemical smog*, the reason for which we shall see shortly. Los Angeles air pollution is the classic example of photochemical smog. It is no longer possible in many cases to classify sharply the chemical nature of a region's air pollution, since both the widespread use of automobiles and the generation of electric power by burning of coal and oil are present in large urban areas.

The formulation of a descriptive and predictive mechanism for chemical reactions in air pollution requires identification of all the important reactions contributing to the chemical dynamics. Similarly, thorough investigation of a specific reaction is achieved only when the reaction rate constant has been carefully determined and the reaction mechanism properly specified. Because of the large number of important reactions that take place in the atmosphere, the rapid rates of many of them, and the low concentrations of most reactants (e.g. free radicals), the experimental investigation of air pollution chemistry is an enormously large and difficult task. Much has been accomplished, however, in recent years. Our objective in this chapter is to elucidate the basic chemical processes which take place in urban atmospheres as well as to point out those areas that are in need of further study. We begin in Sec. 4.1 with a discussion of photochemical reactions in urban atmospheres. We then study individually the reactions of the oxides of nitrogen and hydrocarbons in Secs. 4.2 and 4.3. Drawing on the material in Secs. 4.1 to 4.3, we consider in Sec. 4.4 the chemistry of photochemical smog. Section 4.5 is devoted to the chemistry of the oxides of sulfur in air pollution. The physics and chemistry of atmospheric aerosols are reviewed briefly in Sec. 4.6. Finally, measurement methods for gaseous air pollutants are discussed in Sec. 4.7.

We have endeavored to present here a reasonably complete treatment of the current level of understanding of air pollution chemistry. The chapter is designed to be relatively self-contained. However, because of the vast amount of study that has been devoted to atmospheric chemical reactions, a large number of references are cited throughout the chapter. The interested reader may wish to avail himself of the references cited for detailed treatment of specific topics.

4.1 ATMOSPHERIC PHOTOCHEMICAL REACTIONS

Photochemical reactions can be defined as those which are initiated by the absorption of a photon by an atom, molecule, free radical, or ion. The primary step of a photochemical reaction may be written

$$A + h\nu \rightarrow A^*$$

where A^* is an excited state of A and $h\nu$ indicates that exactly one photon is absorbed per molecule (the Stark-Einstein law of photochemical equivalence). The excited molecule A^* may subsequently partake in

$$\text{Dissociation: } A^* \xrightarrow{1} B_1 + B_2 + \cdots$$

$$\text{Direct reaction: } A^* + B \xrightarrow{2} C_1 + C_2 + \cdots$$

$$\text{Fluorescence: } A^* \xrightarrow{3} A + h\nu$$

$$\text{Collisional deactivation: } A^* + M \xrightarrow{4} A + M$$

The first two reactions lead to chemical change, whereas the last two return the molecule to its original state. Either of the first two may be the primary process resulting from light absorption.

The primary quantum yield for a specific primary process is defined as the ratio of the number of molecules of A* reacting by that process to the number of photons absorbed. Since the total number of A* molecules formed equals the total number of photons absorbed, the primary quantum yield ϕ for a specific process, say dissociation, is just the fraction of the A* molecules that react by dissociation. If the four processes above represent all the possible paths for A*, the sum of the quantum yields for the four processes must equal 1:

$$\sum_{i=1}^{4} \phi_i = 1$$

The rate of formation of A* is equal to the rate of absorption and can be written

$$\frac{d[A^*]}{dt} = k_a[A]$$

where [A] denotes the concentration of A and k_a, the rate constant with units of time^{-1}, is called the specific absorption rate; k_a is normally taken to be independent of [A]. The rate of formation of C_1 in step 2 can be expressed as

$$\frac{d[C_1]}{dt} = \phi_2 \, k_a[A]$$

where ϕ_2 is the quantum yield of step 2. If we assume that A* immediately participates in one of the four steps, the total rate of consumption of A* must equal the total rate of production of A*, the absorption rate. Of course, some of the consumption reactions for A* may be a return to A.

The rate of absorption of radiation by a particular molecule depends strongly on the wavelength distribution and the intensity of the radiation. In the lower atmosphere radiation is attenuated by scattering and absorption, and, as a result, the solar spectral intensity distribution received at the ground varies with solar zenith angle (0° for radiation perpendicular to the surface, 90° for sunrise or sunset). Figure 4.1 shows typical distributions of solar spectral intensity for zenith angles of 0 and 80° at ground level in mid-latitudes. Thus, the major variable which affects photochemical reactions in the lower atmosphere is the intensity change which occurs as a function of time of day, latitude, and time of year. The maximum intensity and the number of hours when the intensities are near these maxima are important because they place an upper limit on the rate of photochemical reactions. In the United States the maximum noonday intensity does not vary substantially with latitude during the summer months. In the 3000- to 4000-Å region the maximum total intensity is 2×10^{16} photons cm^{-2} sec^{-1} and remains near this value for 4 to 6 hr. In

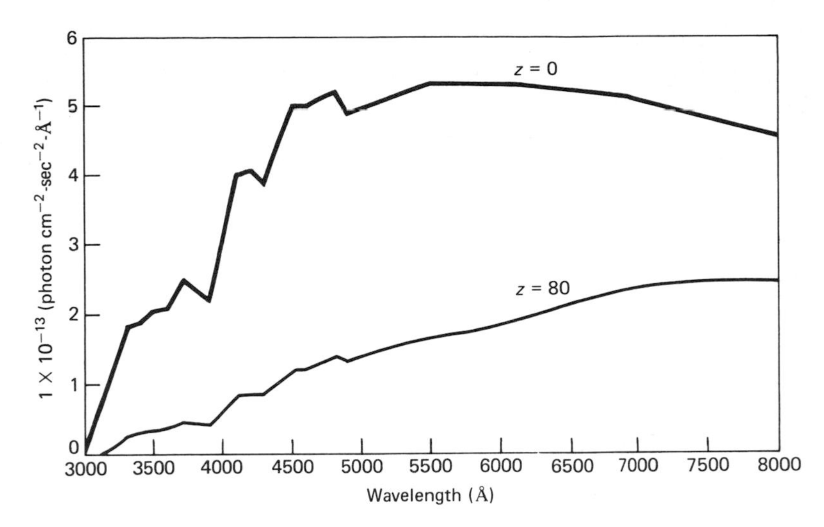

FIGURE 4.1
Typical distributions of solar spectral intensity for zenith angles of 0 and 80° at ground level in mid-latitudes. (*Source: Cadle and Allen,* 1970. *Copyright* 1970 *by the American Association for the Advancement of Science.*)

contrast, the winter values vary from 0.7×10^{16} to 1.5×10^{16} as a function of latitude. The time near this maximum is reduced to 2 to 4 hr.

The photochemistry of the lower atmosphere is dominated by the fact that virtually no solar radiation of wavelength less than 2900 Å reaches the troposphere. We have already seen in the preceding chapter that oxygen and ozone in the upper atmosphere effectively absorb all the radiation of wavelengths below 2900 Å. Therefore, the absorbing species of interest from an air pollution viewpoint are those that absorb in the portion of the spectrum from 3000 to 7000 Å. Table 4.1 lists the absorbing species in this category, which we shall now consider briefly. For deeper coverage of the photochemistry of air pollution the reader may consult the text of Leighton (1961). Table 4.2 summarizes the important aspects of light absorption by lower atmospheric species, and Table 4.3 presents the primary processes for the species in Table 4.2.

Photochemical reactions are of importance in air pollution because of the products (mainly free radicals) that result from them. These products then initiate or participate in a large number of other reactions responsible for the conversion of primary to secondary pollutants. Notice that none of the main primary air pollutants, SO_2, NO, CO, and hydrocarbons (except aldehydes), are important absorbers of radiation at the wavelengths prevalent in the lower atmosphere. Only NO_2, which

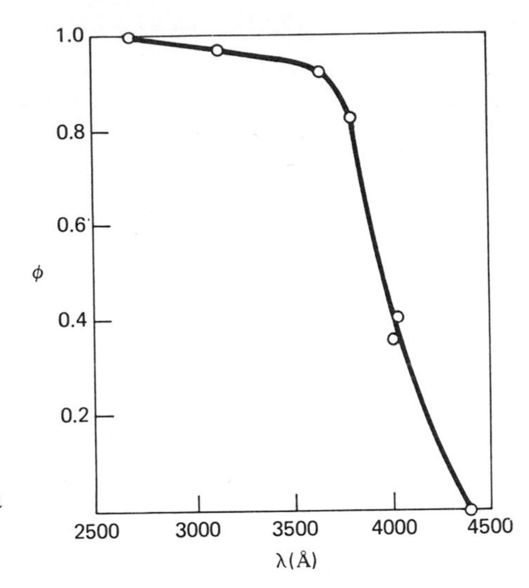

FIGURE 4.2
Primary quantum yield of NO_2 as a function of wavelength. (*Source: Schuck and Stephens*, 1969.)

can be viewed as both a primary and secondary pollutant, is an important absorber. This is indeed an interesting situation, since, as we shall see, the conversion of SO_2, NO, and hydrocarbons to products depends on photochemical reactions but not on the direct dissociation of these species themselves.

Table 4.1 ABSORBING AND NONABSORBING SPECIES OF INTEREST IN AIR POLLUTION

Absorbers	Nonabsorbers
Oxygen	Nitrogen
Ozone	Water
Nitrogen dioxide	Carbon monoxide
Sulfur dioxide	Carbon dioxide
Nitric acid and alkyl nitrates	Nitric oxide
Nitrous acid, alkyl nitrites, and nitro compounds	Sulfur trioxide and sulfuric acid
Aldehydes	Hydrocarbons
Ketones	Alcohols
Peroxides	Organic acids
Acyl nitrites, pernitrites, and nitrates	
Particulate matter	

SOURCE: Leighton (1961).

Table 4.2 PHOTOCHEMICAL REACTIONS IMPORTANT IN AIR POLLUTION

Species	Absorption characteristics
O_2	Absorption by the ground state of O_2 is too weak to lead to dissociation at wavelengths above 3000 Å.
O_3	Ozone absorbs strongly in the region 2000 to 3200 Å and weakly in the region 4500 to 7000 Å. In both cases the result is $O_3 + h\nu \rightarrow O_2 + O$; between 2000 and 3200 Å both products are electronically excited, whereas in the latter case both are in their normal electronic states.
NO_2	Nitrogen dioxide is the most important absorbing molecule in the urban atmosphere. It absorbs over the entire visible and ultraviolet range of the solar spectrum in the lower atmosphere. Below 4200 Å, NO_2 dissociates by $NO_2 + h\nu \rightarrow NO + O$. In Fig. 4.2 the primary quantum yield of NO_2 is shown as a function of wavelength, where the quantum yield ϕ is the ratio of molecules dissociating to molecules absorbing. Between 3000 and 3700 Å over 90 percent of the NO_2 molecules absorbing will dissociate into NO and O. Above 3700 Å this percentage drops off rapidly, and above 4200 Å dissociation does not occur. The bond energy between O and NO is 73 kcal/mole. This energy corresponds to the energy contained in wavelengths near 4000 Å. At longer wavelengths, there is insufficient energy to promote bond cleavage. The point at which dissociation fails to occur is not sharp because the individual molecules of NO_2 do not possess a precise amount of ground-state energy prior to absorption. The gradual transition area in Fig. 4.2 (3700 to 4200 Å) indicates a variation in ground-state energy of about 10 kcal/mole. This transition curve can be shifted slightly to longer wavelengths by increasing the temperature and therefore increasing the ground-state energy of the system.
SO_2	Sulfur dioxide has its absorption maximum about 2850 Å; above 3400 Å the absorption is very weak. The estimated bond dissociation energy for the reaction $SO_2 \rightarrow SO + O$ is 135 kcal/mole, corresponding to $\lambda = 2180$ Å. Since this energy is not available in the lower atmosphere, the primary photochemical process following absorption of sunlight by SO_2 is the formation of either of the two excited states, $^1SO_2$, or $^3SO_2$, the *singlet* and *triplet* states, respectively.
Nitric acid (HNO_3) and alkyl nitrates ($RONO_2$)	Most of the experimental evidence points to the primary process $RONO_2 + h\nu \rightarrow RO\cdot + NO_2$, where RO• is an alkoxyl radical. The absorption rates of alkyl nitrates are small above 3000 Å, so that these processes cannot be expected to be very important in air pollution.
Nitrous acid (HNO_2) and alkyl nitrites (RONO)	Alkyl nitrites dissociate according to $RONO + h\nu \rightarrow RO\cdot + NO$. The nitrites are probably the second most important class of absorbers behind NO_2.
Aldehydes (RCHO)	Photodissociation of aldehydes is possible at wavelengths shorter than 3130 Å. The principal dissociation reaction is $RCHO + h\nu \rightarrow R\cdot + H\dot{C}O$, where $H\dot{C}O$ is the *formyl* radical.
Peroxides (ROOR′)	Alkyl peroxides absorb weakly in the range 3000 to 7000 Å. The process most consistent with observations is $ROOR' + h\nu \rightarrow RO\cdot + R'O\cdot$.

Table 4.3 SUMMARY OF PHOTOCHEMICAL PRIMARY PROCESSES IN AIR POLLUTION

Absorber	Primary process	k_a at $z = 45°$, hr^{-1}	ϕ	Rate at $z = 45°$ and unit conc., pphm/hr
NO_2	$NO_2^* \rightarrow NO + O$	21	0.9–1	~20
O_3 (4500–7500 Å)	$O_3^* \rightarrow O_2 + O$	1	1	1
(2900–3500 Å)	$O_3^* \rightarrow O_2 + O$†	0.3	1	0.3
SO_2	$^1SO_2 \rightarrow {}^3SO_2$‡	0.4	<1	<0.4
Alkyl nitrites	$RCH_2ONO^* \nearrow RCH_2O\cdot + NO$; $\searrow RCHO + HNO$	5.5	<1	≤5.5
Alkyl nitrates	$RONO_2^* \rightarrow RO\cdot + NO_2$	0.006	1	0.006
Olefinic aldehydes	$RCH{=}CHCHO^* \rightarrow$	~1.5	~0.1	~0.15
Formaldehyde	$HCHO^* \nearrow H + H\dot{C}O$; $\searrow H_2 + CO$	0.5	0.4	0.2
Aliphatic aldehydes	$RCHO^* \rightarrow R\cdot + H\dot{C}O$	0.2	0.2–0.7	0.04–0.14
Nitrous acid	$HONO^* \rightarrow OH\cdot + NO$	...	...	...
Hydrogen peroxide	$H_2O_2^* \rightarrow 2OH\cdot$	0.08	1	0.08

SOURCE: Leighton (1961).

† This oxygen atom is in a higher energy state than that in the reaction immediately above.

‡ $^1SO_2$ and $^3SO_2$ denote the singlet and triplet states of excited SO_2, the singlet being more energetic.

4.2 REACTIONS OF NITROGEN OXIDES IN THE URBAN ATMOSPHERE

The oxides of nitrogen, NO and NO_2, play an important role in air pollution chemistry. As we have noted, the principal source of nitrogen oxides in the urban atmosphere is combustion processes. Most of the NO_X formed in combustion is NO. However, NO_2 is formed to some extent from the NO in hot combustion exhaust gases by

$$2NO + O_2 \rightarrow 2NO_2$$

Thus, small amounts of NO_2 are always present in an urban atmosphere which contains significant amounts of NO. Aside from its effects on visibility and human health, NO_2 plays a central role in the formation of so-called photochemical air pollution and may also be important in the oxidation of SO_2 to sulfate. In fact, if SO_2 and NO_X are present together, the usual slow conversion of SO_2 to H_2SO_4 is substantially accelerated. Even the small amounts of NO_2 present in the atmosphere are sufficient to trigger the complex series of reactions producing photochemical smog. The name "photochemical smog" was a result of the discovery by Haagen-Smit (Haagen-Smit, 1952; Haagen-Smit and Fox, 1956) in the early 1950s of the importance of the photolysis of NO_2 in the overall conversion of primary to secondary pollutants in this type of air pollution.

4.2.1 Basic Photochemical Cycle of NO_2, NO, and O_3

It has now been established beyond any doubt that the three most important reactions involving NO_2, NO, and air in the presence of sunlight are

$$NO_2 + h\nu \xrightarrow{1} NO + O$$
$$O + O_2 + M \xrightarrow{2} O_3 + M$$
$$O_3 + NO \xrightarrow{3} NO_2 + O_2$$

where M is any third body (usually either N_2 or O_2). Let us consider the dynamics of a system in which only these three reactions are taking place. Let us assume that known initial concentrations of NO and NO_2, $[NO]_0$ and $[NO_2]_0$, are placed in air. In a reactor of constant volume and at constant temperature the time rate of change of the concentration of NO_2 after the irradiation begins is given by

$$\frac{d[NO_2]}{dt} = -k_1[NO_2] + k_3[O_3][NO] \qquad (4.1)$$

Treating $[O_2]$ as constant, there are four species in the system: NO_2, NO, O, and O_3. We could write the dynamic equations for NO, O, and O_3 just as we have done for NO_2. For example, the equation for [O] is

$$\frac{d[O]}{dt} = k_1[NO_2] - k_2[O][O_2][M] \qquad (4.2)$$

However, if we were to evaluate the right-hand side numerically we would find that it is very close to zero. Physically, this means that the oxygen atom is so reactive that it disappears by reaction 2 almost as fast as it is formed by reaction 1. In dealing with highly reactive species such as the oxygen atom, it is customary to assume that the rate of formation is exactly equal to the rate of disappearance and to set the right-hand side of the rate equation equal to zero. In this case,

$$k_1[NO_2] = k_2[O][O_2][M] \qquad (4.3)$$

This assumption is called the *steady-state approximation*. The steady-state oxygen atom concentration in this system is then given by

$$[O] = \frac{k_1[NO_2]}{k_2[O_2][M]} \qquad (4.4)$$

Note that [O] is not constant; rather it varies with $[NO_2]$ in such a way that at any instant a balance is achieved between its rate of production and loss. What this really means is that the oxygen atom concentration adjusts to changes in the NO_2 concentration many orders of magnitude faster than the NO_2 concentration changes. Thus, on a time scale of the NO_2 changes, [O] always appears to satisfy (4.4).

However, from (4.1) and (4.2) we see that these three reactions will eventually reach a steady state in which all concentrations become constant. At that point reactions 1 to 3 represent a cycle in which NO_2 is destroyed and re-formed so fast that a steady state is maintained.

Let us compute the steady-state concentrations of NO, NO_2, and O_3 [the steady-state concentration of oxygen atoms is already given by (4.4)]. The steady-state ozone concentration is given by

$$[O_3] = \frac{k_1[NO_2]}{k_3[NO]} \tag{4.5}$$

We now need to compute $[NO_2]$ and [NO]. These are obtained from conservation of nitrogen,

$$[NO] + [NO_2] = [NO]_0 + [NO_2]_0 \tag{4.6}$$

and the stoichiometric reaction of O_3 with NO,

$$[O_3]_0 - [O_3] = [NO]_0 - [NO] \tag{4.7}$$

Solving for $[O_3]$ we obtain the relation for the ozone concentration formed at steady state by irradiating any mixture of NO, NO_2, O_3, and excess O_2 (in which only reactions 1 to 3 are important):

$$[O_3] = -\frac{1}{2}\left([NO]_0 - [O_3]_0 + \frac{k_1}{k_3}\right) + \frac{1}{2}\left[\left([NO]_0 - [O_3]_0 + \frac{k_1}{k_3}\right)^2 + \frac{4k_1}{k_3}([NO_2]_0 + [O_3]_0)\right]^{1/2} \tag{4.8}$$

If $[O_3]_0 = [NO]_0 = 0$, (4.8) reduces to

$$[O_3] = \frac{1}{2}\left\{\left[\left(\frac{k_1}{k_3}\right)^2 + \frac{4k_1}{k_3}[NO_2]_0\right]^{1/2} - \frac{k_1}{k_3}\right\} \tag{4.9}$$

Using a typical value of $k_1/k_3 = 0.01$ ppm we can compute the ozone concentration attained as a function of the initial concentration of NO_2 with $[O_3]_0 = [NO]_0 = 0$:

$[NO_2]_0$, ppm	$[O_3]$, ppm
0.1	0.027
1.0	0.095

If, on the other hand, $[NO_2]_0 = [O_3]_0 = 0$, then $[O_3] = 0$. This is clear since with no NO_2 there is no means to produce ozone. Thus, the maximum steady-state ozone concentrations would be achieved with an initial charge of pure NO_2. The concentrations of ozone attained in urban atmospheres are usually greater than those in the sample calculation. However, as noted in Chap. 2, most of the NO_X emitted is in the form of NO and not NO_2. Thus, with emissions of principally NO, the concentration of ozone reached, if governed solely by reactions 1 to 3, would be far too low to account for the actual observed concentrations. It must be concluded that reactions other than 1 to 3 are important in urban air in which relatively high ozone concentrations occur.

Although NO_2 is the primary energy absorber of those pollutants present in the atmosphere, and reactions 1 to 3 are the most important of those involving the oxides of nitrogen in daytime urban air, other reactions take place in the NO_X-air system. These other reactions, together with reactions 1 to 3, are presented in Table 4.4 with reported rate constant values at 298°K.[1] (We will generally use units of ppm and minutes for rate constants. Conversion factors between these and other units are given in Table 4.5.)

Given the mechanism in Table 4.4, it is interesting to explore the dynamic behavior of a given mixture of NO_2, NO, and dry air (note that we have not yet introduced water vapor into the system) which is irradiated (exposed to radiation of intensity and spectral composition typically observed at the earth's surface) for a

Table 4.4 REACTIONS AND RATE CONSTANTS FOR THE PHOTOLYSIS OF NITROGEN DIOXIDE IN N_2 AND O_2

No.	Reaction	Rate constant at 298°K	Reference
1	$NO_2 + h\nu \rightarrow NO + O$	Depends on light intensity	
2	$O + O_2 + M \rightarrow O_3 + M$	2.33×10^{-5} ppm^{-2} min^{-1}	Johnston (1968)
3	$O_3 + NO \rightarrow NO_2 + O_2$	2.95×10^{1} ppm^{-1} min^{-1}	Johnston and Crosby (1954)
4	$O + NO_2 \rightarrow NO + O_2$	1.38×10^{4} ppm^{-1} min^{-1}	Schuck et al. (1966)
5	$O + NO_2 + M \rightarrow NO_3 + M$	4.50×10^{-3} ppm^{-2} min^{-1}	Schuck et al. (1966)
6	$NO_3 + NO \rightarrow 2NO_2$	1.48×10^{4} ppm^{-1} min^{-1}	Schott and Davidson (1958)
7	$O + NO + M \rightarrow NO_2 + M$	2.34×10^{-3} ppm^{-2} min^{-1}	Kaufman (1958)
8	$2NO + O_2 \rightarrow 2NO_2$	7.62×10^{-10} ppm^{-2} min^{-1}	Glasson and Tuesday (1963)
9	$NO_3 + NO_2 \rightarrow N_2O_5$	4.43×10^{3} ppm^{-1} min^{-1}	Schott and Davidson (1958)
10	$N_2O_5 \rightarrow NO_3 + NO_2$	1.38×10^{1} min^{-1}	Mills and Johnston (1951)
11	$NO_2 + O_3 \rightarrow NO_3 + O_2$	0.46×10^{-1} ppm^{-1} min^{-1}	Ghormley et al. (1973)

[1] Rate constant values to be presented should not be interpreted as absolute values. Rather, each value is an experimental determination which is subject to some degree of error. It would therefore be more precise to present each rate constant as $k_i \pm \kappa$, where κ_i is, say, the 90 percent confidence interval on the measurement.

certain period of time. In order to answer this question, we could integrate the ordinary differential rate equations for the concentrations of the species in the mechanism. (If these ordinary differential equations are nonlinear, this integration must normally be carried out on a computer.) Once we have integrated the rate equations, we can examine the concentrations of any desired species as a function of time.

In our previous analysis of reactions 1 to 3 we assumed that oxygen atoms obeyed the steady-state approximation. In the analysis of the mechanism in Table 4.4 we again make the steady-state approximation for oxygen atoms and treat the remaining six species with differential equations. Letting R_i denote the rate of reaction i, for example, $R_1 = k_1[NO_2]$, $R_2 = k_2[O][O_2][M]$, etc., the differential equations governing NO_2, NO, O_3, NO_3, O_2, and N_2O_5 are

$$\frac{d[NO_2]}{dt} = -R_1 + R_3 - R_4 - R_5 + 2R_6 + R_7 + 2R_8 - R_9 + R_{10} - R_{11} \quad (4.10)$$

$$\frac{d[NO]}{dt} = R_1 - R_3 + R_4 - R_6 - R_7 - 2R_8 \quad (4.11)$$

$$\frac{d[O_3]}{dt} = R_2 - R_3 - R_{11} \quad (4.12)$$

$$\frac{d[NO_3]}{dt} = R_5 - R_6 - R_9 + R_{10} + R_{11} \quad (4.13)$$

$$\frac{d[O_2]}{dt} = -R_2 + R_3 + R_4 - R_8 + R_{11} \quad (4.14)$$

$$\frac{d[N_2O_5]}{dt} = R_9 - R_{10} \quad (4.15)$$

Table 4.5 CONVERSION FACTORS FOR CONCENTRATIONS AND RATE CONSTANTS (RELATIVE TO AIR AT 1 ATM AND 298°K)

Concentrations	
moles/l × 2.445 × 10^7	= ppm
moles/l × 6.0 × 10^{20}	= molecules/cm^3
molecules/cm^3 × 4.08 × 10^{-14}	= ppm
Rate constants	
Bimolecular	
l mole^{-1} sec^{-1} × 2.45 × 10^{-6}	= ppm^{-1} min^{-1}
cm^3 molecule^{-1} sec^{-1} × 1.47 × 10^{15}	= ppm^{-1} min^{-1}
Termolecular	
l^2 mole^{-2} sec^{-1} × 1.005 × 10^{-13}	= ppm^{-2} min^{-1}

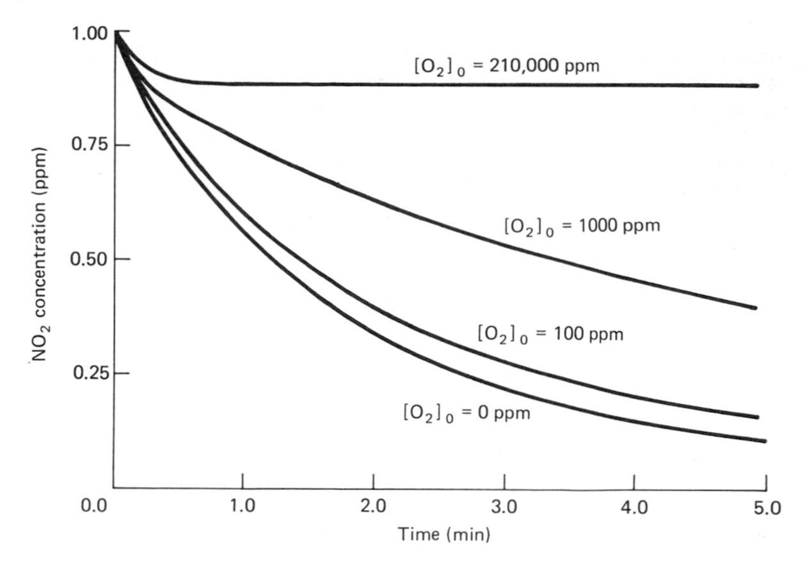

FIGURE 4.3
Nitrogen dioxide concentration as a function of time in a system initially comprising NO_2, N_2, and O_2. The curves are based on the numerical solution of (4.10) to (4.16), using the rate constants in Table 4.4 and $k_1 = 0.4\ \text{min}^{-1}$.

where the steady-state concentration of oxygen atoms is given by

$$[O] = \frac{k_1[NO_2]}{k_2[O_2][M] + k_4[NO_2] + k_5[NO_2][M] + k_7[NO][M]} \tag{4.16}$$

In order to analyze the concentration-time behavior of a system with given initial amounts of NO_2, NO, N_2, and O_2, one would solve (4.10) to (4.16) numerically, using the rate constant values in Table 4.4 as well as an estimated value of k_1, which depends on the light intensity. Figure 4.3 shows the results of such an integration for the NO_2 concentration at varying initial concentrations of O_2 with $[NO]_0 = [O_3]_0 = 0$.

Measurement of Light Intensity

Although the rate constants for the thermal reactions in Table 4.4 vary only with temperature, k_1 depends on the intensity of the radiation, either solar or artificial, used to drive the reaction. A problem of some importance is the measurement of radiation intensity and, consequently, k_1 in an experimental system. Actual physical measurement of UV intensity by various spectroradiometric techniques would be desirable from the standpoint of comparing the sources of artificial sunlight used in the laboratory with one another and with the ultraviolet portion of the solar spectrum. However, such absolute measurements require specialized apparatus which is complex and expensive. Since NO_2 is the main absorbing species in these systems and since the first-order rate constant for decay of NO_2, $k_1 = \phi k_a$, is the parameter which depends on the light intensity, it has been reasoned that a method which determines k_1 directly avoids the need for elaborate measurements of the radiation spectrum. The system of photolysis of NO_2 in N_2, first suggested by Tuesday (1961), has been used for this purpose. From measurements of the rate of decay of NO_2 with time, one would like to calculate the rate constant k_1 directly. Then, having determined the value of k_1 which characterizes the artificial lighting, one can carry out the desired experiments with other mixtures of reactant species in the same vessel[1].

In principle, we can determine the value of k_1 for any system by comparing the measured NO_2 decay as a function of time with that obtained by integration of the governing rate equations. When the measured and predicted curves can be brought into agreement, the value of k_1 employed in obtaining the predicted curves is the estimate of k_1 in the real system.

Schuck and Stephens (1969) and Holmes et al. (1973) considered whether one might be able to determine k_1 directly from measured NO_2 decay curves in an irradiated system of NO_2, NO, N_2, and O_2. If possible, this would circumvent the need to solve the reaction rate equations in a trial-and-errror fashion to find k_1. In order to proceed with the analysis we make several simplifying assumptions.

First, we neglect reactions 8 and 11. Reaction 8 is slow and at the customary levels of NO can be neglected. Similarly, the concentration of ozone attained in such a system is quite low, so that reaction 11 can be dismissed. We assume that in addition to oxygen atoms, O_3, NO_3, and N_2O_5 are in steady state. The steady-state approximation applied to these three species yields $R_2 - R_3 = 0$, $R_5 - R_6 = 0$, and $R_9 - R_{10} = 0$, respectively. Substituting these relations and that derived for oxygen atoms, namely $R_1 = R_2 + R_4 + R_5 + R_7$, into (3.10), we obtain

$$\frac{d[NO_2]}{dt} = -2k_4[O][NO_2] \tag{4.17}$$

Substituting (4.16) into (4.17) and rearranging we get

$$\frac{-2k_1}{d \ln [NO_2]/dt} = 1 + \frac{k_5[M]}{k_4} + \frac{k_7[M][NO]}{k_4[NO_2]} + \frac{k_2[M][O_2]}{k_4[NO_2]} \tag{4.18}$$

Equation (4.18) seems complex, but it can be readily interpreted. In the absence of any reactions except 1 and 4 all oxygen atoms formed would react with NO_2 to form NO and O_2, giving an overall quantum yield of 2; that is two molecules of NO_2 disappear for each photon absorbed. Thus,

$$-\frac{d \ln [NO_2]}{dt} = 2k_1 \tag{4.19}$$

This result holds if the experiment were run without N_2 at low pressure ($M = 0$). In the actual situation, however, various species compete with NO_2 for the oxygen atoms. The effect of this competition is expressed in the last three terms of (4.18). These are the ratios of oxygen atom reaction rates by other pathways to the rate of reaction 4.

We can integrate (4.18) analytically under the assumption that most of the NO_X is either NO_2 or NO:

$$[NO] = [NO]_0 + [NO_2]_0 - [NO_2] \tag{4.20}$$

[1] The assumption implicit in such laboratory experiments is that artificial radiation which drives reaction 1 at a rate comparable to that observed with natural sunlight will also allow the experimenter to simulate other processes which occur in the atmosphere. This assumption is not completely valid, since other species present can also undergo photolysis and the rates of these processes, in general, have a wavelength dependence different from that of NO_2. To a good approximation, however, the rate of NO_2 dissociation can serve as a useful measure of UV intensity.

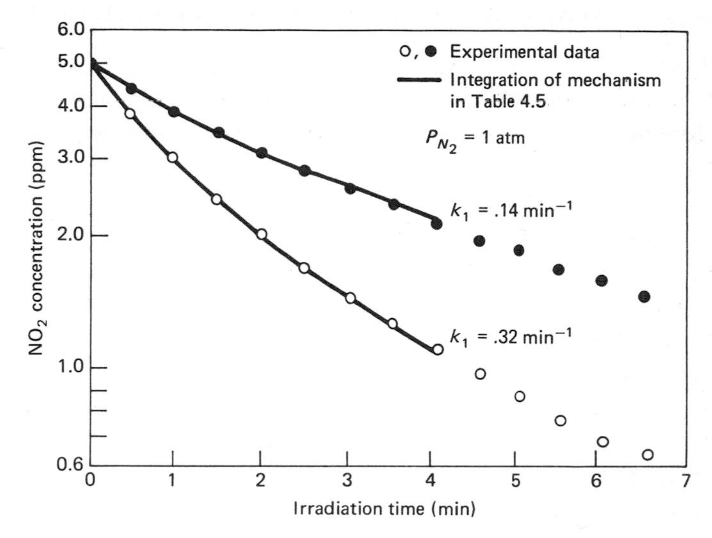

FIGURE 4.4
Nitrogen dioxide concentration as a function of time in a system initially comprising 5 ppm NO_2 in N_2. Experimental data and the predictions of the mechanism in Table 4.5 are shown for two light intensities. (*Source*: *Holmes et al.*, 1973. *Copyright* 1973 *by the American Chemical Society. Reprinted by permission of the copyright owner.*)

The result of the integration is

$$k_1 = \frac{1}{2t}\left[(1+\alpha_1-\alpha_2)\ln\frac{[NO_2]_0}{[NO_2]} + \alpha_2\left(\frac{[NO_2]_0}{[NO_2]}-1\right) + (\alpha_2[NO]_0 + \alpha_3[O_2])\frac{[NO_2]_0-[NO_2]}{[NO_2]_0[NO_2]}\right] \quad (4.21)$$

where

$$\alpha_1 = \frac{k_5[M]}{k_4} \qquad \alpha_2 = \frac{k_7[M]}{k_4} \qquad \alpha_3 = \frac{k_2[M]}{k_4}$$

Equation (4.21) may be simplified if the initial concentrations of NO and O_2 are zero:

$$k_1 = \frac{1}{2t}\left[(1+\alpha_1-\alpha_2)\ln\frac{[NO_2]_0}{[NO_2]} + \alpha_2\left(\frac{[NO_2]_0}{[NO_2]}-1\right)\right] \quad (4.22)$$

Equation (4.22) can be used to compute k_1 from the measured $[NO_2]$ versus time data in an irradiated system of NO_2 in N_2. Figure 4.4 shows NO_2 concentrations as a function of time for two different light intensities. The points represent experimental data taken in a laboratory reactor with artificial radiation. The curves in Fig. 4.4 represent (4.22) using the values of k_1 shown. Thus, the validity of (4.22) and hence the assumptions leading to (4.22) are established for an NO_2-N_2 system at ppm concentrations of NO_2.

4.2.2 Additional Reactions in the System of NO_X, H_2O, CO, and Air

As we have seen, when a mixture of NO_2 and NO is irradiated in air, a large number of "secondary" reactions can occur subsequent to NO_2 photolysis. These secondary reactions are thermal rather than photolytic in nature, since the energy contained in the reactants and that produced by their collision, rather than light energy, drives the reaction.[1]

One of the characteristics of urban atmospheres containing oxides of nitrogen, as noted, is the formation of substantial quantities of ozone. We have seen that the reactions which take place in a dry mixture of NO, NO_2, and air are insufficient to explain the occurrence of ozone at levels typically observed. The next logical step in the analysis of atmospheric reactions involving oxides of nitrogen is to consider the role of water vapor and carbon monoxide, two species virtually always present in the urban atmosphere. Let us now see if the additional reactions involving H_2O and CO that take place are capable of providing an explanation of the formation of substantial quantities of ozone.

Ozone, which forms as a product of NO_2 photolysis by the reactions

$$NO_2 + h\nu \xrightarrow{k_1} NO + O$$

$$O + O_2 + M \xrightarrow{k_2} O_3 + M$$

is destroyed rapidly in the presence of NO by the reaction

$$O_3 + NO \xrightarrow{k_3} NO_2 + O_2$$

If NO_2 and O_3 reach sufficiently high concentration levels, the reaction

$$O_3 + NO_2 \xrightarrow{k_{11}} NO_3 + O_2$$

becomes more important than the O_3-NO reaction. Any NO_3 formed by reaction 11 before the maximum concentration of NO_2 is reached will react rapidly with NO

[1] The rate constant for a thermal reaction is conventionally written in the *Arrhenius form* $k = k_0 \exp(-E_A/RT)$, where E_A is the *activation energy* [measured in kilocalories (kcal) per mole], R is the universal gas constant, and T is the temperature. From the functional dependence of k on T we see that, depending on the value of the activation energy, the rate constant for a reaction has the potential of varying enormously with temperature. Most of the inorganic and organic reactions involving air pollutants for which activation energies have been determined have values of E_A less than 5 kcal (for example, see Johnston et al., 1970). For instance, the NO-O_3 reaction, which contributes substantially to NO and O_3 loss rates and the NO_2 formation rate, has an activation energy of 2.5 kcal. Increasing the temperature from 77 to 95°F (25 to 35°C) increases the rate constant from 29.4 to 33.8 ppm^{-1} min^{-1}, a 15 percent change which, by itself, should have only moderate effect on observed atmospheric conversion rates. The activation energy of the NO_2-O_3 reaction, on the other hand, is 7.0 kcal, and a 10°C increase in temperature results in a 46 percent increase in the rate constant. The only reaction of those listed in Table 4.4 having a high activation energy is reaction 10 for which $E_A = 21$ kcal/mole.

to regenerate NO_2 by the reaction

$$NO_3 + NO \xrightarrow{k_6} 2NO_2$$

After the NO_2 peak, a time after which NO has reached low levels, NO_3 will react primarily with NO_2, presumably leading to an equilibrium concentration of N_2O_5:

$$NO_3 + NO_2 \underset{k_{10}}{\overset{k_9}{\rightleftharpoons}} N_2O_5$$

In the presence of water, however, N_2O_5 can hydrolyze to form nitric acid:

$$N_2O_5 + H_2O \xrightarrow{k_{12}} 2HNO_3$$

Nitric acid may oxidize NO by a process with the following overall stoichiometry:

$$2HNO_3 + NO \rightarrow 3NO_2 + H_2O$$

A possible reaction sequence to account for this (Gray et al., 1972) is

$$HNO_3 + NO \xrightarrow{k_{13}} HNO_2 + NO_2$$

$$HNO_3 + HNO_2 \xrightarrow{k_{14}} 2NO_2 + H_2O$$

Nitrous acid is known to form by the reaction

$$NO + NO_2 + H_2O \xrightarrow{k_{15}} 2HNO_2$$

We have seen that, upon absorption of radiation, HNO_2 photolyzes by

$$HNO_2 + h\nu \xrightarrow{k_{16}} NO + OH\cdot$$

The rate of photolysis of HNO_2 is about one-tenth that of NO_2. The importance of HNO_2 photolysis in urban atmospheres is a consequence of the high reactivity of the OH radical (particularly with hydrocarbons).[1]

[1] Although measurements of atmospheric HNO_2 concentrations have not been reported, we can estimate roughly the equilibrium concentration of HNO_2 that would be established under typical atmospheric conditions. Using equilibrium constants determined by Wayne and Yost (1951) and concentrations of NO (0.2 ppm), NO_2 (0.1 ppm), and humidity (63 percent at 25°C) which are typically observed in, say, Los Angeles just before sunrise, we estimate the equilibrium concentration of HNO_2 to be about 0.026 ppm or about one-fourth of the NO_2 concentration. At these low concentrations HNO_2 will form at a rate of about 0.10 ppm/hr. (If the concentration of NO were 0.40 ppm and NO_2 were 0.20 ppm, the rate of nitrous acid production would rise to about 0.40 ppm/hr, and the equilibrium concentration would be about 0.05 ppm.) Thus, nitrous acid may well be an important source of free radicals in a polluted atmosphere early in the morning.

Subsequent reactions involving hydroxyl radicals are

$$OH\cdot + NO_2 \xrightarrow{k_{17}} HNO_3$$

$$OH\cdot + NO \xrightarrow{k_{18}} HNO_2$$

Until recently there has been little direct evidence to substantiate the hypothesis that the hydroperoxyl radical HO_2 is an important oxidizer of NO in urban atmospheres. Wilson and Ward (1970) observed that irradiation of NO ($\sim$1 ppm) in the presence of large amounts of CO ($\sim$500 ppm) in air results in the complete oxidation of NO to NO_2 and the formation of ozone. The mechanism generally assumed to explain this observation is as follows: Water in the air reacts with NO and traces of NO_2 to form nitrous acid, which subsequently photolyzes to form OH radicals. The following reactions then occur:

$$OH\cdot + CO \xrightarrow{k_{19}} CO_2 + H\cdot$$

$$H\cdot + O_2 + M \xrightarrow{k_{20}} HO_2\cdot + M$$

$$HO_2\cdot + NO \xrightarrow{k_{21}} OH\cdot + NO_2$$

$$HO_2\cdot + HO_2\cdot \xrightarrow{k_{22}} H_2O_2 + O_2$$

$$H_2O_2 + h\nu \xrightarrow{k_{23}} 2OH\cdot$$

Reactions 19 to 21 provide a means to explain the oxidation of NO to NO_2 in a reaction system consisting of oxides of nitrogen, air, water, and carbon monoxide.

Table 4.6 REACTIONS INVOLVING H_2O AND CO IN THE NO_X–AIR SYSTEM

No.	Reaction	Rate constant at 298°K	Reference
12	$N_2O_5 + H_2O \rightarrow 2HNO_3$	2.5×10^{-3} ppm^{-1} min^{-1}	Johnston et al. (1970)
13	$HNO_3 + NO \rightarrow HNO_2 + NO_2$	2.5×10^{-4} ppm^{-1} min^{-1}	Jaffe and Ford (1967)
14	$HNO_3 + HNO_2 \rightarrow 2NO_2 + H_2O$	0.2 ppm^{-1} min^{-1}	Estimated
15	$NO + NO_2 + H_2O \rightarrow 2HNO_2$	4.3×10^{-6} ppm^{-2} min^{-1}	Wayne and Yost (1951)
16	$HNO_2 + h\nu \rightarrow NO + OH\cdot$	Depends on light intensity	
17	$OH\cdot + NO_2 \rightarrow HNO_3$	1.5×10^{4} ppm^{-1} min^{-1}	Morley and Smith (1972)
18	$OH\cdot + NO \rightarrow HNO_2$	1.2×10^{4} ppm^{-1} min^{-1}	Demerjian et al. (1973)
19	$OH\cdot + CO \rightarrow CO_2 + H\cdot$	250 ppm^{-1} min^{-1}	Demerjian et al. (1973)
20	$H\cdot + O_2 + M \rightarrow HO_2\cdot + M$	Very fast	
21	$HO_2\cdot + NO \rightarrow NO_2 + OH\cdot$	700 ppm^{-1} min^{-1}	Davis et al. (1972)
22	$HO_2\cdot + HO_2\cdot \rightarrow H_2O_2 + O_2$	5.3×10^{3} ppm^{-1} min^{-1}	Johnston et al. (1970)
23	$H_2O_2 + h\nu \rightarrow 2OH\cdot$	Depends on light intensity	

We have seen that the three basic inorganic reactions (reactions 1 to 3) are incapable alone of providing an explanation for the observed conversion of NO to NO_2 and the subsequent appearance of O_3. What reactions 19 to 21 give, in effect, is an alternate path by which NO can be converted to NO_2, without the consumption of a molecule of ozone. If we then consider (4.5) to predict roughly the prevailing ozone concentration at any ratio of NO_2 to NO, we see that when $[NO_2]/[NO]$ is approximately ten or greater, observed concentrations of ozone can be predicted. In an urban atmosphere, however, CO concentrations do not reach the levels required for reactions 19 to 21 to proceed at the rates necessary to explain the rate of oxidation of NO commonly observed. Thus, we must still look to additional reactions for an explanation of observed behavior in urban atmospheres containing oxides of nitrogen. Nevertheless, reactions 19 to 21 do provide a key to understanding the required process, that is a means by which NO is converted to NO_2 without the consumption of a molecule of ozone. We will see that reactions involving hydrocarbons offer such a means, mainly through chain reactions of free radicals derived from the hydrocarbons.

Table 4.6 summarizes rate constants for reactions 12 to 23.

4.3 REACTIONS OF HYDROCARBONS IN THE URBAN ATMOSPHERE

Although the irradiation of a system containing NO, CO, H_2O, and air results in the oxidation of NO and production of ozone, the addition of any of several organic species to the system substantially accelerates the photooxidation process. Organics that enhance the oxidation rate include olefins, aldehydes, ketones, most paraffins and aromatics, and the longer-chain acetylenes. These species enter the atmosphere in several ways:

1 Auto exhaust contains large amounts of unburned and partially burned gasoline.
2 The filling of gas tanks displaces air saturated with gasoline into the atmosphere.
3 Organic solvents used in metal working plants, dry cleaning, and as carriers for paints evaporate into the air.
4 Organic products escape to the atmosphere from chemical manufacturing plants such as petroleum refineries.

The contribution from motor vehicles is generally the highest, for example, being about 66 percent of the total organics and 86 percent of the reactive organics found in the Los Angeles atmosphere.[1]

In 1967 Altshuller et al. (1971) measured the hydrocarbon composition of the atmosphere at two locations in the Los Angeles basin. They found that atmospheric organics consist of about

53%	paraffins (exluding methane)
16%	olefins
20%	alkyl benzenes
11%	acetylene

In this section we first examine the mechanisms and products of the oxidation of each class of hydrocarbons (paraffins, olefins, aromatics, and acetylenes) by atomic oxygen, hydroxyl radicals, and ozone. We then discuss the reactions of aldehydes and ketones that can occur in the atmosphere. Finally, we describe the reactions of free radicals, the products of hydrocarbon oxidation reactions.

4.3.1 Mechanisms of Oxidation of Hydrocarbons in Air Pollution

Elucidation of the mechanisms of hydrocarbon-oxidant reactions that contribute to the formation of secondary air pollutants has proved to be an arduous task. The low concentrations of the reactants, the typically rapid rates of reaction, and the short lifetimes and low concentrations of the products of the primary oxidation have, in general, precluded definitive experimental investigations. In this section we discuss the reactions of the three most common classes of hydrocarbons (paraffins, olefins,

[1] Eccleston and Hurn (1970) measured the composition of exhaust emissions from eight automobiles in an effort to determine the effect of switching from leaded to unleaded gasolines. The average emissions from cars using regular leaded gasoline were

38.34%	paraffins
36.45%	olefins
13.34%	aromatics
1.41%	oxygenates
10.61%	acetylenes

Exhaust analysis of the cars run on an unleaded, high-olefin fuel showed a moderate increase in aromatic emissions and almost equal decrease in olefinic emissions, when compared with emissions from cars using regular leaded fuel:

35.92%	paraffins
31.34%	olefins
21.06%	aromatics
1.86%	oxygenates
10.13%	acetylenes

Thus, switching from leaded to unleaded gasoline resulted in an increase in combined olefinic and aromatic emissions and reduced paraffinic emissions.

and aromatics) with O, OH, and O_3, the oxidants thought to be the most important in the atmosphere.[1,2]

4.3.1.1 Oxygen atom oxidation reactions Oxygen atoms form as a result of NO_2 photolysis and are generally thought to be a species that initiates the reactions leading to smog formation in smog chambers. However, although oxygen atoms react rapidly with olefins, reactions of aromatics and acetylenes with oxygen atoms proceed slowly, at rates about one to two orders of magnitude slower than for olefins.

Two electronic states of the oxygen atom are important in atmospheric chemistry: the unexcited and first-excited electronic states, triplet-P oxygen atoms $O(^3P)$, and singlet-D oxygen atoms $O(^1D)$, respectively. The dissociation of NO_2 upon light absorption in the wavelength range 2900–4300 Å yields an oxygen atom in the triplet-P state. It is this form to which our attention has been focused. The singlet-D oxygen atom, which is much more reactive than the ground-state triplet-P oxygen atom, can be formed by ozone photolysis in the wavelength range 2900–3500 Å (see Table 4.2). A potentially important reaction in which $O(^1D)$ may participate is that reaction with a water molecule to form a hydroxyl radical,

$$O(^1D) + H_2O \rightarrow 2OH\cdot$$

a REACTIONS WITH PARAFFINS The reaction of atomic oxygen with paraffins probably results in the abstraction of a hydrogen atom (Leighton, 1961).

$$RH + O \longrightarrow R\cdot + OH\cdot$$

The result of the reaction is an alkyl radical and an OH radical.

[1] Acetylenes are generally less reactive than the other three classes and their reactions will not be discussed here.

[2] Two electronically excited singlet molecular oxygen species, $O_2(^1\Delta_g)$ and $O_2(\Sigma_g{}^+)$, can be produced in the atmosphere by light absorption by O_2,

$$O_2 + h\nu \rightarrow O_2(^1\Delta_g) \quad \sim 12{,}700 \text{ Å}$$
$$\rightarrow O_2(^1\Sigma_g{}^+) \quad \sim 7600 \text{ Å}$$

by electronic energy transfer from electronically excited NO_2 molecules formed by absorption of light of wavelengths greater than 4000 Å (not enough energy to dissociate NO_2),

$$NO_2{}^* + O_2 \rightarrow NO_2 + O_2(^1\Delta_g)$$

and by ozone photolysis in the wavelength range 2900–3060 Å (Kummler et al., 1969),

$$O_3 + h\nu \rightarrow O_2(^1\Delta_g) + O$$

It has been suggested that singlet molecular oxygen might be important in hydrocarbon oxidation reactions (Khan et al., 1967), although the role of these species is still uncertain.

b REACTIONS WITH OLEFINS Oxygen atoms generally add to olefins, forming an excited epoxide, which then decomposes to an alkyl and an acyl radical:

$$\mathrm{O} + \mathrm{(R_1)(R_2)C{=}C(R_3)(R_4)} \longrightarrow \mathrm{(R_1)(R_2)C\overset{O}{—}C(R_3)(R_4)} \longrightarrow$$

$$\mathrm{R_1{-}C^{\bullet}(R_2)(R_3) + R_4C^{\bullet}{=}O}$$

or

$$\mathrm{R_1C^{\bullet}{=}O + R_2{-}C^{\bullet}(R_3)(R_4)} \quad \text{, etc.}$$

This reaction thus results in the formation of two free radicals.

(*c*) REACTIONS WITH AROMATICS The mechanism for oxygen atom attack on aromatics is not yet known. Among the products that have been observed from the reaction chain initiated by oxygen atom attack on aromatics are peroxides, acids, and alcohols. The reaction of atomic oxygen with benzene is very slow. With monosubstituted aromatics atomic oxygen might attack either the alkyl side chain (as it would a paraffin) or the ring. There is little evidence to indicate which of the two mechanisms predominates, but studies of toluene and NO_X in smog chambers have led to the observation of benzaldehyde and peroxybenzoylnitrate as products (Heuss and Glasson, 1968). The fact that these products are formed indicates that the toluene-oxidant reaction does not result exclusively in chain opening.

In Table 4.7 rate constants for selected atomic oxygen–hydrocarbon reactions are summarized.

4.3.1.2 Hydroxyl radical oxidation reactions Hydroxyl radicals enter the atmospheric system as a result of HNO_2 photolysis and, as we shall see later, as products of the degradation reactions of free radicals. Reactions of OH with hydrocarbons are very similar to those of atomic oxygen, with two exceptions:

1 The reactions of OH with a given hydrocarbon are generally very much faster than those of atomic oxygen.

Table 4.7 RATE CONSTANTS OF SELECTED ATOMIC OXYGEN–HYDROCARBON REACTIONS

Hydrocarbon	Rate constant at 298°K, $ppm^{-1}\ min^{-1}$†
Olefins	
Ethylene $H_2C{=}CH_2$	1190
Propylene $H_2C{=}CHCH_3$	6810
1-Butene $H_2C{=}CHCH_2CH_3$	6810
Isobutene $(H_3C)_2C{=}CH_2$	3.0×10^4
Cyclopentene	2.35×10^4
Cyclohexene	2.06×10^4
Alkanes	
Methane CH_4	0.0176
Ethane C_2H_6	1.37
Propane C_3H_8	12.3
n-Butane C_4H_{10}	32.4
n-Pentane C_5H_{12}	85.5
Cyclopentane	147
Cyclohexane	177
Aromatics	
Benzene	30.8
Toluene (—CH_3)	107
Aldehydes	
Formaldehyde HCHO	220
Acetaldehyde CH_3CHO	544

† A complete survey of rate constant values and references as of 1970 is given by Johnston et al. (1970). Values presented here are representative measurements.

2 Hydrogen abstraction reactions do not result in chain branching (more than one free radical) since the OH becomes H_2O, whereas atomic oxygen becomes OH.

a REACTIONS WITH PARAFFINS The paraffin-OH reaction results in hydrogen abstraction, an alkyl radical and water being formed:

$$RH + OH\cdot \longrightarrow R\cdot + H_2O$$

The rate of reaction of an individual paraffin with OH generally increases with the number of hydrogen atoms on the molecule, especially secondary and tertiary hydrogens.

b REACTION WITH OLEFINS Hydroxyl radicals react with olefins by addition at the double bonds. Morris et al. (1971) observed OH adducts directly in the reactions of OH with C_2H_4 and C_3H_6.

$$OH\cdot + CH_3CH{=}CH_2 \longrightarrow CH_3\dot{C}HCH_2OH \quad \text{or} \quad CH_3\underset{\displaystyle |\atop \displaystyle OH}{C}H\dot{C}H_2$$

Recently, the rate constants for OH attack on several olefins were measured by Morris and Niki (1971). Their results indicate that, for a given olefin, the rate constant for the OH-olefin reaction is about 10 times greater than for the corresponding atomic oxygen–olefin reaction.

c REACTIONS WITH AROMATICS It is current speculation that hydroxyl radicals abstract alpha hydrogen atoms from branched aromatics in the same manner as we have assumed for atomic oxygen:

$$C_6H_5CH_2CH_3 + OH\cdot \longrightarrow C_6H_5\dot{C}HCH_3 + H_2O$$

Rate constants for selected hydroxyl radical–hydrocarbon reactions are presented in Table 4.8.

4.3.1.3 Ozone oxidation reactions Ozone begins to form in significant amounts in the atmosphere when the NO_2 concentration reaches a level about 25 times that of the NO concentration. Ozone is not nearly as strong an oxidizing agent as O or OH. However, concentrations of ozone of about 0.25 ppm or greater are not uncommon in polluted atmospheres; at these concentrations ozone and olefins react at appreciable rates.

a REACTIONS WITH PARAFFINS AND AROMATICS Ozone does not react at a significant rate with either paraffins or aromatics. Several paraffin-ozone rate constants have been determined; these have been summarized by Peters and Wingard (1970).

b REACTIONS WITH OLEFINS The reactions of ozone with olefins in the gas phase have been studied extensively during the past two decades, and rate constants for the ozonolysis reaction are available for a large number of olefins (see Table 4.10). However, the mechanism of the olefin-ozone reaction in the gas phase has not yet been established, and the initial decomposition products have not yet been conclusively identified. Although the exact mechanism is still unresolved, in the liquid phase ozone is thought to add to the olefin double bond to form a molozonide intermediate, the molozonide then decomposing into an aldehyde and a diradical (or *zwitterion*). For propylene, for example, this reaction is

$$C_3H_6 + O_3 \longrightarrow \underset{\underset{\diagdown O \diagup}{O \quad\; O}}{CH_3HC{-}{-}CH_2} \longrightarrow \begin{cases} HCHO + CH_3\dot{C}H{-}OO\cdot \\ \\ CH_3CHO + H_2\dot{C}OO\cdot \end{cases}$$

Table 4.8 RATE CONSTANTS OF SELECTED HYDROXYL RADICAL-HYDROCARBON REACTIONS

Hydrocarbon	Rate constant at 298°K, $ppm^{-1}\ min^{-1}$†
Alkanes	
Methane	16.5
Ethane	443
Propane	1800
n-Butanc	5700
Cyclohexane	1.2×10^4
Olefins	
Ethylene	7550
Propylene	2.5×10^4‡
Acetylene	1470
Aldehydes	
Formaldehyde	2.35×10^4‡
Acetaldehyde	2.35×10^4‡

† A complete survey of rate constant values and references for the OH-alkane reactions as of 1970 is given by Johnston et al. (1970). Values presented here are representative measurements.

‡ A survey of rate constants for OH reactions with propylene, formaldehyde, and acetaldehyde is presented by Niki et al. (1972).

Ozonolysis in the gas phase may proceed by this same mechanism or by allylic oxidation, which would lead to the same decomposition products:

$$C_3H_6 + O_3 \longrightarrow \underset{\underset{\quad\;+\;\;-}{O\text{—}O\text{—}O}}{\overset{}{CH_3CH\text{—}CH_2}} \longrightarrow \begin{cases} HCHO + CH_3\dot{C}H \\ \qquad\qquad\quad\;\; | \\ \qquad\qquad\quad\;\; OO\cdot \\ \\ CH_3CHO + H_2\dot{C}OO\cdot \end{cases}$$

Table 4.9 MECHANISMS OF OZONE REACTIONS WITH ETHYLENE, PROPYLENE, AND ISOBUTYLENE

Olefin	Mechanism
Ethylene	$C_2H_4 + O_3 \longrightarrow HCHO + \overset{\overset{O\nearrow O}{\|}}{C}H_2$
Propylene	$C_3H_6 + O_3 \longrightarrow HCHO + CH_3\overset{\overset{O\nearrow O}{\|}}{C}H$
	$CH_3\overset{\overset{O\nearrow O}{\|}}{C}H \longrightarrow CH_3\overset{\overset{O}{\|}}{C}\cdot + OH\cdot$
Isobutylene	$C_4H_8 + O_3 \longrightarrow HCHO + CH_3\overset{\overset{O\nearrow O}{\|}}{C}CH_3$
	$C_4H_8 + O_3 \longrightarrow CH_3\overset{\overset{O}{\|}}{C}CH_3 + \overset{\overset{O\nearrow O}{\|}}{C}H_2$
	$CH_3\overset{\overset{O\nearrow O}{\|}}{C}CH_3 \xrightarrow{O_2} (CH_3)_2CO + O_3$
	$CH_3\overset{\overset{O\nearrow O}{\|}}{C}CH_3 \longrightarrow CH_3\overset{\overset{O}{\|}}{C}\cdot + CH_3O\cdot$
	$\overset{\overset{O\nearrow O}{\|}}{C}H_2 + O_2 \longrightarrow HCHO + O_3$

Mechanisms of ozone reactions with ethylene, propylene, and isobutylene are summarized in Table 4.9, and rate constants for ozone-olefin reactions are given in Table 4.10.

The products of ozone attack on olefins are thus an aldehyde and an unstable intermediate called a zwitterion. Possible decomposition reactions of the zwitterion are

$$HRCOO \longrightarrow \begin{cases} RO\cdot + H\dot{C}O \\ R\dot{C}O + OH\cdot \\ ROH + CO \end{cases}$$

In addition, the zwitterion might participate in reactions with O_2, NO, and NO_2, for example,

$$HRCOO + NO \longrightarrow RCHO + NO_2$$

4.3.2 Mechanisms of Oxidation of Oxygenated Hydrocarbons in Air Pollution

Seizinger and Dimitriades (1972) have shown that aldehydes and ketones are present in the exhaust of automobile engines fueled by simple hydrocarbons. As indicated earlier, oxygenates constitute only about 1.5 percent of the hydrocarbons found in auto exhaust. Thus, the most important source of atmospheric oxygenates may be the oxidation of hydrocarbons and the decomposition of free radicals in the atmosphere. Of the oxygenates produced in this manner, aldehydes seem to form more

Table 4.10 RATE CONSTANTS OF SELECTED OZONE–OLEFIN REACTIONS

Olefin	Rate constant at 298°K, $ppm^{-1}\ min^{-1}$†
Ethylene	2.96×10^{-3}
Propylene	0.016
1-Butene	0.0147
Isobutene	0.034
Cyclohexene	0.044
Butadiene $H_2C{=}CHCH{=}CH_2$	0.0134
2,3-Dimethyl-2-butene	1.1

† A complete survey of rate constant values and references as of 1970 is given by Johnston et al. (1970). Values presented here are representative measurements.

readily than ketones.[1] In this section, therefore, we focus our attention on the reactions of aldehydes.

4.3.2.1 Photolysis Aldehydes photodissociate in sunlight at wavelengths greater than 3000 Å in chain-initiating reactions:

$$RCHO + h\nu \longrightarrow R\cdot + H\dot{C}O$$

Leighton (1961) estimated that the rate of this photolysis reaction is about one one-hundredth the rate of NO_2 photodissociation. In the case of formaldehyde, a second primary photodissociation is possible (Calvert et al., 1972):

$$HCHO + h\nu \longrightarrow H_2 + CO$$

This reaction pathway, which does not directly result in the formation of free radicals, has a probability of occurrence about equal to that for the chain-initiating reaction.

4.3.2.2 Oxygen atom oxidation reactions The typical rate constant for the atomic oxygen–aldehyde reactions falls between those of the atomic oxygen–olefin and atomic oxygen–paraffin reactions. The reaction is

$$O + RCHO \longrightarrow \underset{\underset{O}{\|}}{RC\cdot} + OH\cdot$$

resulting in the formation of an acyl radical.

4.3.2.3 OH oxidation reactions Hydroxyl radicals abstract a hydrogen atom from aldehydes, forming an acyl radical and water:

$$OH\cdot + RCHO \longrightarrow \underset{\underset{O}{\|}}{RC\cdot} + H_2O$$

The chain transfer reaction takes place about as rapidly as the reaction of OH with propylene. The rate constants for the reaction of OH with formaldehyde and acetaldehyde have been determined by Morris and Niki (1971). They are both about

[1] Ketones are produced only when reaction occurs at a carbon atom bonded to at least two other carbon atoms in the molecule being oxidized, e.g.

$$O_3 + \overset{H}{\underset{R_1}{>}}C{=}C\overset{R_2}{\underset{R_3}{<}} \longrightarrow \underset{\underset{O\!-\!O\cdot}{|}}{R_1C\cdot} + \underset{\underset{O}{\|}}{R_2CR_3} \quad \text{or} \quad \underset{\underset{O}{\|}}{R_1CH} + R_2\!-\!\overset{+}{\dot{C}}(R_3)\!-\!OO\cdot$$

Leighton (1961) supported this view, concluding that the atmospheric processes that result in the formation of aldehydes apparently do not generate ketones in comparable amounts.

23,000 ppm^{-1} min^{-1} as compared with 25,000 ppm^{-1} min^{-1} for the propylene-OH reactions. Thus, the reactions of OH radicals with aldehydes may serve as an important mechanism for removal of aldehydes from the atmosphere.

4.3.3 Free-radical Reactions

As we have just seen, the hydrocarbon oxidation reactions yield the following classes of free radicals:

$R\cdot$	alkyl radical
$R\dot{C}O$	acyl radical
$RO\cdot$	alkoxyl radical (including $OH\cdot$)

We will now consider the reactions of these species, as well as others that may appear as a result, in the atmosphere. Some of these radicals, notably alkyl and acyl, are extremely reactive, combining immediately with molecular oxygen to form peroxy-radicals:[1]

$ROO\cdot$	peroxyalkyl radical (including $HO_2\cdot$)
$R\overset{\overset{O}{\parallel}}{C}OO\cdot$	peroxyacyl radical

Subsequent reactions of the peroxyacyl radical produce

$$R\overset{\overset{O}{\parallel}}{C}O\cdot \quad \text{acylate radical}$$

These six radical classes enter into reactions of interest with O_2, NO, NO_2, and each other. Unfortunately, the rate constants for each class of reactions are largely unknown. A detailed study of atmospheric free-radical reactions is beyond our scope. Thus, in this section we shall merely summarize the most likely reactions that free radicals within each of the six classes might undergo in the atmosphere.

[1] If the alkyl radical is a product of highly exothermic reaction, it is possible that the O_2 might abstract a hydrogen at a position alpha to the free radical, forming an olefin rather than adding at the site, e.g.

$$CH_3CH_2\dot{C}HCH_3 + O_2 \longrightarrow CH_3CH_2CH{=}CH_2 + HO_2\cdot$$

or

$$CH_3CH{=}CHCH_3 + HO_2\cdot$$

This reaction is observed in high-temperature flames (that is, 425°C) (Benson, 1968). At ambient temperatures, however, probably no more than 1 percent of the $R\cdot$-O_2 reactions proceed in this manner, the remaining 99 percent resulting in the formation of the conventional peroxyalkyl radicals. Even so, this type of reaction may be a source of olefins in the oxidation of paraffinic systems and, as a consequence, merits further experimental investigation.

4.3.3.1 Reactions with O_2 As mentioned, both R· and R$\dot{C}$O react rapidly with O_2 to yield their peroxy counterparts:

$$R\cdot + O_2 \longrightarrow ROO\cdot$$

$$R\dot{C}O + O_2 \longrightarrow \underset{\displaystyle}{R\overset{\displaystyle O}{\overset{\|}{C}}OO\cdot}$$

In addition, alkoxy radicals also react with O_2 to give an aldehyde and a hydroperoxyl radical:

$$RO\cdot + O_2 \rightarrow R'CHO + HO_2\cdot$$

The rate constant for this reaction has been determined by Heicklen (1968) as 4.4×10^3 ppm^{-1} min^{-1}, where R is CH_3 or C_2H_5. Reactions of other radicals with O_2 are generally less important than these three.

4.3.3.2 Reactions with NO The primary mechanism by which NO is thought to be oxidized to NO_2 without the consumption of ozone is through reaction with peroxy radicals:

$$NO + \begin{cases} ROO\cdot \\ R\overset{\displaystyle O}{\overset{\|}{C}}OO\cdot \end{cases} \longrightarrow NO_2 + \begin{cases} RO\cdot \\ R\overset{\displaystyle O}{\overset{\|}{C}}O\cdot \end{cases}$$

The rate of this type of reaction probably decreases as the size of the R group increases, because large R groups have a greater number of vibrational degrees of freedom over which to distribute the energy of the free radical. Demerjian et al. (1973) estimated that the rate constants for the CH_3O_2-NO and $CH_3\underset{\displaystyle O}{\underset{\|}{C}}O_2$-NO reactions are 910 and 470 ppm^{-1} min^{-1}, respectively. [Spicer et al. (1971), however, found the CH_3O_2-NO reaction to be quite slow.] Nitric oxide may also react with RO· to give alkyl nitrates (recall reaction 18 in Table 4.6):

$$RO\cdot + NO \longrightarrow RONO$$

4.3.3.3 Reactions with NO_2 As NO is oxidized and NO_2 accumulates during smog formation, the reaction of peroxy radicals with NO_2 becomes increasingly important. If the peroxy radical is an acyl peroxy radical, stable products, peroxyacyl nitrates (PAN), will form:

$$R\underset{\displaystyle O}{\underset{\|}{C}}OO\cdot + NO_2 \longrightarrow R\underset{\displaystyle O}{\underset{\|}{C}}OONO_2$$

Formylperoxynitrates, however, are thought to be extremely unstable, if, indeed, they form at all; they have never been observed in smog chamber studies. Similarly, alkylperoxynitrates, $ROONO_2$, formed by

$$ROO\cdot + NO_2 \longrightarrow ROONO_2$$

have yet to be observed in laboratory smog chambers, and it is not known whether or not these species are stable. Nitrogen dioxide may also react with $RO\cdot$ to give alkyl nitrates (recall reaction 17 in Table 4.6):

$$RO\cdot + NO_2 \longrightarrow RONO_2$$

4.3.3.4 Radical–radical reactions The population of peroxy radicals in the smog system is limited by radical-radical recombination reactions. These reactions may be direct recombinations to form a peroxide and oxygen,

$$ROO\cdot + ROO\cdot \longrightarrow ROOR + O_2$$

or disproportionation to form an aldehyde (or ketone), an alcohol, and oxygen:

$$ROO\cdot + ROO\cdot \longrightarrow R'\overset{\displaystyle O}{\overset{\|}{C}}H + R''CH_2OH + O_2$$

4.3.3.5 Radical decomposition reactions The acylate radical is quite unstable and loses CO_2, forming an alkyl radical:

$$R\overset{\displaystyle O}{\overset{\|}{C}}O\cdot \longrightarrow R\cdot + CO_2$$

4.4 PHOTOCHEMICAL SMOG

Photochemical smog is the designation given to the particular mixture of reactants and products that exists when hydrocarbons and oxides of nitrogen occur together in an urban atmosphere in the presence of sunlight. Hydrocarbons usually occur with oxides of nitrogen in an urban atmosphere (since a major fraction of each results from the same type of source, namely motor vehicles). The understanding of photochemical smog therefore involves a consideration of the reactions which occur when both nitrogen oxides and hydrocarbons are present in air. Having discussed these reactions in some detail in Secs. 4.2 and 4.3, we are now in a position to consider photochemical smog.

The formation of photochemical smog takes place in an extremely complicated system, one in which meterology, continuous emissions of pollutants, and chemical reactions all play important roles. In order to isolate the effects of chemistry from other variables in an atmospheric environment, investigators have simulated the

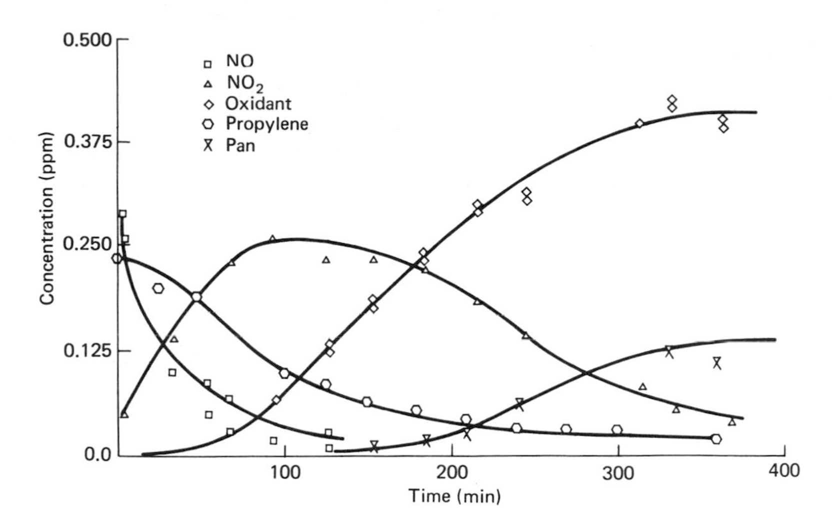

FIGURE 4.5
Experimental data on the photolysis of an initial mixture of 0.25 ppm propylene, 0.26 ppm NO, and 0.05 ppm NO_2 in air. Experiment conducted by S. L. Kopczynski of the U.S. Environmental Protection Agency.

chemical reaction processes by irradiating the primary pollutants (NO_X and hydrocarbons) in laboratory reactors, called *smog chambers*. Both in the laboratory and in the atmosphere ultraviolet irradiation of air containing ppm concentrations of hydrocarbons and oxides of nitrogen leads to (1) oxidation of NO to NO_2, (2) oxidation of hydrocarbons, and (3) formation of ozone.

The general features of the concentration-time behavior of a hydrocarbon-NO_X-air system in an irradiated laboratory reactor are shown in Fig. 4.5. Although the experimental data shown in Fig. 4.5 involve only the single hydrocarbon propylene (C_3H_6), propylene photooxidation in the presence of NO_X manifests the major characteristics of photochemical smog and has been used extensively in laboratory studies. Soon after the irradiation begins, NO is oxidized to NO_2. (Recall that most of the NO_X is emitted initially as NO.) The concentration of NO reaches a low value which persists throughout the latter stages of the reaction. The concentration of NO_2 reaches a maximum when its rate of formation (from NO) is just balanced by depletion reactions to form nitric acid and organic nitrates. At about the time the NO_2 concentration reaches a maximum, measurable ozone formation begins to take place. The propylene is continually converted to organic products.

Figure 4.6 shows the concentrations of NO, NO_2, hydrocarbons, and ozone measured at the Los Angeles County Air Pollution Control District station in down-

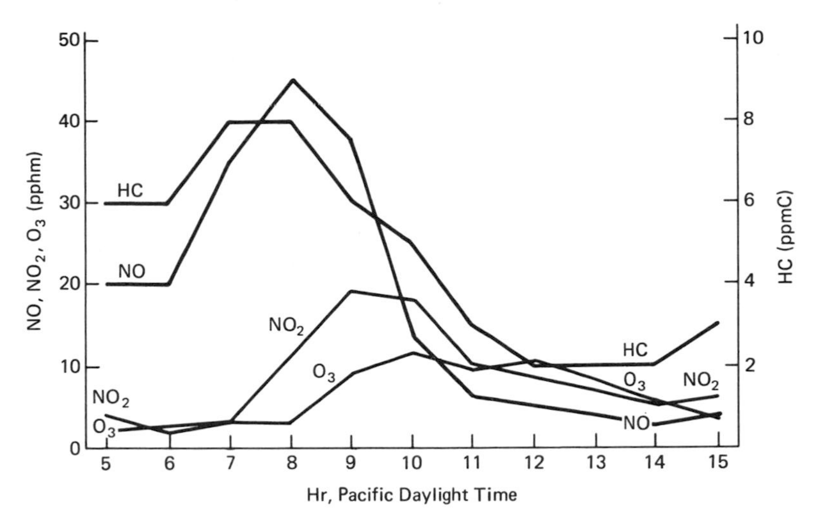

FIGURE 4.6
Measured concentrations of total hydrocarbons, NO, NO_2, and O_3 at the downtown Los Angeles monitoring station of the Los Angeles County Air Pollution Control District on Sept. 29, 1969.

town Los Angeles on Sept. 29, 1969. Although the behavior shown is the result of not only chemistry but also meteorology and continuous source emissions, the resemblance of Fig. 4.6 to Fig. 4.5, in which only chemical reaction effects appear, is obvious. At 6 A.M., hydrocarbons and NO begin increasing because of local automobile traffic. The NO_2 maximum is reached by 9 A.M. The maximum ozone concentration in downtown Los Angeles on that day was about 0.12 ppm, reached at 10 A.M.

Ozone has traditionally been used as the measure of the severity of a photochemical smog episode. An important question from the standpoint of abatement strategy is the effect of different initial concentrations of hydrocarbons and NO_X on the maximum concentration (or the dosage) of ozone achieved over a certain period of irradiation. Presumably the best abatement strategy would be to reduce source emissions of hydrocarbons and NO_X in such a way that initial conditions in the atmosphere (in early morning) favor minimum ozone formation. In the study of the effect of initial concentrations on ozone formation it has been observed that the peak amount of ozone produced over a fixed time of irradiation increases rapidly, goes through a maximum, and finally decreases as one performs a series of controlled experiments in which the initial hydrocarbon concentration is held constant and the initial concentration of NO is increased. This phenomenon is illustrated in Fig. 4.7.

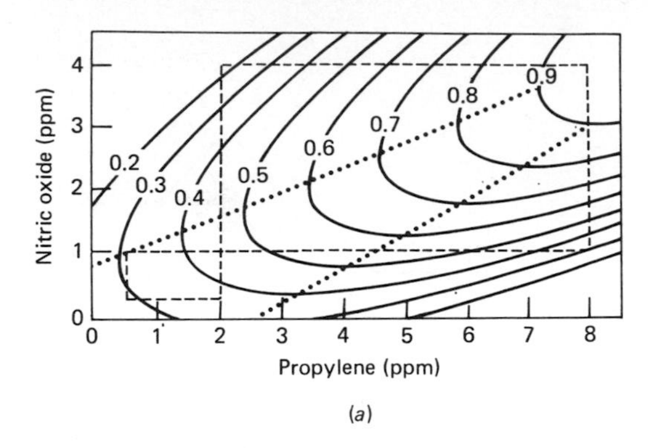
Nitric oxide (ppm)
Propylene (ppm)
0.2
0.3
0.4
0.5
0.6
0.7
0.8
0.9

(a)

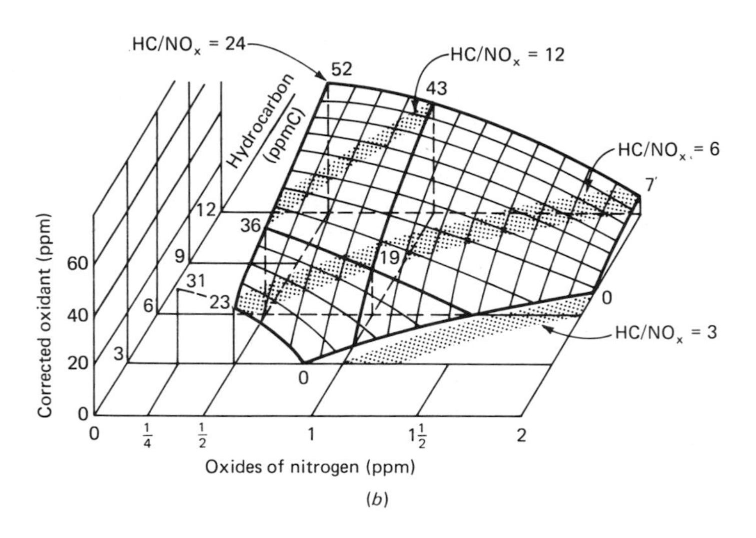
HC/NO$_x$ = 24
HC/NO$_x$ = 12
HC/NO$_x$ = 6
HC/NO$_x$ = 3
52
43
36
31
23
19
7
0
Hydrocarbon (ppmC)
Corrected oxidant (ppm)
Oxides of nitrogen (ppm)

(b)

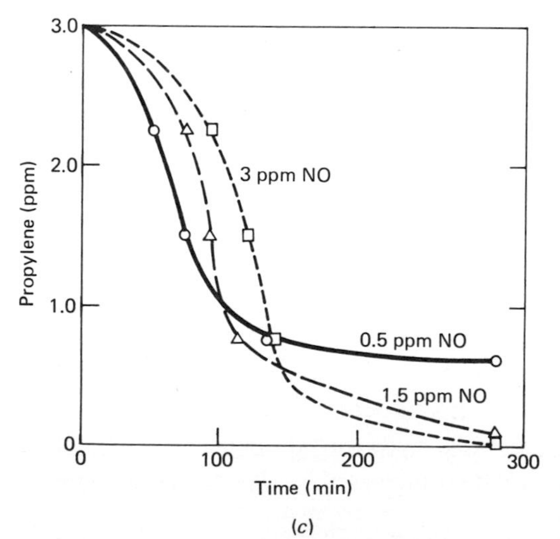
Propylene (ppm)
Time (min)
3 ppm NO
0.5 ppm NO
1.5 ppm NO

(c)

Figure 4.7*a* shows the effect of varying initial concentrations of NO and propylene on the maximum oxidant (primarily ozone) concentration achieved over a fixed time of irradiation. Combined reduction of both initial propylene and NO concentrations reduces the oxidant yield, while reduction of either propylene or NO initial concentration, holding the other fixed, may increase or decrease oxidant yield. The dotted lines connect the NO and propylene minima. Within the region bounded by these lines, oxidant reduction is obtained from both initial NO and propylene reductions. The rectangle in the lower left-hand corner indicates the region of atmospheric levels. The results shown in Fig. 4.7*a* suggest an inhibitory effect of NO on the maximum amount of ozone produced at a fixed hydrocarbon level. Results such as these have led to the suggestion that control of NO_X emissions in areas with photochemical smog might actually lead to increased ozone levels.

Although the behavior shown in Fig. 4.7*a* has been verified in many studies [see, for example, Fig. 4.7*b*, due to Korth et al. (1964)], there is evidence that the inhibitory effect of NO on ozone formation is not a true inhibition but only a delay. Results of the type shown in Fig. 4.7*a* are generally based on irradiation experiments of a few hours duration. When longer-period irradiations are carried out (5 hr or more), there is evidence that the inhibitory effect of initial NO concentrations is no longer observed. Irradiations of this duration are, of course, more relevant for atmospheric simulations (an urban area is exposed to sunlight for about 12 hr).

Figure 4.7*c* represents a set of three experiments at the same temperature and light intensity in a smog chamber. In each case the initial concentration of propylene was 3 ppm, whereas the three initial concentrations of NO_X were 0.5, 1.5, and 3.0 ppm. During the first hour the fastest rate of propylene disappearance occurred with the least NO_X (0.5 ppm), and the slowest rate occurred with the most NO_X (3 ppm). However, after 2 hr the rate of propylene oxidation decreased significantly in the $[NO_X]_0 = 0.5$ ppm case, whereas that in the 3 ppm case continued at a finite rate until the virtual disappearance of the propylene at the end of 5 hr. If Fig. 4.7*c* is replotted to show the total propylene consumed at the end of 1, 2, and, 3 hr as a function of $[NO_X]_0$, the 1-hr curve shows that increasing $[NO_X]_0$ leads to a decrease

FIGURE 4.7
Effect of initial reactant ratios on ozone concentrations in photochemical smog. (*a*) Effect of varying initial concentrations of propylene and NO on the maximum oxidant concentration achieved over a fixed time of irradiation. (*Source: Romanovsky et al.*, 1967.) (*b*) Final oxidant concentrations as a function of initial NO and hydrocarbon concentrations for automobile exhaust. (*Source: Korth et al.*, (1964.) (*c*) Propylene concentrations as a function of time for three smog chamber experiments with $[C_3H_6]_0 = 3$ ppm and $[NO_X]_0 =$ 0.5, 1.5, and 3.0 ppm. (*Source: Johnston*, 1973.)

in the propylene oxidized. However, the 3-hr curve shows quite the opposite behavior, with the least amount of hydrocarbon consumed for $[NO_X]_0 = 0.5$ ppm.

In this section our objective is to explain on a chemical basis the observed behavior of the photochemical smog system. We have already discussed all the relevant chemical reactions in Secs. 4.2 and 4.3, so that here we wish to see what the combined effect of these reactions is and, in fact, if these reactions can provide an explanation for the behavior exhibited in Figs. 4.5 to 4.7. Although SO_2 may also be present in photochemical smog, it is not a "necessary" constituent. For this reason we do not consider the role of SO_2 in photochemical smog in this section; SO_2 chemistry will be discussed in the next section.

4.4.1 Qualitative Kinetic Mechanism for Photochemical Smog

Based on Secs. 4.3 and 4.3, a framework can be provided for the discussion of kinetic mechanisms with the highly generalized kinetic scheme given below:

$$NO_2 + h\nu \xrightarrow{1} NO + O$$

$$O + O_2 + M \xrightarrow{2} O_3 + M$$

$$O_3 + NO \xrightarrow{3} NO_2 + O_2$$

$$O + \text{hydrocarbons} \xrightarrow{4} \text{stable products} + \text{radicals}$$

$$O_3 + \text{hydrocarbons} \xrightarrow{5} \text{stable products} + \text{radicals}$$

$$\text{Radicals} + \text{hydrocarbons} \xrightarrow{6} \text{stable products} + \text{radicals}$$

$$\text{Radicals} + NO \xrightarrow{7} \text{radicals} + NO_2$$

$$\text{Radicals} + NO_2 \xrightarrow{8} \text{stable products}$$

$$\text{Radicals} + \text{radicals} \xrightarrow{9} \text{stable products}$$

We are not proposing this mechanism as one for quantitative prediction but rather only for discussion of the qualitative features of the overall photochemical smog system. We have chosen the three most important inorganic reactions (reactions 1 to 3), the three types of hydrocarbon oxidation reactions (reactions 4 to 6), and three types of free-radical reactions (reactions 7 to 9).

The photochemical smog system consists of chain reactions, wherein free radicals once produced may participate in many reactions prior to extinction. In the terminology of chain-reaction processes, the reactions in the smog system can be classified as one of four types:

1 Initiation reactions: those that provide free radicals to induce chain reactions, for example $NO_2 + h\nu \rightarrow NO + O$†
2 Branching reactions: those in which there is a net increase in radical species, for example, $O + HC \rightarrow$ products + radicals, where more than one radical results
3 Propagation reactions: those in which there is no net gain or loss of radicals but merely a change in identity, e.g., radical + HC → product + radical
4 Termination reactions: those that remove free radicals through the formation of stable end products, e.g., reactions 8 and 9.

Let us now examine the behavior of this highly generalized mechanism to see if it is capable of predicting the qualitative features of the photochemical smog system, in particular the rapid rate of conversion of NO to NO_2 and the effect of varying initial reactant concentrations on ozone formation. In the discussion that follows, we denote the total concentration of radicals as [R] and that of hydrocarbons as [HC]. We can write a material balance on the total concentration of free radicals in the following manner:

$$\frac{d[R]}{dt} = a[NO_2] + b[HC][R] - c[NO_2][R] - d[R]^2$$

where the terms on the right-hand side have the following significance:

$a[NO_2]$ = production of oxygen atoms by reaction 1
$b[HC][R]$ = net rate of production of radicals from reactions 4–6
$c[NO_2][R]$ = net depletion of radicals by reaction 8
$d[R]^2$ = net depletion of radicals by reaction 9

The symbols a, b, c, d represent the combination of appropriate rate constants and stoichiometric coefficients.

We can identify two limiting cases of interest:

1 Near $t = 0$ when $[R] = 0$, that is,

$$\frac{d[R]}{dt} = a[NO_2]$$

From this we can see that initially the concentration of free radicals increases at a rate proportional to the NO_2 concentration. Presumably, most of these radicals are oxygen atoms.

† This reaction is not the only initiation reaction that occurs in the generation of smog. Photolysis of nitrous acid, alkyl nitrites, and aldehydes also provides free radicals. For simplicity, and because NO_2 photolysis is the most important initiation reaction, we have omitted others in this generalized mechanism.

2 When the total radical population reaches a steady state (ss), $d[R]/dt = 0$. At this point

$$O = a[NO_2] + b[HC][R] - c[NO_2][R] - d[R]^2$$

For case 2 we can examine the effect on $[R]_{ss}$ of the relative concentrations of NO_2 and HC. When $[NO_2] \gg [HC]$,

$$O \cong a[NO_2] - c[NO_2][R]_{ss}$$

or $[R]_{ss}$ is a constant a/c, dependent only on the photolysis rate of NO_2 and the rate of termination with NO_2. When the rate of production by chain branching, $b[HC][R]$, equals the termination rate with NO_2, $c[NO_2][R]$, the steady-state radical concentration is

$$[R]_{ss} = \left(\frac{a[NO_2]}{d}\right)^{1/2}$$

Finally, when $[HC] \gg [NO_2]$,

$$O \cong b[HC][R] - d[R]^2$$

so that $[R]_{ss} = b[HC]/d$.

Summarizing, at steady state the limits of total free-radical concentrations are

$$\underset{\text{(high } NO_2)}{\frac{a}{c}} \leq [R]_{ss} \leq \underset{\text{(low } NO_2)}{\frac{b[HC]}{d}}$$

Using these relationships, Johnston et al. (1970) estimated that actual concentrations may be

$\frac{c[NO_2]}{b[HC]}$	10	1	0.1
$[R]_{ss}$, ppm	1.1×10^{-4}	2×10^{-3}	3.6×10^{-2}

Leighton (1961) estimated that in photooxidation of an olefin such as propylene NO is converted to NO_2 at a rate of 0.04 ppm min^{-1} by type 7 reactions. This rate is consistent with the steady-state radical concentrations just calculated. For example, using $[R] = 10^{-3}$ ppm, $[NO] = 0.1$ ppm, and a rate constant for reaction 7 of 500 ppm^{-1} min^{-1}, the rate of conversion of NO to NO_2 is 0.05 ppm min^{-1}.

Aside from predicting the rapid conversion of NO to NO_2, a mechanism must also be capable of simulating the effect of different initial reactant ratios on product formation in the smog system. Let us consider the two cases of low and high NO_2 concentrations. The photochemical smog process is initiated by NO_2, with the rate of increase of the total radical population depending on the initial concentration of NO_2. The maximum concentration of NO_2 occurs when the rate of conversion of NO to NO_2 just equals the rate of termination of radicals by NO_2. If this maximum concentration of NO_2 is low (i.e. low total initial NO_X), the concentration of radicals reaches very high values. The radicals rapidly convert NO to NO_2, allowing ozone to form. As the total initial NO_X, and hence the NO_2 concentration at its maximum, is increased, the amount of ozone formed should increase up to a point. That point is the one at which radical removal by NO_2 is at least as fast as chain branching, that is, $b[HC] \cong c[NO_2]$. As the amount of initial NO_X, and hence the NO_2, continues to increase the steady-state radical concentration continues to decrease, as we have seen. The result is that the rate of ozone formation slows down.

We can also examine the effect of initial reactant ratios on the maximum amount of ozone formed during an irradiation. We consider first an experiment in which the concentration of initial NO_X ($NO + NO_2$) is large compared with initial HC. Note that the relative amounts of NO and NO_2 serve only to govern the time needed to convert the NO to NO_2 and should not affect the ultimate maximum ozone concentration attained. In this case the hydrocarbon is expended before all the NO is converted to NO_2. Since an appreciable amount of NO remains after the hydrocarbon is depleted, no significant ozone concentration can be reached. We consider next the opposite limit, namely $[NO_X]_0 \ll [HC]_0$. Now the NO is rapidly converted to NO_2 with little expenditure of HC. As [NO] becomes small, ozone can accumulate. However, because of the large concentration of HC, the reaction of O_3 and HC will prevent $[O_3]$ from becoming too large. Finally, the intermediate case of $[HC]_0 \cong [NO_X]_0$ can be expected to yield the largest ozone concentration. Thus, if one considers the maximum ozone concentration achieved, over a fixed time of irradiation, as a function of $[HC]_0/[NO_X]_0$, one gets a curve which exhibits a maximum. These effects have been illustrated in Fig. 4.7.

We have seen that the simplified mechanism presented early in this section can account for several of the qualitative features of the photochemical smog system. However, for quantitative discussion of concentration-time behavior in smog chamber systems and in the atmosphere we need to include the reactions cited in Secs. 4.2 and 4.3. Therefore, we shall now expand somewhat on the simple, qualitative mechanism presented at the beginning of this section. In particular, we shall consider the role of free radicals produced by hydrocarbon and aldehyde oxidations. Our objective is to explain the rapid conversion of NO to NO_2 and the subsequent formation of ozone in photochemical smog.

In the photochemical smog system the large number of hydrocarbons and free radicals present makes it desirable to consider, as we have done in Sec. 4.3, reactions by species class rather than by individual compound. In order to represent the important organic reactions in photochemical smog as concisely as possible, we consider the four classes of organics: olefins, aromatics, paraffins, and aldehydes. Likely products from reactions of these species with O, O_3, and OH are

$$\text{Olefin} + \begin{cases} O & \longrightarrow & R\cdot + R\dot{C}O \\ O_3 & \longrightarrow & R\dot{C}O + RO\cdot + \text{aldehyde} \\ OH\cdot & \longrightarrow & R\cdot \end{cases}$$

$$\text{Aromatic} + \begin{cases} O & \longrightarrow & R\cdot + OH\cdot \\ O_3 & \longrightarrow & \text{not important} \\ OH\cdot & \longrightarrow & R\cdot + H_2O \end{cases}$$

$$\text{Paraffin} + \begin{cases} O & \longrightarrow & R\cdot + OH\cdot \\ O_3 & \longrightarrow & \text{not important} \\ OH\cdot & \longrightarrow & R\cdot + H_2O \end{cases}$$

$$\text{Aldehyde} + \begin{cases} O & \longrightarrow & R\dot{C}O + OH\cdot \\ O_3 & \longrightarrow & \text{not important} \\ OH\cdot & \longrightarrow & R\dot{C}O + H_2O \end{cases}$$

Since OH has an odd number of electrons and oxygen atoms and ozone each have even numbers of electrons, when these species attack hydrocarbons (with even numbers of electrons) the OH must produce an odd number of free radicals (normally one) whereas the oxygen atom and ozone must produce an even number of free radicals (normally either zero or two). Thus, the number and type of free-radical products in each step shown above can be specified with some certainty.

As we have seen, the reactions of O, O_3, and OH with hydrocarbons yield the following classes of free radicals: R, RCO, RO, where R can be a hydrogen atom or an alkyl group. We must now trace the most likely reactions of these species with the others in the system. Let us consider R, RCO, and OH as typical products. Both R and RCO will most probably react with O_2 by

$$R\cdot + O_2 \longrightarrow RO_2\cdot$$

$$R\dot{C}O + O_2 \longrightarrow R\overset{\overset{\displaystyle O}{\|}}{C}OO\cdot$$

These peroxy radicals will undergo a variety of reactions, the most important of which are with NO and NO_2:

$$RO_2\cdot + NO \longrightarrow NO_2 + RO\cdot$$

$$RO_2\cdot + NO_2 \longrightarrow \begin{cases} RNO_2 + O_2 \\ RO\cdot + NO_3 \\ ROONO_2 \end{cases}$$

$$R\overset{\overset{\displaystyle O}{\|}}{C}OO\cdot + NO \longrightarrow NO_2 + R\overset{\overset{\displaystyle O}{\|}}{C}O\cdot$$

$$R\overset{\overset{\displaystyle O}{\|}}{C}OO\cdot + NO_2 \longrightarrow R\overset{\overset{\displaystyle O}{\|}}{C}OONO_2$$

The two new radicals formed will then probably react with O_2:

$$RO\cdot + O_2 \longrightarrow RCHO + HO_2\cdot$$

$$R\overset{\overset{\displaystyle O}{\|}}{C}O\cdot + O_2 \longrightarrow RO_2\cdot + CO_2$$

The likely history of typical alkyl and acyl radicals in chain propagation reactions can thus be depicted as

$$R\dot{C}O \xrightarrow{O_2} R\overset{\overset{\displaystyle O}{\|}}{C}OO\cdot \xrightarrow{NO} R\overset{\overset{\displaystyle O}{\|}}{C}O\cdot \xrightarrow{O_2} RO_2\cdot$$

$$R\cdot \xrightarrow{O_2} RO_2\cdot \xrightarrow{NO} RO\cdot \xrightarrow{O_2} HO_2\cdot \xrightarrow{NO} OH\cdot \xrightarrow{HC} R\cdot$$

We see that, during the lifetimes of R and RCO, many molecules of NO can be converted to NO_2 (of course, each step in the sequence competes with other propagation and termination reactions). This is essentially the process we have been seeking to depict, namely the conversion of NO to NO_2 in significant quantities without the consumption of ozone.

We can now summarize the chemistry of photochemical smog. In the early morning there are large inputs of hydrocarbons and nitric oxide from automobiles and fixed sources. As a result of nighttime oxidation of NO, there is always a residual amount of NO_2 present at sunrise. As the sun rises, the NO_2 present photolyzes to provide oxygen atoms. The sequence of inorganic reactions in Tables 4.4 and 4.6 begins to generate, in addition to oxygen atoms, ozone and hydroxyl radicals. Each of these three species is capable of oxidizing hydrocarbons to form a number of free radicals, most of which end up as peroxy radicals. These peroxy radicals (ROO· and $R\overset{\overset{\displaystyle O}{\|}}{C}OO\cdot$) efficiently convert NO to NO_2. The key to smog chemistry is that one free radical formed, for example, as the result of the reaction of an oxygen atom with a hydrocarbon will participate in many propagation steps, i.e. the conversion of NO to NO_2, before extinction. A typical history of one such radical might be reaction with NO to give NO_2 and another radical. This radical then combines with O_2 to replenish the oxygen lost to NO. Then the radical participates in a reaction with another NO molecule to generate NO_2, etc., resulting in many molecules of NO converted to NO_2 for each free radical formed. This process involving oxygen-containing free radicals provides the alternative path for oxidation of NO to NO_2 and subsequent accumulation of ozone. The termination steps in the chain reaction explain the existence of many of the oxygen-containing organic products found in polluted air. For more detailed discussion of photochemical smog chemistry we refer the reader to Leighton (1961), Stephens (1966, 1969), Altshuller and Bufalini (1971), and Demerjian et al. (1973).

Hydrocarbon Reactivity in Photochemical Smog

In general, the term *hydrocarbon reactivity* in photochemical smog refers to the potential of a hydrocarbon to promote formation of products when irradiated in a system of nitrogen oxides and air. Because photochemical smog is manifested in such a variety of ways, e.g. eye irritation, plant damage, ozone formation, it is difficult to find one measure upon which the reactivity of a hydrocarbon may be expressed that includes all deleterious aspects of photochemical smog. From a chemical standpoint, reactivity may be expressed in terms of reaction rates and product yields, for example:

Reaction Rate Measures

1. Rate of NO_2 formation
2. Rate of hydrocarbon consumption
3. Rate of ozone formation

Product Yield Measures

1. Maximum concentration of ozone achieved
2. Total ozone dosage over a fixed time of irradiation
3. Maximum concentration of PAN achieved

From the standpoint of effects, reactivity may be defined in terms of the amount of eye irritation experienced (say by a panel of judges), the amount of plant damage incurred, or the degree of visibility limitation resulting. Aside from the subjectivity of eye irritation, a major problem in selecting a particular reactivity measure is that the measures are not necessarily interrelatable. Thus, ethylene is consumed at a rate much slower than that of tetramethylethylene [$(CH_3)_2C{=}C(CH_3)_2$], yet photooxidation of ethylene yields formaldehyde, a powerful eye irritant, as its chief end product, and tetramethylethylene yields mainly acetone, which is not an eye irritant. Thus, it is clear that there does not exist one single reactivity scale that includes all the deleterious effects of photochemical smog. Because air quality standards relating to photochemical smog are basically directed toward oxidant formation, the reactivity measure which has been most studied is that of the amount of oxidant (ozone) formed upon irradiation of the hydrocarbon in a mixture of air and NO_X.

For the normal constituents of automobile exhaust, it has been substantially confirmed that, regardless of the reactivity criterion (oxidant production, NO_2 formation rate, hydrocarbon consumption rate, or eye irritation), the constituents may be listed in the following order of decreasing reactivity (Altshuller and Cohen, 1963; Glasson and Tuesday, 1970):

1. Olefins, internally double-bonded
2. Alkyl aromatics, di- and tri-
3. Ethylene
4. Monoalkyl aromatics
5. C_5 and greater paraffins
6. C_1 to C_5 paraffins

Based on the wealth of experimental data available, it appears that the only hydrocarbons which can be considered to have negligible photochemical reactivity are methane, ethane, propane, acetylene, and benzene.

The photochemical reactivity of 45 organic solvents has been studied by Levy et al. (1971) with respect to four reactivity measures: NO_2 formation rate, oxidant and aldehyde production, and eye irritation. Their results indicate that many solvents considered heretofore as unreactive must be reclassified as reactive.

It would be highly desirable to relate the reactivity of a given organic species directly to its molecular structure. For internally double-bonded olefins, for example, reactivity would be expected to increase with the degree of substitution of the double-bonded carbon atoms, since the ease of formation and stability of free radicals formed by, for example, oxygen atom attack increase with degree of substitution. Thus, the reactivity of olefins would be expected to adhere to the ordering

$$R_2C{=}CR_2 > R_2C{=}CRH > RHC{=}CRH > H_2C{=}CRH > H_2C{=}CH_2$$

Following this idea, Niki et al. (1972) demonstrated that reactivity based on the rate of formation of NO_2 correlates well with the OH-rate constant for the particular organic.

The most difficult problem in assessing the reactivity of atmospheric hydrocarbon mixtures is accounting for the synergism that occurs when several hydrocarbons are present simultaneously. Thus, the rates of hydrocarbon consumption in a system of two hydrocarbons may be vastly different from the rates when the two are irradiated separately. For example, it is observed that when propylene is mixed with a less reactive hydrocarbon, such as ethylene, the rate of consumption of propylene per unit concentration is decreased from the pure propylene rate and is increased when propylene is mixed with a more reactive hydrocarbon, such as trans-2-butene. When a relatively nonreactive hydrocarbon like ethylene is mixed with a more reactive hydrocarbon like propylene, its rate of consumption is substantially increased over the pure component value. It is also observed in such a system that, when the more reactive hydrocarbon is depleted, the rate of oxidation of the less reactive hydrocarbon decreases abruptly. These observations are consistent with the free-radical mechanism outlined in this section. From a practical point of view, the quantitative estimation of mixture reactivities from individual hydrocarbon reactivities remains an important problem in assessing the effect of emission composition changes on oxidant formation in photochemical smog.

Role of Carbon Monoxide in Photochemical Smog

Carbon monoxide has been shown to accelerate the oxidation of NO to NO_2 when present in concentrations of 100 ppm or more (Westberg et al., 1971). The effect of CO is explained by reactions 19 to 21 in Table 4.6; accordingly, the oxidation of CO by OH (reaction 19), the first step in the chain, has received a great deal of attention. However, because ambient concentrations of CO are often of the order of only 5 to 30 ppm, CO probably has little effect on smog formation (Dodge and Bufalini, 1972).

A simplified example demonstrates the acceleration in the rate of oxidation of NO in the presence of CO due to an increase in the ratio of the rate of radical formation to the rate of radical termination. The four most important reactions in smog involving OH are (it is sufficient to examine only these reactions since CO reacts with no species other than OH)

$$OH\cdot + NO_2 \longrightarrow HNO_3 \qquad k = 15{,}000 \text{ ppm}^{-1}\text{ min}^{-1}$$

$$OH\cdot + NO \longrightarrow HNO_2 \qquad k = 12{,}000 \text{ ppm}^{-1}\text{ min}^{-1}$$

$$OH\cdot + HC \xrightarrow{O_2} 1RO_2\cdot \qquad k = 25{,}000 \text{ ppm}^{-1}\text{min}^{-1} \text{ for propylene}$$

$$OH\cdot + CO \xrightarrow{O_2} 1HO_2\cdot \qquad k = 250 \text{ ppm}^{-1}\text{min}^{-1}$$

If propylene, NO, and NO_2 are all present at a concentration of 1 ppm,

$$R = \frac{\text{rate of radical formation}}{\text{rate of radical termination}} = \frac{25{,}000}{15{,}000 + 12{,}000} \cong 0.9$$

With 100 ppm of CO added to the system,

$$R = \frac{25{,}000 + 25{,}000}{15{,}000 + 12{,}000} = 1.8$$

Thus, the addition of 100 ppm of CO to this system results in a doubling of the ratio of the rate of radical formation to the rate of radical termination. The effect of CO is even more striking in a system containing a hydrocarbon of lower reactivity such as *n*-butane for which the OH–*n*-butane rate constant is 3800 $ppm^{-1}\ min^{-1}$. In that case, if $[n\text{-butane}] = [NO] = [NO_2] = 1$ ppm

$$R = \frac{3800}{15{,}000 + 12{,}000} = 0.14$$

Upon addition of 100 ppm of CO, R increases to

$$R = \frac{3800 + 25{,}000}{15{,}000 + 12{,}000} = 1.1$$

The ratio of the rate of radical formation to the rate of termination has, therefore, increased eightfold upon addition of the CO. Thus, the effect of high concentrations of CO on the rate of smog formation would be expected to be proportionately greater for systems containing hydrocarbons of low reactivity than for those containing very reactive hydrocarbons.

4.4.2 Abatement of Photochemical Smog

The final question to be considered with regard to photochemical smog is to what levels must atmospheric hydrocarbon and nitrogen oxides concentrations be reduced so that air quality standards for ozone are not violated. This question can be studied chemically in a laboratory reactor with hydrocarbon and NO_X mixtures typical of an urban atmosphere (primarily that of Los Angeles). A series of experiments of this type were carried out by Dimitriades (1972), who irradiated mixtures of automobile exhaust and NO_X at a variety of hydrocarbon-NO_X ratios. The initial concentrations were chosen as those typical of current problem atmospheres in early morning as well as those anticipated as a result of current controls. Reactivity was measured and expressed in terms of five smog manifestations: (1) rate of NO_2 formation, (2) oxidant dosage, (3) peroxyacetyl nitrate (PAN) dosage, (4) formaldehyde dosage, and (5) NO_2 dosage. In addition, reactivity was expressed in terms of the same units used to express the California air quality standards for oxidant and NO_2, namely the time during which oxidant exceeded 0.1 ppm and the time during which NO_2 exceeded 0.25 ppm.

In contrast to the results shown in Fig. 4.7, the oxidant dependence on NO_X as measured by Dimitriades did not show a maximum within the range of hydrocarbon-NO_X ratios used. The propylene-NO_X system, for example, shows such a maximum at comparable HC/NO_X ratios. Therefore, it appears that the use of simple hydrocarbon systems, for example propylene, is somewhat unrealistic in simulating the

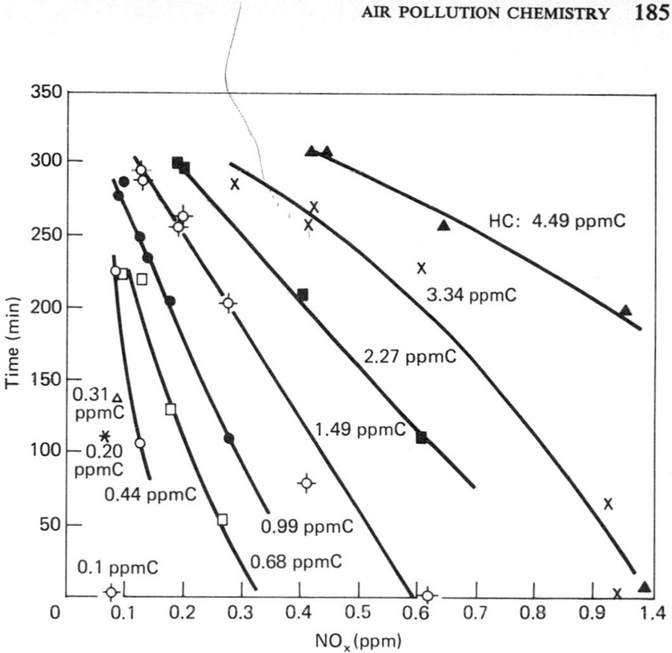

FIGURE 4.8
Time during which oxidant exceeded 0.1 ppm as a function of NO_X at various initial hydrocarbon levels for irradiation of auto exhaust. (*Source: Dimitriades,* 1972. *Copyright* 1972 *by the American Chemical Society. Reprinted by permission of the copyright owner.*)

atmosphere. The dependence of other reactivity manifestations on the HC/NO_X ratio was generally in agreement with those reported by others. For example, at low NO_X levels, the NO_2 dosage is nearly proportional to NO_X and is slightly inhibited by HC at higher HC levels, the formaldehyde dosage is nearly proportional to HC level and slightly inhibited by NO_X, and the rate of NO_2 formation is nearly proportional to HC and varies similarly with NO_X in the lower NO_X level range.

Figure 4.8 shows the time during which oxidant exceeded 0.1 ppm as a function of initial NO_X at various initial hydrocarbon levels. Hydrocarbon concentrations are expressed as ppm of carbon (ppm C). [A typical atmospheric level of hydrocarbons in ppm C was 3.05 ppm C (nonmethane) measured at Riverside, California, on Oct. 24, 1968.] We note from Fig. 4.8 that decreasing hydrocarbon at constant NO_X always decreased the oxidant dosage and that decreasing NO_X at constant hydrocarbon always increased the oxidant dosage. From this figure there can be determined combinations of initial hydrocarbon and NO_X that lead to an oxidant yield equal to the California standard of 0.1 ppm of oxidant for 1 hr. The result is Fig. 4.9. Points above the line *aeb* in Fig. 4.9 represent values of $[HC]_0$ and $[NO_X]_0$

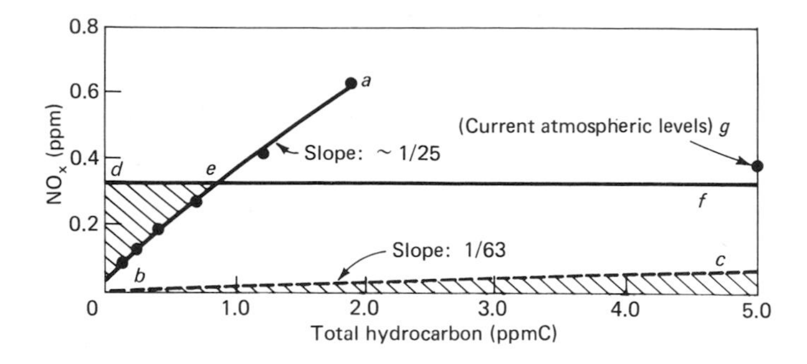

FIGURE 4.9
Regions representing combinations of total hydrocarbons and NO_X corresponding to oxidant and NO_2 yields meeting the California standards. (*Source: Dimitriades,* 1972. *Copyright* 1972 *by the American Chemical Society. Reprinted by permission of the copyright owner.*)

that meet the California oxidant standard. At extremely low $[NO_X]_0$ levels it is to be expected that the oxidant will begin to decrease as $[NO_X]_0$ approaches zero, since with zero $[NO_X]_0$ no oxidant would be expected. Assuming that oxidant dependence on $[NO_X]_0$ for $[HC]_0$ at 5 ppm C has a maximum at $[NO_X]_0 = 0.4$ ppm (see Fig. 4.8) and the oxidant yield is proportional to $[NO_X]_0$ within the range 0 to 0.4 ppm, then the $[NO_X]_0$ level corresponding to the California oxidant standard is estimated to be 0.08 ppm, a $[HC]_0/[NO_X]_0$ ratio of 62. Applying this argument to the other curves in Fig. 4.8, one obtains ratios of 61, 64, 60, and 67. These results suggest that the California oxidant standard can be met when $[HC]_0/[NO_X]_0 \geq 63$, that is, that region lying below line *bc* in Fig. 4.9. The NO_2 standard is represented by line *def.* Thus, the shaded regions in Fig. 4.9 correspond to those sets of initial concentrations for which both the oxidant and NO_2 standards are met. Current Los Angeles atmospheric levels are indicated by point *g*.

From Fig. 4.9 it is possible to estimate the direction and extent to which current atmospheric levels must be modified to achieve a certain level of air quality. Because the data upon which Fig. 4.9 is based do not include effects of meteorology and continuous source emissions, Fig. 4.9 provides only a semiquantitative indication of necessary reductions. From Fig. 4.9, we see that in order to reach the shaded region bounded by $[NO_X]_0 \leq 0.33$ ppm and $[HC]_0 \leq 2.5 \times [NO_X]_0$ it is necessary to reduce current Los Angeles NO_X emissions by 15 percent and to reduce HC emissions by at least a factor of 5. To reach the shaded region defined by $[NO_X]_0 \leq 0.08$ ppm and $[HC]_0 \geq 62 \times [NO]_0$, at least an 80 percent NO_X reduction, but no HC reduction, is required.

4.5 REACTIONS OF SULFUR OXIDES IN THE URBAN ATMOSPHERE

Sulfur oxides, in particular sulfur dioxide, have been the subject of much study with respect to atmospheric chemistry. In spite of this study, however, an understanding of the chemistry of sulfur oxides in the atmosphere is still far from complete. Most evidence suggests that the eventual fate of atmospheric SO_2 is oxidation to sulfate, although the detailed mechanism of SO_2 oxidation in the atmosphere is basically incomplete. One of the problems that complicates the understanding of atmospheric sulfur oxide processes is that reaction paths may be both homogeneous and heterogeneous. In Chap. 2 we delineated two processes for conversion of SO_2 to sulfate: catalytic and photochemical. It is certainly an oversimplification to assume that SO_2 is oxidized by one or the other of these two paths in any one circumstance. Nevertheless, under overcast, high-humidity, and high-particulate-concentration conditions it is likely that catalytic (heterogeneous) oxidation is the primary process for SO_2 conversion. In the presence of sunny, NO_X-hydrocarbon polluted atmospheres, it appears that the existence of a photochemical reaction path may be necessary to account for observed rates, particularly if the availability of particles containing catalysts and acid-neutralizing compounds is insufficient to produce observed rates by the catalytic route.

In this section we have summarized current understanding of atmospheric oxidation processes for SO_2, both heterogeneous (catalytic) and homogeneous routes. There are a number of reviews dealing with atmospheric reactions of SO_2, for example, Urone and Schroeder (1969) and Bufalini (1971), to which the reader may refer for more detailed coverage.

4.5.1 Catalytic Oxidation of SO_2

In clean air, SO_2 is very slowly oxidized to SO_3 by homogeneous reactions. However, the rate of SO_2 oxidation in a power plant plume has been observed to be 10 to 100 times the clean-air photooxidation rate (see, for example, Gartrell et al., 1963). Such a rapid rate of oxidation is similar to that expected of oxidation in solution in the presence of a catalyst.[1]

Sulfur dioxide dissolves readily in water droplets and can be rapidly oxidized to sulfuric acid by dissolved oxygen in the presence of metal salts, such as iron and manganese. The overall reaction can be expressed as

$$2SO_2 + 2H_2O + O_2 \xrightarrow{\text{catalyst}} 2H_2SO_4$$

[1] The rate of the direct reaction of SO_2 with O_2,

$$2SO_2 + O_2 \longrightarrow 2SO_3$$

is too slow at room temperature to be of importance in the atmospheric oxidation of SO_2.

Catalysts for this reaction include several metal salts, such as sulfates and chlorides of manganese and iron, which usually exist in air as suspended particulate matter. At high humidities these particles act as condensation nuclei or undergo hydration to become solution droplets. The oxidation process then proceeds by absorption of both SO_2 and O_2 by the liquid aerosol with subsequent chemical reaction in the liquid phase. The oxidation slows considerably when the droplet becomes highly acidic, because of the decreased solubility of SO_2. However, if sufficient ammonia is present, the oxidation process is not impeded by the accumulation of H_2SO_4. Measurements of particulate compositions in urban air often show large concentrations of ammonium sulfate.

Early experiments by Junge and Ryan (1958), Johnstone and Coughanowr (1958), and Johnstone and Moll (1960) confirmed the basic catalytic mechanism and the effect of solution acidity on the rate of SO_2 oxidation. Recent laboratory results on the catalytic oxidation of SO_2 in aerosol drops containing metal salts were reported by Cheng et al. (1971). An aerosol stabilizing technique was developed in which aerosol particles were deposited on inert beads in a fluidized bed. The beads with the deposited aerosol particles were then packed into a flow reactor in which the catalytic oxidation of SO_2 occurred as an SO_2, and a humid air mixture flowed through the reactor. After a brief induction period during which the solubility of SO_2 decreased as the solution became more acidic, the rate of SO_2 oxidation reached a steady state. Because H_2SO_4 in dilute concentration undergoes complete dissociation to HSO_4^- and H^+, the added H^+ concentration diminishes the solubility of SO_2. When the solution acid concentration exceeds a certain level, the high H^+ concentration prevents further dissociation of H_2SO_4, and the solubility of SO_2 becomes constant.

The individual steps in the liquid-phase catalytic oxidation of SO_2 are:

1 Gas-phase diffusion of SO_2 to the drop
2 Diffusion of SO_2 from the drop surface to the interior
3 Catalytic reaction in the interior

At steady-state conditions, the overall rate of SO_2 conversion is limited by the slowest of the above three steps. If gas-phase diffusion of SO_2 to the drop is the controlling step, then the rate of SO_2 conversion should depend on the gas velocity in the system. If liquid-phase diffusion of SO_2 controls, then conversion rates can be expected to be independent of the type of catalyst. In varying the gas flow rate through their reactor, Cheng et al. found that the overall rate of SO_2 conversion was unaffected. In addition, it was found that the oxidation depended strongly on the type of catalyst. Thus, the rate-controlling step is the chemical reaction itself. Similar conclusions were reached by Foster (1969).

Table 4.11 summarizes the steady-state conversion rates measured by Cheng et al. (1971). Catalytic effectiveness is seen to follow the order $MnSO_4 > MnCl_2$

$> CuSO_4 > NaCl$. A similar study by Matteson et al. (1969) with metal salt catalysts revealed catalytic effectiveness in the order $MnCl_2 > CuCl_2 > FeCl_2 > CoCl_2$. From the studies of Matteson et al. and Cheng et al., one may draw two general conclusions. First, catalytic oxidation of SO_2 is fostered in basic or neutral solutions and inhibited in acidic solutions.[1] Second, relative humidity plays an important role in the rate of oxidation. If the water concentration drops below that necessary to maintain a catalyst solution, the rate of oxidation decreases sharply. Although the conversion of SO_2 proceeded even at very low relative humidities (less than 40 percent) in the experiment of Cheng et al., it did so slowly. Above about 70 percent relative humidity, at which transition from solid crystals surrounded by a layer of water to an actual solution drop takes place, the rate of conversion increased dramatically.

Although the detailed mechanisms of SO_2 oxidation in solution are still a matter of speculation (see, for example, Matteson et al., 1969), most data show that the catalytic oxidation of SO_2 can be represented as a first-order reaction based on the gas-phase concentration of SO_2 with a rate constant dependent on the catalyst type and relative humidity. For example, Cheng et al. determined the following first-order rate expression for $MnSO_4$:

$$R_{SO_2} = 0.67 \times 10^{-2}[SO_2]$$

Table 4.11 EFFECT OF CATALYST ON SO_2 OXIDATION

Catalyst	Weight, mg	Mean residence time, min	Influent SO_2 conc., ppm	Fraction conversion	Effectiveness factor†
NaCl	0.36	1.7	14.4	0.069	1.0
$CuSO_4$	0.15	1.7	14.4	0.068	2.4
$MnCl_2$	0.255	0.52	3.3	0.052	3.5
$MnSO_4$	0.51	0.52	3.3	0.365	12.2

SOURCE: Cheng et al. (1971).

† The catalytic effectiveness of the various materials was compared with that of NaCl. Thus, effectiveness factor = 1.0 × (ratio of weight of catalyst in the reactor) × (ratio of reactor mean residence time) × (ratio of reaction conversion of SO_2 in the reactor). The effectiveness factor for $MnSO_4$, for example, is

$$1.0 \times \frac{0.36}{0.51} \times \frac{1.7}{0.52} \times \frac{0.365}{0.069} = 12.2$$

[1] Water solutions of species in high oxidation states tend to be acidic, with acidity increasing with oxidation state. Therefore, solutions of SO_3 are more acidic than SO_2, and those of Fe^{+3} are more so than Fe^{+2}. For solutions of metal ions in the same oxidation state, acidity tends to increase as the ionic radius decreases. Ionic radii of the metals we are considering fall in the order $Cu^{+2} > Mn^{+2} > Fe^{+2} > Co^{+2} > Ni^{+2}$. Thus, on the basis of acidity alone, we would expect the catalytic effectiveness to follow this same order.

where $R_{SO_2} = \mu g SO_2 \ min^{-1}/mg$ of $MnSO_4$ converted, $[SO_2]$ is the gas-phase concentration of SO_2 in μg^{-3}, and the constant factor is that for drops containing 500 ppm of $MnSO_4$. The factor can be altered for other catalysts using Table 4.11.

In summary, the gross features of catalytic oxidation of SO_2 in droplets containing metal ions are reasonably well understood. Nevertheless, predictions of the rate of oxidation of SO_2 in particulate matter containing a variety of metal compounds, and perhaps also ammonia, at different relative humidities, are difficult to make. Additional measurements of SO_2 oxidation rates under plume conditions will be valuable.

4.5.2 Photochemical Oxidation of SO_2

In the presence of air, SO_2 is slowly oxidized to SO_3 when exposed to solar radiation. If water is present, the SO_3 is rapidly converted to sulfuric acid. Since virtually no radiation of wavelength shorter than 2900 Å reaches the earth's surface, and since the dissociation of SO_2 into SO and O is possible only for wavelengths of absorbed light below 2180 Å (corresponding to a bond dissociation energy of 135 kcal), the primary photochemical processes following absorption by SO_2 in the lower atmosphere involve activated SO_2 molecules and not direct dissociation. Thus, the conversion of SO_2 to SO_3 in clean air is a result of a several-step reaction sequence involving excited SO_2 molecules, oxygen, and oxides of sulfur other than SO_2. In the presence of reactive hydrocarbons and oxides of nitrogen the rate of conversion of SO_2 to SO_3 increases markedly. In addition, oxidation of SO_2 in systems of this type is frequently accompanied by considerable aerosol formation. The elucidation of the chemical reaction mechanism prevailing in irradiated systems of SO_2, hydrocarbons, NO_X, air, and water is one of the most important current problems in air pollution chemistry. We divide our discussion here into photochemical oxidation of SO_2 in pure air and photochemical oxidation of SO_2 in the presence of hydrocarbons and oxides of nitrogen.

4.5.2.1 Photochemical oxidation of SO_2 in air The absorption spectrum for SO_2 shows two bands above 2900 Å, the first a weak absorption with transition to the first excited (triplet) state at 3840 Å and the second a strong absorption to the second excited (singlet) state at 2940 Å. The former is denoted $^3SO_2$, and the latter is written $^1SO_2$. Thus, we have

$$SO_2 + h\nu \underset{2}{\overset{1}{\rightleftharpoons}} {}^1SO_2 \qquad (2900 \text{ to } 3400 \text{ Å})$$

$$\underset{4}{\overset{3}{\rightleftharpoons}} {}^3SO_2 \qquad (3400 \text{ to } 4000 \text{ Å})$$

The more energetic singlet state decays either to ground-state SO_2 or to the less energetic triplet state by the following reactions:

$$^1SO_2 + M \xrightarrow{5} SO_2 + M$$
$$\xrightarrow{6} {}^3SO_2 + M$$

Okuda et al. (1969) showed that the major, if not the exclusive, chemical entity in the urban photochemistry of SO_2 is the triplet state $^3SO_2$. Apparently the major role of the singlet species $^1SO_2$ is the generation of the triplet molecule by reaction 6.

The mechanism by which SO_2 is oxidized to SO_3 in air is presumably based on subsequent reactions of the two excited states, $^1SO_2$ and $^3SO_2$, with other molecules in the system. First, $^3SO_2$ may be returned to ground-state SO_2 by quenching with other atmospheric species:

$$^3SO_2 + M \xrightarrow{7} SO_2 + M$$

Rate constants for $M = N_2$, O_2, CO, CO_2, and CH_4 are all very similar (0.68 to 1.2 $\times 10^8$ l mole^{-1} sec^{-1}). For $M = H_2O$ and O_3 the rate constants are 8.9×10^8 and 11×10^8, respectively (Sidebottom et al., 1972).[1]

Considerable study has been devoted to elucidating the mechanism of SO_2 photolysis in air. A careful survey of the possible reactions by Bufalini (1971) and Sidebottom et al. (1972) resulted in the conclusion that potentially the most important oxidation step for $^3SO_2$ from among those involving O_2 only is

$$^3SO_2 + O_2 \xrightarrow{8} SO_3 + O$$

A number of experimental studies on the photooxidation of SO_2 in air have been reported. Gerhard and Johnstone (1955), for example, studied the rate of photooxidation of SO_2 in air at concentrations from 5 to 30 ppm SO_2. They found essentially a linear increase in H_2SO_4, corresponding to 0.1 to 0.2 percent conversion per hour of SO_2. Table 4.12 summarizes these results and those of two other investigators. These data indicate that a typical rate of conversion of SO_2 in clean air by photochemical oxidation is about 0.1 percent per hour. Based on reactions 1 to 8, a maximum theoretical rate of SO_2 oxidation of about 2 percent per hour is predicted (Sidebottom et al., 1972).

[1] When M is a hydrocarbon molecule containing double bonds the rate constant increases markedly ($\sim 2 \times 10^{11}$ l mole^{-1} sec^{-1}). Because of the abnormally high quenching rate constant for olefinic hydrocarbons, about one or two excited SO_2 molecules out of 10^4 can react with olefins even though the olefin may be present only at the pphm level. The marked influence of SO_2 on aerosol formation in smog chamber experiments may be attributable in part to this reaction.

4.5.2.2 Photochemical oxidation of SO_2 in mixtures of hydrocarbons, oxides of nitrogen, and air Once hydrocarbons and oxides of nitrogen are introduced into a mixture of SO_2 and air, the rate of SO_2 oxidation upon irradiation increases markedly over that observed in clean air. Table 4.13 presents several values of the measured rate of SO_2 oxidation in such mixtures. In contrast to the situation in an industrial fog containing metal ions, the existence of a catalytic mechanism is unlikely in this case in spite of the fact that pronounced aerosol formation takes place in such systems.

The reactions taking place in a system of SO_2, hydrocarbons, NO_X, and air are probably the least well understood of all those in atmospheric chemistry. There are potentially a number of paths for oxidation of SO_2 in such a system.

Table 4.14 summarizes several SO_2 oxidation reactions which might be important in the atmosphere. Reactions 1, 2, and 3 are probably unimportant in the gas phase, although when water droplets are present in a mixture of SO_2, O_3, and air, both

Table 4.12 MEASURED RATES OF SO_2 PHOTOOXIDATION IN CLEAN AIR

Investigators	Initial [SO_2], ppm	Relative humidity	Rate of SO_2 oxidation,† %/hr
Gerhard and Johnstone (1955)	5–30	32–91	0.102–0.198
Urone et al. (1968)	10–20	50	0.084
Urone et al. (1971)	1000	0	0.023
		50	0.028
Katz and Gale (1971)	3.2	0	0.28
		50	1.0

† At light intensities equivalent to noonday sunlight.

Table 4.13 MEASURED RATES OF SO_2 PHOTOOXIDATION IN AIR CONTAINING HYDROCARBONS AND OXIDES OF NITROGEN

Investigators	Initial [SO_2], ppm	Hydrocarbons	Rate of SO_2 oxidation, %/hr
Renzetti and Doyle (1960)	0.2–0.6	2-Methyl-2-butene	48–294
Urone et al. (1968)	15–21	*n*-Hexane	1.8–12
Wilson and Levy (1970)	0–0.75	1-Butene	15–30
Wilson et al. (1972)	0–0.75	1-Butene	
		1-Heptene	
		2,2,4-Trimethylpentane	
		Toluene	

SO_2 and O_3 are observed to disappear rapidly, indicating that reaction 3 proceeds efficiently in the liquid phase. Rate constants for reactions 4 and 5 are not yet available, and the values given in Table 4.14 represent very crude estimates. Reactions 6 and 7 involving peroxy radicals may be important in SO_2 oxidation, as are the similar reactions in NO oxidation. Cox and Penkett (1971) have reported that SO_2 is oxidized by a product of ozone-olefin reactions, presumably the diradical (zwitterion) intermediate, reaction 8.

Experimentally, Wilson et al. (1972) have found that lower NO_2 peaks occur in all hydrocarbon-NO_X reaction systems to which SO_2 is added. This observation may be related to the reaction of SO_2 with a reaction product in the O_3-NO_2 system. Ozone formation was found to be decreased in systems of olefins and paraffins to which SO_2 is added. Removal of NO_X by SO_2 lowers the NO_2 concentration leading to lower ozone concentrations. However, in an experiment conducted with toluene, NO_X and SO_2, Wilson and Levy found that the ozone levels observed were higher than those observed in the absence of SO_2. A convincing explanation for this observation has not yet been put forward.

In summary, SO_2 interacts strongly with many species present in photochemical smog. The most striking effect is the greatly enhanced tendency to form aerosol

Table 4.14 ATMOSPHERIC REACTIONS INVOLVING SO_2, NO_X, AND HYDROCARBONS

No.	Reaction	Rate constant at 298°K	Reference
1	$O + SO_2 + M \rightarrow SO_3 + M$	2.71×10^{-4} ppm^{-2} min^{-1}	Mulcahy et al. (1967)
2	$SO_2 + NO_2 \rightarrow SO_3 + NO$	1.3×10^{-14} ppm^{-1} min^{-1}	Calvert (1973)
3	$SO_2 + O_3 \rightarrow SO_3 + O_2$	Slow in gas phase	
4	$SO_2 + NO_3 \rightarrow SO_3 + NO_2$	$\leq 2.5 \times 10^{-5}$ ppm^{-1} min^{-1}	Calvert (1973)
5	$SO_2 + N_2O_5 \rightarrow SO_3 + 2NO_2$	$\leq 1.7 \times 10^{-8}$ ppm^{-1} min^{-1}	Calvert (1973)
6	$SO_2 + HO_2 \rightarrow SO_3 + OH$	0.44 ppm^{-1} min^{-1}($\Delta H = -19$)	Davis et al. (1972)
7	$SO_2 + RO_2 \rightarrow SO_3 + RO$	0.57† ppm^{-1} min^{-1}	
	$\rightarrow RO_2SO_2$	Unmeasured	
8	$SO_2 + R_2CO_2 \rightarrow R_2CO + SO_3$	Unmeasured‡	Cox and Penkett (1971)
9	$^3SO_2 + O_2 \rightarrow SO_2 + O(^3P)$	>235 ppm^{-1} min^{-1}	Badcock et al. (1971)
10	$SO_2 + RH \rightarrow RSO_2H$	Depends on hydrocarbon	Badcock et al. (1971)
	$\rightarrow SO_2H + R$		
11	$SO_2 + R \rightarrow RSO_2$	$\sim 4.4 \times 10^2$ ppm^{-1} min^{-1}(CH_3)	Calvert (1973)
12	$SO_2 + RO \rightarrow ROSO_2$	$\sim 4.4 \times 10^2$ ppm^{-1} min^{-1}(OH)	Calvert (1973)

† RO_2 and $RCOO_2$ are estimated to oxidize NO about 30 percent faster than does HO_2. Assuming the same 30 percent difference holds for SO_2, a rate constant of 0.57 ppm^{-1} min^{-1} is obtained based on that given for reaction 6.

‡ Cox and Penkett (1971) observed SO_3 to form in a system of O_3, olefin, and SO_2. For every 100 ozone molecules consumed approximately 8 SO_2 molecules were oxidized, from which it might be inferred that 8 out of every 100 biradicals reacted with SO_2. At the concentration levels of the experiment the SO_2 concentration was four times the O_3 concentration. Thus, at equivalent levels, one might expect 2 out of every 100 biradicals to react with SO_2 at equivalent O_3 and SO_2 levels.

when SO_2 is present. We have not discussed the chemistry of aerosol formation in such systems because little is actually known about possible reactions. Clearly, more work is needed on this aspect of air pollution chemistry.

4.6 AEROSOL PROCESSES IN THE URBAN ATMOSPHERE

It has long been recognized that atmospheric reaction processes are frequently accompanied by aerosol formation. In spite of this early recognition, however, neither the available smog chamber data nor existing atmospheric data are presently adequate to identify fully the physical and chemical mechanisms which govern aerosol formation in the atmosphere. The experimental difficulties associated with the physical and chemical characterization of submicron aerosols are primarily responsible for this inadequacy.

Historically, aerosol research has focused mainly on aerosol physics, namely growth mechanisms, diffusion, and size spectra (see, for example, Hidy and Brock, 1970). Scant information is available even at this time on the chemical processes occurring in atmospheric aerosol particles.

The central problems of urban atmospheric aerosol analysis are identifying the formation mechanisms and predicting the chemical composition and rates of growth of aerosols. With respect to formation mechanisms, aerosols can be classed as primary (those introduced into the atmosphere in particulate form, such as dust and smoke) and secondary (produced in the atmosphere by condensation and chemical reaction, such as Aitken nuclei). Both chemical and physical mechanisms play important roles in aerosol formation and growth. Chemical reactions provide species convertible from the gas to the particulate (liquid or solid) phase and may take place in the particles themselves. Physical processes such as nucleation, condensation, absorption, adsorption, and coagulation are primarily responsible for determining physical properties, i.e., number concentration, size distribution, optical properties, settling properties, etc., of the formed aerosols.

Viewed from the standpoint of air pollution control, the objective of aerosol measurements is to relate both gaseous and particulate source emissions to urban aerosols which are responsible for visibility impairment and possible health effects. The goal, then, is to account in as much detail as possible for the origin, atmospheric concentration, and fate of atmospheric particles. Knowledge of the following elements is required, in part, to achieve this goal:

1 *Aerosol sources:* automobile and industrial emissions, natural aerosol production, etc.

2 *Chemical and physical rate processes:* gas-gas reactions, gas-particle reactions, nucleation, condensation, coagulation, growth by absorption, adsorption, etc.

3 *Removal mechanisms:* settling, washout, etc.

4.6.1 Aerosol Source Inventory

We consider first the question of an aerosol "source inventory" for an urban area. The compilation of an aerosol source inventory is more difficult than the compilation of a gaseous emission inventory for the following reasons:

1 Of the aerosols of anthropogenic origin, a significant quantity is formed in the atmosphere (secondary) and cannot be accounted for in a primary emissions inventory.

2 Aerosols are distributed not only with respect to chemical composition but also with respect to size.

3 Primary sources can be either anthropogenic or natural.

Apparently, the first large-scale program aimed at understanding the behavior and properties of urban aerosols was the 1969 Pasadena Smog Experiment. During the experimental period of about three weeks (August and September 1969), the aerosol size distribution, along with other physical, chemical, and meteorological parameters, was measured in six intensive experimental periods of about one day each. The experimental data and analysis are reported in Aerosol Measurements in Los Angeles Smog (1971) and in the collection of papers edited by Hidy (1972). A second major aerosol characterization study aimed at the Los Angeles aerosol was carried out during 1972.

Although an aerosol emissions inventory, in the sense of the spatial and temporal distribution of aerosol emissions from sources, was not compiled in either of these two studies, a concerted effort was made to identify the *major* sources which contribute to the Los Angeles aerosol (Hidy and Friedlander, 1971; Miller et al., 1972). For example, Hidy and Friedlander (1971) estimated the relative importance of major natural and anthropogenic sources of the Los Angeles aerosol from available data on their chemical composition. The fractional contribution of each source was estimated from data on emissions of gases and particles, in combination with the use of tracer elements from sources.

The results of these studies may be summarized as follows: The total number concentration, N, was found to average about 10^4 cm^{-3}, with a range from 10^3 to 10^5 cm^{-3}. Number concentrations, however, for particles exceeding 0.5 μm in diameter average only 10 to 100 cm^{-3}. As Hidy and Friedlander note, the natural background level near Los Angeles for particles >0.5 μm is 1 to 10 cm^{-3}, so that the majority of the anthropogenic particles are produced in a diameter range >0.5 μm. The average mass concentration of >100 μg/m^3 in Los Angeles is associated principally with particles >5 μm in diameter, and visibility reduction is associated principally with particles in the range 0.2 to 2 μm.

Table 4.15 shows the composition of the Los Angeles aerosol. The contribution of the marine aerosol was estimated to be about 10 $\mu g/m^3$. Table 4.16 presents estimates for the contribution of the four sources of aerosol, primary and secondary, natural and anthropogenic. The mineral-dust concentration was obtained by assuming that all the silicon originated in mineral dust. All the chloride and sodium were assumed to be of marine origin. All the nitrate and sulfate were assumed to be generated by chemical reactions involving NO_2 and SO_2 produced originally by man-made sources. The estimate of natural organic aerosol is due to Junge; that for secondary anthropogenic aerosol was based on an assumed 1 percent conversion of reactive gas-phase organics to aerosol. The total aerosol concentration estimated was 93 to 108 $\mu g/m^3$, somewhat lower than the range for the 1960s.

In subsequent studies, Friedlander and his colleagues (Miller et al., 1972; Friedlander, 1973) developed a novel means for estimating the contribution of various sources to the atmospheric aerosol measured at any point in an airshed. The method is essentially based on a mass balance for various individual elements which are known to come primarily from certain well-defined sources. Air pollution sources of given types, whether natural or man-made, emit a characteristic set of chemical elements in approximately fixed proportions. If the sources in a polluted region are known, the chemical-element balance method enables the estimation of the contribution from each source from measurements of the elemental concentrations at a given point. Several dozen chemical elements have been identified in urban atmospheres.

Table 4.15 COMPOSITION OF THE LOS ANGELES AEROSOL

Substance	Arithmetic average concentration, $\mu g/m^3$	Year measured or estimated
Total suspended particulate	119†	1966
Benzene soluble	15.2	1966
Sulfate	14	1965
Nitrate	13	1965
Lead	5.6	1963
	2.0	1963
Silicon	2.5–6	?
Chloride	3	1965
Sodium	3 (?)	Calculated assuming $Cl^-/Na^+ = 1$ (Junge, 1963, p. 326). Excludes soil dust
Ammonium	0.7	1965

SOURCE: Hidy and Friedlander (1971).

† Based on sample equilibration for 24 hr at 75°F and 50 percent or less relative humidity.

Many of these elements, such as sodium, chlorine, silicon, and aluminum, are associated with the natural background, but certain species can be attributed to particular anthropogenic sources. Examples include:

Lead: automobile exhaust
Vanadium: fuel-oil fly ash
Nickel: fuel-oil fly ash
Zinc: tire dust
Barium: diesel fuel exhaust

It is important to stress that an ordinary source inventory based merely on estimates of emission rates from various anthropogenic sources is inadequate for aerosols because it is uncertain how much material is emitted from sources such as the soil and

Table 4.16 ESTIMATED CONCENTRATIONS OF LOS ANGELES AEROSOL PARTICLES BY SOURCE (ANNUAL AVERAGE)

Source	Mass concentration, $\mu g/m^3$	
Natural background		
Primary		14–26
Dust rise by wind	8–20	
Sea salt $\{Na^+, Cl^-\}$	3	
Spores, pollen, etc.	Unknown	
Secondary		4–7
Vegetation (organic vapors)	3–6	
Biological (soil bacterial action, decay of organics) —NH_3, NO_x, S...	0.7	
Man-made		
Primary		36
Motor vehicles	15	
Organic solvent use	6	
Petroleum	1.3	
Aircraft	4	
Combustion of fuels	5	
Other	5	
Secondary		39
Reactive hydrocarbon vapors	11	
NO_3^-	13.5	
SO_4^{2-}	14.4	
Total accounted for		93–108
Measured total		119

SOURCE: Hidy and Friedlander (1971).

the sea. In addition, the fraction which remains airborne is often unknown. The mass balance method circumvents these difficulties by essentially employing certain characteristic elements, such as the five listed above, as fingerprints for specific types of sources. In some cases, particularly for the elements carbon, sulfur, and nitrogen, it is necessary to have some information on the combined state in order to ascertain whether an element is present in the gas or particulate phase.

In summary, the development of an aerosol source inventory first requires the analysis of particulate monitoring data to determine the fractional contribution of various types of sources (natural and anthropogenic) to the atmospheric aerosol. Then, a regular emissions inventory can be assembled, based on the major anthropogenic sources. The emissions inventory should specify, if possible, the size range and chemical constituency of the effluent.

4.6.2 Dynamic Processes Affecting Aerosols

As formed, aerosols exhibit differing size and chemical composition distributions, depending on the method of generation. Subsequently, in the atmosphere, a number of dynamic processes affect particle size distributions and chemical composition. Within a given size range, the following processes may alter the number of particles:

1 Production of particles by homogeneous or heterogeneous nucleation[1] or dispersion of fine particles
2 Growth of particles by heterogeneous gas reactions on the surface of particles
3 Growth of particles by homogeneous gas-phase reactions and subsequent absorption of the reaction products on the particles
4 Brownian coagulation
5 Turbulent coagulation
6 Scavenging of smaller particles by larger ones during their fall
7 Gravitational settling
8 Impaction on obstacles on the earth's surface
9 Washout

The presence of gaseous pollutants influences only processes 1, 2, and 3. The rate at which processes 4 and 5 take place depends on the number concentration of particles and does not alter the total mass of aerosol present. Process 6 can generally be neglected, whereas processes 7 to 9 result in an overall decrease in aerosol mass by removing primarily the larger-size particles. A detailed treatment of each of these nine processes is well beyond our scope. An up-to-date treatise in this area is that of Hidy and Brock

[1] In *homogeneous* nucleation, particles are formed in the vapor phase by molecular clustering without the aid or intervention of foreign nuclei. In *heterogeneous* nucleation, foreign particles are present in the vapor and act as nuclei for the growing particles.

(1970) to which we refer the interested reader. In this subsection we shall concentrate only on those processes of greatest importance for urban aerosols, that is, homogeneous and heterogeneous nucleation, particle growth through chemical reactions, and coagulation.

4.6.2.1 Homogeneous nucleation When the supersaturation level of a vapor in a gas reaches a certain point (in the absence of foreign particles), tiny clusters of the condensed vapor will grow with time rather than reevaporate. Such a situation represents homogeneous nucleation. In order to achieve a molecular cluster consisting of, say, j molecules, it is necessary to overcome an energy barrier which depends on j. By considering the thermodynamics of an idealized spherical cluster of j molecules of a droplet in its own vapor, the conditions under which the droplet will grow or reevaporation will take place may be determined. The critical radius r^* above which molecular clusters are stable and grow is given by (the derivation of this relation is given on page 258 of Hidy and Brock)

$$r^* = \frac{2\sigma v_L}{kT \ln p/p_0}$$

where σ is the surface tension of the drop, v_L is the volume of a molecule in the liquid phase, and p and p_0 are the actual vapor pressure and that above a planar surface, respectively.

The process of particle growth can be represented by

$$e_i + e_j \underset{\text{splitting}}{\overset{\text{collision}}{\rightleftharpoons}} e_{i+j}$$

in which e_i and e_j represent clusters containing i and j molecules, respectively. We shall consider all collisions and splitting to be binary processes only. A balance on all clusters of size k yields:

Gain:

1 Collisions between particles whose indices add to k
2 Splitting of clusters such that at least one product is a cluster of size k

Loss:

3 Collisions of a cluster of size k with any other cluster
4 Splitting of a cluster of size k

The total rate of generation of clusters of size k, B_k, can be written

$$B_k = \frac{1}{2}\sum_{i+j=k}^{M} N_{ij} + \sum_{i=1}^{M-1} S_{ik} - \sum_{i=1}^{M-1} N_{ik} - \frac{1}{2}\sum_{i+j=k}^{M} S_{ij} \qquad k = 1, 2, \ldots, M$$

where N_{ij} equals the number of collisions per unit time per unit volume between classes i and j, S_{ij} equals the number of splittings per unit time per unit volume into classes i and j, and M is the number of molecules in the largest cluster. Thus, the first and third terms account for collisions and the second and fourth terms represent evaporation.

In so-called *coagulation theory* it is customary to neglect the second and fourth terms, thereby assuming that collisions are irreversible. If we represent the collision rate N_{ij} with the rate equation

$$N_{ij} = b_{ij} n_i n_j$$

where n_i is the number of i clusters per unit volume and b_{ij} is a collision rate constant, the rate of change of the number density of k clusters as a result of coagulation processes only is

$$\frac{dn_k}{dt} = \frac{1}{2} \sum_{\substack{j=1 \\ i+j=k}}^{i=k-1} b_{ij} n_i n_j - n_k \sum_{i=1}^{M} b_{ik} n_i \tag{4.23}$$

Nucleation theory, on the other hand, considers only processes involving single molecules, and so that the total rate of generation of clusters of size k as a result of nucleation processes only is

$$B_k = N_{1,k-1} + S_{i,k+1} - N_{1k} - S_{1k}$$

Over the time evolution of the growth of an aerosol, both nucleation and coagulation theories are relevant. In the early stages of growth, the concentration of single molecules is much greater than that of clusters. Also, during this early period the evaporation of a single molecule is much more likely than the splitting of a cluster. The main growth processes are therefore described by nucleation theory during this stage. In the late stages of growth, the concentration of gas molecules has been reduced (unless supersaturation is maintained by external means). Evaporation and splitting are secondary in importance to collisions in their contribution to growth dynamics. Coagulation theory is thus relevant at this stage.

4.6.2.2 Heterogeneous nucleation In any but the most carefully controlled laboratory situations foreign particles in the gas as well as molecular clusters act as centers for condensation. When foreign centers are present, the stable formation of clusters takes place at much lower supersaturations than if no foreign particles are present. Nucleation to form clouds of water droplets, for example, takes place at supersaturations rarely exceeding a few percent. The analysis of vapor condensation on foreign material is much more complex than that of homogeneous nucleation. The important factors include (1) the type of particle, (2) its shape, and (3) its chemical properties, in other words, those factors involved in a detailed consideration of surface chemistry.

4.6.2.3 Urban aerosol dynamics Perhaps the most comprehensive situation of urban aerosol formation and growth is that which takes place in the presence of hydrocarbons, oxides of nitrogen, and sulfur dioxide. Husar and Whitby (1973) proposed the following mechanism for photochemical aerosol formation:

The driving force for the gas-particle conversion is provided by a gaseous photochemical reaction or chain of reactions. The gaseous reaction(s) produces a supply of molecular species (or radicals) which, upon collision with each other, agglomerate and form molecular clusters, i.e., homogeneously nucleate. If suitable aerosol particles or ions are present, the monomers or radicals deposit preferentially on the existing surfaces and thus the nucleation is heterogeneous. The growth rate of the newly formed particles is controlled by the diffusion rate; i.e., the collision rate of the condensable species. The condensation[1] itself may be physical (governed by supersaturation) or chemical (if the condensable species react with each other upon collision). If the concentration of the droplets is sufficiently high, they may interact by coagulation.

Thus, in the atmosphere the mode of nucleation of aerosols from gaseous pollutants is believed to be, in most cases, heterogeneous. However, in laboratory systems in which the air is carefully filtered to remove all particles an aerosol is still observed to form. For example, in an initially particle-free chamber, Husar and Whitby found that the chronological evolution of a photochemical aerosol may be characterized by:

1 Initial rapid increase in total number concentration by homogeneous nucleation
2 Coagulation when the aerosol concentration reaches a sufficiently high level (maximum concentration when production rate equals coagulation rate)
3 The attainment of a steady-state surface area sufficient to accommodate all the vapor (the nucleation rate then diminishes); decay of number concentration by coagulation; gas-particle conversion rate associated with the steady-state surface areas

The degree to which homogeneous nucleation is important in the atmosphere is still unknown.

In summary, the available evidence seems to indicate that aerosol formation and growth in photochemical smog are basically a result of the condensation of gas-phase species (presumably higher-molecular-weight organic species) onto existing nuclei, such as sea-salt particles and lead aerosol from automobile exhaust. Coagulation processes may be important for the very small particles; however, within the 0.1 to 1 μm range, the most important growth mechanism appears to be diffusion of gaseous species to the particles, followed by absorption, as shown below.

[1] In this context, the term "condensation" is used to designate a diffusion-controlled growth process, regardless of the nature of the accommodation process, i.e., physical or chemical.

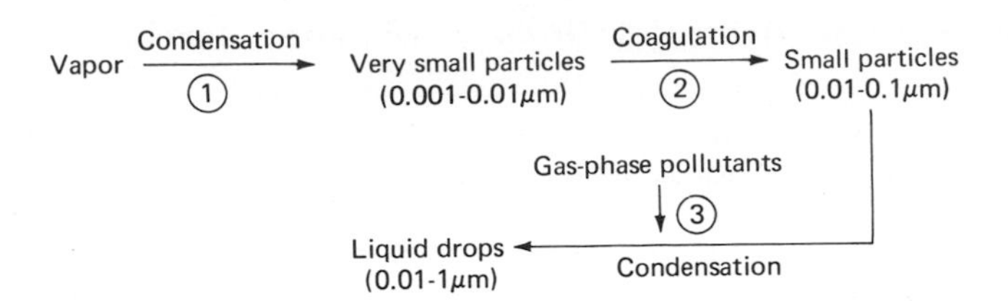

Dynamics of Aerosol Size and Chemical Composition Distributions

It should be evident, even from our brief discussion, that the dynamics of a population of small particles can depend on a number of different physical and chemical mechanisms. It is of considerable interest to be able to describe mathematically the dynamics of an aerosol population as a result of these mechanisms. We introduced the size-composition probability density function (pdf) in Chap. 2. In principle, this function embodies all the desired information about a distribution of particles. Thus, the elucidation of the general differential equation which governs the dynamics of this function will provide the means to describe the changes in an aerosol population.

In Chap. 2 we defined the size-composition pdf as $g(n_1, n_2, \ldots, n_K)$ for K species. By its definition

$$\int \cdots \int g(n_1, \ldots, n_K)\, dn_1 \cdots dn_K = 1$$

Given $n_1, n_2, \ldots, n_K$, the size of a particle can be computed by

$$v = n_1\bar{v}_1 + n_2\bar{v}_2 + \cdots + n_K\bar{v}_K$$

where $\bar{v}_i$ is the partial molar volume of species i. (Thus, if we wish to include particle size, say D_p, formally in the argument of g we need only specify any K-1 molar quantities, since the total volume is fixed once all K are given.)

We shall now derive a general dynamic equation governing $g(n_1, n_2, \ldots, n_K, t)$ which includes the following processes:

1 Condensation
2 Coagulation
3 Chemical reaction among species in the particles

Thus, from a physical point of view, we are considering the time evolution of a population of particles in a system in which there are no appreciable variations of g from point to point in the system. Consequently, we are neglecting any spatial dependence of g and, therefore, convection, diffusion, and sedimentation processes.

The K-dimensional vector space S of $(n_1, n_2, \ldots, n_K)$ may be taken as an orthogonal coordinate system. Since each particle has a set of values $(n_1, n_2, \ldots, n_K)$ associated with it, there is a one-to-one correspondence between the particles and the points in S space. Motion of a particle in S space may be a result of two different phenomena: (1) The particle can gain mass by condensation of vapor (if the vapor is of species 1, the point in S space moves in the n_1 direction with all other coordinates remaining fixed) and (2) chemical reactions may take place in the particle (if, for example, 2 moles of species 1 combine to form 1 mole of species 2, the corresponding point in S space will decrease by two units in the n_1 direction and increase one unit in the n_2 direction). Motion by either of these two processes can be called *drifting*.

The number of points in a small volume of S space may change with time in three ways: (1) by drifting, (2) by extinction (simply disappearing), and (3) by creation (appearing). Extinction of a point occurs when a particle collides and coalesces with another particle. Simultaneously, a new point appears in another part of S space, representing the new particle that has the sum of the masses of the two colliding particles. This process can be called *collection*. (The creation and extinction of points associated with the processes of homogeneous nucleation and evaporation could also be considered but will be neglected here.)

Two intuitive conservation laws can be formulated (Berry, 1965). First, under drifting alone the number of points in S space (which corresponds to the number of particles) is conserved. Second, under collection alone the total mass of particles is conserved. Hence, the total rate of change of the particle concentration can be expressed as the sum of two terms:

$$\left.\frac{\partial(N_\infty g)}{\partial t}\right|_{\text{total}} = \left.\frac{\partial(N_\infty g)}{\partial t}\right|_{\text{drifting}} + \left.\frac{\partial(N_\infty g)}{\partial t}\right|_{\text{collection}} \tag{4.24}$$

Let us consider first the second term on the right-hand side of (4.24), namely the change in the number density of particles due to coagulation. The basic representation of coagulation processes was already presented as (4.23). Equation (4.23) was written assuming a discrete number M of sizes. Actually, M is very large, and so it is best to assume a continuous distribution of sizes or, equivalently, a continuous distribution of chemical compositions. Therefore, we have

$$\left.\frac{\partial(N_\infty g)}{\partial t}\right|_{\text{collection}} = \frac{1}{2} N_\infty^2 \int_0^{n_1} \cdots \int_0^{n_K} b(v', v - v') g(\mathbf{n}') g(\mathbf{n} - \mathbf{n}')\, d\mathbf{n}' - N_\infty^2 \int_0^\infty \cdots \int_0^\infty b(v,v') g(\mathbf{n}) g(\mathbf{n}')\, d\mathbf{n}' \tag{4.25}$$

where the collision parameter $b(v,v')$ incorporates all the information on the probability of two particles colliding and sticking. In (4.25) we have assumed that b is independent of chemical composition and depends only on the volumes of the two particles. A great deal of effort has gone into the prediction of b for various physical situations, such as, for example, brownian-motion coagulation (the particles' motion is due to thermal agitation by molecules of the gas) and turbulent coagulation (particle motion is due to the turbulent velocity field). A discussion of the functional form of b in these different situations may be found in Hidy and Brock (1970).

Equation (4.25) is written in its most general form. Ordinarily we deal with particles of uniform chemical composition. In that case $g(n_1, \ldots, n_K, t)$ reduces simply to $g(v,t)$, where $N_\infty g(v,t)\, dv$ is the number of particles per unit volume having volume in the range v to $v + dv$ at time t. The total number of collisions between particles of volume $(m, m + dm)$ and $(p, p + dp)$ to give $(v, v + dv)$ is given by $N_\infty^2 b(m,p) g(m,t) g(p,t)\, dm\, dp\, dt$. Thus, the rate of change of the number of particles per unit volume by coagulation is

$$\frac{\partial[N_\infty g(v,t)]}{\partial t} = \frac{1}{2} N_\infty^2 \int_0^v \int_0^m b(m,p) g(m,t) g(p,t)\, dp\, dm - g(v,t)\, dv N_\infty^2 \int_0^\infty b(m,v) g(m,t)\, dm \tag{4.26}$$

Since $v = m + p$, the first term on the right-hand side can be simplified to

$$\frac{dv}{2} \int_0^v b(m, v - m) g(m,t) g(v - m, t)\, dm$$

Thus, (4.26) becomes

$$\frac{\partial[N_\infty g(v,t)]}{\partial t} = \frac{N_\infty^2}{2} \int_0^v b(m, v - m) g(m, t) g(v - m, t)\, dm - N_\infty^2 g(v,t) \int_0^\infty b(m,v) g(m,t)\, dm \tag{4.27}$$

which is the continuous version of (4.23), as well as the special case of (4.25) when only the volume dependence of g is of interest.

Having determined the form of the second term on the right-hand side of (4.24), let us now turn to the first term on the right-hand side, i.e. that due to drifting (condensation and chemical reaction). Let us consider an arbitrary volume Ω with surface area A in S space. We have noted that under drifting alone the number of points in S space is conserved. Thus, the change in the number of points in the volume Ω is equal to the net number of points drifting across A into Ω per unit time. Expressed mathematically,

$$\frac{\partial}{\partial t}\int_{\Omega} N_{\infty} g \, d\Omega = -\int_{A} N_{\infty} g \mathbf{u} \cdot \mathbf{i} \, dA \tag{4.28}$$

where $\mathbf{u}$ is the local particle velocity in S space and $\mathbf{i}$ is the outwardly directed unit normal to the surface of the volume A. Equation (4.28) merely states that the accumulation of particles in the volume Ω is attributable to the net flux of particles through the surface of Ω. The velocity $\mathbf{u}$ is not a true physical velocity but represents the rate of movement of particles along the n_i coordinates i.e.

$$\mathbf{u} = \left(\frac{\partial n_1}{\partial t}, \frac{\partial n_2}{\partial t}, \ldots, \frac{\partial n_K}{\partial t}\right)$$

A K-dimensional generalization of the divergence theorem can be invoked to express the surface integral on the right-hand side of (4.28) in terms of a volume integral

$$\int_{A} N_{\infty} g \mathbf{u} \cdot \mathbf{i} \, dA = \int_{\Omega} \nabla \cdot (N_{\infty} g \mathbf{u}) \, d\Omega \tag{4.29}$$

where ∇ is the gradient operator,

$$\nabla = \left(\frac{\partial}{\partial n_1}, \frac{\partial}{\partial n_2}, \ldots, \frac{\partial}{\partial n_K}\right)$$

Combining (4.28) and (4.29), we have, for arbitrary Ω,

$$\frac{\partial(N_{\infty} g)}{\partial t} = -\nabla \cdot (N_{\infty} g \mathbf{u}) \tag{4.30}$$

Thus, the drifting contribution to (4.24) is, rewriting (4.30),

$$\left.\frac{\partial(N_{\infty} g)}{\partial t}\right|_{\text{drifting}} = -\sum_{i=1}^{K} \frac{\partial}{\partial n_i}\left(N_{\infty} g \frac{\partial n_i}{\partial t}\right) \tag{4.31}$$

As we noted, the change in the molar quantities of species in a particle can result from condensation of gaseous species on the particle or chemical reaction within the particle. If the rates of condensation and reaction are expressed as $I_i(\mathbf{n},t)$ and $R_i(\mathbf{n},t)$, respectively, then the change in moles in a particle can be written

$$\frac{\partial n_i}{\partial t} = I_i(\mathbf{n},t) + R_i(\mathbf{n},t) \tag{4.32}$$

Substituting (4.32) into (4.31) and then using (4.31) and (4.26) in (4.24), we obtain the complete dynamic equation governing the composition pdf $g(n_1, n_2, \ldots, n_K)$:

$$\begin{aligned}\frac{\partial(N_{\infty} g)}{\partial t} = &-\sum_{i=1}^{K} \frac{\partial}{\partial n_i}[I_i(\mathbf{n},t)N_{\infty} g] - \sum_{i=1}^{K} \frac{\partial}{\partial n_i}[R_i(\mathbf{n},t)N_{\infty} g] \\ &+ \frac{1}{2} N_{\infty}^2 \int_0^{n1} \cdots \int_0^{nK} b(v', v - v') g(\mathbf{n}',t) g(\mathbf{n} - \mathbf{n}', t) \, d\mathbf{n}' \\ &- N_{\infty}^2 g(\mathbf{n},t) \int_0^{\infty} \cdots \int_0^{\infty} b(v,v') g(\mathbf{n}',t) \, d\mathbf{n}'\end{aligned} \tag{4.33}$$

Equation (4.33) provides the means, in principle, to determine the effect of condensation, chemical reaction, and coagulation on a population of chemically heterogeneous particles. As one might suspect, in its general form (4.33) is virtually impossible to solve analytically. In fact, most studies of aerosol dynamics have neglected any consideration of the chemical composition of the particles and have instead concentrated on the size distribution alone. In addition, condensation and coagulation are rarely considered simultaneously. For a population of chemically homogeneous particles growing by condensation alone

$$\frac{\partial g(v,t)}{\partial t} = - \frac{\partial}{\partial v} [I(v,t)g(v,t)] \qquad (4.34)$$

For a population of chemically homogeneous particles undergoing only coagulation, we have already derived the applicable equation (4.27).

We shall not discuss here the determination of the functions and parameters in (4.33) that must be specified in any given situation [that is, $I_i(\mathbf{n},t)$, $R_i(\mathbf{n},t)$, and $b(v,v')$]. The condensation rate I_i depends on the gas involved, its vapor pressure and molecular weight, the size of the particle, the diffusion coefficient for the species in the vapor phase, etc. The collision frequency $b(v,v')$ depends on the physical mechanism responsible for coagulation, i.e. brownian motion, turbulent mixing, etc. A detailed treatment of these functions can be found in Hidy and Brock (1970).

4.7 MEASUREMENT METHODS FOR GASEOUS AIR POLLUTANTS

The measurement of the concentrations of gaseous air pollutants presents a number of special problems not routinely encountered in the analysis of gas compositions. The basic source of difficulty is the extremely low concentration levels involved, levels almost always below 1 ppm by volume. A second source of problems in measurement is the interferences which can exist when two or more pollutants are present in the same sample. As a result of these inherent difficulties in measurement, only very recently have techniques been developed which offer promise for rapid, accurate measurement of pollutant concentrations at ambient levels. In this section we discuss these analytical techniques.

Analytical instruments fall into two classes. The first class includes instruments which may be used to make a direct measurement of some *physical* property of the species under observation. An example of such an instrument is the infrared spectrometer, which detects the amount of infrared radiation absorbed by specific pollutants. The second class of instruments, and by far the most common, is the chemicophysical. In this case the pollutant being measured first undergoes chemical transformation; the product of the reaction is then measured by using an appropriate analytical technique. As an example, the classic "wet" chemical techniques are all chemicophysical. In these, a sample of the pollutant is collected for a period of time in a liquid absorber. The sample is then reacted with another reagent, resulting in a change in color or in the formation of another product. The intensity of the color or the concentration of the products is related to the concentration of the original pollutant. Many newer types of instrumentation, such as chemiluminescent devices (which are based on measuring the light emitted when the pollutant being monitored reacts with another reagent), are also chemicophysical.

A number of criteria may be specified for evaluating the appropriateness and utility of an analytical technique for a particular application. These include:

1 The accuracy,[1] specificity, and ease of calibration of the instrument
2 The precision[1] or reproducibility of the measurements
3 The volume of gas required to make a determination
4 The response time of the instrument

In the discussion that follows, we first examine these criteria individually. We then use them in evaluating the quality of measurements to be expected from analytical instruments which either have been employed extensively in the past, or have recently been developed, to measure NO_2, NO, O_3, hydrocarbons, CO, and SO_2.

4.7.1 Accuracy and Primary Standards

The accuracy of a method, that is, the extent to which the observed or measured value and the "true" value agree, depends on both the specificity of the technique and the accuracy of calibration. Lack of specificity, or the presence of interferences due to species other than the one desired, is the greatest shortcoming of all wet methods. For example, in the measurement of ozone by the potassium iodide (KI) method (see Subsec. 4.7.4), peroxides, NO_2, and peroxyacyl nitrates, in addition to ozone, all give a positive response, whereas SO_2 and dust give a negative response relative to the value observed without the interference. Thus, the accuracy of the indicated ozone concentration is, at best, questionable when these other species are present.

The accuracy of calibration procedures depends upon the availability of primary standards. The well-established wet chemical techniques fare very well in this regard, but suitable primary standards are still being sought for many of the gas-phase

[1] *Accuracy* refers to the extent to which a given measurement agrees with the true but unknown value of the quantity being measured. *Precision* refers to the extent to which a given set of measurements agrees with the mean of the observations. Thus, if in a given experiment we observe y_i and the true but unknown value of the observed quantity is η_i, we might write $y_i = \eta_i + b_i + \varepsilon_i$, where b_i is a bias in the measurement (due perhaps to interference by another species) and ε_i is a random error, say, normally distributed with zero mean and variance σ^2. If b_i is large relative to η_i, y_i is inaccurately determined, whereas if σ is large, with the result that ε_i is large, y_i is imprecisely determined. Thus, accuracy and precision are measures of bias and scatter, respectively.

monitoring techniques. For instance, as yet no primary standard for ozone has been developed, and ozone sources such as air ozonizers must be calibrated by potassium iodide titration or other techniques before each use as an ozone source. Nitric oxide and many hydrocarbons, on the other hand, can be obtained in prepurified gas cylinders and can thus serve as precision gas standards for those species.

O'Keefe and Ortman (1966) developed permeation tubes as calibration standards. They found that liquefied gases, such as SO_2, NO_2, and hydrocarbons, sealed in teflon tubing, would permeate through the walls of the tubing at a constant rate which is dependent upon the surface area of the tubing and the temperature of the surroundings. The success of this technique depends on the fact that the equilibrium vapor pressure of a liquid is a constant when temperature is held constant. Under these conditions, the permeation rate of the gas through the tube is constant, and the absolute amount of gas released by the tube over a period of time can be calculated from the change in weight of the tube. Permeation tubes have been used successfully for SO_2, H_2S, many hydrocarbons, anhydrous NH_3, anhydrous HF, $COCl_2$ (phosgene), and organic mercury compounds (see Saltzman et al., 1971). Unfortunately, it has been found that NO_2 permeation tubes are seriously affected by relative humidity and past exposure. Thus, permeation tubes are not suitable as a primary standard for NO_2 unless they have been recently calibrated.

The frequency of recalibration must be established for each instrument in the laboratory. Loss of calibration can occur as a result of aging and decomposition of chemical reagents or because of changes in ambient operating conditions (i.e., temperature, relative humidity, pressure). The need for recalibration also depends on the sensitivity of the plumbing, the stability of the electronic components, and the conditions under which the instrument is used. Recalibration should be performed sufficiently often to assure the accuracy of the results. Any observed drift between calibrations should, of course, be noted. Finally, the calibration experiments should be performed over the full concentration ranges of the species observed during an experiment (e.g., concentrations of NO_2 vary from 0.005 to 2 ppm).

4.7.2 Precision

Precision refers to the extent to which a given set of measurements agrees with the mean of the observations. Although the precision of each instrument must be determined individually, it has been found, in general, that both wet chemical techniques and instrument-based techniques such as chemiluminescence produce highly reproducible results (Maugh, 1972). It is, of course, important that the precision of the measurements be reported with the results. Care should be taken to establish the precision as a function of concentration.

4.7.3 Sampling Procedures

The reliability of a measurement depends to a large degree on the technique used to sample the bulk gas. The sampling-tube material should be selected to minimize the heterogeneous loss of pollutants during sampling. Further, the residence time in the sampling line should be as short as possible to minimize the possibility of chemical reaction along the path. The overall accuracy of a measurement depends upon careful calibration of the airflow rate into the monitoring equipment, for the signal output of the instrument must be corrected for the total volume of gas being determined.

One consideration related to sampling is that it is often desirable to minimize the total volume of air required for the analytical instruments, particularly when dealing with laboratory reaction systems. Finally, knowledge of the response time of the measuring systems is necessary for interpreting the data relative to real time. Response time is the time from the moment the pollutant enters the sampling instrument to the moment that the measured value is obtained. The measured value, in turn, is taken to be a specified percentage of the final or true value, such as 90 percent. Thus, if it requires 2 min to achieve 90 percent of full response, 2 min is the response time.

Measurements are generally made on a continuous, discrete, or average time basis. The first and most desirable type of measurement is *continuous* and is carried out with instruments having very fast response times, such as infrared spectrometers. A second class of measurements, *discrete*, includes those analyses in which a sample is injected into the instrument. Gas-chromatographic analysis is an example of a discrete procedure. The sensitivity of discrete measuring instruments to short-term fluctuations and sharp peak concentrations depends upon the rapidity with which determinations can be made. In the limit, as the time between measurements goes to zero, discrete measurements, of course, become the same as continuous measurements. Monitoring devices which require a pollutant to be collected over a finite period of time in order to obtain measurable concentrations yield *integrated* or averaged results. Most wet techniques operate on this principle. Such methods suffer from the obvious disadvantage that they are insensitive to short-term variations in concentration.

4.7.4 Analytical Techniques Available for Measuring Air Pollutants

Because of the wide variety of analytical instruments presently available for monitoring air pollutants, there is often uncertainty concerning which procedures are "best." Although recommended procedures are published regularly by the United States government in *The Federal Register*, review articles on current trends in instrumentation suggest that techniques now in use will soon be replaced by more rapid, more reliable, more efficient, and certainly more expensive techniques (Maugh, 1972). A logical goal in upgrading analytical devices would be to develop a single instrument capable of measuring many or all of the pollutants simultaneously. Microwave spectrometers, which measure the rotational energy of polar molecules at discrete frequencies, appear promising, with commercial availability in the near future (Maugh, 1972). Because of the uncertainties concerning analytical procedures, it is worthwhile to examine briefly the techniques most often used to measure the standard pollutants and to discuss the advantages and disadvantages of techniques now becoming available.

Nitrogen dioxide Nitrogen dioxide has generally been determined by the Saltzman method (Saltzman, 1954). In this technique the sample is bubbled through an absorbing reagent at a rate of about 0.4 l/min for 10 min in order to concentrate the NO_2. Reaction of NO_2 with the absorbing solution results in a color change which can be related to the NO_2 concentration. The technique is quite reproducible but suffers interferences from ozone and PAN. The primary sources of error in this method are chemical interferences, inefficiency of collection, and the aging of reagents.

No substantial improvement over the Saltzman method has, as yet, been made for determining NO_2. Hodgeson et al. (1972) suggested reducing the NO_2 to NO, using a gold catalyst and measuring the resulting NO by chemiluminescence (see the following subsection). The problem with this approach is that the gold reduces all the higher oxides of nitrogen (i.e., nitrates, PAN, nitrite, NO_2) to NO, and so the technique lacks specificity. One promising approach for measuring NO_2 which is currently being developed involves exciting NO_2 to a higher energy state with a tunable dye laser. The NO_2 concentration can then be determined by measuring the fluorescence of the excited molecules.

Nitric oxide Until recently, there was no technique available for determining nitric oxide concentrations directly. The pollutant gas is passed over an oxidizing agent such as dichromate which oxidizes NO to NO_2. Total oxides of nitrogen are then determined as NO_2 by the Saltzman method, and NO is calculated by subtracting the NO_2 reading made at the same time from total NO_X. In addition to all the difficulties involved with the Saltzman method, the technique depends on the efficiency of oxidation of NO. Efficiencies of only 70 to 85 percent are not uncommon, and thus measurements of NO by this method are unreliable.

Stedman et al. (1972) and Hodgeson et al. (1972) have shown that NO can be determined very accurately by *chemiluminescence*. In this technique the pollutant sample is exposed to a large excess of ozone. The ozone reacts rapidly with the NO to form NO_2 in an excited state:

$$NO + O_3 \longrightarrow NO_2^* + O_2$$

Collapse of the excited NO_2 to the ground state is accompanied by the emission of light, which is then measured and related to the initial NO.

$$NO_2^* \longrightarrow NO_2 + \text{light}$$

This technique is sensitive to 10^{-3} ppm of NO, has an overall response time of 10 sec and requires a flow rate of air of only 20 cm^3/min. Interferences can be caused by collisional deactivation of the excited NO_2 (this problem is reduced by operating the system at low pressures) and by the reaction of other species present with ozone, the product of which will also chemiluminesce. Stedman et al. investigated the possibility of interferences from SO_2, NO_2, Cl_2, H_2O, CH_4, C_2H_2, C_2H_4, and C_3H_6 at a concentration of 100 ppm in air but found that these species did not emit a detectable amount of light in the region of the NO_2^* luminescence. An interesting feature of this instrument is that it can be used as an ozone detector as well by subjecting an air sample to an excess of NO. In view of its accuracy, response time, sensitivity, stability, and versatility, this instrument is being rapidly adopted as the standard measuring technique for NO.

Ozone Traditionally, ozone determination has been based on reaction of the species with potassium iodide (KI)

$$2KI + O_3 + H_2O \longrightarrow I_2 + O_2 + 2KOH$$

and measurement of the amount of iodine which is liberated by the reaction. Under well-controlled laboratory conditions, this method can be quite accurate, and the technique remains as one important method for calibrating ozone sources. However, automatic monitors for ozone based on the KI reaction suffer from numerous difficulties. Several species, including NO_2, peroxides, and PAN, give a positive response for ozone in varying degrees, but the presence of SO_2 and reducing dusts results in a serious negative interference. Although some of the interfering species such as SO_2 can be removed by chemical scrubbing before the pollutant gases reach the ozone monitor and others can be corrected for mathematically (for example NO_2), the KI method must be regarded, at best, as a method for determining the total *oxidant* concentration in a sample. An additional difficulty with the method is that the oxidant must be collected in a concentrating solution by bubbling gas at the rate of 1 to 2 l/min for 15 mins; full development of the color requires an additional 30 to 60 min. Thus, in summary, the KI method provides a nonspecific, averaged oxidant concentration with poor response time.

Two techniques have recently been developed as commercial products for determining ozone. The first method is based on the chemiluminescence of the ozone-ethylene reaction (Warren and Babcock, 1970). An air sample containing ozone is exposed to a large excess of ethylene in a reaction vessel. The light released during the reaction is then measured, using a photomultiplier tube, and related to the ozone concentration.

Ozone can be determined by its chemiluminescent reaction with many species besides ethylene. We have already mentioned that the reaction of ozone with NO has been used successfully by Stedman et al. (1972). Kummer et al. (1971) pointed out that ozone will undergo chemiluminescent reactions with many olefins and sulfides, including tetramethylethylene and dimethylsulfide, and that these reactions result in a much greater intensity of emitted light than the ozone-ethylene reaction.

Chemiluminescent methods for ozone appear to be free of interferences from other species, and they are sensitive and fast. However, they suffer the disadvantage of having to be operated at very low pressures in the reaction vessel (0.5 to 5 torr). Tanks of gaseous reagents must also be used in operating the equipment.

Hydrocarbons Most routine hydrocarbon monitoring is done with the *flame ionization detector*. The principle of operation of the flame ionization detector (FID) is that the sample gas is ionized and then the electric conductivity of the resulting gas is measured. In this apparatus, enough hydrogen is mixed with the sample gas to maintain a steady hydrogen-oxygen flame. Relatively few ions are produced in the H_2–O_2 flame, but if hydrocarbons are present the conductivity of the flame is greatly increased. The response of the instrument increases as the number of CH groups in the hydrocarbons in the sample increases. Thus, the FID measures only the the total amount of organically bound carbon atoms in the sample and does not distinguish among individual species. The limiting threshold of the FID to, say, propane is 0.003 ppm. The FID is insensitive to the presence of NO, NO_2, CO, CO_2, SO_2, and H_2O.

Measurements of the concentrations of individual hydrocarbon species are obtained by gas chromatography, in which a mixture of inert (carrier) gas and the sample is passed at a constant rate through a bed of high-porosity solids. As the mixture travels through the bed, individual components of the sample are preferentially adsorbed on the solid. As a result, some molecules spend relatively more time attached to the solid phase than do others and therefore require more time to pass through the column. The degree of separation achieved is expressed by the partition ratio for each species, which is the ratio of the equilibrium concentration of the species in the solid phase to that in the mobile phase. A complex mixture injected into the column emerges from the column as a series of bands of individual components. Thus, the column serves as a means of separating the species. After leaving the column, the identification is made by a detector, which might range from a high-resolution mass spectrometer to a thermal conductivity cell. In addition, compounds can be characterized by their retention time on the column. Resolution in a gas chromatograph can be extremely fine, and this method is the preferred way to carry out hydrocarbon measurements.

Carbon monoxide The basic procedure for measuring atmospheric CO concentrations is by *nondispersive infrared spectrometry* (NDIR). In this instrument the absorption spectra of molecules present in the sample are measured in the infrared wavelength range (2.5 to 25 μm). The infrared spectrum of CO is quite distinctive. In the NDIR, the removal of infrared radiation by the absorption bands of CO is used to differentially cool one side of a two-sided sample cell. The uneven heating leads to the displacement of a membrane separating the two sides of the cell. The displacement is measured electronically and related to a prior calibration to yield the quantity of CO in the sample.

There are three major infrared absorbing gases in air, CO, CO_2, and H_2O, and thus interferences can exist when both CO_2 and H_2O are present (especially H_2O). The precision that can be achieved for CO through NDIR is about ± 1 ppm, although without proper filtering of the radiation, errors as large as ± 5 ppm can result because of humidity variations.

An alternative technique for measuring CO is based on the reaction of mercuric oxide with CO at 210°C to produce CO_2 and mercury:

$$HgO + CO \longrightarrow CO_2 + Hg$$

The mercury vapor is then measured through its absorption of radiation from a mercury lamp. The range of applicability of this method is from 0.025 to 10 ppm of CO.

Sulfur dioxide The two most common procedures for measuring atmospheric SO_2 concentrations are wet chemical in nature. The West-Gaeke method is based on converting the SO_2 to pararosaniline methyl sulfonic acid, a reddish compound, the concentration of which is determined spectrophotometrically. The hydrogen peroxide method involves absorption of SO_2 in H_2O_2 in which it is oxidized to sulfuric acid. The quantity of H_2SO_4 is determined by titration. The limit of detection by each method is about 0.005 ppm. The West-Gaeke method does not suffer from interference of other species as does the hydrogen peroxide method. Both methods are reviewed in Selected Methods for the Measurement of Air Pollutants (1965).

REFERENCES

Aerosol Measurements in Los Angeles Smog, vol. I, *U.S. Environ. Protection Agency, Rep. APTD*–0630, 1971.

ALTSHULLER, A. P., and J. J. BUFALINI: Photochemical Aspects of Air Pollution: A Review, *Environ. Sci. Technol.*, **5**:39 (1971).

ALTSHULLER, A. P., and I. R. COHEN: Structural Effects on the Rate of NO_2 Formation in the Photooxidation of Organic Compounds-NO Mixtures in Air, *Int. J. Air Water Pollut.*, **7**:1043 (1963).

ALTSHULLER, A. P., W. A. LONNEMAN, F. D. SUTTERFIELD, and S. L. KOPCZYNSKI: Hydrocarbon Composition of the Atmosphere of the Los Angeles Basin—1967, *Environ. Sci. Technol.*, **5**:1009 (1971).

BENSON, S. W.: Some Current Views of the Mechanism of Free Radical Oxidations, *Adv. Chem.*, **76**:143 (1968).

BERRY, E. X.: Cloud Droplet Growth by Collection: A Theoretical Formulation, Ph.D. dissertation, Univ. of Nevada, Reno, 1965.

BUFALINI, M.: The Oxidation of Sulfur Dioxide in Polluted Atmospheres: A Review, *Environ. Sci. Technol.*, **5**: 685 (1971).

CADLE, R. D., and E. R. ALLEN: Atmospheric Photochemistry, *Science*, **167**:243 (1970).

CALVERT, J. G.: Interactions of Air Pollutants, presented at National Academy of Sciences, Conference on Health Effects of Air Pollutants, Washington, D.C., Oct. 3–5, 1973.

CALVERT, J. G., J. A. KERR, K. L. DEMERJIAN, and R. D. MCQUIGG: Photolysis of Formaldehyde as a Hydrogen Atom Source in the Lower Atmosphere, *Science*, **175**:751 (1972).

CHENG, R. T., M. CORN, and J. O. FROHLIGER: Contribution to the Reaction Kinetics of Water Soluble Aerosols and SO_2 in Air at ppm Concentrations, *Atmos. Environ.*, **5**:987 (1971).

COX, R. A., and S. A. PENKETT: Oxidation of Atmospheric SO_2 by Product of the Ozone-Olefin Reaction, *Nature*, **230**:321, April 2, 1971.

DAVIS, D. D., W. A. PAYNE, and L. J. STIEF: The Hydroperoxyl Radical in Atmospheric Chemical Dynamics: Reaction with Carbon Monoxide, *Science*, **179**:280 (1973).

DAVIS, D. D., W. WONG, W. A. PAYNE, and L. J. STIEF: A Kinetics Study to Determine the Importance of HO_2 in Atmospheric Chemical Dynamics: Reaction with CO, presented at the Symposium on Sources, Sinks, and Concentrations of CO and CH_4 in the Earth's Environment, St. Petersburg Beach, Fla., August, 1972.

DEMERJIAN, K. L., J. A. KERR, and J. G. CALVERT: The Mechanism of Photochemical Smog Formation, in J. N. PITTS, JR. and R. L. METCALF (eds.), "Advances in Environmental Science," vol. 3, Wiley Interscience. New York, 1973.

DIMITRIADES, B.: Effects of Hydrocarbon and Nitrogen Oxides on Photochemical Smog Formation, *Environ. Sci. Technol.*, **6**:253 (1972).

DODGE, M. C., and J. J. BUFALINI: The Role of Carbon Monoxide in Polluted Atmospheres, *Adv. Chem.* **113**:232 (1972).

ECCLESTON, B. H., and R. W. HURN: Comparative Emissions from Some Leaded and Prototype Lead-free Automobiles, *U.S. Dept. Interior*, *Bureau of Mines*, TN23.U7 No. 7390 622.06173, 1970.

FOSTER, P. M.: The Oxidation of Sulfur Dioxide in Power Station Plumes, *Atmos. Environ.*, **3**:157 (1969).

FRIEDLANDER, S. K.: Chemical Element Balances and Identification of Air Pollution Sources, *Environ. Sci. Technol.*, **7**: 235 (1973).

GARTRELL, F. E., F. W. THOMAS, and S. B. CARPENTER: Atmospheric Oxidation of SO_2 in Coal Burning Power Plant Plumes, *Am. Ind. Hygiene Assoc. J.*, **24**:113 (1963).

GERHARD, E. R., and H. F. JOHNSTONE: Photochemical Oxidation of Sulfur Oxide in Air, *Ind. Eng. Chem.*, **47**:972 (1955).

GHORMLEY, J. A., R. L. ELLSWORTH, and C. J. HOCHANANDEL: Reaction of Excited Oxygen Atoms with Nitrous Oxide. Rate Constants for Reaction of Ozone with Nitric Oxide and with Nitrogen Dioxide, *J. Phys. Chem.*, **77**:1341 (1973).

GLASSON, W. A., and C. S. TUESDAY: The Atmospheric Thermal Oxidation of Nitric Oxide, *J. Am. Chem. Soc.*, **85**:2901 (1963).

GLASSON, W. A., and C. S. TUESDAY: Hydrocarbon Reactivities and the Kinetics of the Atmospheric Photooxidation of Nitric Oxide, *J. Air Pollut. Control Assoc.*, **20**:239 (1970).

GRAY, D., E. LISSI, and J. HEICKLEN: The Reaction of Hydrogen Peroxide with Nitrogen Dioxide *J. Phys. Chem.*, **76**:1919 (1972).

HAAGEN-SMIT, A. J.: Chemistry and Physiology of Los Angeles Smog, *Ind. Eng. Chem.* **44**:1423 (1952).

HAAGEN-SMIT, A. J., and M. FOX: Ozone Formation in Photochemical Oxidation of Organic Substances, *Ind. Eng. Chem.*, **48**:1484 (1956).

HEICKLEN, J.: Gas Phase Reactions of Alkylperoxy and Alkoxy Radicals, *Adv. Chem.*, **76**:23 (1968).

HEUSS, J. M., and W. A. GLASSON: Hydrocarbon Reactivity and Eye Irritation, *Environ. Sci. Technol.*, **2**:1109 (1968).

HIDY, G. M. (ed.): "Aerosols and Atmospheric Chemistry," Academic Press, New York, 1972.

HIDY, G. M., and J. R. BROCK: "Dynamics of Aerocolloidal Systems," Pergamon Press, New York, 1970.

HIDY, G. M., and S. K. FRIEDLANDER: The Nature of the Los Angeles Aerosol, in H. M. ENGLUND and W. T. BEERY, (eds.), "Proceedings of the Second International Clean Air Congress," Academic Press, New York, 1971.

HODGESON, J. A., K. A. REHME, B. E. MARTIN, and R. K. STEVENS: Measurements for Atmospheric Oxides of Nitrogen and Ammonia by Chemiluminescence, presented at the Air Pollution Control Association Meeting, Miami Beach, Fla., June, 1972.

HOLMES, J. R., R. J. O'BRIEN, J. H. CRABTREE, T. A. HECHT, and J. H. SEINFELD: Measurement of Ultraviolet Radiation Intensity in Photochemical Smog Studies, *Environ. Sci. Technol.*, **7**:519 (1973).

HUSAR, R. B., and K. T. WHITBY: Growth Mechanisms and Size Spectra of Photochemical Aerosols, *Environ. Sci. Technol.*, **7**:241 (1973).

JAFFE, S., and H. W. FORD: The Photolysis of Nitrogen Dioxide in the Presence of Nitric Acid at 3660Å and 25°, *J. Phys. Chem.*, **71**:1832 (1967).

JOHNSTON, H. S.: Gas Reaction Kinetics of Neutral Oxygen Species, NSRDS–NBS20, National Bureau of Standards, Washington, D.C., 1968.

JOHNSTON, H. J.: A Review of the CMVE Panel Report, National Academy of Sciences, Washington, D.C. (1973).

JOHNSTON, H. S., and J. H. CROSBY: Kinetics of the Fast Gas Phase Reaction Between O_3 and NO, *J. Chem. Phys.*, **22**:689 (1954).

JOHNSTON, H. S., J. N. PITTS, JR., J. LEWIS, L. ZAFONTE, and T. MOTTERSHEAD: Atmospheric Chemistry and Physics, Project Clean Air, Task Force Assessments, vol. 4, University of California, Berkeley, 1970.

JOHNSTONE, H, F., and D. R. COUGHANOWR: Absorption of Sulfur Dioxide from Air, *Ind. Eng. Chem.*, **50**:1169 (1958).

JOHNSTONE, H. F., and A. J. MOLL: Formation of Sulfuric Acid in Fogs, *Ind. Eng. Chem.*, **52**:861 (1960).

JUNGE, C. E.: "Air Chemistry and Radioactivity," Academic Press, New York, 1963.

JUNGE, C. E., and T. G. RYAN: Study of the SO_2 Oxidation in Solution and its Role in Atmospheric Chemistry, *Q. J. Roy. Meteorol. Soc.*, **84**:46 (1958).

KATZ, M.: Photochemical Reactions of Atmospheric Pollutants, *Can. J. Chem. Eng.*, **48**:3 (1970).

KATZ, M., and S. B. GALE: Mechanism of Photooxidation of Sulfur Dioxide in Atmosphere, in H. M. ENGLUND and W. T. BEERY (eds.), "Proceedings of the Second International Clean Air Congress," Academic Press, New York, 1971.

KAUFMAN, F.: The Air Afterglow and Its Use in the Study of Some Reactions of Atomic Oxygen, *Proc. Roy. Soc.*, (*London*), *Ser. A*, **247**:123 (1958).

KHAN, A. U., J. N. PITTS, JR., and E. B. SMITH: Singlet Oxygen in the Environmental Sciences: The Role of Singlet Molecular Oxygen in the Production of Photochemical Smog, *Environ. Sci. Technol*, **1**: 656 (1967).

KORTH, M. W., A. H. ROSE, JR., and R. C. STAHMAN: Effects of Hydrocarbon to Oxides of Nitrogen Ratios on Irradiated Auto Exhaust. Pt. I, *J. Air Pollut. Control Assoc.*, **14**:168 (1964).

KUMMER, W. A., J. N. PITTS, JR., and R. P. STERN: Chemiluminescent Reactions of Ozone with Olefins and Sulfides, *Environ. Sci. Technol.*, **5**:1045 (1971).

KUMMLER, R. H., M. H. BORTNER, and T. BAURER: The Hartley Photolysis of Ozone as a Source of Singlet Oxygen in Polluted Atmospheres, *Environ. Sci. Technol.*, **3**:248 (1969).

LEIGHTON, P. A.: "Photochemistry of Air Pollution," Academic Press, New York, 1961.

LEVY, A., S. E. MILLER, and F. SCOFIELD: The Photochemical Smog Reactivity of Solvents, in H. M. ENGLUND and W. T. BEERY (eds.), "Proceedings of the Second International Clean Air Congress," Academic Press, New York, 1971.

MATTESON, M. J., W. STOBER, and H. LUTHER: Kinetics of the Oxidation of Sulfur Dioxide by Aerosols of Manganese Sulfate, *Ind. Eng. Chem. Fund.*, **8**:677, (1969).

MAUGH, T. H., II: Air Pollution Instrumentation: A Trend Toward Physical Methods, *Science*, **177**:685 (1972).

MILLER, M. S., S. K. FRIEDLANDER, and G. M. HIDY: A Chemical Element Balance for the Pasadena Aerosol, *J. Colloid Interface Sci.*, **39**:165 (1972).

MILLS, R. L., and H. S. JOHNSTON: Decomposition of N_2O_5 in the Presence of NO, *J. Am. Chem. Soc.*, **73**:938 (1951).

MORLEY, C., and I. W. M. SMITH: Rate Measurements of Reactions of OH by Resonance Absorption. I, *J. Chem. Soc., Faraday Trans.* II, **68**:1016 (1972).

MORRIS, E. D., JR., and H. NIKI: Reactivity of Hydroxyl Radicals with Olefins, *J. Phys. Chem.*, **75**:3640 (1971).

MORRIS, E. D., JR., D. H. STEDMAN, and H. NIKI: Mass Spectrometric Study of the Reactions of the Hydroxyl Radical with Ethylene, Propylene, and Acetaldehyde in a Discharge-flow System, *J. Am. Chem. Soc.*, **93**:3570 (1971).

MULCAHY, M. F. R., J. R. STEVEN, and J. C. WARD: The Kinetics of Reaction between Oxygen

Atoms and Sulfur Dioxide: An Investigation by Electron Spin Resonance Spectrometry, *J. Phys. Chem.*, **71**:2124 (1967).

NIKI, H., E. E. DABY, and B. WEINSTOCK: Mechanisms of Smog Reactions, *Adv. Chem.*, **113**:16 (1972).

O'KEEFE, A. E., and G. O. ORTMAN: Primary Standards for Trace Gas Analysis, *Anal. Chem.*, **38**:760 (1966).

OKUDA, S., T. N. RAO, D. H. SLATER, and J. G. CALVERT: Identification of the Photochemically Active Species in SO_2 Photolysis within the First Allowed Absorption Band, *J. Phys. Chem.*, **73**:4412 (1969).

PETERS, L. K., and L. B. WINGARD, JR.: The Ozone Oxidation of Ethylene as It Pertains to Air Pollution, University of Pittsburgh, Pittsburgh, Pa., July 24, 1970.

RENZETTI, N. A., and G. J. DOYLE: Photochemical Aerosol Formation in Sulfur Dioxide–Hydrocarbon Systems, *Int. J. Air Water Pollut.*, **2**:327 (1960).

ROMANOVSKY, J. C., R. M. INGELS, and R. J. GORDON: Estimation of Smog Effects in the Hydrocarbon–Nitric Oxide System, *J. Air Pollut. Control Assoc.*, **17**:454 (1967).

SALTZMAN, B. E.: Colormetric Microdetermination of Nitrogen Dioxide in the Atmosphere, *Anal. Chem.*, **26**:1949 (1954).

SALTZMAN, B. E., W. R. BURG, and G. RAMASWAMY: Performance of Permeation Tubes as Standard Gas Sources, *Environ. Sci. Technol.*, **5**:1121 (1971).

SCHOTT, G., and N. DAVIDSON: Shock Waves in Chemical Kinetics: The Decomposition of N_2O_5 at High Temperatures, *J. Am. Chem. Soc.*, **80**:1841 (1958).

SCHUCK, E. A., and E. R. STEPHENS: Oxides of Nitrogen, in J. N. PITTS, JR., and R. METCALF (eds.), "Advances in Environmental Sciences," vol. I, Wiley, New York, 1969.

SCHUCK, E. A., E. R. STEPHENS, and P. R. SCHROCK: Rate Constant Ratios During Nitrogen Dioxide Photolysis, *J. Air Pollut. Control Assoc.*, **16**:695 (1966).

SEIZINGER, D. E., and B. DIMITRIADES: Oxygenates in Exhaust from Simple Hydrocarbon Fuels, *J. Air Pollut. Control Assoc.*, **22**:47 (1972).

Selected Methods for the Measurement of Air Pollutants, *U.S. Dept. Health, Education and Welfare Publ. AP*–11, 1965.

SIDEBOTTOM, H. W., C. D. BADCOCK, G. E. JACKSON, J. G. CALVERT, G. W. REINHARDT, and E. K. DAMON: Photooxidation of Sulfur Dioxide, *Environ. Sci. Technol.*, **6**:72 (1972).

SPICER, C. W., A. VILLA, H. A. WIEBE, and J. HEICKLEN: The Reactions of Methylperoxy Radicals with NO and NO_2, *Pennsylvania State Univ., Rept. CAES* 223–71, 1971.

STEDMAN, D. H., E. E. DABY, F. STUHL, and H. NIKI: Analysis of Ozone and Nitric Oxide by a Chemiluminescent Method in Laboratory and Atmospheric Studies of Photochemical Smog, *J. Air Pollut. Control Assoc.*, **22**:260 (1972).

STEPHENS, E. R.: Reactions of Oxygen Atoms and Ozone in Air Pollution, *Int. J. Air Water Pollut.* **10**:649 (1966).

STEPHENS, E. R.: Chemistry of Atmospheric Oxidants, *J. Air Pollut. Control Assoc.*, **19**:181 (1969).

TUESDAY, C. S.: The Atmospheric Photooxidation of *trans*-2-Butene and Nitric Oxide, in "Chemical Reactions in the Lower and Upper Atmosphere," pp. 1–49, Interscience, New York, 1961.

URONE, P., and W. H. SCHROEDER: SO_2 in the Atmosphere: A Wealth of Monitoring Data, but Few Reaction Rate Studies, *Environ. Sci. Technol.*, **3**:436 (1969).

URONE, P., W. H. SCHROEDER, and S. R. MILLER: Reactions of Sulfur Dioxide in Air, in H. M. ENGLUND and W. T. BEERY (eds.), "Proceedings of the Second International Clean Air Congress," Academic Press, New York, 1971.

URONE, P., H. LUTSEP, C. M. NOYES, and J. F. PARCHER: Static Studies of Sulfur Dioxide in Air, *Environ. Sci. Technol.*, **2**:611 (1968).

WARREN, G. I., and G. BABCOCK: Portable Ethylene Chemiluminescence Ozone Monitor, *Rev. Sci. Instrum.*, **41**:280 (1970).

WAYNE, L. G., and D. M. YOST: Kinetics of the Rapid Gas Phase Reaction Between NO, NO_2 and H_2O, *J. Chem. Phys.*, **19**:41 (1951).

WESTBERG, K., N. COHEN, and K. W. WILSON: Carbon Monoxide: Its Role in Photochemical Smog Formation, *Science*, **171**:1013 (1971).

WILSON, W. E., JR., and A. LEVY: A Study of Sulfur Dioxide in Photochemical Smog. I. Effect of SO_2 and Water Vapor Concentration in the 1-Butene/NO_x/SO_2 System, *J. Air Pollut. Control Assoc.*, **20**: 385 (1970).

WILSON, W. E., JR., and G. F. WARD: The Role of Carbon Monoxide in Photochemical Smog, presented at the 160th Meeting of the American Chemical Society, Chicago, Ill., September, 1970.

WILSON, W. E., JR., A. LEVY, and D. B. WIMMER: A Study of Sulfur Dioxide in Photochemical Smog. II. Effect of SO_2 on Oxidant Formation in Photochemical Smog, *J. Air Pollut. Control Assoc.*, **22**:27 (1972).

WILSON, W. E., JR., E. L. MERRYMAN, and A. LEVY: A Literature Survey of Aerosol Formation and Visibility Reduction in Photochemical Smog, American Petroleum Institute, Project EF–2, Aug. 1, 1969.

PROBLEMS

4.1 You wish to estimate the rate of conversion of NO to NO_2 by reaction 8 in Table 4.4,

$$2NO + O_2 \longrightarrow 2NO_2$$

at NO and O_2 levels typical of an urban atmosphere. In the atmosphere, the concentration of O_2 is so high that it changes imperceptibly because of this reaction. Thus, the product $2k_8[O_2]$ may be taken as an effective second-order rate constant, k_8'. For air, $[O_2] = 0.21 \times 10^6$ ppm can be used. Compute the half-life of 0.5 ppm of NO if the the only reaction contributing to its loss is the above reaction. What do you conclude about the importance of this reaction in the conversion of NO to NO_2 in an urban atmosphere?

4.2 Develop a means of converting relative humidity to parts per million of water in the atmosphere.

4.3 Consider the consecutive reactions $A \xrightarrow{k_1} B \xrightarrow{k_2} C$. Given pure species A at time zero, solve the first-order rate equations to determine the concentrations of A, B, and C at any subsequent time in a constant-volume batch reactor at constant temperature. Now let $k_2/k_1 \gg 1$, so that one can employ the pseudo-steady-state approximation for B. Derive a quantitative expression for the error committed in using this approximation.

4.4 You wish to estimate the concentrations of some of the more highly reactive species in Tables 4.4 and 4.6. Under conditions prevailing in the atmosphere, the species can be divided into accumulating and steady-state species:

1 Accumulating species: NO, NO_2, HNO_3, HNO_2, H_2O, CO, H_2O_2
2 Steady-state species: O, O_3, NO_3, OH, HO_2

Consider two situations: high NO, low NO_2 and low NO, high NO_2:

1	*2*
$[NO] = 0.40$ ppm	$[NO] = 0.04$ ppm
$[NO_2] = 0.04$ ppm	$[NO_2] = 0.40$ ppm

(*a*) Estimate the concentrations of oxygen atoms and ozone under both conditions. Assume that $k_1/k_3 = 0.01$ ppm.
(*b*) Estimate the concentration of NO_3 in the system. Assume that the relative humidity is 50 percent.
(*c*) Estimate the concentrations of OH and HO_2 radicals. Assume that $[CO] = 20$ ppm.

4.5 You wish to compute the rate of conversion of SO_2 by catalytic oxidation for conditions typical of a natural fog in an urban atmosphere. Assume that

$[SO_2] = 0.1$ ppm
Fog concentration $= 0.2$ g H_2O/m^3 air

Half the fog droplets contain a catalyst capable of oxidizing SO_2 to H_2SO_4. The catalyst concentration within these droplets is equivalent to 500 ppm $MnSO_4$. Compute the rate of conversion of SO_2 to H_2SO_4 under such conditions.

4.6 An important mechanism for the growth of aerosol particles is by diffusion of gaseous species to the particle surface followed by absorption into the particle. Consider a droplet which has an initial radius of r_0 and which grows by absorption of species A in the gas phase. Because the rate of change of size of the droplet is small compared with the characteristic time for vapor-phase diffusion, assume that the concentration of species A in the vicinity of the particle is governed by the steady-state form of the molecular diffusion equation

$$\frac{d^2 c_A}{dr^2} + \frac{2}{r}\frac{dc_A}{dr} = 0$$

Far from the particle, the concentration of A has a constant value,

$$c_A = c_{A\infty} \qquad r \to \infty$$

and at the particle surface the flux of A is equal to the rate of absorption, which depends on the concentration of A just adjacent to the surface,

$$D\frac{dc_A}{dr} = k_s c_A \qquad r = r_0$$

(*a*) Determine $c_A(r)$ and sketch the form of $c_A/c_{A\infty}$ as a function of r/r_0 for $k_s r_0/D \gg 1$, $=1$, and $\ll 1$. How do you interpret these three cases?

(*b*) Using the values $r_0 = 0.5$ μm, $D = 0.1$ cm²/sec, $k_s = 1$ cm/sec, estimate the time required for the drop to grow to a radius of 0.6 μm. Assume that the density of the drop remains 1 g/cm³ and that $c_{A\infty}$ is $10^{-4}M^{-1}$, where M is the molecular weight of A.

4.7 It is often useful to examine the dynamic behavior of the total number density $N_\infty(t)$ in a coagulating aerosol. An equation governing N_∞ can be obtained by integrating (4.27) over v from 0 to ∞. If $b(m, v) = b_0$ and $N_\infty(0) = N_0$, show that

$$\frac{1}{N_\infty(t)} - \frac{1}{N_0} = \frac{b_0}{2}t$$

[This relation predicts that a plot of $N_\infty(t)^{-1}$ versus t should be a straight line, if b is constant and the aerosol is initially monodisperse. Actually, the straight-line relation has been found to hold in some cases for polydisperse aerosols. The reason is that the decay of the total number density is rather insensitive to the actual details of the statistics of interaction.]

4.8 In the classic theory of coagulation by brownian motion it is assumed that the number concentration n of particles in a field satisfies the molecular diffusion equation with a constant diffusivity D. Each particle has a "radius of influence" R equal to its own radius. Any particle entering another's radius of influence constitutes a coagulation event, and only binary collisions are possible. Consider a uniform dispersion of n_0 spherical particles per unit volume and imagine one of the particles to be stationary and all others moving relative to that particle. If the rate at which other particles arrive at the sphere of radius R surrounding the fixed particle can be determined, it is possible to compute the collision frequency b for use in (4.27) corresponding to coagulation by brownian motion.

The number density of particles satisfies

$$\frac{\partial n(r, t)}{\partial t} = D\frac{1}{r^2}\frac{\partial}{\partial r}\left(r^2\frac{\partial n}{\partial r}\right) \quad \text{(A)}$$

subject to

$$\begin{aligned} n &= n_0 \quad && \text{for} \quad r > R, t = 0 \\ n &= 0 && \text{at} \quad r = R, t > 0 \\ n &= n_0 && \qquad r \to \infty \end{aligned}$$

(*a*) Show that the solution of Eq. (A) is

$$n(r, t) = n_0\left(1 - \frac{R}{r} + \frac{R}{r}\operatorname{erf}\frac{r-R}{2\sqrt{Dt}}\right)$$

(*b*) Show that the flux of particles at $r = R$, and, therefore, the collision frequency, is

$$4\pi D n_0 R\left(1 + \frac{R}{\sqrt{\pi D t}}\right)$$

5

MICROMETEOROLOGY

With this chapter we begin a study of the physical behavior of air pollutants in the atmosphere. Our ultimate objective is to be able to predict the rate of dispersion of contaminants once they are introduced into the atmosphere. To achieve this objective we must study the nature of airflow in the lowest layers of the atmosphere, since it is in these layers that pollutants are most often emitted. As noted in Chap. 3, the study of fluid mechanical and transport phenomena taking place on horizontal scales of tens of kilometers and in the lowest layers of the atmosphere is termed *micrometeorology*.

In this chapter we will concentrate on air motion in the lowest layers of the atmosphere. Such air motion, taking place adjacent to a solid boundary of variable temperature and roughness, is virtually always turbulent. This atmospheric turbulence is responsible for the transport of heat, water vapor, and pollutants from the surface to the atmosphere as a whole. Our objective in this chapter will be to understand the basic phenomena that influence atmospheric turbulence.[1] In Chap. 6 we will consider the dispersion of pollutants as a result of atmospheric turbulence.

[1] The subject of atmospheric turbulence has received a great amount of attention. Our treatment here will, of necessity, be rather limited. The interested reader may pursue this area to greater depth in Lumley and Panofsky (1964), Monin and Yaglom (1971), and Plate (1971).

5.1 BASIC EQUATIONS OF ATMOSPHERIC FLUID MECHANICS

We wish to derive the equations which govern the fluid density, temperature, and velocities in the lowest layers of the atmosphere. These equations will form the basis from which we can subsequently explore the processes that influence atmospheric turbulence.

The equations of continuity and motion for a compressible, newtonian fluid in a gravitational field are[1]

$$\frac{\partial \rho}{\partial t} + \frac{\partial}{\partial x_i}(\rho u_i) = 0 \tag{5.1}$$

and

$$\rho\left(\frac{\partial u_i}{\partial t} + u_j \frac{\partial u_i}{\partial x_j}\right) = \frac{\partial}{\partial x_k}\left[\mu\left(\frac{\partial u_i}{\partial x_k} + \frac{\partial u_k}{\partial x_i}\right) - \left(p + \frac{2}{3}\mu\frac{\partial u_j}{\partial x_j}\right)\delta_{ik}\right] - \rho g \delta_{3i} \tag{5.2}$$

where $u_i(x_1,x_2,x_3,t)$ is the fluid velocity component in direction i, μ is the fluid viscosity, and δ_{ij} is the Kronecker delta, defined by $\delta_{ij} = 1$ if $i = j$, and $\delta_{ij} = 0$ if $i \neq j$. In (5.1) and (5.2) we have taken the x_3 axis as vertically upward. If $\mu = 0$, (5.1) and (5.2) reduce to the Euler equations of motion introduced in Chap. 3 in connection with the derivation of the relation for the geostrophic wind speed. We have not included a term accounting for the Coriolis acceleration in (5.2), since we shall be interested only in processes taking place on limited spatial and temporal scales over which the air motion is not influenced substantially by the rotation of the earth.

The energy equation is

$$\rho\left(\frac{\partial U}{\partial t} + u_j \frac{\partial U}{\partial x_j}\right) = k\frac{\partial^2 T}{\partial x_j\,\partial x_j} - p\frac{\partial u_j}{\partial x_j} + \Phi + Q \tag{5.3}$$

where U is the internal energy per unit mass ($= \hat{C}_V T$ for a perfect gas), k is the thermal conductivity (assumed constant), Φ is the heat generated per unit volume and time as a result of viscous dissipation, and Q represents the heat generated by any sources in the fluid.

[1] Throughout this book when dealing with the equations of fluid mechanics and diffusion we will employ the so-called summation convention, in which a repeated subscript in a term indicates summation over the three components of that term. Thus,

$$\frac{\partial u_k}{\partial x_k} \equiv \frac{\partial u_1}{\partial x_1} + \frac{\partial u_2}{\partial x_2} + \frac{\partial u_3}{\partial x_3}$$

We will usually employ the 1, 2, 3 notation for coordinate directions and velocity components, although on occasion we may use the x, y, z and u, v, w notations when it seems more appropriate.

For an ideal gas, p and ρ are related by the equation of state:

$$p = \frac{\rho RT}{M_a} \tag{5.4}$$

Equations (5.1) to (5.4) represent six equations for the six unknowns u_1, u_2, u_3, p, ρ, and T. These equations can therefore be solved, in principle, subject to appropriate boundary and initial conditions to yield velocity, pressure, density, and temperature profiles in an ideal gas. Because of the highly coupled nature of (5.1) to (5.4), these equations are virtually impossible to solve analytically. However, we can exploit certain aspects characteristic of the lower atmosphere to simplify these equations.

When the atmosphere is at rest ($u_i = 0$), (5.2) and (5.3) become

$$\frac{\partial p_e}{\partial x_1} = \frac{\partial p_e}{\partial x_2} = 0 \qquad \frac{\partial p_e}{\partial x_3} = -g\rho_e \tag{5.5}$$

$$\frac{\partial^2 T_e}{\partial {x_3}^2} = 0 \tag{5.6}$$

where the subscript e denotes equilibrium values and where we have assumed there are no heat sources ($Q = 0$). In writing (5.6), we have also assumed that the atmosphere is horizontally homogeneous. It follows from (5.6) that at equilibrium in the absence of heat sources, the temperature varies linearly with height,

$$T_e = T_0\left(1 - \frac{x_3}{H}\right)$$

where T_0 is the surface temperature and H is the height at which T_e becomes zero [chosen simply to satisfy the two necessary boundary conditions of (5.6)]. Therefore, the pressure, density, and temperature in an equilibrium atmosphere are governed by

$$\begin{aligned} \frac{\partial p_e}{\partial x_3} &= -g\rho_e \\ p_e &= \frac{\rho_e RT_e}{M_a} \\ T_e &= T_0\left(1 - \frac{x_3}{H}\right) \end{aligned} \tag{5.7}$$

Upon integration, we obtain

$$\begin{aligned} p_e &= p_0\left(1 - \frac{x_3}{H}\right)^{gHM_a/RT_0} \\ \rho_e &= \rho_0\left(1 - \frac{x_3}{H}\right)^{gHM_a/RT_0 - 1} \end{aligned} \tag{5.8}$$

where $p_0 = \rho_0 RT_0/M_a$, the surface state. Thus, when the atmosphere is at rest, the ratio of the equilibrium values at any height to the corresponding surface values are related by

$$\frac{p_e}{p_0} = \left(\frac{\rho_e}{\rho_0}\right)^n \tag{5.9}$$

where

$$\frac{n}{n-1} = \frac{gHM_a}{RT_0}$$

The lapse rate Λ, as expressed in $T_e = T_0 - \Lambda x_3$, is given by

$$\Lambda = \frac{T_0}{H} = \frac{gM_a}{R}\frac{n-1}{n}$$

In the special case in which $n = \gamma = \hat{C}_p/\hat{C}_V$, the atmosphere is adiabatic. As we know, in this case (since $R/M_a = \hat{C}_p - \hat{C}_V$ for a perfect gas)

$$\Lambda = \Gamma = \frac{g}{\hat{C}_p} \tag{5.10}$$

We shall now consider the description of the atmosphere when there is motion. In doing so we shall consider only a shallow layer adjacent to the ground, in which case we can make some rather important simplifications in the equations of continuity, motion, and energy. The approximations we make are called the *Boussinesq approximations*. The conditions for their validity have been examined by Spiegel and Veronis (1960) and further clarified by Calder (1968).

We can express the equilibrium profiles of pressure, density, and temperature in terms of functions of x_3 only as follows:

$$\begin{aligned} p_e &= p_0 + p_m(x_3) \\ \rho_e &= \rho_0 + \rho_m(x_3) \\ T_e &= T_0 + T_m(x_3) \end{aligned} \tag{5.11}$$

We consider only a shallow layer, so that p_m/p_0, ρ_m/ρ_0, and T_m/T_0 are all small compared with unity. When there is motion, we can express the actual pressure, density, and temperature in terms of the sum of the equilibrium values and a small correction due to the motion (denoted by a tilde). Thus, we write

$$\begin{aligned} p &= p_0 + p_m(x_3) + \tilde{p}(x_1,x_2,x_3,t) \\ \rho &= \rho_0 + \rho_m(x_3) + \tilde{\rho}(x_1,x_2,x_3,t) \\ T &= T_0 + T_m(x_3) + \tilde{T}(x_1,x_2,x_3,t) \end{aligned} \tag{5.12}$$

where we assume that the deviations induced by the motion are sufficiently small that the quantities $\tilde{p}/p_0$, $\tilde{\rho}/\rho_0$, and $\tilde{T}/T_0$ are also small compared with unity.

For the equation of continuity we have

$$\frac{\partial u_i}{\partial x_i} = -\frac{1}{\rho}\left(\frac{\partial \rho}{\partial t} + u_j \frac{\partial \rho}{\partial x_j}\right) \equiv -\frac{1}{\rho}\frac{D\rho}{Dt} = -\frac{D}{Dt}\log \rho$$

$$= -\frac{D}{Dt}\log\left[\rho_0\left(1 + \frac{\rho_m + \tilde{\rho}}{\rho_0}\right)\right]$$

$$= -\frac{D}{Dt}\frac{\rho_m + \tilde{\rho}}{\rho_0}\dagger \qquad (5.13)$$

where we have employed the Maclaurin series expansion for $\log(1 + x)$ for $x \ll 1$. However, we see that the right-hand side of (5.13) is small compared with unity, so that to a first-order approximation the continuity equation can be written as

$$\frac{\partial u_i}{\partial x_i} = 0 \qquad (5.14)$$

which, of course, is the continuity equation for an incompressible fluid.

We now subtract the reference equilibrium state (5.5) from the equation of motion (5.2). Using (5.14), we obtain

$$\frac{\partial u_i}{\partial t} + u_j \frac{\partial u_i}{\partial x_j} = -\frac{1}{\rho}\frac{\partial \tilde{p}}{\partial x_i} + \frac{\mu}{\rho}\frac{\partial^2 u_i}{\partial x_j\, \partial x_j} - \frac{g\tilde{\rho}}{\rho}\delta_{i3} \qquad (5.15)$$

Let us examine the two terms $-(1/\rho)(\partial \tilde{p}/\partial x_i)$ and $-(g\tilde{\rho}/\rho)\delta_{i3}$ appearing in (5.15). We can write the first of these terms as

$$-\frac{1}{\rho}\frac{\partial \tilde{p}}{\partial x_i} = \frac{\rho_0 - \rho_e}{(\rho_0 + \rho_e)(\rho_0 - \rho_e)}\frac{\partial \tilde{p}}{\partial x_i}$$

$$= \frac{1}{\rho_0^{\,2} - \rho_e^{\,2}}\left(\rho_0 \frac{\partial \tilde{p}}{\partial x_i} - \rho_e \frac{\partial \tilde{p}}{\partial x_i}\right)$$

$$\cong \frac{1}{\rho_0}\frac{\partial \tilde{p}}{\partial x_i} - \frac{\rho_e}{\rho_0^{\,2}}\frac{\partial \tilde{p}}{\partial x_i}$$

The second term on the right-hand side is small compared with the first, and so

$$-\frac{1}{\rho}\frac{\partial \tilde{p}}{\partial x_i} \cong \frac{1}{\rho_0}\frac{\partial \tilde{p}}{\partial x_i} \qquad (5.16)$$

† D/Dt denotes the substantial derivative and is defined by

$$\frac{D}{Dt} \equiv \frac{\partial}{\partial t} + u_j \frac{\partial}{\partial x_j}$$

Next we consider the term $-(g\tilde{\rho}/\rho)\delta_{i3}$, which expresses the vertical acceleration on a fluid element as a result of density fluctuations. Assuming that the fluctuations in density from the surface value are small, we can expand ρ in a Taylor series about ρ_0 as follows:

$$\rho = \rho_0\left(1 + \frac{p_m}{p_0} - \frac{T_m}{T_0}\right) + \rho_0\left(\frac{\tilde{p}}{p_0} - \frac{\tilde{T}}{T_0}\right) \tag{5.17}$$

In the atmosphere the relative magnitude of pressure deviations from the reference pressure as a result of motion, that is, $\tilde{p}$, is small compared with temperature fluctuations,

$$\frac{\tilde{p}}{p_0} \ll \frac{\tilde{T}}{T_0}$$

Thus, (5.17) becomes

$$\tilde{\rho} = -\rho_0 \frac{\tilde{T}}{T_0}$$

i.e., the deviations in density from the reference state can be attributed solely to temperature deviations.

The final form of the approximate equation of motion is therefore

$$\frac{\partial u_i}{\partial t} + u_j \frac{\partial u_i}{\partial x_j} = -\frac{1}{\rho_0}\frac{\partial \tilde{p}}{\partial x_i} + \frac{\mu}{\rho_0}\frac{\partial^2 u_i}{\partial x_j\,\partial x_j} + \frac{g\tilde{T}}{T_0}\delta_{i3} \tag{5.18}$$

We now consider the energy equation (5.3). First, the contribution of viscous dissipation Φ to the energy balance is negligible in the atmosphere. Upon subtracting (5.6) from (5.3) we obtain the equation governing the temperature fluctuations,

$$\rho\hat{C}_V\left[\frac{\partial(T_e + \tilde{T})}{\partial t} + u_j \frac{\partial}{\partial x_j}(T_e + \tilde{T})\right] = k\frac{\partial^2 \tilde{T}}{\partial x_j\,\partial x_j} - p\frac{\partial u_j}{\partial x_j} + Q \tag{5.19}$$

where we have retained the term $p(\partial u_j/\partial x_j)$ since, although $\partial u_j/\partial x_j \cong 0$, p is large so that $p(\partial u_j/\partial x_j)$ is the same order of magnitude as the other terms in the equation. In fact, we can determine the order of $p(\partial u_j/\partial x_j)$ as follows: From the continuity equation and the ideal-gas law,

$$\begin{aligned}
p\frac{\partial u_j}{\partial x_j} &= -\frac{p}{\rho}\frac{D\rho}{Dt} \\
&= -\frac{RT}{M_a}\frac{D\rho}{Dt} \\
&= \frac{\rho_0 R}{M_a}\frac{D}{Dt}(T_e + \tilde{T}) - \frac{Dp_e}{Dt} \\
&= \frac{\rho_0 R}{M_a}\frac{D}{Dt}(T_e + \tilde{T}) + u_3 g\rho_0
\end{aligned} \tag{5.20}$$

where in the last step we have employed (5.5). Using the relation $\hat{C}_p - \hat{C}_V = R/M_a$, and upon substituting (5.20) into (5.19), we obtain the approximate form of the energy equation:

$$\rho_0 \hat{C}_p\left(\frac{\partial \tilde{T}}{\partial t} + u_j \frac{\partial \tilde{T}}{\partial x_j}\right) + \rho_0 \hat{C}_p u_3\left(\frac{\partial T_e}{\partial x_3} + \frac{g}{\hat{C}_p}\right) = k \frac{\partial^2 \tilde{T}}{\partial x_j \, \partial x_j} + Q \qquad (5.21)$$

This equation holds regardless of the choice of reference equilibrium atmosphere. Usually, however, the equilibrium reference condition is chosen to be the adiabatic case, in which

$$\frac{\partial T_e}{\partial x_3} = -\frac{g}{\hat{C}_p}$$

Let us assume that at some initial time $\tilde{T} = 0$ relative to the adiabatic atmosphere. Then, from (5.21), we see that if $Q = 0$ the condition of $\tilde{T} = 0$ is preserved for $t > 0$ even though there may be motion of the air. Also, the equation of motion (5.18) reduces to the usual form of the Navier-Stokes equation for the dynamics of an incompressible fluid under the influence of a motion-induced pressure fluctuation $\tilde{p}$ with no contribution from buoyancy forces since $\tilde{T} = 0$. Therefore, for an atmosphere with no sources of heat and initially having an adiabatic lapse rate, the temperature profile is unaltered if the atmosphere is set in motion. As a result, the adiabatic condition can be envisioned as one in which a large number of parcels are rising and falling, a sort of "convective" equilibrium. Thus, we have been able to derive the relation for the adiabatic lapse rate here from the full equations of continuity, motion, and energy, in contrast with the derivation presented in Chap. 3, which was based on thermodynamic arguments.

With the choice of the adiabatic equilibrium condition, (5.21) becomes

$$\rho_0 \hat{C}_p\left(\frac{\partial \tilde{T}}{\partial t} + u_j \frac{\partial \tilde{T}}{\partial x_j}\right) = k \frac{\partial^2 \tilde{T}}{\partial x_j \, \partial x_j} + Q \qquad (5.22)$$

which is the classic form of the heat conduction equation for an incompressible fluid with constant physical properties.

In summary, the complete set of equations for the lowest layer of the atmosphere is (5.14), (5.18), and (5.21) [or (5.22)]. The set consists of five equations for the five unknowns u_1, u_2, u_3, $\tilde{p}$, and $\tilde{T}$. The ideal-gas equation of state is no longer required as it has been incorporated into the equations. The Boussinesq approximations have led to a considerable simplification of the original equations. First of all, the incompressible form of the continuity equation can be used, together with a nearly incompressible form of the equation of motion. In (5.18) the density enters as ρ_0 in every term except that representing the acceleration due to buoyancy forces.[1] Finally, the energy equation is just the usual heat conduction equation with T replaced by $\tilde{T}$.

We introduced in Chap. 3 the potential temperature θ and noted that $d\theta/dx_3$ is a measure of the departure of the actual temperature profile from the adiabatic. Thus we expect that a definite relationship should exist between $\tilde{T}$ and θ. From (3.14) we see that

$$\frac{1}{\theta}\frac{\partial \theta}{\partial x_i} = \frac{1}{T}\left(\frac{\partial T}{\partial x_i} + \frac{g}{\hat{C}_p}\delta_{3i}\right) \tag{5.23}$$

where the quantity in parentheses is the difference between the actual and the adiabatic lapse rates, i.e.,

$$\frac{1}{\theta}\frac{\partial \theta}{\partial x_i} = \frac{1}{T}\frac{\partial \tilde{T}}{\partial x_i} \tag{5.24}$$

As noted in obtaining (3.15), since, in magnitude, θ is quite close to T, we can replace (5.24) by

$$\frac{\partial \theta}{\partial x_i} \cong \frac{\partial \tilde{T}}{\partial x_i} \tag{5.25}$$

Now, using (5.25), we can rewrite (5.22) as

$$\rho_0 \hat{C}_p\left(\frac{\partial \theta}{\partial t} + u_j \frac{\partial \theta}{\partial x_j}\right) = k \frac{\partial^2 \theta}{\partial x_j \, \partial x_j} + Q \tag{5.22$'$}$$

Although the ρ_0 and T_0 in (5.18) and (5.22) refer to the constant surface values, equations of precisely the same form can be derived in which ρ_0 and T_0 are replaced by ρ_e and T_e, the reference profiles. These equations written in that form will be useful later when we consider the dynamics of potential temperature in the atmosphere.

5.2 TURBULENCE

Equations (5.14), (5.18), and (5.22) govern the fluid velocity and temperature in the lower atmosphere. Although these equations are at all times valid, their solution is impeded by the fact that atmospheric flow is turbulent (as opposed to laminar). It is difficult to define turbulence; instead we can cite a number of the characteristics of

[1] Because of the variation of air density with temperature, motion can arise solely as a result of buoyancy effects induced by temperature nonuniformities. Since the variation of density with temperature leads to the last term on the right-hand side of (5.18), the equations of motion and energy are not uncoupled, as in the case of forced convection. This situation is called, by contrast, *free convection*, since flow can arise without the imposition of external pressure gradients.

turbulent flows.[1] Turbulent flows are irregular and random, so that the velocity components at any location vary randomly with time. Since the velocities are random variables, their exact values can never be predicted precisely. Thus (5.14), (5.18), and (5.22) become partial differential equations whose dependent variables are random functions. We cannot, therefore, expect to solve any of these equations exactly; rather, we must be content to determine some convenient statistical properties of the velocities and temperature. The random fluctuations in the velocities result in rates of momentum, heat, and mass transfer in turbulence that are many orders of magnitude greater than the corresponding rates due to pure molecular transport. Turbulent flows are dissipative in the sense that there is a continuous conversion of kinetic to internal energy. Thus, unless energy is continuously supplied, turbulence will decay. The usual source of energy for turbulence is shear in the flow field, although in the atmosphere buoyancy can also be a source of energy.

A particular turbulent flow, say that produced in a laboratory, can be envisioned as one of an infinite ensemble of flows with identical macroscopic boundary conditions. Let us consider a situation of turbulent pipe flow. If the same pipe and pressure drop is used each time the turbulent flow experiment is repeated, the velocity field would always be different no matter how carefully the conditions of the experiment were reproduced. The *mean* or *average* velocity, say as a function of radial position in the pipe, could be determined, in principle, only by averaging the readings made over an infinite ensemble of identical experiments. Figure 5.1 shows a hypothetical record of the ith velocity component at a certain location in a turbulent flow. The specific features of a second velocity record taken under the same conditions would be different but there might well be a decided resemblance in some of the characteristics of the record. In practice it is usually not possible to repeat measurements under identical conditions (particularly in the atmosphere). To compute the mean value of u_i at location $\mathbf{x}$ and time t we would need to average the values of u_i at $\mathbf{x}$ and time t from all the similar velocity records. This ensemble mean is denoted by $\langle u_i(t,\mathbf{x})\rangle$. If the ensemble mean does not change with time t, we can substitute a time average for the ensemble average. The time-average velocity is defined by

$$\bar{u}_i = \lim_{T\to\infty} \frac{1}{T}\int_{t_0}^{t_0+T} u_i(t)\,dt$$

In practice, u_i is usually not a strictly stationary function, that is, one whose statistical properties are independent of time. Rather, the flow may change with time. However, we still wish to define a mean velocity; this is done by defining

$$\bar{u}_i(t) = \frac{1}{T}\int_{t-T/2}^{t+T/2} u_i(t')\,dt'$$

[1] Turbulence is a characteristic of flows and not of fluids themselves.

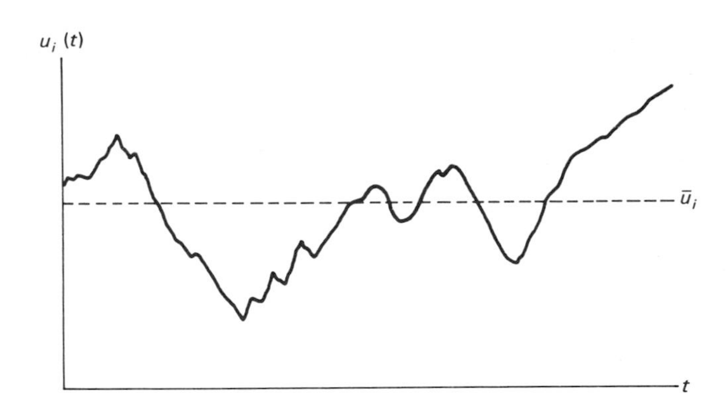

FIGURE 5.1
Typical record of the velocity in direction i at a point in a turbulent flow.

Clearly, $\bar{u}_i(t)$ will depend on the averaging interval T. We need to choose T large enough so that an adequate number of fluctuations are included, but yet not so large that important macroscopic features of the flow would be masked. For example, if T_1 and T_2 are time scales associated with fluctuations and macroscopic changes in the flow, respectively, we would want $T_2 \gg T \gg T_1$.

It is customary to represent the instantaneous value of the wind velocity as the sum of a mean and fluctuating component, $\bar{u}_i + u'_i$. The mean values of the velocities tend to be smooth and slowly varying. The fluctuations $u'_i = u_i - \bar{u}_i$ are characterized by extreme spatial and temporal variations. In spite of the severity of fluctuations, it is observed experimentally that turbulent spatial and temporal inhomogeneities still have considerably greater sizes than molecular scales. The viscosity of the fluid prevents the turbulent fluctuations from becoming too small. Because the smallest scales (or eddies) are still many orders of magnitude larger than molecular dimensions, the turbulent flow of a fluid is described by the basic equations of continuum mechanics in Sec. 5.1. In general, the largest scales of motion in turbulence (the so-called big eddies) are comparable to the major dimensions of the flow and are responsible for most of the transport of momentum, heat, and mass. Large scales of motion have comparatively long time scales, whereas the small scales have short time scales and are often statistically independent of the large-scale flow. A physical picture that is often used to describe turbulence involves the transfer of energy from the larger to the smaller eddies which ultimately dissipate the energy as heat.

5.3 EQUATIONS FOR THE MEAN QUANTITIES

What we seek, in principle, is a description of a turbulent flow at all points in space and time. Unfortunately, in the equations of motion and energy, the dependent variables u_i, p, and T are random variables, making the equations virtually impossible to solve. To proceed we decompose the velocities, temperature, and pressure into a mean and a fluctuating component.

$$\begin{aligned} u_i &= \bar{u}_i + u_i' \\ \theta &= \bar{\theta} + \theta' \\ p &= \bar{p} + p' \end{aligned} \tag{5.26}$$

(Note that $\tilde{T}$ and θ differ only by a constant, and T and $\tilde{T}$ differ only by T_e. Thus, θ' can be regarded as a fluctuation in T, $\tilde{T}$, or θ.) By definition, the mean of a fluctuating quantity is zero, i.e.

$$\bar{u}_i' = \bar{\theta}' = \bar{p}' = 0 \tag{5.27}$$

Thus, a term of the form $\overline{u_i u_j}$ can be written $\bar{u}_i \bar{u}_j + \overline{u_i' u_j'}$, where the mean of the product of two fluctuations is not necessarily (and usually is not) equal to zero. If $\overline{u_i' u_j'} \neq 0$, u_i' and u_j' are said to be correlated.

Our objective is to determine equations for $\bar{u}_i$, $\bar{\theta}$, and $\bar{p}$. To obtain these equations we first substitute (5.26) into (5.14), (5.18), and (5.22). We then average each term in the resulting equations with respect to time. The result, employing (5.27), is

$$\frac{\partial \bar{u}_i}{\partial x_i} = 0 \tag{5.28}$$

$$\frac{\partial \bar{u}_i}{\partial t} + \bar{u}_j \frac{\partial \bar{u}_i}{\partial x_j} + \overline{u_j' \frac{\partial u_i'}{\partial x_j}} = -\frac{1}{\rho_0}\frac{\partial \bar{p}}{\partial x_i} + \nu_0 \frac{\partial^2 \bar{u}_i}{\partial x_j \, \partial x_j} + \frac{g}{T_0}\bar{\theta}\delta_{i3} \tag{5.29}$$

$$\rho_0 \hat{C}_p\left(\frac{\partial \bar{\theta}}{\partial t} + \bar{u}_j \frac{\partial \bar{\theta}}{\partial x_j} + \overline{u_j' \frac{\partial \theta'}{\partial x_j}}\right) = k \frac{\partial^2 \bar{\theta}}{\partial x_j \, \partial x_j} \tag{5.30}$$

It is customary to employ the relation

$$\frac{\partial u_i'}{\partial x_i} = 0$$

obtained by subtracting (5.28) from (5.14), to transform the third terms on the left-hand side of (5.29) and (5.30) to $\partial \overline{u_i' u_j'}/\partial x_j$ and $\partial \overline{u_j' \theta'}/\partial x_j$. Then (5.29) and (5.30) are written in the form

$$\frac{\partial}{\partial t}(\rho_0 \bar{u}_i) + \frac{\partial}{\partial x_j}(\rho_0 \bar{u}_i \bar{u}_j) = -\frac{\partial \bar{p}}{\partial x_i} + \frac{\partial}{\partial x_j}\left(\mu \frac{\partial \bar{u}_i}{\partial x_j} - \rho_0 \overline{u_i' u_j'}\right) + \frac{g}{T_0}\bar{\theta}\delta_{i3} \tag{5.31}$$

$$\rho_0 \hat{C}_p\left(\frac{\partial \bar{\theta}}{\partial t} + \bar{u}_j \frac{\partial \bar{\theta}}{\partial x_j}\right) = \frac{\partial}{\partial x_j}\left(k \frac{\partial \bar{\theta}}{\partial x_j} - \rho_0 \hat{C}_p \overline{u_j' \theta'}\right) \tag{5.32}$$

These equations, now time-averaged, contain only smoothly varying average quantities, so that the difficulties associated with the stochastic nature of the original equations have been alleviated. However, a new difficulty has arisen. We note the emergence of new dependent variables $\overline{u_i' u_j'}$, $i, j = 1, 2, 3$, and $\overline{u_j' \theta'}$, $j = 1, 2, 3$. When the equations are written in the form of (5.31) and (5.32), we can see that $\rho_0 \overline{u_i' u_j'}$ represents a new contribution to the total stress tensor and that $\rho_0 \hat{C}_p \overline{u_j' \theta'}$ is a new contribution to the heat flux vector, i.e.

$$\bar{\tau}_{ij} = -\left(\mu \frac{\partial \bar{u}_i}{\partial x_j} - \rho_0 \overline{u_i' u_j'}\right) \tag{5.33}$$

$$\bar{q}_j = -\left(k \frac{\partial \bar{\theta}}{\partial x_j} - \rho_0 \hat{C}_p \overline{u_j' \theta'}\right) \tag{5.34}$$

Let us consider the terms $\rho_0 \overline{u_i' u_j'}$. These terms, called the *Reynolds stresses*, indicate that the velocity fluctuations lead to a transport of momentum from one volume of fluid to another. Let us consider the physical interpretation of the Reynolds stresses. To do this, we envision the situation of a steady mean wind in the x_1 direction near the ground. Let the x_2 and x_3 directions be the horizontal direction perpendicular to the mean wind and the vertical direction, respectively. Then, a sudden increase or gust in the mean wind would result in a positive u_1', whereas a lull would lead to a negative u_1'. Left- and right-hand swings of the wind direction from its mean direction can be described by positive and negative u_2', respectively, and upward and downward vertical gusts by positive and negative u_3'. If the mean value of a product such as $u_1' u_2'$ is not to vanish, as we time-average over T, we must find a high frequency of terms of the same sign, either positive or negative, indicating that, say, positive values of u_1' are more likely to be found with positive values of u_2' than negative values. Since there is probably no reason to associate gusts or lulls with the wind having a tendency to swing in any particular direction, we would find $\overline{u_1' u_2'} = 0$. However, since the air needed to sustain a gust must come from faster moving air from above, we would expect positive values of u_1' to be correlated with negative values of u_3'. Similarly, a lull will result when air is transported upward rather than forward, so that we would expect negative u_1' to be associated with positive u_3'. As a result of both effects, $\overline{u_1' u_3'}$ will not vanish and the Reynolds stress $-\rho \overline{u_1' u_3'}$ will play an important role in the transport of momentum.

Equations (5.28), (5.31), and (5.32) have as dependent variables $\bar{u}_i$, $\bar{p}$, $\bar{\theta}$, $\overline{u_i' u_j'}$, and $\overline{u_j' \theta}$. We thus have 14 dependent variables $(\overline{u_i' u_j'} = \overline{u_j' u_i'})$ and only five equations. In general, one possible means for circumventing this problem is to generate equations which, in essence, are conservation equations for the new dependent variables. We can derive such an equation for the variables $\overline{u_i' u_j'}$, for example, by first subtracting (5.29) from (5.18), leaving an equation for u_i'. We then multiply this equation by u_j' and average over all terms. Although we can derive the desired equation for $\overline{u_i' u_j'}$, we have unfortunately at the same time generated still more dependent variables $\overline{u_i' u_j' u_j'}$. This problem, arising in the description of turbulence, is called the *closure problem*, for which no general solution has yet been found. At present, we must rely on models and estimates based on intuition and experience to obtain a closed set of equations. Since mathematics by itself will not provide a solution, we must resort to dimensional analysis and quasi-physical models for the Reynolds stresses and the turbulent heat fluxes. The next section is devoted to the most popular empirical models for the turbulent momentum and energy fluxes, the so-called mixing-length models.

5.4 MIXING–LENGTH MODELS FOR TURBULENT TRANSPORT

As we have just seen, the closure problem is the fundamental impediment to obtaining solutions for the mean velocities in turbulent flows. In order to progress at all, from a purely mathematical point of view, we must obtain a closed set of equations. The simplest approach to closing the equations is based on an appeal to a physical picture of the actual nature of turbulent momentum transport.

We can envision the turbulent fluid as comprising lumps of fluid which, for a short time, retain their integrity before being destroyed. These lumps or eddies transfer momentum, heat, and material from one location to another, conceptually in much the same way as molecular motion is responsible for transport in gases. Thus, it is possible to imagine an eddy, originally at one level in the fluid, breaking away and conserving some or all of its momentum until it mixes with the mean flow at another level.

Let us first consider turbulent momentum transport, i.e. the Reynolds stresses. We assume a steady turbulent shear flow in which $\bar{u}_1 = \bar{u}_1(x_2)$ and $\bar{u}_2 = \bar{u}_3 = 0$. The equation for $\bar{u}_1$ is

$$\frac{d}{dx_2} \rho \overline{u_1' u_2'} = \mu \frac{d^2 \bar{u}_1}{dx_2^2} - \frac{\partial \bar{p}}{\partial x_1} \tag{5.35}$$

The mean flux of x_1 momentum in the x_2 direction due to the turbulence is $\rho \overline{u_1' u_2'}$. Let us see if we can derive an estimate for this flux.

We can assume that the fluctuation in u_1 at any level x_2 is due to the arrival at that level of a fluid lump or eddy which originated at some other location where the mean velocity was different from that at x_2. We illustrate this idea in Fig. 5.2, in which a fluid lump which is at $x_2 = x_2 + l_\alpha$ at $t - \tau_\alpha$ arrives at x_2 at t. Let u_{1_α}' be the fluctuation in u_1 at x_2 at time t due to the αth eddy. If the eddy maintains its x_1 momentum during its sojourn, the fluctuation in u_1 at x_2 can be written

$$u_{1_\alpha}' = u_{1_\alpha}(x_2,t) - \bar{u}_1(x_2) \tag{5.36}$$

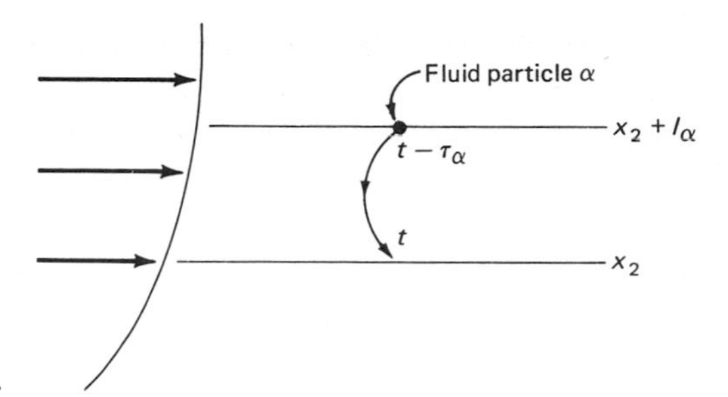

FIGURE 5.2
Eddy transfer in a turbulent shear flow.

Suppose that this eddy originated at the level $x_2 + l_\alpha$ at time $t - \tau_\alpha$ with a velocity equal to the mean velocity at that level, namely $\bar{u}_1(x_2 + l_\alpha)$. As long as the x_1 momentum of the eddy is conserved,

$$u_{1_\alpha}(x_2,t) = \bar{u}_1(x_2 + l_\alpha,\ t - \tau_\alpha) \tag{5.37}$$

Substituting (5.37) into (5.36) and expanding $\bar{u}_1(x_2 + l_\alpha, t - \tau_\alpha)$ in a Taylor series about the point (x_2,t), we get

$$u'_{1_\alpha} = l_\alpha \frac{\partial \bar{u}_1}{\partial x_2} - \tau_\alpha \frac{\partial \bar{u}_1}{\partial t} + \frac{1}{2} l_\alpha{}^2 \frac{\partial^2 \bar{u}_1}{\partial x_2{}^2} + \frac{1}{2} \tau_\alpha{}^2 \frac{\partial^2 \bar{u}_1}{\partial t^2} + \cdots \tag{5.38}$$

First, we note that, since the flow has been assumed to be steady, $\bar{u}_1$ does not vary with time. Thus, (5.38) becomes

$$u'_{1_\alpha} = l_\alpha \frac{d\bar{u}_1}{dx_2} + \frac{1}{2} l_\alpha{}^2 \frac{d^2 \bar{u}_1}{dx_2{}^2} + \cdots \tag{5.39}$$

Let L_e be the maximum distance over which an eddy maintains its integrity, that is, $L_e > |l_\alpha|$ for nearly all eddies. Let L be a characteristic length scale of the $\bar{u}_1$ field, say, given by

$$L = \frac{d\bar{u}_1/dx_2}{d^2\bar{u}_1/dx_2{}^2} \tag{5.40}$$

Then, as long as

$$L \gg L_e \tag{5.41}$$

we can truncate second- and higher-order terms in (5.39), leaving

$$u'_{1_\alpha} = l_\alpha \frac{d\bar{u}_1}{dx_2} \tag{5.42}$$

We can now consider the turbulent flux $\rho\overline{u_1'u_2'}$. First, we multiply (5.42) by u_2', the velocity fluctuation in the x_2 direction at (x_2,t) associated with the αth eddy, and average the resulting equation over an ensemble of eddies which pass the point x_2, say,

$$\lim_{N\to\infty} \frac{1}{N} \sum_{\alpha=1}^{N} u_{2_\alpha}' l_\alpha$$

Then, assuming that this ensemble eddy average converges to the time average (which it should for a steady flow), we obtain

$$\overline{u_1'u_2'} = \overline{u_2' l} \frac{d\bar{u}_1}{dx_2} \tag{5.43}$$

The term $\overline{u_2' l}$ represents the correlation (negative, if $d\bar{u}_1/dx_2 > 0$) between the fluctuating x_2 velocity at x_2 and the distance of travel of the eddy. As the eddy travels, we expect $\overline{u_2' l}$ to decrease, since u_2' is the velocity in the x_2 direction at $x_2 + l$, and the particle is getting farther and farther away from $x_2 + l$ as it moves. Assume l and u_2' become uncorrelated at L_e, where L_e is a measure of the eddy size. We can then estimate the order of $\overline{u_2' l}$ as $-L_e \hat{u}_2$, where $\hat{u}_2$ is the turbulent intensity, $(\overline{u_2'^2})^{1/2}$. Employing this relation in (5.43), we obtain

$$\overline{u_1'u_2'} = -cL_e \hat{u}_2 \frac{d\bar{u}_1}{dx_2} \tag{5.44}$$

where c is a constant of proportionality.

A reasonable definition of the *mixing length* L_e is the integral length scale

$$L_e = \int_0^\infty \frac{\overline{u_2'(x_2 + l)u_2'(x_2)}}{\overline{u_2'^2}} \, dl \tag{5.45}$$

The integrand is expected to vanish for sufficiently large l, so that L_e is a measure of the maximum distance in the fluid over which the velocity fluctuations are correlated, or, in some sense, of the eddy size. The experimental determination of L_e simply involves measuring the velocities at two points separated by larger and larger distances.

The turbulent flux can, therefore, be written

$$\rho\overline{u_1'u_2'} = -c\rho L_e \hat{u}_2 \frac{d\bar{u}_1}{dx_2} \tag{5.46}$$

Based on (5.46) we define an *eddy viscosity* or *turbulent momentum diffusivity* K_M by

$$\rho\overline{u_1'u_2'} = -\rho K_M \frac{d\bar{u}_1}{dx_2} \tag{5.47}$$

where $K_M = cL_e \hat{u}_2$.

We can extend the mixing-length concept to the turbulent heat flux. We consider the same shear flow as above, in which buoyancy effects are, for the moment, neglected. The mean vertical turbulent heat flux is $\rho \hat{C}_p \overline{u_2'\theta'}$. By analogy to the definition of the eddy viscosity, we can define an eddy diffusivity for heat transfer by

$$\rho \hat{C}_p \overline{u_2' \theta'} = -\rho \hat{C}_p K_T \frac{d\bar{\theta}}{dx_2} \tag{5.48}$$

Equations (5.47) and (5.48) provide a solution to the closure problem inasmuch as the turbulent fluxes have been related directly to the mean velocity and potential temperature. However, we have essentially exchanged our lack of knowledge of $\rho \overline{u_1' u_2'}$ and $\rho \hat{C}_p \overline{u_2'\theta'}$ for K_M and K_T, respectively. In general, both K_M and K_T are different for transport in different coordinate directions and are a function of location in the flow field. The variation of these coefficients with flow properties is usually determined from experimental data.

The result of the mixing-length idea used to derive the expressions (5.47) and (5.48) is that the turbulent momentum and energy fluxes are related to the gradients of the mean quantities. Substitution of these relations into (5.31) and (5.32) leads to advection diffusion equations for the mean quantities. Thus, except for the fact that K_M and K_T vary with position and direction, the models for turbulent transport are analogous to those for molecular transport of momentum and energy. The use of a diffusion equation model implies that the length scale of the transport process is much smaller than the characteristic length over which the mean profiles are changing [such as, for example, L, given by (5.40)]. In molecular diffusion in a gas at normal densities, the length scale of the diffusion process, the mean free path of a molecule, is many orders of magnitude smaller than the distances over which the mean properties of the gas vary. In turbulence, on the other hand, the motions responsible for the transport of momentum, heat, and material are usually roughly the same size as the characteristic length scale for changes in the mean fields of velocity, temperature, and concentration. Thus, in general, a diffusion model for turbulent transport, as exemplified by (5.47) and (5.48), is inapplicable in turbulence. In the atmosphere, for example, the characteristic vertical dimension of eddies is of the same order as the distance above the ground as is the characteristic scale of changes in the velocity profile. We must conclude, therefore, that expressions such as (5.47) and (5.48) do not possess a firm, rigorous basis, and the success in using them depends on two factors. First, they should ideally be employed in situations in which the length scale for changes in the mean properties is considerably greater than that of the eddies responsible for transport, that is (5.41). Second, the values and functional forms of K_M and K_T should be determined from experiments in situations similar to those in which (5.47) and (5.48) are to be applied.

5.5 VARIATION OF WIND WITH HEIGHT IN THE ATMOSPHERE

The atmosphere near the surface of the earth can be divided into three layers, as shown in Fig. 3.15: the free atmosphere, the Ekman layer, and the surface layer.[1] The Ekman layer and the surface layer constitute the so-called planetary boundary layer. The Ekman layer extends to a height of from 300 to 500 m depending on the type of terrain, with the greater thickness corresponding to the more disturbed terrain.

In the Ekman layer, the wind direction tends to turn clockwise with increasing height in the Northern Hemisphere (counterclockwise in the Southern Hemisphere). The wind speed in the Ekman layer generally increases rapidly with height; however, the rate lessens as the free atmosphere is approached. The exact distribution of the wind speed depends on many parameters, particularly the vertical distribution of the horizontal pressure gradient as well as the atmospheric stability.

The layer immediately adjacent to the surface, typically up to 30 to 50 m from the ground, is called the surface layer. Within this layer, the vertical turbulent fluxes of momentum and heat are assumed constant with respect to height, and indeed they define the extent of this region.

In this section we consider the prediction of the variation of wind with height in the surface and Ekman layers. Most of our attention will be devoted to the surface layer, the region in which pollutants are usually first released.

One other item should be discussed before we begin, and that is the question of smooth versus rough surfaces. In meteorological applications, the surface features leading to roughness are usually so closely distributed (for example, grass, crops, bushes, etc.) that only the height of the roughness elements and not their spacing is important. Thus, we characterize a particular rough surface by a single length parameter ε. Whether the surface is "smooth" or "rough" depends on the comparison of ε with the depth of the laminar sublayer. In general, a surface is called *smooth* if the roughness elements are sufficiently small to allow the establishment of a laminar sublayer in which they are submerged. On the other hand, a *rough* surface is one in which the roughness elements are high enough to prevent the formation of a laminar sublayer, so that the flow is turbulent down to the roughness elements. The depth of the laminar sublayer, and hence the classification of the surface as smooth or rough, depends on the Reynolds number of the flow.

[1] Immediately adjacent to the ground surface, a laminar sublayer can be identified in which molecular viscosities become important. However, the thickness of this layer is typically less than a centimeter. Therefore, for all practical purposes, it can be ignored in the present discussion.

5.5.1 Mean Velocity in the Surface Layer in Adiabatic Conditions

Let us consider the steady, two-dimensional turbulent flow of air in the surface layer parallel to the ground at $x_3 = 0$. We assume that $\bar{u}_1 = \bar{u}_1(x_3)$ and $\bar{u}_2 = 0$. Our object is to determine $\bar{u}_1(x_3)$ when the vertical temperature profile is adiabatic. Since in this case $\theta = 0$ we need consider only the x_1 component of the time-averaged equation of motion,

$$\frac{d}{dx_3}\,\rho\overline{u_1'u_3'} = \mu\frac{d^2\bar{u}_1}{dx_3{}^2} - \frac{\partial\bar{p}}{\partial x_1} \tag{5.49}$$

which we can write in the more concise form

$$\frac{\partial\bar{\tau}_{13}}{\partial x_3} = \frac{\partial\bar{p}}{\partial x_1} \tag{5.50}$$

where $\bar{\tau}_{13}$ is the total shear stress. If $\partial\bar{p}/\partial x_1$, the pressure gradient in the direction of the mean wind, is independent of x_3, we can integrate (5.50) to give

$$\bar{\tau}_{13} = \tau_0 + x_3\frac{\partial\bar{p}}{\partial x_1} \tag{5.51}$$

where τ_0 is the value of $\bar{\tau}_{13}$ as $x_3 \to 0$. In most atmospheric situations, and as we shall assume here, $\partial\bar{p}/\partial x_1$ is small, and, provided x_3 is not too large, $\bar{\tau}_{13}$ is approximately equal to τ_0 in the surface layer.

Let us now see what can be determined about the functional dependence of $\bar{u}_1(x_3)$ employing dimensional analysis. The term $\bar{u}_1$ and hence $d\bar{u}_1/dx_3$ should depend on τ_0, ν, ρ, and x_3. We thus have five quantities involving three dimensions (mass, length, and time). We invoke the Buckingham π theorem:

Let $B_1, B_2, \ldots, B_m$ be the m variables in a physical problem. Their functional relationship may be written $F(B_1,B_2,\ldots,B_m) = 0$. If k fundamental dimensions are required to define these variables, then the above relation may be written in terms of $m - k$ dimensionless groups, $F_1(\pi_1,\pi_2,\ldots,\pi_{m-k}) = 0$.

In this case $m = 5$ and $k = 3$, and so there are only two independent dimensionless groups relating the five variables. The π method does not tell us *what* the groups are, only *how many* exist. It will be useful to have a characteristic velocity from among the variables. For this we choose $\sqrt{\tau_0/\rho}$, denote it u_*, and call it the *friction velocity.* As our two groups we select

$$\pi_1 = \frac{d\bar{u}_1}{dx_3}\frac{x_3}{u_*} \qquad \pi_2 = \frac{x_3 u_*}{\nu}$$

The π method tells us that

$$F_1(\pi_1,\pi_2) = 0$$

or, alternatively, that

$$\frac{d\bar{u}_1}{dx_3} = \frac{u_*}{x_3} F_2\left(\frac{x_3 u_*}{\nu}\right) \tag{5.52}$$

This is as far as dimensional analysis will bring us. We now need to add some physical insight. We note first that $x_3 u_*/\nu$ is essentially a Reynolds number. Typical values of u_* and ν for the atmosphere are 100 cm/sec and 0.1 cm^2/sec, respectively. Thus,

$$\frac{x_3 u_*}{\nu} = \begin{cases} 10^5 & x_3 = 1 \text{ m} \\ 10^7 & x_3 = 100 \text{ m} \end{cases}$$

The large values indicate a fully turbulent region. We expect that $d\bar{u}_1/dx_3$ should be independent of ν in this region, since the thickness of the laminar sublayer, proportional to ν/u_*, is of the order of 0.01 cm. Thus, we can set $F_2(x_3 u_*/\nu) = \text{const} = a$, and

$$\frac{d\bar{u}_1}{dx_3} = a\frac{u_*}{x_3} \tag{5.53}$$

Upon integration,

$$\frac{u_1(x_3)}{u_*} = a \ln x_3 + \text{const}$$

which may be represented in the dimensionless form

$$\frac{\bar{u}_1(x_3)}{u_*} = a \ln \frac{u_* x_3}{\nu} + \text{const} \tag{5.54}$$

The constant of integration would in principle be determined by the condition that $\bar{u}_1 = 0$ at $x_3 = 0$. Unfortunately, this boundary condition cannot be satisfied by a finite constant in (5.54). Thus, we must employ a somewhat different boundary condition. We choose as the condition that the velocity gradient increase without limit as $x_3 \to 0$. This is automatically satisfied by (5.53). The constant in (5.54) was evaluated experimentally by Nikuradse for smooth surfaces and he found

$$\frac{\bar{u}_1(x_3)}{u_*} = \frac{1}{\kappa} \ln \frac{u_* x_3}{\nu} + 5.5 \tag{5.55}$$

where the constant a has been written as $1/\kappa$ with the experimental value of $\kappa = 0.4$. We find from (5.55) that $\bar{u}_1$ vanishes at $x_3 = \nu/9u_*$, so that (5.55) holds only for values of x_3 greater than this, approximately 10^{-4} cm.

For a rough surface, there is no laminar sublayer. Thus, $\bar{u}_1$ should depend on τ_0, ρ, ε, and x_3. By dimensional analysis, we reason that

$$\frac{\bar{u}_1(x_3)}{u_*} = \frac{1}{\kappa} \ln \frac{x_3}{\varepsilon} + \text{const} \tag{5.56}$$

which is usually written

$$\frac{\bar{u}_1(x_3)}{u_*} = \frac{1}{\kappa} \ln \frac{x_3}{z_0} \qquad x_3 \geq z_0 \tag{5.57}$$

where the integration constant z_0 is called the *roughness* length. Clearly, z_0 should be related to the height of the roughness elements ε. By experiment it has been found that $z_0 \cong \varepsilon/30$. The criteria for smooth or rough flow regimes have been determined experimentally to be

$$\frac{u_* z_0}{\nu} \begin{cases} <0.13 & \text{smooth flow} \\ >2.5 & \text{rough flow} \end{cases}$$

Values of z_0 for typical surfaces are given in Table 5.1.

The logarithmic law (5.57) is determined by the parameters κ, z_0, and u_*. Of the two variables, z_0 is a property of the roughness, whereas u_* must somehow be measured. Commonly, the velocity is measured at some reference height, say 10 m. Substituting this measurement into (5.57) allows calculation of u_* and subsequent specification of $\bar{u}_1$ at all values of x_3. A better way of obtaining u_* would be a direct measurement of the surface shear stress, but this requires elaborate experimental equipment. Thus, in micrometeorological studies it is usual to infer the shear at the ground from measured profiles of wind velocity distribution, a procedure that works satisfactorily in neutrally stratified boundary layers (Plate, 1971).

5.5.2 Effects of Temperature on the Surface Layer

In our study of atmospheric turbulence we have to this point neglected any effects of buoyancy. In the atmosphere, however, buoyancy plays an important role in maintaining (or suppressing) the energy of the turbulence. Consequently, we must

Table 5.1 ROUGHNESS LENGTHS FOR VARIOUS SURFACES

Surface	z_0, cm
Very smooth (ice, mud flats)	0.001
Snow	0.005
Smooth sea	0.02
Level desert	0.03
Lawn, grass up to 1 cm high	0.1
Lawn, grass up to 5 cm high	1–2
Lawn, grass up to 50 cm high	4–9
Fully grown root crops	14

SOURCES: Sutton (1953), Priestley (1959), and Pasquill (1962).

examine the effect of temperature stratification on the nature of turbulence in the surface layer.

The diurnal changes in solar radiation set up a cycle of heating and cooling of the atmospheric boundary layer which is strongly reflected in the wind field (recall Fig. 3.8). At night the air is stably stratified because the ground is colder than the air. As the sun rises, on a clear day, solar radiation heats up the ground faster than the air. Soon after dawn the near laminar flow of the nighttime stable air gives way to turbulent flow. As height increases, the effect of shear stresses at the surface in maintaining turbulence decreases and the effect of buoyancy increases. The warm thermals of air cause vigorous mixing aloft. The thickness of the layer of convective influence increases during the day as surface heating continues. Late in the afternoon the air reaches the same temperature as the ground, and the temperature profile becomes adiabatic since there is no heat flux from the ground. Near evening, the temperature of the air exceeds that of the ground, and the resulting heat flux to the ground causes a stably stratified temperature profile. The stable layer builds in thickness throughout the night just as the unstable layer grew during the day. Wind speed is often very low at night, and under these circumstances shear is virtually nonexistent and stratification becomes dominant.

This typical behavior serves to point out that the atmosphere is seldom adiabatic. Thus, it becomes essential in predicting velocities in the surface layer to consider the effects of temperature on the type of turbulence to be expected.

All that we essentially want to do is to examine the differential equations for the dynamics of the kinetic energy of turbulence and for the dynamics of temperature fluctuations. We will not attempt to derive or discuss in any detail these two equations, as our interest is only in the physical interpretation of the terms in the equations. For complete treatment of the dynamics of atmospheric turbulence we refer the reader to Monin and Yaglom (1971, sec. 6).

Let us consider a shear flow that is steady and homogeneous in the x_1x_2 plane with the only nonzero mean velocity $\bar{u}_1(x_3)$. The kinetic energy of the turbulence is given by $\frac{1}{2}\overline{u_i'u_i'} = \frac{1}{2}(\overline{u_1'u_1'} + \overline{u_2'u_2'} + \overline{u_3'u_3'})$. A measure of the effect of the turbulence on temperature fluctuations is the mean-square fluctuation $\overline{\theta'^2}$. The dynamic equations governing $\frac{1}{2}\overline{u'u_i'}$ and $\overline{\theta'^2}$ in this situation reduce to

$$0 = \underset{①}{-\overline{u_1'u_3'}\frac{\partial \bar{u}_1}{\partial x_3}} + \underset{②}{\frac{g}{T_e}\overline{u_3'\theta'}} - \underset{③}{\frac{\partial}{\partial x_3}\left(\frac{1}{2}\overline{u_i'u_i'u_3'} + \frac{1}{\rho}\overline{p'u_3'}\right)} - \underset{④}{\nu\overline{\frac{\partial u_i'}{\partial x_j}\frac{\partial u_i'}{\partial x_j}}} \tag{5.58}$$

$$0 = \underset{⑤}{-\overline{u_3'\theta'}\frac{\partial \bar{\theta}}{\partial x_3}} - \underset{⑥}{\frac{\partial}{\partial x_3}\left(\frac{1}{2}\overline{\theta'^2u_3'}\right)} - \underset{⑦}{\alpha\overline{\frac{\partial \theta'}{\partial x_j}\frac{\partial \theta'}{\partial x_j}}} \tag{5.59}$$

The terms in these two equations can be interpreted as follows:

① Production of turbulent kinetic energy by shear stresses
② Production of turbulent kinetic energy by buoyancy (if this term is negative, it represents *loss* of kinetic energy by buoyancy)
③ Turbulent flux of kinetic energy
④ Dissipation of kinetic energy by molecular viscosity
⑤ Production of fluctuations by the mean temperature gradient
⑥ Turbulent flux of mean-square temperature fluctuations
⑦ Decay of mean-square temperature fluctuations due to molecular conductivity

These are the basic equations used in the description of atmospheric turbulence. The key feature of interest in this discussion is the buoyant production of turbulent kinetic energy, i.e. term ②. In order to have a means of assessing the importance of this term, let us consider the ratio of terms ② and ①.

$$\frac{(g/T_e)\overline{u_3'\theta'}}{\overline{u_1'u_3'}\,\partial\bar{u}_1/\partial x_3}$$

This ratio is called the *flux Richardson number* and is denoted by Rf. We can then rewrite (5.58):

$$0 = -\overline{u_1'u_3'}\frac{\partial\bar{u}_1}{\partial x_3}(1 - \text{Rf}) - \frac{\partial}{\partial x_3}\left(\frac{1}{2}\overline{u_i'u_i'u_3'} + \frac{1}{\rho}\overline{p'u_3'}\right) - \nu\overline{\frac{\partial u_i'}{\partial x_j}\frac{\partial u_i'}{\partial x_j}} \tag{5.60}$$

In our situation $\partial\bar{u}_1/\partial x_3 > 0$, and so $\overline{u_1'u_3'} < 0$ as explained in Sec. 5.3. Thus, the term $\overline{u_3'\theta'}$ governs the sign of Rf and thus whether kinetic energy is produced or destroyed by the buoyancy. We stress that buoyancy can lead not only to kinetic energy production but also to its destruction. This can be explained readily by the sign of the term $\overline{u_3'\theta'}$.

CASE 1 $\overline{u_3'\theta'} > 0$, $\text{Rf} < 0$ Positive values of u_3' occur with positive values of θ'. The actual mean profile is $\bar{\theta}(x_3) = \bar{T}(x_3) - T_e$, as shown in Fig. 5.3. Consider a parcel of air which experiences an upward displacement with $u_3' > 0$. Its temperature will change adiabatically if the fluctuation is rapid. At the new level, the temperature fluctuation θ' is the difference between the parcel's temperature $\bar{\theta}(x_3)$ and that of the surroundings $\bar{\theta}(x_3 + l)$. The parcel's *actual* temperature T decreases in accordance with the adiabatic relation, but its temperature θ, $\theta = T - T_e$, *relative to an adiabatic profile*, remains constant. Thus, in this case, $\theta' > 0$. The production of

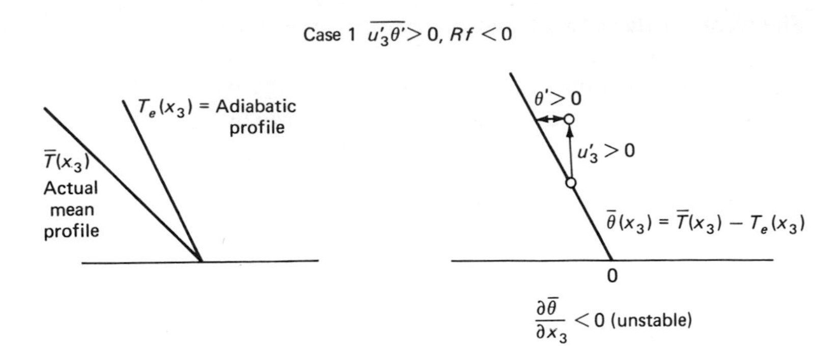

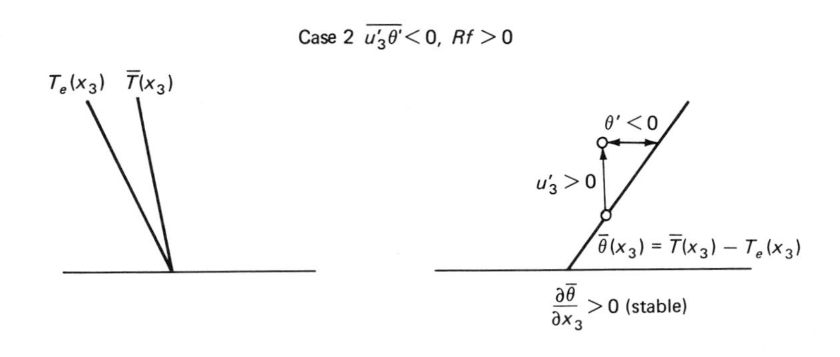

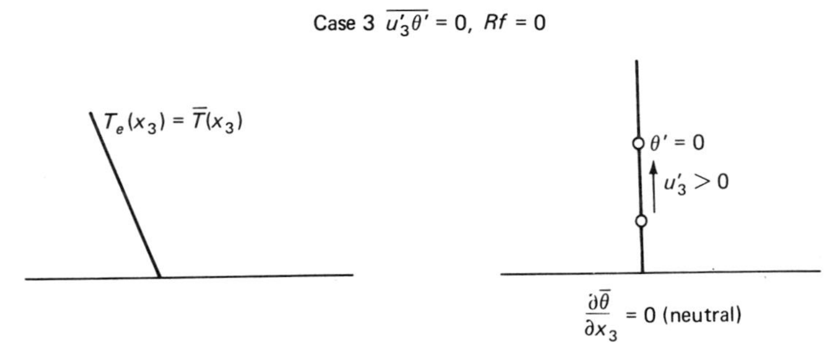

FIGURE 5.3
Relationship between the sign of Rf and atmospheric stability.

turbulent kinetic energy is *increased* in (5.60). Since the actual mean temperature profile must be as shown in Fig. 5.3 (case 1) we know that this situation occurs under *unstable* conditions ($\partial\bar{\theta}/\partial x_3 < 0$).

CASE 2 $\overline{u_3'\theta'} < 0$, $\mathrm{Rf} > 0$ Positive values of u_3' occur with negative values of θ'. Consider a parcel of air which experiences an upward displacement with $u_3' > 0$. Its temperature will change adiabatically and, as noted in case 1, remain at $\bar{\theta}(x_3)$. Its temperature difference at the new level, θ', will therefore be negative since $\partial\bar{\theta}/\partial x_3 > 0$. Turbulent kinetic energy is lost, since $\mathrm{Rf} > 0$ in (5.60). This situation occurs under *stable* conditions ($\partial\bar{\theta}/\partial x_3 > 0$).

CASE 3 $\overline{u_3'\theta'} = 0$, $\mathrm{Rf} = 0$ In this case, there are no contributions to the turbulent kinetic energy from temperature fluctuations. This is recognized as the adiabatic, or *neutral*, case ($\partial\bar{\theta}/\partial x_3 = 0$).

From (5.60) it is evident that, if $\mathrm{Rf} = 1$, turbulent energy is consumed by buoyancy forces as fast as it is generated by shear stresses. Thus, $\mathrm{Rf} = 1$ represents a theoretical limit beyond which, that is, $\mathrm{Rf} > 1$, atmospheric turbulence is completely suppressed. (Actually, experimental observations have shown that turbulence cannot be maintained for values of Rf greater than about 0.2. We will discuss this point below.)

The flux Richardson number is a function of the distance from the ground. To illustrate the dependence of Rf on x_3, we consider the case (near-neutral conditions) in which the velocity profile is logarithmic, i.e.

$$\frac{\partial\bar{u}_1}{\partial x_3} = \frac{u_*}{\kappa x_3}$$

By definition of the surface layer, the Reynolds stress $-\rho\overline{u_1'u_3'}$ is constant and equal to $\rho u_*{}^2$, and the vertical mean turbulent heat flux $\bar{q}_3 = \rho\hat{C}_p\overline{u_3'\theta'}$ is constant. We can then write Rf:

$$\mathrm{Rf} = -\frac{\kappa g x_3 \bar{q}_3}{\rho\hat{C}_p T_0 u_*{}^3} \tag{5.61}$$

We see that Rf is essentially a dimensionless length,

$$\mathrm{Rf} = \frac{x_3}{L} \tag{5.62}$$

where L, called the *Monin-Obukhov* length, is given by[1]

$$L = \frac{-\rho \hat{C}_p T_0 u_*^3}{\kappa g \bar{q}_3} \tag{5.63}$$

We see that L is simply the height above the ground at which the production of turbulence by both mechanical and buoyancy forces is equal. The Monin-Obukhov length, like Rf, provides a measure of the stability of the surface layer, i.e.

$$L > 0 \qquad \text{stable } (\bar{q}_3 < 0)$$
$$L < 0 \qquad \text{unstable } (\bar{q}_3 > 0)$$
$$L = \infty \qquad \text{neutral } (\bar{q}_3 = 0)$$

Although Rf is a convenient measure of the stability condition of the atmosphere, its measurement is difficult since both heat and momentum fluxes must be determined simultaneously. In order to obtain a more convenient form of Rf from the point of view of measurement, we employ the definitions of an eddy viscosity and thermal conductivity by (5.47) and (5.48), so that Rf may be written

$$\text{Rf} = \frac{K_T g}{K_M T_0} \frac{\partial \bar{\theta}/\partial x_3}{(\partial \bar{u}_1/\partial x_3)^2} \tag{5.64}$$

Aside from K_T and K_M, Rf now involves quantities which can be measured rather easily. In order to isolate the ratio K_T/K_M we define the *gradient Richardson number* Ri by

$$\text{Ri} = \frac{g}{T_0} \frac{\partial \bar{\theta}/\partial x_3}{(\partial \bar{u}_1/\partial x_3)^2} \tag{5.65}$$

Thus,

$$\text{Rf} = \frac{K_T}{K_M} \text{Ri} \tag{5.66}$$

[1] The Monin-Obukhov length can be obtained directly from dimensional analysis. The turbulence characteristics in the surface layer are governed in general by the following variables: (g/T_0), ρ, ν, k, τ (or u_*), $\bar{q}_3$, z_0, and x_3. We assume that (1) molecular effects can be neglected, that is, ν and k, and (2) variations in the roughness length z_0 only shift the profiles but do not affect their form. Then, the variables are (g/T_0), ρ, τ (or u_*), $\bar{q}_3$ (or $\bar{q}_3/\rho \hat{C}_p$), and x_3. We have five variables in four dimensions (mass, length, time, and temperature), so that by the Buckingham π theorem we have only one dimensionless group from among these variables. That group is

$$\text{Rf} = \frac{x_3}{L}$$

where κ is just a dimensionless proportionality factor.

Both Rf and Ri have significance in determining atmospheric conditions. The stability of the surface layer at any height is described by Rf, which is identically equal to zero in neutral conditions. A small absolute value of Rf indicates that the atmosphere is in a near-neutral condition, where a logarithmic velocity profile is valid. A small absolute value of Rf can arise in either of two ways. First, L can be large, implying that the heat flux $\bar{q}_3$ is small or u_* is large. Second, x_3 can be small. Thus, even in a flow significantly deviating from neutral conditions, there exists a layer close to the ground in which $|\mathrm{Rf}|$ is small and the flow resembles that in neutral stratification. Because the logarithmic law was used in deriving (5.62), this equation is valid only in the limit of small $|\mathrm{Rf}|$. When $|\mathrm{Rf}|$ is large, we still might expect Rf to depend only on x_3/L, although not in as simple a fashion as (5.62).

In unstable conditions, $\mathrm{Rf} < 0$. As Rf becomes more negative (as a result of $\bar{q}_3$ or x_3 increasing), the effect of mechanical generation of turbulence becomes less and less important compared with buoyant production of turbulent energy. A point is reached, at about $\mathrm{Rf} \cong -0.03$, where the flow becomes totally dominated by buoyancy effects and essentially becomes a free convection flow.

Under stable conditions, $\mathrm{Rf} > 0$. As Rf becomes more positive (as a result of $\bar{q}_3$ or x_3 decreasing), the effect of buoyancy is to suppress the mechanically generated turbulence. As Rf increases, a point is reached where the turbulence should be suppressed. A first guess at this critical value of Rf was unity, although experimental data seem to indicate that the critical flux Richardson number is no larger than 0.2 (Plate, 1971).

Theoretically we have obtained (5.66) relating Rf, Ri, and K_T/K_M. Since Rf is related to x_3/L, it should therefore be possible to correlate both Ri and K_T/K_M with x_3/L. Experimental data on Ri and K_T/K_M versus x_3/L collected by Plate (1971) from many sources are shown in Figs. 5.4 and 5.5. For Ri the x_3/L dependency is well confirmed, and excellent agreement is achieved between field and laboratory data. The correlation of K_T/K_M with x_3/L is less successful. The reason is that K_T and K_M refer to dynamically different quantities, i.e. heat and momentum, and differences in stability lead to different modes of transport of these two quantities.

Measurements of wind velocity components, such as depicted in Fig. 5.1, are important in characterizing atmospheric turbulence. Certain statistical properties of the turbulence can be extracted from such records. The *intensity* of turbulence is related to $\overline{u_i'^2}$, or $\sigma_{u_i}^2$ (no summation), the variance of the velocity distribution of the ith component about its mean value. The σ_{u_i} values bear a direct relation to the diffusing power of the atmosphere. Two other useful properties are the standard deviations of the fluctuations in the horizontal direction of the wind, σ_θ, and the vertical direction of the wind, σ_ϕ. It is important to realize that σ_{u_i}, σ_θ, and σ_ϕ depend on the sampling and averaging times inherent to a velocity record such as that shown in Fig. 5.1.

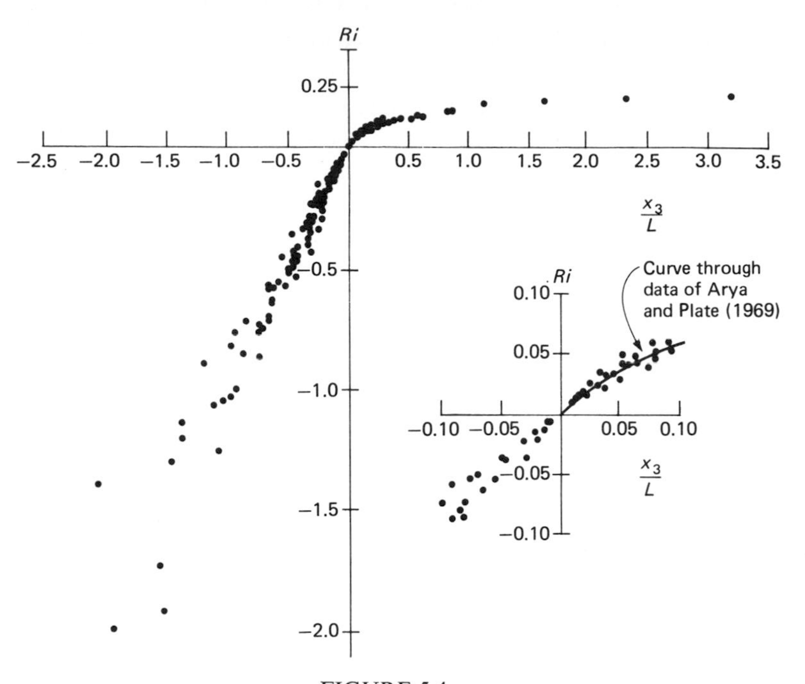

FIGURE 5.4
Ri as a function of x_3/L. (*Source: Plate*, 1971.)

In Fig. 5.6 is presented a general summary of the vertical variation of the standard deviation of the horizontal and of the vertical wind directions with height and stability. The construction of these figures is based on actual data. Figure 5.6*a* shows the variation of the horizontal wind direction standard deviation for a sampling time of about 10 min as measured up to about 130 m. For a given stability condition, values of σ_θ for sampling times greater than a minute or so will always be greater when the wind is light than when it is strong. The curve in Fig. 5.6*a* representing very stable conditions (which by their nature are associated with light winds) exhibits this behavior. The three branches represent limits to σ_θ, with the central curve that for typical inversion conditions. In all cases, the large values of σ_θ at the surface do not decrease very rapidly with height. Often the standard deviation of the crosswind velocity fluctuation σ_2 (if the mean wind is directed along the x_1 axis) can be related to σ_θ. The following are general statements concerning the variation of σ_{u_2} with wind speed, stability, and height (Slade, 1968):

1 At a given height during neutral conditions, σ_{u_2} is proportional to wind speed.
2 For a given wind speed and height, σ_{u_2} is greater during unstable than during stable conditions.

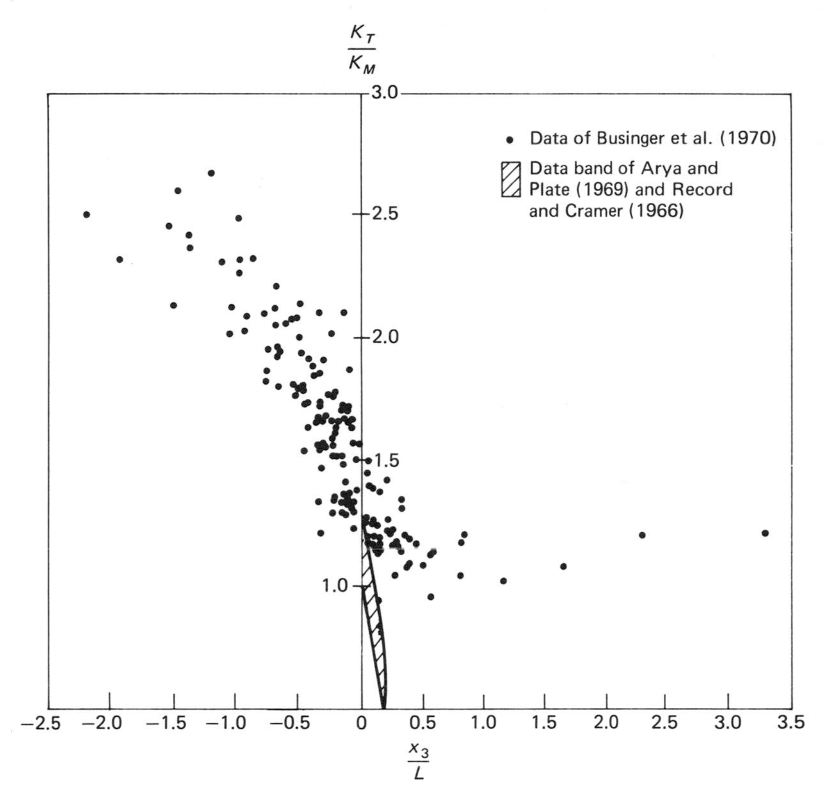

FIGURE 5.5
K_T/K_M as a function of x_3/L. (*Source: Plate*, 1971.)

3 For a given stability condition at a given height, σ_{u_2} increases with wind speed and surface roughness, most markedly during stable conditions.
4 The value of σ_{u_2} does not change appreciably with height during any stability condition.

Figure 5.6*b* presents vertical profiles of vertical direction fluctuations for averaging times of a few seconds. The effect of wind speed on vertical fluctuations is different from that on the horizontal. Light winds at low levels are associated with large values of σ_ϕ during unstable conditions and small values during stable conditions. Mechanically induced vertical turbulence decreases with height whereas buoyancy-induced turbulence increases with height. The standard deviation of the vertical velocity fluctuations σ_{u_3} are related to σ_ϕ. General statements regarding σ_{u_3} are (Slade, 1968):

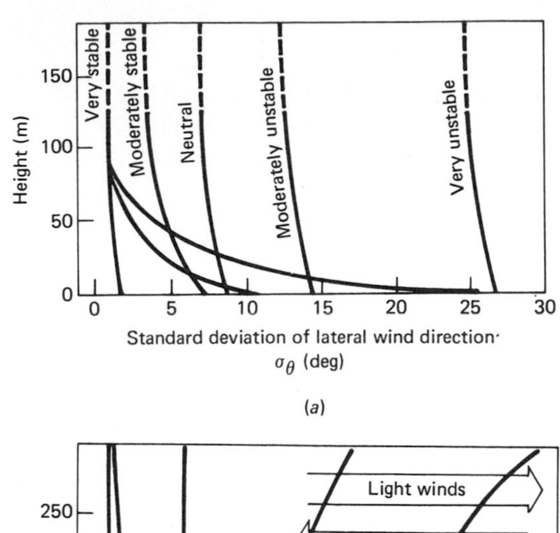

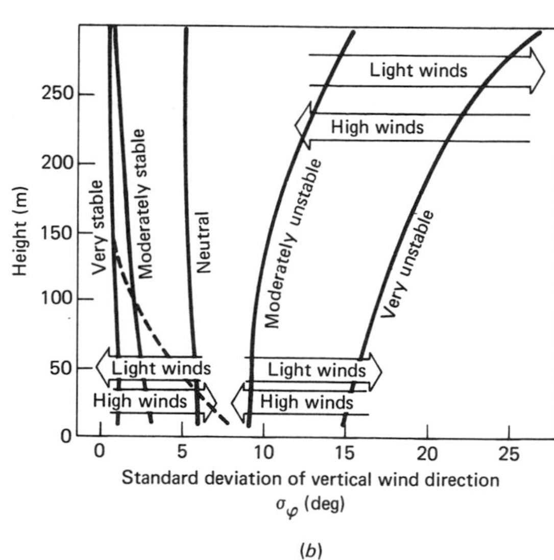

FIGURE 5.6
(*a*) Vertical variation of the lateral wind direction standard deviation σ_θ as a function of stability. (*b*) Vertical variation of the vertical wind direction standard deviation σ_ϕ as a function of stability. (*Source: Slade*, 1968.)

1 Under neutral conditions, σ_{u_3} is proportional to wind speed and constant with height.
2 In a stable atmosphere, σ_{u_3} decreases with height.
3 Under stable conditions, σ_{u_3} increases markedly with height.

Although the along-wind component of turbulence has not received as much experimental attention as the other two components, some general characteristics of σ_{u_1} are:

1 At a fixed height, σ_{u_1} is proportional to wind speed.
2 At a fixed height, σ_{u_1} increases with increasing instability.
3 σ_{u_1} is generally independent of height during neutral and unstable conditions but decreases with height during stable conditions.

The capacity of the atmosphere for diffusing heat and matter is related not only to the standard deviation of the wind direction but also to the range of frequencies of the fluctuations which contribute to the standard deviation. Although two records may have the same standard deviation, this could result in one case by a few oscillations of long period and in the other by many shorter-period oscillations. Long-period oscillations tend to transport parcels of pollutant in their entirety, whereas short-period fluctuations tend to tear the parcels apart. Thus, the distribution of turbulent energy among the various spatial and temporal scales is of great interest in understanding the structure of a turbulent flow. Such information is embodied in time and space *spectra* of the turbulence, basically representations of the turbulence as waves (time) and eddies (space). In Appendix A some elements of the spectral representation of turbulence are outlined briefly.

5.5.3 Wind Profiles in the Nonadiabatic Surface Layer

In stable and unstable conditions the velocity profiles of the atmospheric surface layer deviate from the logarithmic law. In this section we will outline briefly the forms of velocity profiles in these conditions. Since the stratified-boundary-layer conservation equations cannot be solved (because of the closure problem), we must resort to empirical profiles, based largely on dimensional analysis.

Suppose we are interested in the height dependence of some mean property $\bar{f}$ (say velocity) in the surface layer. In carrying out a dimensional analysis, six quantities must be considered: (g/T_0), ρ_0, u_*, $(\bar{q}/\rho C_p)$, x_3, $\bar{f}$. From the π theorem, we have six quantities involving four fundamental units; two independent dimensionless groups thus result. Let us choose x_3/L and $\bar{f}(x_3)/f_0$ as these two groups, where f_0 is a group with the dimensions of f, formulated from (g/T_0), ρ, u_*, and $(\bar{q}/\rho C_p)$. Thus,

$$\frac{\bar{f}(x_3)}{f_0} = F\left(\frac{x_3}{L}\right)$$

where F is some universal function. For example, if $\bar{f}(x_3) = \bar{u}(x_3)$, we choose f_0 as u_*/κ, and if $\bar{f}(x_3) = \bar{\theta}(x_3)$ we choose f_0 as $-(1/\kappa u_*)(\bar{q}/\rho \hat{C}_p) = T_*$. Let us denote the group x_3/L by ζ. Note that $\zeta = \text{Rf}$ only in the surface layer. Therefore, we can in theory represent the dependence on height of mean velocity and temperature (and their gradients) by universal functions of ζ, for example,

$$\frac{\partial \bar{u}_1}{\partial x_3} = \frac{u_*}{\kappa L} g(\zeta) \qquad (5.67)$$

$$\frac{\partial \bar{\theta}}{\partial x_3} = \frac{T_*}{L} g_1(\zeta) \qquad (5.68)$$

These equations are generalizations of the logarithmic layer equations to the case of a thermally stratified layer. We remind the reader that the adiabatic temperature profile in a *stagnant* layer is the familiar 1°C/100 m decrease. However, in the presence of a *mean wind* in the x direction with a logarithmic profile, the neutral temperature profile is given by

$$\frac{\partial \bar{\theta}}{\partial x_3} = \frac{T_*}{\gamma x_3}$$

We have not taken the space to derive this result, although it is obtained in the same manner as the logarithmic velocity law.

Substituting (5.67) and (5.68) into the defining equations for the eddy coefficients K_M and K_T, we obtain

$$K_M = \frac{\kappa u_* L}{g(\zeta)} \tag{5.69}$$

and

$$K_T = \frac{\kappa u_* L}{g_1(\zeta)} \tag{5.70}$$

In neutral conditions, $g(\zeta) = \zeta^{-1} = L/x_3$ and

$$K_M = \kappa u_* x_3 \tag{5.71}$$

The basic problem is to determine the forms of $g(\zeta)$ and $g_1(\zeta)$. We know the forms under neutral conditions, and so we look for distinct forms for stable and unstable conditions which will approach each other as $|\zeta| \to 0$. We consider first the velocity profile and rewrite (5.67):

$$\frac{\partial \bar{u}_1}{\partial x_3} = \frac{u_*}{\kappa L} g(\zeta) = \frac{u_* \zeta}{\kappa x_3} g(\zeta) = \frac{u_*}{\kappa x_3} \phi(\zeta) \tag{5.72}$$

As $\bar{q} \to 0$, $\zeta \to 0$, we must obtain

$$\frac{\partial \bar{u}_1}{\partial x_3} = \frac{u_*}{\kappa x_3}$$

and so we need $\lim_{\zeta \to 0} \phi(\zeta) = 1$. In effect, for $x_3 \ll |L|$ the layer is essentially adiabatic. Thus, as we noted, L is a measure of the thickness of a layer in which thermal effects are unimportant. When $x_3 \ll |L|$ (near neutral conditions)

$$\frac{x_3}{|L|} \cong |\mathrm{Rf}|$$

At low levels $x_3/|L| \cong 0$ and the effect of buoyancy can be neglected. The level to which the logarithmic law is valid depends on the magnitude of Rf. Over rough ground with strong winds, the so-called dynamic sublayer (adiabatic) may extend to 10 m, whereas over smooth ground with strong surface heating it may extend only up to 1 m.

For ζ close to zero, we can expand $\phi(\zeta)$ in a power series:

$$\phi(\zeta) = 1 + \beta_1\zeta + \beta_2\zeta^2 + \cdots \tag{5.73}$$

Usually the available data permit evaluation only of one coefficient, and so we truncate after the linear term. Substituting (5.73) into (5.72) and integrating we obtain

$$\bar{u}_1(x_3) = \frac{u_*}{\kappa}\left(\ln\frac{x_3}{z_0} + \beta\frac{x_3 - z_0}{L}\right) \tag{5.74}$$

It is important to note that $\zeta \geq 0$ and $\zeta \leq 0$ may correspond to different values of the parameter β. However, although its magnitude must be established from experimental data, the sign of β can be determined. Under stable conditions ($L > 0$, $\zeta > 0$) vertical turbulent momentum exchange is suppressed and the velocity profile increases more rapidly with x_3 than in the adiabatic case. Under unstable conditions ($L < 0$, $\zeta < 0$), on the other hand, intense turbulent mixing leads to equalization of the velocity, and so $\bar{u}_1(x_3)$ increases more slowly with height than in the adiabatic case. Thus, $\beta[(x_3 - z_0)/L]$ must be >0 for $L > 0$ and <0 for $L < 0$. Therefore, β must be >0 for both $\zeta > 0$ and $\zeta < 0$. The value of β yielding the best fit to data depends on the range of ζ values considered. For unstable stratification, values of β from 3 to 6 have been obtained; for stable stratification, typical values range from 5 to 7 (Plate, 1971).

Although (5.74) holds for fairly small values of $|\zeta|$, the cases of large negative (unstable) or positive (stable) require other formulas, the theory underlying which are discussed in detail by Monin and Yaglom (1971, sec. 7.4).

For unstable stratification, one form often used is

$$\phi(\zeta) = (1 - \beta'\zeta)^{-0.25}$$

where β' is about 15 (Plate, 1971). For stable stratification, the form

$$\phi(\zeta) = 4.2\zeta^{0.3}$$

valid for $0.1 \leq \zeta \leq 0.2$ has been used.

Figure 5.7 summarizes the dimensionless function $\phi(\zeta)$ as a function of ζ, and Fig. 5.8 shows $\kappa\bar{u}/u_*$ as a function of $\ln(x_3/z_0)$.

5.5.4 Wind Speed Variation in the Planetary Boundary Layer

It has been found that the mean wind speed variation with height in the entire atmospheric boundary layer can often be described quite well by the empirical-power-law profile

$$\frac{\bar{u}_1(x_3)}{u_G} = \left(\frac{x_3}{h_G}\right)^{\alpha} \tag{5.75}$$

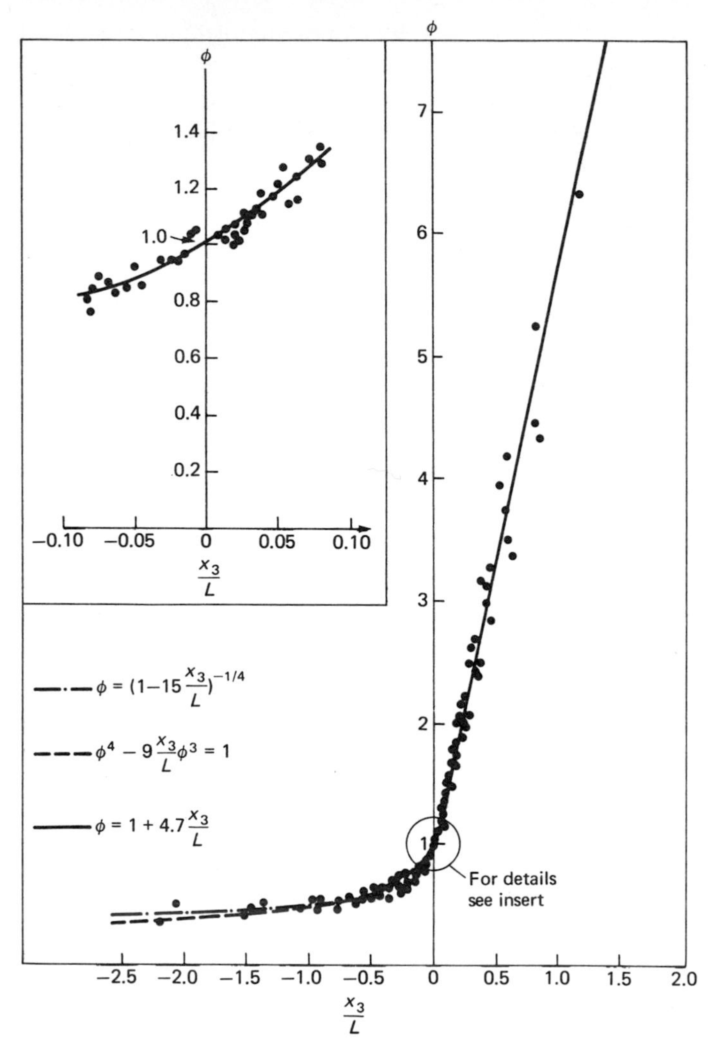

FIGURE 5.7
$\phi(\zeta)$ as a function of ζ. (*Source: Plate*, 1971.)

where u_G is the geostrophic wind speed and h_G the thickness of the planetary boundary layer. The exponent α is a number which varies from 0.1 to 0.4, depending on the roughness of the ground surface as well as on atmospheric stratification. "Best estimates" of this parameter, based on experiments made under fair weather conditions, are given in Table 5.2.

Two comments are in order concerning the utility of the power law. First, this law is derived empirically from data collected in neutrally stratified atmospheres.

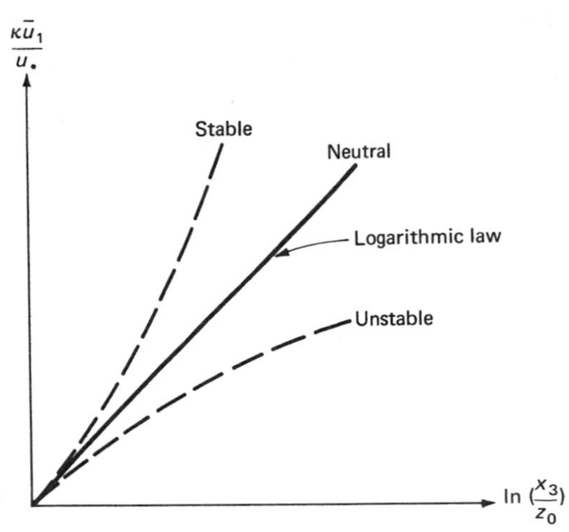

FIGURE 5.8
Surface-layer wind profiles under diabatic conditions.

Strictly speaking, it is not applicable under diabatic conditions. Second, this equation is found to fit the data quite well for all altitudes below the geostrophic height; it can therefore be used for the entire planetary boundary layer including the surface layer. In practice, because of the difficulties involved in the determination of u_G and h_G, (5.75) is usually replaced by

$$\frac{\bar{u}_1(x_3)}{\bar{u}_1(h)} = \left(\frac{x_3}{h}\right)^{\alpha} \tag{5.76}$$

where $\bar{u}_1(h)$ is the wind speed at reference height h within the planetary boundary layer (e.g. at 10 m).

Figure 5.9 shows empirical-power-law profiles over different terrain, and Fig. 5.10 shows the exponent α and the height of the boundary layer, h_G, as functions of the roughness length z_0 and the type of terrain. The rougher the terrain, (i.e., the larger the surface obstructions) the thicker will be the affected layer of air, and the

Table 5.2 ESTIMATES FOR α IN (5.75)

	Type of terrain		
	Open country	Suburbs	Urban
α	0.16†,‡	0.28†	0.40†

† Davenport (1965).
‡ Shellard (1965).

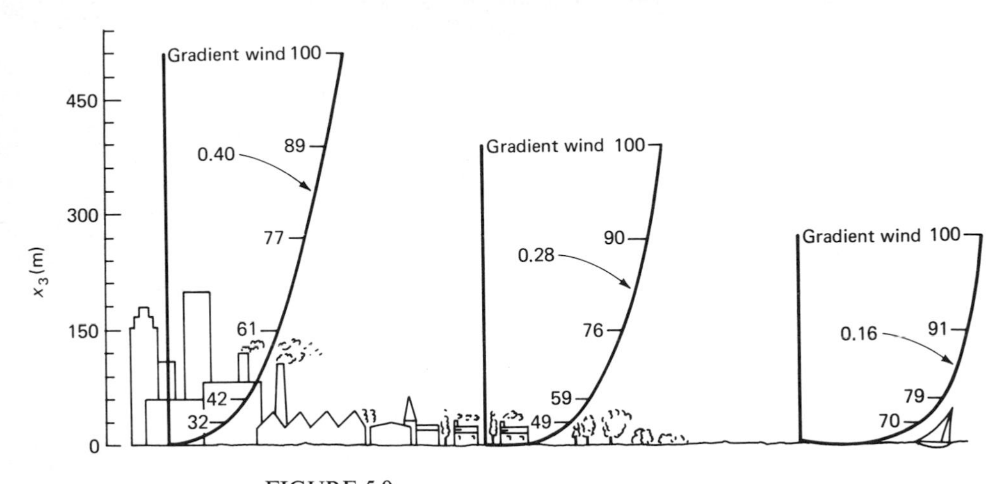

FIGURE 5.9
Form of wind profiles over different terrain. Each profile is shown with the corresponding value of α in (5.75). (*Source: Davenport*, 1965.)

more gradual will be the increase of velocity with height. Thus, as the roughness increases, the exponent α increases. The stability limits of α in (5.76) are:

$$\alpha = \begin{cases} 0.83 & \text{very stable} \\ 1/7 & \text{neutral} \\ 0.02 & \text{very unstable} \end{cases}$$

Finally, we will find in Chap. 6 that we are often confronted with the problem of solving a partial differential equation for the mean concentration of a pollutant in the atmosphere. The use of a velocity profile such as (5.74) renders these equations impossible to solve analytically. Therefore, for diffusion calculations it is desirable to use a profile in the form (5.76).

5.6 METEOROLOGICAL MEASUREMENTS IMPORTANT IN AIR POLLUTION

There are several *primary* meteorological measurements necessary in assessing atmospheric transport and mixing characteristics as related to the dispersion of air pollutants. Those related to the wind are:

1 Wind direction
2 Wind speed
3 Wind turbulence

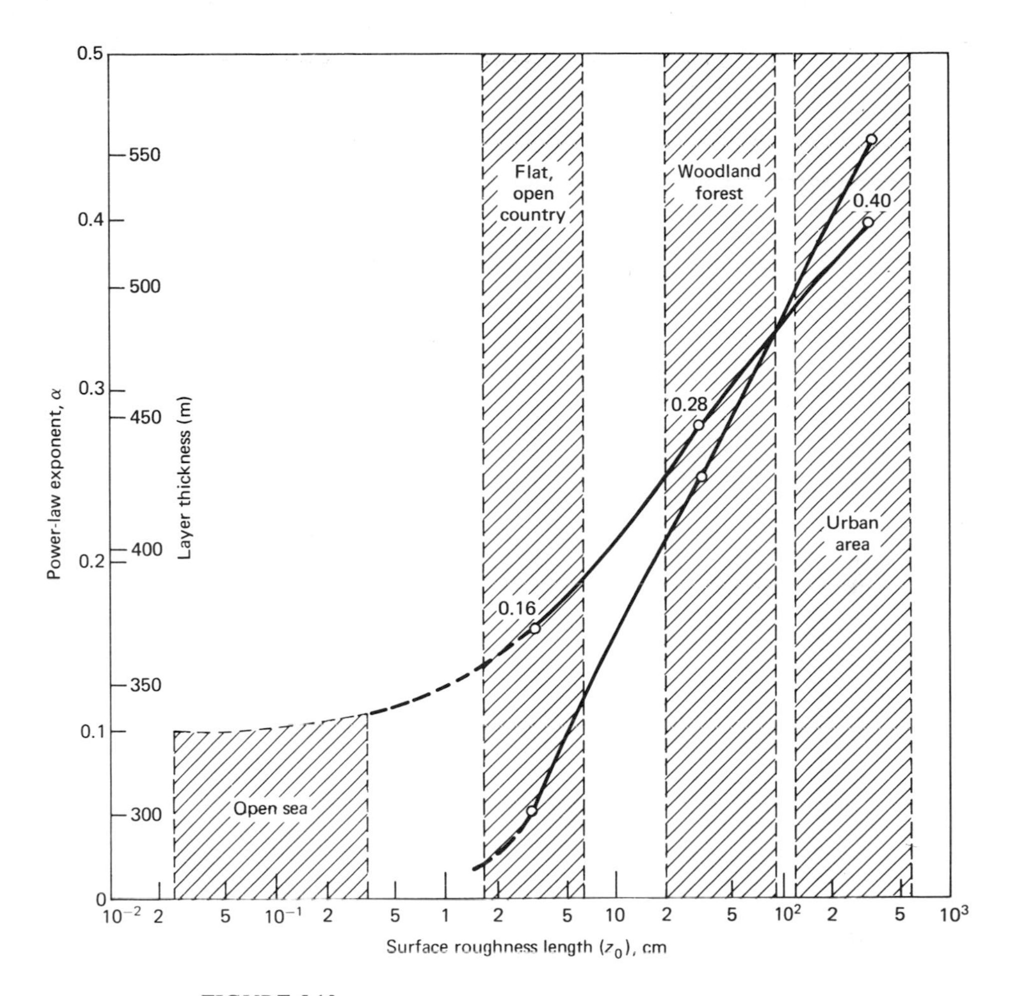

FIGURE 5.10
Exponent in the power law (5.75) and the height h_G of the boundary layer as functions of the roughness length and type of terrain. (*Source: Davenport*, 1965.)

Surface wind direction is meaured by a *wind vane*, essentially a flat plate on a horizontal rod which aligns itself with the wind direction. Wind direction aloft is determined by *pilot balloons* (less than 1 m in diameter, equipped to carry a small radio for tracking) and *tetroons* (a constant-volume balloon in the shape of a tetrahedron, which is ballasted so as to remain at a nearly constant elevation).

Surface wind speed is recorded by a class of instruments called *anemometers*. Typical anemometers are:

1 Rotation anemometer: a small windmill

2 Cup anemometer: three conical cups on a freely rotating vertical shaft

3 Hot-wire anemometer: an electrically heated element the temperature of which is related to the velocity of the passing wind

Wind speed measurements aloft are made by balloons and tetroons.

Figure 5.1 shows a typical record of one wind component which might be obtained by an anemometer. From such a record a mean velocity $\bar{u}_i$ can be obtained. Wind turbulence refers to the properties of the fluctuations about the mean in a record such as in Fig. 5.1. These fluctuations are primarily responsible for the spreading of a cloud of pollutants in the atmosphere. It is important to measure all three components of the wind velocity since the vertical fluctuations are often the most important in transporting pollutants. The vertical velocity fluctuations provide a direct measure of the stability of the atmosphere. In order to obtain a record such as shown in Fig. 5.1 a sensitive, rapidly responding wind vane and anemometer are required. Vertical fluctuations can be measured by:

1 Bivane: a wind vane constructed so as to meaure wind direction fluctuations in vertical as well as horizontal planes

2 *u-v-w* anemometer (refers to the three wind components): a system of three propellers mounted at the ends of orthogonal shafts

The standard elevation for wind instruments is 10 m over level, open terrain, where "open terrain" is usually interpreted as that where the distance to any obstacle is 10 times or more the height of the obstacle.

Other primary meteorological measurements are temperature profile and radiation. Temperature profiles are measured by *wiresonde* (a temperature sensor carried aloft by a kite balloon), *dropsonde* (a temperature sensor lowered by a parachute), and *radiosonde* (pressure, temperature, and humidity sensors carried aloft by a balloon). Solar radiation intensity is measured by an *actinometer*, a device in which two metallic strips, one of which is absorbing and one reflecting, lie side by side under a glass dome. The amount of differential bending is an indication of the radiation intensity. Also, the net heating or cooling at or near the surface can be measured by two black metal strips, one facing up, the other down. The differential heating or cooling establishes an emf between the strips.

Other *secondary* measurements often reported include visibility, humidity, and precipitation.

REFERENCES

ARYA, S. P. S., and E. J. PLATE: Modeling of the Stably Stratified Atmospheric Boundary Layer, *J. Atmos. Sci.*, **26**:656 (1969).

BUSINGER, J. A., J. C. WYNGAARD, Y. IZUMI, and E. F. BRADLEY: Flux Profile Relationships in the Atmospheric Surface Layer, *J. Atmos. Sci.*, **28**:181 (1971).

CALDER, K. L.: In Clarification of the Equations of Shallow-layer Thermal Convection for a Compressible Fluid Based on the Boussinesq Approximation, *Q. J. Roy. Meteorol. Soc.*, **94**:88 (1968).

DAVENPORT, A. G.: The Relationship of Wind Structure to Wind Loading, in Wind Effects on Buildings and Structures, National Physical Laboratory, Symposium 16, Her Majesty's Stationery Office, London, 1965.

HINZE, J. O.: "Turbulence," McGraw-Hill, New York, 1959.

LUMLEY, J. L., and H. A. PANOFSKY: "The Structure of Atmospheric Turbulence," Wiley Interscience, New York, 1964.

MONIN, A. S., and A. M. YAGLOM: "Statistical Fluid Mechanics," M.I.T., Cambridge, Mass., 1971.

PASQUILL, F.: "Atmospheric Diffusion," Van Nostrand, New York, 1962.

PLATE, E. J.: Aerodynamic Characteristics of Atmospheric Boundary Layers, U.S. Atomic Energy Commission, Oak Ridge, Tenn., 1971.

PRIESTLEY, C. H. B.: "Turbulent Transfer in the Lower Atmosphere," University of Chicago Press, Chicago, 1959.

RECORD, F. A., and H. E. CRAMER: Turbulent Energy Dissipation Rates and Exchange Processes Above a Nonhomogeneous Surface, *Q. J. Roy. Meteorol. Soc.*, **92**:519 (1966).

SHELLARD, H. C.: The Estimation of Design Wind Speeds, in Wind Effects on Buildings and Structures, National Physical Laboratory, Symposium 16, Her Majesty's Stationery Office, London, 1965.

SLADE, D. H. (ed.): Meteorology and Atomic Energy—1968, U.S. Atomic Energy Commission, Washington, D.C., 1968.

SPIEGEL, E. A., and G. VERONIS: On the Boussinesq Approximation for a Compressible Fluid, *Astrophys. J.*, **131**:442 (1960).

SUTTON, O. G.: "Micrometeorology," McGraw-Hill, New York, 1953.

PROBLEMS

5.1 An interpretation of the potential temperature θ is afforded by considering an adiabatic process. Changes in entropy can be related to those in temperature and pressure by

$$dS = \left(\frac{\partial S}{\partial T}\right)_p dT + \left(\frac{\partial S}{\partial p}\right)_T dp$$

(*a*) Show that

$$dS = \frac{\hat{C}_p}{T}\,dT - \frac{R}{M_a}\frac{dp}{p}$$

(*b*) From this result, show that in an adiabatic process ($dS = 0$), $d\theta = 0$, and that

$$\theta = \text{const} \times \frac{T}{p^{(\gamma-1)/\gamma}}$$

5.2 In Chap. 3 we discussed the Ekman spiral, that is, the variation of wind direction with altitude in the planetary boundary layer. The analytical form of the Ekman spiral can be derived by considering a two-dimensional horizontal wind field (no vertical component), the two components of which satisfy

$$\frac{\partial}{\partial z}\left(K_M \frac{\partial \bar{u}}{\partial z}\right) - \frac{1}{\rho}\frac{\partial p}{\partial x} + f\bar{v} = 0$$

$$\frac{\partial}{\partial z}\left(K_M \frac{\partial \bar{v}}{\partial z}\right) - \frac{1}{\rho}\frac{\partial p}{\partial y} - f\bar{u} = 0$$

where K_M is a constant eddy viscosity and f is the Coriolis parameter. At some height the first terms in each of these equations are expected to become negligible, leading to the geostrophic wind field. Take the x axis to be oriented in the direction of the geostrophic wind, in which case

$$f\bar{u}_G = -\frac{1}{\rho}\frac{\partial p}{\partial y}$$

Show that the solutions for $\bar{u}(z)$ and $\bar{v}(z)$ are

$$\bar{u}(z) = \bar{u}_G(1 - e^{-\alpha z}\cos \alpha z)$$

$$\bar{v}(z) = \bar{u}_G e^{-\alpha z}\sin \alpha z$$

where $\alpha = \sqrt{f/2K_M}$. In what way is this solution oversimplified?

5.3 The outer boundary of the planetary boundary layer can be defined as the point at which the component in the Ekman spiral disappears. From the solution of Prob. 5.2, it is seen that this occurs when $\alpha z_G = \pi$.

(*a*) Using representative values of f and K_M, estimate the depth of the planetary boundary layer.

(*b*) The magnitude of the total turbulent stress at the surface is given by

$$\tau_0 = \frac{\rho K_M}{\sqrt{2}}\left(\left.\frac{\partial \bar{u}}{\partial z}\right|_{z=0} + \left.\frac{\partial \bar{v}}{\partial z}\right|_{z=0}\right)$$

Show that for a planetary boundary layer

$$\tau_0 = \frac{z_G}{\sqrt{2\pi}} f\rho\bar{u}_G$$

Estimate the magnitude of τ_0 in dynes per square centimeter.

(*c*) The surface layer can be estimated as that layer in which τ_0 changes by only 10 percent of its value at the surface. In the planetary boundary layer the turbulent stress terms in the equations of motion are of the same order as the Coriolis acceleration terms, both about 0.1 in cgs units. Estimate $\partial\tau/\partial z$, and from this estimate the thickness of the surface layer.

5.4 Consider the prediction of the diurnal atmospheric temperature profile under stagnant conditions. In this circumstance it is necessary to consider spatial variations in temperature in only the vertical direction. If it is assumed that absorption of radiation by the atmosphere can be neglected, the potential temperature θ satisfies

$$\frac{\partial \theta}{\partial t} = \frac{\partial}{\partial z}\left(K \frac{\partial \theta}{\partial z}\right) \qquad \text{(A)}$$

assuming that K may be taken as constant. It may also be assumed that, at sufficiently high altitudes, the temperature profile should approach the adiabatic lapse rate, i.e.

$$\theta \to 0 \quad \text{as} \quad z \to \infty \tag{B}$$

The ground ($z = 0$) temperature is governed by solar heating during the day and radiational cooling during the night. Therefore $\theta(0,t)$ may be expressed as

$$\theta(0,t) = A \cos \omega t \tag{C}$$

where A is the amplitude of the diurnal surface temperature variation and $\omega = 7.29 \times 10^{-5}\ \text{sec}^{-1}$.

(*a*) Show that a solution which satisfies Eqs. (A) to (C) is

$$\theta(z,t) = Ae^{-\beta z} \cos (\omega t - \beta z)$$

where $\beta = \sqrt{\omega/2K}$. This is the so-called long-time solution or steady-state solution which represents the temperature dynamics corresponding to the influence of the surface forcing function (C).

(*b*) Show that the elevation H which marks either the base or top of an inversion is found from

$$\sin (\omega t - \beta H) - \cos (\omega t - \beta H) = \frac{g}{A\beta\hat{C}_p} e^{\beta H}$$

Can there be more than one inversion layer?

(*c*) Consider the evolution of the temperature profile from an initial profile

$$\theta(z,0) = f(z) \tag{D}$$

Show that the solution of (A) to (D) is

$$\theta = Ae^{-\beta z} \cos (\omega t - \beta z) - \frac{A}{\pi}\int_0^\infty e^{-\eta t} \sin\left(z\sqrt{\frac{\eta}{K}}\right)\frac{\eta}{\eta^2 + \omega^2}\, d\eta$$
$$+ \frac{1}{2\sqrt{\pi K t}}\int_0^\infty f(\eta)[e^{-(z-\eta)^2/4Kt} - e^{-(z+\eta)^2/4Kt}]\, d\eta$$

(*d*) Plot the steady-state solution for $K = 10^4\ \text{cm}^2/\text{sec}$ and $A = 4°\text{C}$. Note that a surface inversion is predicted at night (take $t = 0$ to be midnight), followed by weak, elevated, probably multiple inversions during the day.

6
ATMOSPHERIC DIFFUSION

In the preceding two chapters we have discussed the chemistry of air pollutants and the properties of the lowest layers of the atmosphere. Clearly, each of these subjects is of importance in understanding the behavior of air pollutants in the atmosphere. Our ultimate goal, with respect to the atmospheric aspects of air pollution, is to describe mathematically the spatial and temporal history of contaminants released into the atmosphere. To achieve this goal we must be able to describe the physical and chemical behavior of pollutants in the turbulent atmospheric boundary layer. This chapter is devoted to such description.

6.1 GENERAL DESCRIPTION OF TURBULENT DIFFUSION

In this section we will define the turbulent diffusion problem and illustrate the difficulties inherent in the exact description of the behavior of species in turbulent fluid.[1] The treatment is divided into two parts, corresponding to the two basic ways of

[1] It is common to refer to the behavior of gases and particles in turbulent fluid as turbulent "diffusion" or, in this case, as atmospheric "diffusion," although, as pointed out in Chap. 5, the processes responsible for dispersion are not diffusion in the sense of ordinary molecular diffusion.

describing turbulent diffusion. The first is the *eulerian* approach in which the behavior of species is described relative to a fixed coordinate system. The eulerian description is the most common way of treating heat and mass transfer in a flowing system. The second approach is the *lagrangian* in which concentration changes are described relative to the moving fluid. As we will see, the two approaches yield different types of mathematical relationships for the pollutant concentrations. Each of the two modes of expression is, of course, a valid description of turbulent diffusion; however, each has inherent associated difficulties which render exact solution for the species concentrations impossible.

The essential results presented in this chapter may be understood without a knowledge of turbulence theory. Nevertheless, several of the derivations of results require certain elements from the theory of turbulence. In Appendix A some basic concepts from the statistical description of turbulence are summarized, and we urge the reader interested in the underlying theory of turbulent diffusion to consult that appendix.

6.1.1 Eulerian Approach

Let us consider N species in a fluid. The concentration of each must, at each instant, satisfy a material balance taken over a volume element. Thus, any accumulation of material over time, when added to the net amount of material convected into the volume element, must be balanced by an equivalent amount of material that is produced by chemical reaction in the element, that is, emitted into it by sources, and that enters by molecular diffusion. Expressed mathematically, the concentration of each species, c_i, must satisfy the continuity equation

$$\frac{\partial c_i}{\partial t} + \frac{\partial}{\partial x_j} u_j c_i = D_i \frac{\partial^2 c_i}{\partial x_j \, \partial x_j} + R_i(c_1, \ldots, c_N, T) + S_i(\mathbf{x},t) \qquad i = 1, 2, \ldots, N \tag{6.1}$$

where u_j is the jth component of the fluid velocity, D_i is the molecular diffusivity of species i in the carrier fluid, R_i is the rate of generation of species i by chemical reaction (which depends in general on the fluid temperature T), and S_i is the rate of addition of species i at location $\mathbf{x} = (x_1,x_2,x_3)$ and time t.

In addition to the requirement that the c_i satisfy (6.1), the fluid velocities u_j and the temperature T, in turn, must satisfy the Navier-Stokes and energy equations (5.18) and (5.22), which themselves are coupled through the u_j, c_i, and T with the total continuity equation and the ideal-gas law (we restrict our attention to gaseous systems). In general, it is necessary to carry out a simultaneous solution of the coupled equations of mass, momentum, and energy conservation to account properly for the changes in u_j, T, and c_i and the effects of the changes of each of these on each other. In dealing with atmospheric pollutants, however, since species occur at parts-per-million concentrations, it is quite justifiable to assume that the presence of pollutants does not affect the meteorology to any detectable extent; thus, the equations of continuity (6.1) can be solved independently of the coupled momentum and energy equations.[1] Consequently, the fluid velocities u_j and the temperature T can be considered independent of the c_i. From this point on we will not explicitly indicate the dependence of R_i on T.

The complete description of pollutant behavior rests with the solution of (6.1). Unfortunately, because the flows of interest are turbulent, we can never solve (6.1) exactly. We first note that, since the fluid is turbulent, the fluid velocities u_j are random variables in space and time. As was done in Chap. 5, it is customary to represent the wind velocities u_j as the sum of a deterministic and stochastic component, $\bar{u}_j + u'_j$. We note that the definition of $\bar{u}_j$ and u'_j, given the records from a network of wind stations such as that depicted in Fig. 5.1, is far from a trivial undertaking but is, unfortunately, beyond our scope.

To illustrate the importance of the definition of the deterministic and stochastic velocity components $\bar{u}_j$ and u'_j, let us suppose a puff of pollutant of known concentration distribution $c(\mathbf{x},t_0)$ at time t_0. In the absence of chemical reaction and other sources, and assuming molecular diffusion to be negligible, the concentration distribution at some later time is described by the *advection equation*,

$$\frac{\partial c}{\partial t} + \frac{\partial}{\partial x_j}(u_j c) = 0$$

If we solve this equation with $u_j = \bar{u}_j$ and compare the solution with observations we would find in reality that the material spreads out *more* than predicted. This extra spreading is, in fact, what is referred to as *turbulent diffusion* and results from the influence of the random component u'_j which we have ignored. Now let us solve this equation with the precise velocity field u_j. We should then find that the solution agrees exactly with the observations (assuming, of course, that molecular diffusion is negligible), implying that if we knew the velocity field precisely at all locations and times there would be no such phenomenon as turbulent diffusion. Thus, turbulent diffusion is an artifact of our lack of complete knowledge of the true velocity field. Consequently, one of the fundamental tasks in turbulent diffusion theory is to define the deterministic and stochastic components of the velocity field.

Replacing u_j by $\bar{u}_j + u'_j$ in (6.1) gives

[1] Two effects could, in principle, serve to invalidate this assumption. Least likely is that sufficient heat would be generated by chemical reactions to influence the temperature. More likely, however, is that a polluted layer would become so concentrated that absorption, reflection, and scattering of radiation by the pollutants would result in alterations of the fluid behavior. We will not consider either of these effects here.

$$\frac{\partial c_i}{\partial t} + \frac{\partial}{\partial x_j}[(\bar{u}_j + u'_j)c_i] = D_i \frac{\partial^2 c_i}{\partial x_j \, \partial x_j} + R_i(c_1, \ldots, c_N) + S_i(\mathbf{x},t) \tag{6.2}$$

Since the u'_j are random variables, the c_i resulting from the solution of (6.2) must also be random variables; i.e., because the wind velocities are random functions of space and time, the airborne pollutant concentrations are themselves random variables in space and time. Thus, the determination of the c_i, in the sense of being a specified function of space and time, is not possible, just as it is not possible to determine precisely the value of any random variable in an experiment. We can at best derive the probability density functions satisfied by the c_i, for example, the probability that at some location and time the concentration of species i will lie between two closely spaced values. Unfortunately, the specification of the probability density function for a random process as complex as atmospheric diffusion is almost never possible. Instead, we must adopt a less desirable but more feasible approach, the determination of certain statistical properties of the c_i, most notably the *mean* $\langle c_i \rangle$.

The mean concentration can be interpreted in the following way. Let us suppose an experiment in which a puff of pollutant is released at a certain time and concentrations are measured downwind at subsequent times. We would measure $c_i(\mathbf{x},t)$, which would exhibit random characteristics because of the wind. If it were possible to repeat this experiment under identical conditions, we would again measure $c_i(\mathbf{x},t)$, but because of the randomness in the wind field we could not reproduce the first $c_i(\mathbf{x},t)$. Theoretically we could repeat this experiment an infinite number of times. We would then have a so-called ensemble of experiments. If at every location $\mathbf{x}$ and time t we averaged all the concentration values over the infinite number of experiments, we would have computed the theoretical mean concentration $\langle c_i(\mathbf{x},t) \rangle$.[1] Experiments like this cannot, of course, be repeated under identical conditions, and so it is virtually impossible to measure $\langle c_i \rangle$. Thus, a measurement of the concentration of species i at a particular location and time is more suitably envisioned as one sample from a hypothetically infinite ensemble of possible concentrations. Clearly, an individual measurement may differ considerably from the mean $\langle c_i \rangle$.

It is convenient to express c_i as $\langle c_i \rangle + c'_i$, where, by definition, $\langle c'_i \rangle = 0$. Averaging (6.2) over an infinite ensemble of realizations of the turbulence yields the equation governing $\langle c_i \rangle$, namely

$$\frac{\partial \langle c_i \rangle}{\partial t} + \frac{\partial}{\partial x_j}(\bar{u}_j \langle c_i \rangle) + \frac{\partial}{\partial x_j}\langle u'_j c'_i \rangle$$

$$= D_i \frac{\partial^2 \langle c_i \rangle}{\partial x_j \, \partial x_j} + \langle R_i(\langle c_1 \rangle + c'_1, \ldots, \langle c_N \rangle + c'_N) \rangle + S_i(\mathbf{x},t) \tag{6.3}$$

[1] We have used different notation for the mean values of the velocities and the concentrations, that is, $\bar{u}_j$ versus $\langle c_i \rangle$, in order to emphasize the fact that the mean fluid velocities are normally determined by a process involving temporal and spatial averaging, whereas the $\langle c_i \rangle$ always represent the theoretical ensemble averages.

Let us consider the case of a single inert species, that is, $R = 0$. We note that (6.3) contains dependent variables $\langle c \rangle$ and $\langle u'_j c' \rangle$, $j = 1, 2, 3$. We thus have more dependent variables than equations. Again, this is the closure problem of turbulence. For example, if we were to derive an equation for the $\langle u'_j c' \rangle$ by subtracting (6.3) from (6.2), multiplying the resulting equation by u'_j and then averaging, we would obtain

$$\frac{\partial}{\partial t}\langle u'_j c' \rangle + \frac{\partial}{\partial x_k}(\bar{u}_k \langle u'_j c' \rangle) + \langle u'_j u'_k \rangle \frac{\partial \langle c \rangle}{\partial x_k}$$

$$= \frac{\partial}{\partial x_k}\left(D \frac{\partial}{\partial x_k}\langle u'_j c' \rangle - \langle u'_j u'_k c' \rangle\right) \qquad j = 1, 2, 3 \qquad (6.4)$$

Although we have derived the desired equations, we have at the same time generated new dependent variables, $\langle u'_j u'_k c' \rangle$, $j, k = 1, 2, 3$. If we generate additional equations for these variables, we find that still more dependent variables appear. The closure problem becomes even worse if a nonlinear chemical reaction is occurring. If the single species decays by a second-order reaction, then the term $\langle R \rangle$ in (6.3) becomes $-k(\langle c \rangle^2 + \langle c'^2 \rangle)$, where $\langle c'^2 \rangle$ is a new dependent variable. If we were to derive an equation for $\langle c'^2 \rangle$ we would find the emergence of new dependent variables $\langle u'_j c'^2 \rangle$, $\langle c'^3 \rangle$, and $\langle \partial c'/\partial x_j \, \partial c'/\partial x_j \rangle$. It is because of the closure problem that an eulerian description of turbulent diffusion will not permit exact solution even for the mean concentrations $\langle c_i \rangle$.

6.1.2 Lagrangian Approach

The lagrangian approach to turbulent diffusion is concerned with the behavior of representative fluid particles.[1] We therefore begin by considering a single particle which is at location $\mathbf{x}'$ at time t' in a turbulent fluid. The subsequent motion of the particle can be described by its trajectory, $\mathbf{X}[\mathbf{x}',t';t]$, that is, its position at any later time t. Let $\psi(x_1,x_2,x_3,t)\, dx_1\, dx_2\, dx_3 = \psi(\mathbf{x},t)\, d\mathbf{x}$ = probability that particle at time t will be in volume element x_1 to $x_1 + dx_1$, x_2 to $x_2 + dx_2$, and x_3 to $x_3 + dx_3$, that is, that $x_1 \le X_1 < x_1 + dx_1$, etc. Thus, $\psi(\mathbf{x},t)$ is the probability density function (pdf) for the particle's location at time t. By definition

$$\int_{-\infty}^{\infty}\int_{-\infty}^{\infty}\int_{-\infty}^{\infty} \psi(\mathbf{x},t)\, d\mathbf{x} = 1$$

The probability density of finding the particle at $\mathbf{x}$ at t can be expressed as the product of two other probability densities:

[1] By a "fluid particle" we mean a volume of fluid large compared with molecular dimensions but small enough to act as a point which exactly follows the fluid. The "particle" may contain fluid of a different composition than the carrier fluid, in which case the particle is referred to as a "marked particle."

The probability density that if the particle is at $\mathbf{x}'$ at t' it will undergo a displacement to $\mathbf{x}$ at t. Denote this probability density $Q(\mathbf{x},t|\mathbf{x}',t')$ and call it the *transition probability density* for the particle.
The probability density that the particle was at $\mathbf{x}'$ at t', $\psi(\mathbf{x}',t')$, integrated over all possible starting points $\mathbf{x}'$. Thus,

$$\psi(\mathbf{x},t) = \int_{-\infty}^{\infty}\int_{-\infty}^{\infty}\int_{-\infty}^{\infty} Q(\mathbf{x},t|\mathbf{x}',t')\psi(\mathbf{x}',t')\,d\mathbf{x}' \tag{6.5}$$

The density function $\psi(\mathbf{x},t)$ has been defined with respect to a single particle. If, however, an arbitrary number m of particles are initially present and the position of the ith particle is given by the density function $\psi_i(\mathbf{x},t)$, it can be shown that the ensemble mean concentration at the point $\mathbf{x}$ is given by

$$\langle c(\mathbf{x},t)\rangle = \sum_{i=1}^{m} \psi_i(\mathbf{x},t) \tag{6.6}$$

By expressing the pdf $\psi_i(\mathbf{x},t)$ in (6.6) in terms of the initial particle distribution and the spatial-temporal distribution of particle sources $S(\mathbf{x},t)$, say in units of particles per volume per time, and then substituting the resulting expression into (6.5), we obtain the following general formula for the mean concentration:

$$\begin{aligned}\langle c(\mathbf{x},t)\rangle &= \int_{-\infty}^{\infty}\int_{-\infty}^{\infty}\int_{-\infty}^{\infty} Q(\mathbf{x},t|\mathbf{x}_0,t_0)\langle c(\mathbf{x}_0,t_0)\rangle\,d\mathbf{x}_0 \\ &\quad + \int_{-\infty}^{\infty}\int_{-\infty}^{\infty}\int_{-\infty}^{\infty}\int_{t_0}^{t} Q(\mathbf{x},t|\mathbf{x}',t')S(\mathbf{x}',t')\,dt'\,d\mathbf{x}'\end{aligned} \tag{6.7}$$

The first term on the right-hand side represents those particles present at t_0, and the second term on the right-hand side accounts for particles added from sources between t' and t.

Equation (6.7) is the fundamental lagrangian relation for the mean concentration of a species in turbulent fluid in which there are sources. The determination of $\langle c(\mathbf{x},t)\rangle$, given $\langle c(\mathbf{x}_0,t_0)\rangle$ and $S(\mathbf{x}',t)$, rests with the evaluation of the transition probability $Q(\mathbf{x},t|\mathbf{x}',t')$. If Q were known for $\mathbf{x}$, $\mathbf{x}'$, t, and t', the mean concentration $\langle c(\mathbf{x},t)\rangle$ could be computed by simply evaluating (6.7). However, there are two substantial problems with using (6.7). First, it holds only when the particles are not undergoing chemical reactions. Second, such complete knowledge of the turbulence properties as would be needed to know Q is generally unavailable except in the simplest of circumstances.

As a slight generalization of (6.7), we can assume that the particles may decay according to a first-order reaction. An important feature of this type of chemical reaction is that each molecule (or particle) decays individually and does not affect the decay of other particles in the field. Because of this feature, we can include first-order decay rather easily in the foregoing development. In the presence of first-order decay, the probability of finding a given particle at location $\mathbf{x}$ at time t, given that it was at $\mathbf{x}'$ at t', is just the product of two probabilities:

The probability $Q(\mathbf{x},t|\mathbf{x}',t')$ that the particle will undergo a displacement from $\mathbf{x}'$ to $\mathbf{x}$ from t' to t
The probability that the particle will not lose its identity by chemical decay during the time interval $t - t'$

These two events are independent as long as the marked particles in no way influence the fluid motion. Let us now compute the second probability.

First-order decay is equivalent to the condition that the probability that a particle will decay in a time interval dt is proportional to dt. Thus, the probability that a particle will not decay in the small time interval dt is $1 - k\,dt$, where k is the first-order decay constant. To include the case in which k may change with time, we first transform the time coordinate t into the dimensionless value

$$\xi(t) = \int_{t'}^{t} k(t'')\,dt'' \tag{6.8}$$

The probability that the particle will not decay in a small time interval is then $1 - d\xi(t)$. If we now divide the interval $\xi(t) - \xi(t')$ into n subintervals of length $d\xi$, the probability that the particle will not decay in any of these subintervals is $(1 - d\xi)^n$. Taking the limit as $d\xi \to 0$, the probability the particle will not decay in $t - t'$ is

$$\begin{aligned}
\lim_{d\xi \to 0} (1 - d\xi)^n &= \lim_{d\xi \to 0} \left[1 - n\,d\xi + \frac{(n\,d\xi)^2}{2} - \cdots\right] \\
&= e^{-\xi(t)} \\
&= \exp\left[-\int_{t'}^{t} k(t'')\,dt''\right]
\end{aligned} \tag{6.9}$$

Thus, (6.7) becomes

$$\begin{aligned}
\langle c(\mathbf{x},t)\rangle &= \int_{-\infty}^{\infty}\int_{-\infty}^{\infty}\int_{-\infty}^{\infty} Q(\mathbf{x},t|\mathbf{x}_0,t_0)\langle c(\mathbf{x}_0,t_0)\rangle \exp\left[-\int_{t_0}^{t} k(t'')\,dt''\right] d\mathbf{x}_0 \\
&+ \int_{-\infty}^{\infty}\int_{-\infty}^{\infty}\int_{-\infty}^{\infty}\int_{t_0}^{t} Q(\mathbf{x},t|\mathbf{x}',t')S(\mathbf{x}',t') \exp\left[-\int_{t'}^{t} k(t'')\,dt''\right] dt'\,d\mathbf{x}'
\end{aligned} \tag{6.10}$$

Unfortunately, there is no convenient way to include nonlinear chemical reactions within the lagrangian approach to turbulent diffusion. Nevertheless, under certain conditions to be specified later it is possible to use (6.10) as the basis of a model in which nonlinear chemical reactions are admissible.

6.1.3 Comparison of Eulerian and Lagrangian Approaches

The techniques for describing the statistical properties of the concentrations of marked particles, such as air pollutants, in a turbulent fluid can be divided into two categories: eulerian and lagrangian. The eulerian methods attempt to formulate the concentration statistics in terms of the statistical properties of the eulerian fluid velocities, i.e. the velocities measured at fixed points in the fluid. A formulation of this type is very useful not only because the eulerian statistics are readily measurable (as determined from continuous time recordings of the wind velocities by a fixed network of anemometers) but also because the mathematical expressions are directly applicable to situations in which chemical reactions are taking place. Unfortunately, the eulerian approaches lead to a serious mathematical obstacle known as the closure problem, for which no generally valid solution has yet been found. A number of approximate solutions have been proposed (see Subsec. 6.2.1) but each leads to an equation which gives accurate results for only a limited class of problems.

By contrast, the lagrangian techniques attempt to describe the concentration statistics in terms of the statistical properties of the displacements of groups of particles released in the fluid. The mathematics of this approach is more tractable than that of the eulerian methods, in that no closure problem is encountered, but the applicability of the resulting equations is limited because of the difficulty of accurately determining the required particle statistics. Moreover, the equations are not directly applicable to problems involving nonlinear chemical reactions.

Having demonstrated that exact solution for the mean concentrations $\langle c_i(\mathbf{x},t)\rangle$ even of inert species in a turbulent fluid is not possible in general by either the eulerian or lagrangian approaches, we now consider what assumptions and approximations can be invoked to obtain practical descriptions of atmospheric diffusion. In Sec. 6.2 we shall proceed from the two basic equations for $\langle c_i\rangle$, (6.3) and (6.10), to obtain the equations commonly used for atmospheric diffusion. A particularly important aspect is the delineation of the assumptions and limitations inherent in each description.

6.2 EQUATIONS GOVERNING THE MEAN CONCENTRATION OF SPECIES IN TURBULENCE

6.2.1 Eulerian Approaches

As we have seen, the eulerian description of turbulent diffusion leads to the so-called closure problem, as illustrated in (6.3) by the new dependent variables $\langle u_j' c_i'\rangle$, $j = 1, 2, 3$, as well as any that might arise in $\langle R_i\rangle$ if nonlinear chemical reactions are occurring. Let us first consider only the case of chemically inert species, that is, $R_i = 0$. The problem is to relate the variables $\langle u_j' c_i'\rangle$ to the $\langle c_i\rangle$ if we wish not to introduce additional differential equations.

The most common means of relating the turbulent fluxes $\langle u_j' c_i' \rangle$ to the $\langle c_i \rangle$ is based on the mixing-length model of Sec. 5.4. In particular, it is assumed that (summation implied over k)

$$\langle u_j' c_i' \rangle = -K_{jk} \frac{\partial \langle c_i \rangle}{\partial x_k} \qquad j = 1, 2, 3 \tag{6.11}$$

where K_{jk} is called the eddy diffusivity. Equation (6.11) is called both mixing-length theory and K theory. Since (6.11) is essentially only a definition of the K_{jk}, which are, in general, functions of location and time, we have, by means of (6.11), replaced the three unknowns $\langle u_j' c_i' \rangle$, $j = 1, 2, 3$, with the six unknowns K_{jk}, $j, k = 1, 2, 3$ $(K_{jk} = K_{kj})$. If the coordinate axes coincide with the principal axes of the eddy diffusivity tensor $\{K_{jk}\}$, then only the three diagonal elements K_{11}, K_{22}, and K_{33} are nonzero, and (6.11) becomes

$$\langle u_j' c_i' \rangle = -K_{jj} \frac{\partial \langle c_i \rangle}{\partial x_j} \dagger \tag{6.12}$$

In using (6.3), two other assumptions are ordinarily invoked, namely:

1 Molecular diffusion is negligible compared with turbulent diffusion,

$$D_i \frac{\partial^2 \langle c_i \rangle}{\partial x_j \, \partial x_j} \ll \frac{\partial}{\partial x_j} \langle u_j' c_i' \rangle$$

2 The atmosphere is incompressible,

$$\frac{\partial \bar{u}_j}{\partial x_j} = 0$$

With these assumptions and (6.12), (6.3) becomes

$$\frac{\partial \langle c_i \rangle}{\partial t} + \bar{u}_j \frac{\partial \langle c_i \rangle}{\partial x_j} = \frac{\partial}{\partial x_j} \left(K_{jj} \frac{\partial \langle c_i \rangle}{\partial x_j} \right) + S_i(\mathbf{x},t) \tag{6.13}$$

This equation is termed the *semiempirical equation of atmospheric diffusion* and will play a very important role in the remainder of this chapter.

Let us return to the case in which chemical reactions are occurring, for which we refer to (6.3). Since R_i is almost always a nonlinear function of the c_i, we have already seen that additional terms of the type $\langle c_i' c_j' \rangle$ will arise from $\langle R_i \rangle$. The crudest approximation we can make regarding $\langle R_i \rangle$ is to replace $\langle R_i(c_1, \ldots, c_N) \rangle$ by $R_i(\langle c_1 \rangle, \ldots, \langle c_N \rangle)$, thereby neglecting the effect of concentration fluctuations on the

† No summation is implied in this term, e.g.

$$\langle u_1' c_i' \rangle = -K_{11} \frac{\partial \langle c_i \rangle}{\partial x_1}$$

We discuss the conditions under which $\{K_{jk}\}$ may be taken as diagonal at the end of Subsec. 6.2.4.

rate of reaction. Invoking this approximation, as well as those inherent in (6.13), we obtain

$$\frac{\partial\langle c_i\rangle}{\partial t} + \bar{u}_j \frac{\partial\langle c_i\rangle}{\partial x_j} = \frac{\partial}{\partial x_j}\left(K_{jj}\frac{\partial\langle c_i\rangle}{\partial x_j}\right) + R_i(\langle c_1\rangle, \ldots, \langle c_N\rangle) + S_i(\mathbf{x},t) \qquad (6.14)$$

We stress that (6.14) is *not* the fundamental equation governing the mean concentrations of reactive substances in turbulence but rather is only an approximate equation which is considerably limited in its applicability to reactive pollutants in the atmosphere. A key question is: Can we develop the conditions under which (6.14) is a valid description of the diffusion of reactive species in turbulent fluid? Unfortunately, this cannot be done for arbitrary R_i. Nevertheless, the conditions of validity of (6.14) can be obtained for second-order decay of a single species, and these give a good indication of the conditions of validity of (6.14) for general R_i.

6.2.2 Conditions for Validity of (6.14) (Lamb, 1973)

Let us consider a two-dimensional flow containing a single species which decays by a second-order reaction. Assuming $\bar{u}_1 = \bar{u} = \text{const}$, $\bar{u}_2 = \bar{v} = 0$, and molecular diffusion can be neglected, the exact equation for $\langle c\rangle$ is

$$\frac{\partial\langle c\rangle}{\partial t} + \bar{u}\frac{\partial\langle c\rangle}{\partial x} + \frac{\partial}{\partial y}\langle v'c'\rangle = -k(\langle c\rangle^2 + \langle c'^2\rangle) \qquad (6.15)$$

where we have ignored the term $(\partial/\partial x)\langle u'c'\rangle$ as small compared with $\bar{u}\partial\langle c\rangle/\partial x$. Our objective is to develop the conditions under which (6.15) can be approximated by

$$\frac{\partial\langle c\rangle}{\partial t} + \bar{u}\frac{\partial\langle c\rangle}{\partial x} = \frac{\partial}{\partial y}\left(K_{yy}\frac{\partial\langle c\rangle}{\partial y}\right) - k\langle c\rangle^2 \qquad (6.16)$$

In the context of the two-dimensional problem considered here, the mixing-length hypothesis holds that the concentration fluctuation at any point (x,y,t) is due to the arrival at that point of a fluid eddy which originated at some other location where the mean concentration $\langle c\rangle$ was different from that at (x,y,t).

Let c_j' be the fluctuation at (x,y,t) due to the jth eddy. Then

$$c_j' = \underset{\text{Arriving eddy}}{c_j(x,y,t)} - \underset{\text{Mean concentration}}{\langle c(x,y,t)\rangle} \qquad (6.17)$$

Let us suppose that this eddy originated on the level $y + l_j$ at time $t - \tau_j$, where the mean concentration was $\langle c(x, y + l_j, t - \tau_j)\rangle$. During the time τ_j required to traverse the distance l_j, the concentration of the diffusing material in the eddy changes because of chemical decay according to (neglecting entrainment effects)

$$\frac{dc_j}{dt} = -kc_j^2 \qquad (6.18)$$

Therefore, the concentration in the eddy on arrival at (x,y,t) is

$$c_j(x,y,t) = \frac{\langle c(x, y + l_j, t - \tau_j)\rangle}{1 + k\tau_j\langle c(x, y + l_j, t - \tau_j)\rangle} \tag{6.19}$$

Let τ_e be the maximum time over which an eddy maintains its integrity, i.e., so that $\tau_e > \tau_j$ for nearly all eddies. Also, let τ_c be a characteristic time scale for the second-order chemical reaction:

$$\tau_c = (k\langle c\rangle_{\max})^{-1} \tag{6.20}$$

Then, for cases in which the chemical decay takes place much more slowly than the turbulent transport, that is, $\tau_c \gg \tau_e$, (6.19) reduces to

$$c_j(x,y,t) = \langle c(x, y + l_j, t - \tau_j)\rangle \tag{6.21}$$

Indeed this is the exact form employed in the usual Prandtl mixing-length hypothesis for an inert species. Substituting (6.21) into (6.17) and expanding $\langle c(x, y + l_j, t - \tau_j)\rangle$ in a Taylor series about the point (x,y,t) we get

$$c_j' = l_j \frac{\partial\langle c\rangle}{\partial y} - \tau_j \frac{\partial\langle c\rangle}{\partial t} + \frac{1}{2} l_j^2 \frac{\partial^2\langle c\rangle}{\partial y^2} + \frac{1}{2}\tau_j^2 \frac{\partial^2\langle c\rangle}{\partial t^2} + \cdots \tag{6.22}$$

Let L and T be a characteristic length and time scale, respectively, of the $\langle c\rangle$ field such that $\partial\langle c\rangle/\partial y \sim \langle c\rangle/L$, $\partial\langle c\rangle/\partial t \sim \langle c\rangle/T$, $\partial^2\langle c\rangle/\partial y^2 \sim \langle c\rangle/L^2$, etc. and let l_e be a length scale, associated with the eddy time scale τ_e, such that l_e is the maximum distance over which an eddy maintains its integrity, that is, $l_e > |l_j|$ for nearly all eddies. Then if $L \gg l_e$ and $T \gg \tau_e$ we can truncate second- and higher-order terms in (6.22), leaving

$$c_j' = l_j \frac{\partial\langle c\rangle}{\partial y} - \tau_j \frac{\partial\langle c\rangle}{\partial t} \tag{6.23}$$

Having determined an expression for c_j', we can now consider $\langle u'c'\rangle$ and $\langle c'^2\rangle$ in (6.15). Multiplying (6.23) by v_j', the velocity fluctuation in the y component of the velocity associated with the jth eddy, and averaging the resulting equation over a large number of eddies which pass the point (x,y), we get

$$\langle v'c'\rangle = \langle v'l\rangle \frac{\partial\langle c\rangle}{\partial y} - \langle v_j'\tau_j\rangle \frac{\partial\langle c\rangle}{\partial t} \tag{6.24}$$

where l is the "mixing length" defined so that $\langle v'l\rangle = \langle v_j' l_j\rangle$. Similarly, the expression for $\langle c'^2\rangle$ is obtained by squaring (6.23) and averaging:

$$\langle c'^2\rangle = \langle l^2\rangle \left(\frac{\partial\langle c\rangle}{\partial y}\right)^2 - 2\langle l_j\tau_j\rangle \frac{\partial\langle c\rangle}{\partial y}\frac{\partial\langle c\rangle}{\partial t} + \langle\tau_j^2\rangle \left(\frac{\partial\langle c\rangle}{\partial t}\right)^2 \tag{6.25}$$

Since the quantities $\langle v'l \rangle$, $\langle v_j'\tau_j \rangle$, $\langle l_j\tau_j \rangle$, $\langle l^2 \rangle$, and $\langle \tau_j^2 \rangle$ involve hypothetical parameters, there is no chance to obtain rigorous analytical expressions for these quantities; rather, it is possible only to estimate their order of magnitude. We consider first $\langle v_j'\tau_j \rangle$ and $\langle l_j\tau_j \rangle$. Although the pair of variables in each of these two terms are not statistically independent, we expect them to be uncorrelated in each case since τ_j is always positive and l_j and v_j' are equally likely to be either positive or negative. Thus

$$\langle v_j'\tau_j \rangle = \langle l_j\tau_j \rangle = 0 \qquad (6.26)$$

Next we consider $\langle v'l \rangle$. We expect this term to be negative since negative values of l_j are correlated with positive values of v_j' in a shear flow in which $\bar{u}$ increases with y. It is reasonable to approximate $\langle v'l \rangle$ by

$$\langle v'l \rangle = -\langle v'^2 \rangle \tau_e \qquad (6.27)$$

Similarly, we find that

$$\langle l^2 \rangle = \langle v'^2 \rangle \tau_e^2 \qquad (6.28)$$

Considering $\langle \tau_j^2 \rangle$, it seems reasonable to assume that the square root of this quantity is roughly equal to the time scale of the energy-containing eddies since it is those eddies which are primarily responsible for the turbulent flux of material. Since τ_e is a good measure of such a time scale,

$$\langle \tau_j^2 \rangle = \tau_e^2 \qquad (6.29)$$

Also, l_e and $\langle l^2 \rangle^{1/2}$ are about equal.

The basic restrictions leading to (6.23) can be expressed as

$$k\langle c \rangle_{\max} \tau_e \ll 1 \qquad (6.30)$$

$$\frac{\langle v'^2 \rangle^{1/2} \tau_e}{L} \ll 1 \qquad (6.31)$$

$$\frac{\tau_e}{T} \ll 1 \qquad (6.32)$$

If (6.30) to (6.32) are satisfied, we can substitute (6.24) to (6.29) into (6.15) to obtain

$$\frac{\partial \langle c \rangle}{\partial t} + \bar{u} \frac{\partial \langle c \rangle}{\partial x} = \frac{\partial}{\partial y}\left(K_{yy} \frac{\partial \langle c \rangle}{\partial y}\right) - k\left[\langle c \rangle^2 + K_{yy}\tau_e\left(\frac{\partial \langle c \rangle}{\partial y}\right)^2 + \tau_e^2\left(\frac{\partial \langle c \rangle}{\partial t}\right)^2\right] \qquad (6.33)$$

where $K_{yy} = \langle v'^2 \rangle^{1/2} \tau_e$.

Let us now compare (6.33) and (6.16), the only difference being the last two terms on the right-hand side of (6.33). It can be shown, however, that under restrictions (6.30) to (6.32) these two terms are in fact negligible compared with the other terms in the equation. First, we rewrite all four terms on the right-hand side of (6.33) and below each its order of magnitude:

$$\frac{\partial}{\partial y}\left(K_{yy}\frac{\partial\langle c\rangle}{\partial y}\right) - k\langle c\rangle^2 - kK_{yy}\tau_e\left(\frac{\partial\langle c\rangle}{\partial y}\right)^2 - k\tau_e^2\left(\frac{\partial\langle c\rangle}{\partial t}\right)^2$$

$$\frac{K_{yy}\langle c\rangle}{L^2} \qquad k\langle c\rangle^2 \qquad \frac{kK_{yy}\tau_e\langle c\rangle^2}{L^2} \qquad \frac{k\tau_e^2\langle c\rangle^2}{T^2}$$

From (6.32), the last term is negligible. Furthermore, since the first and second terms are automatically assumed to be of the same order of magnitude, we must have

$$\frac{K_{yy}\langle c\rangle}{L^2} \sim k\langle c\rangle^2$$

Using this fact, we may express the third term as $k\tau_e\langle c\rangle(k\langle c\rangle^2)$, which by (6.30) is negligible compared with $k\langle c\rangle^2$.

We may conclude from the above that (6.16) is a valid description of turbulent diffusion and chemical reaction as long as (6.30) to (6.32) hold, namely that the reaction processes are slow compared with turbulent transport and the characteristic length and time scales for changes in the mean concentration field are large compared with the corresponding scales for turbulent transport.

Because the eddy time scale τ_e and the length scale $\langle v'^2\rangle^{1/2}\tau_e$ are often quite large in the atmosphere, the above conditions are violated near strong isolated sources. For example, for the lateral turbulent velocity, component τ_e may be about 1 min and $\langle v'^2\rangle^{1/2} \cong 1$ m/sec. Thus, to satisfy the condition that the characteristic length scale of the concentration field be much greater than that of the turbulence, the spatial scale for variations in $\langle c\rangle$, and hence S, must be of the order of 100 to 1000 m. In addition, under these conditions the time scale of the fastest reactions must be no smaller than of the order of about 10 min.

The conclusion we draw at this point is that (6.14) is a valid model *provided it is applied to situations in which chemical reactions are "slow" and the distribution of sources is "smooth."*

6.2.3 Lagrangian Approaches

We now wish to consider the derivation of usable expressions for $\langle c_i(\mathbf{x},t)\rangle$ based on the fundamental lagrangian expression (6.10) or, for an inert, (6.7). As we have seen, the utility of either (6.7) or (6.10) rests on the ability to evaluate the transition probability $Q(\mathbf{x},t|\mathbf{x}',t')$. The first question, then, is: Are there any circumstances under which the form of Q is known?

If the turbulence is stationary and homogeneous, the transition probability Q of a particle depends only upon the displacements in time and space and not on where or when the particle was introduced into the flow. Thus, in that case, $Q(\mathbf{x},t|\mathbf{x}',t) = Q(\mathbf{x}-\mathbf{x}'; t-t')$. In addition, empirical data indicate that Q obeys a multidimensional normal distribution (Monin and Yaglom, 1971),

$$Q(\mathbf{x}-\mathbf{x}'; t-t') = \frac{1}{(2\pi)^{3/2}|\mathbf{P}|^{1/2}} e^{-\boldsymbol{\zeta}^T\mathbf{P}^{-1}\boldsymbol{\zeta}/2} \tag{6.34}$$

where ζ^T is the transpose of the column vector which has elements

$$\zeta_i = x_i - x_i' - \langle x_i - x_i' \rangle \qquad i = 1, 2, 3$$

and where $\mathbf{P}^{-1}$ and $|\mathbf{P}|$ are the inverse and determinant, respectively, of the matrix $\mathbf{P}$ whose elements are

$$P_{ij} = \langle \zeta_i \zeta_j \rangle \tag{6.35}$$

In using (6.34) it is customary to assume that $P_{ij} = 0$ for $i \neq j$, in which case (6.34) becomes, for $\tau = t - t'$,

$$Q(\mathbf{x}-\mathbf{x}'; \tau) = \frac{1}{(2\pi)^{3/2}[P_{11}(\tau)P_{22}(\tau)P_{33}(\tau)]^{1/2}} \exp\left[-\frac{1}{2}\sum_{i=1}^{3}\frac{\zeta_i^2}{P_{ii}(\tau)}\right] \tag{6.36}$$

We assume that the mean displacements $\langle x_i - x_i' \rangle$ are due only to the deterministic velocity, i.e.

$$\langle x_i - x_i' \rangle = \int_{t'}^{t} \bar{u}_i(\mathbf{x}'', t'')\, dt'' \tag{6.37}$$

where $\mathbf{x}''$ is the particle position at t'' if the only velocity were $\bar{u}_i$. The assumption that $\langle \zeta_i \zeta_j \rangle = 0$ can be justified only through correlation of measurements of the diffusion of many particles relative to the fixed coordinate system being used. In general, $\langle \zeta_i \zeta_j \rangle$ will not be zero, but these terms are usually an order of magnitude smaller than the $\langle \zeta_i^2 \rangle$, in which case they can be safely neglected. The diagonal elements $P_{ii}(\tau)$ are often denoted by $\sigma_i^2(\tau)$, since these are the variances of the gaussian distribution (6.36).

We may now substitute (6.36) into (6.7). Several special cases of the resulting equation are enumerated in Table 6.1. The formulas in Table 6.1 are the classic relations for diffusion in stationary, homogeneous turbulence and are called the gaussian puff and plume formulas. Let us consider these relations somewhat further.

The mean concentration of an inert species emitted from an instantaneous point source of strength S g at $t = 0$ at $(x_{1_0}, x_{2_0}, x_{3_0})$ into an unbounded fluid having a mean velocity $\bar{u}_1 = U$ in the x_1 direction, i.e. entry 1 in Table 6.1, is obtained from (6.7) and (6.36):

$$\langle c(\mathbf{x},t) \rangle = \int_{-\infty}^{\infty}\int_{-\infty}^{\infty}\int_{-\infty}^{\infty}\int_{0}^{t} Q(\mathbf{x}-\mathbf{x}'; t-t')S\delta(\mathbf{x}'-\mathbf{x}_0)\delta(t')\, dt'\, d\mathbf{x}' \tag{6.38a}$$

where we see that for the instantaneous point source at $\mathbf{x}_0$ at $t = 0$, $S(\mathbf{x}',t') = S\delta(\mathbf{x}' - \mathbf{x}_0)\delta(t')$. The concentration distribution obeys a three-dimensional joint normal or gaussian distribution. Projected on each axis, the mean concentration will appear as a bell-shaped curve with standard deviation σ_i, depending on the axis. Because we assumed that the off-diagonal terms $P_{ij} = 0$, the concentration distributions in the three directions are independent.

The standard deviation $\sigma_2(t)$ is interpreted as the standard deviation of the concentration distribution relative to the fixed x_2 axis obtained by emitting a very large number of identical puffs. Thus, even though the puff is being convected and is

Table 6.1 EQUATIONS FOR THE MEAN CONCENTRATION OF AN INERT SPECIES IN HOMOGENEOUS, STATIONARY TURBULENCE

Source	Boundary conditions	Equation
1 Instantaneous point source at $\mathbf{x}_0$. Mean wind $\bar{u}_1 = U$. Source strength $= S$ grams	None	$\langle c(\mathbf{x},t)\rangle = \frac{S}{(2\pi)^{3/2}\sigma_1(t)\sigma_2(t)\sigma_3(t)} \times \exp\left[-\frac{(x - x_{1_0} - Ut)^2}{2\sigma_1^2(t)} - \frac{(x_2 - x_{2_0})^2}{2\sigma_2^2(t)} - \frac{(x_3 - x_{3_0})^2}{2\sigma_3^2(t)}\right]$
2 Continuous point source at $\mathbf{x}_0$. Mean wind $\bar{u}_1 = U$. Source strength $= S$ g/sec	None	$\langle c(\mathbf{x})\rangle = \frac{S}{2\pi\sigma_2(x_1 - x_{1_0})\sigma_3(x_1 - x_{1_0})U} \times \exp\left[-\frac{(x_2 - x_{2_0})^2}{2\sigma_2^2(x_1 - x_{1_0})} - \frac{(x_3 - x_{3_0})^2}{2\sigma_3^2(x_1 - x_{1_0})}\right]$
3 Instantaneous point source at $\mathbf{x}_0$. Mean wind $\bar{u}_1 = U$. Source strength $= S$ grams	Reflecting barrier (ground) at $x_3 = 0$	$\langle c(\mathbf{x},t)\rangle = \frac{S}{(2\pi)^{3/2}\sigma_1(t)\sigma_2(t)\sigma_3(t)} \times \exp\left[-\frac{(x_1 - x_{1_0} - Ut)^2}{2\sigma_1^2(t)} - \frac{(x_2 - x_{2_0})^2}{2\sigma_2^2(t)}\right] \times \left\{\exp\left[-\frac{(x_3 - x_{3_0})^2}{2\sigma_3^2(t)}\right] + \exp\left[-\frac{(x_3 + x_{3_0})^2}{2\sigma_3^2(t)}\right]\right\}$
4 Continuous point source at $\mathbf{x}_0$. Mean wind $\bar{u}_1 = U$. Source strength $= S$ g/sec	Reflecting barrier (ground) at $x_3 = 0$	$\langle c(\mathbf{x})\rangle = \frac{S}{2\pi\sigma_2(x_1 - x_{1_0})\sigma_3(x_1 - x_{1_0})U} \times \exp\left[-\frac{(x_2 - x_{2_0})^2}{2\sigma_2^2(x_1 - x_{1_0})}\right] \times \left\{\exp\left[-\frac{(x_3 - x_{3_0})^2}{2\sigma_3^2(x_1 - x_{1_0})}\right] + \exp\left[-\frac{(x_3 + x_{3_0})^2}{2\sigma_3^2(x_1 - x_{1_0})}\right]\right\}$

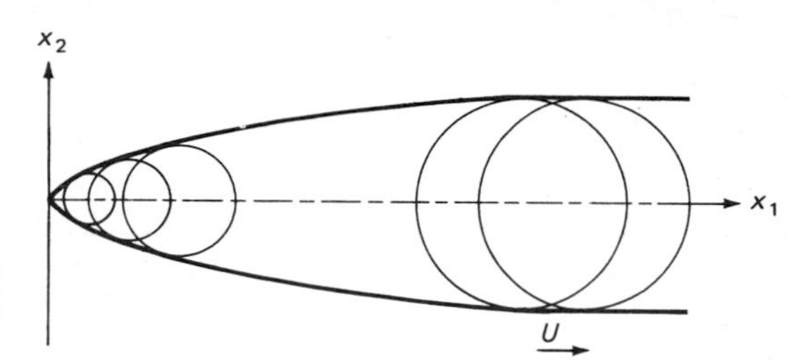

FIGURE 6.1
Visualization of a continuous plume as the superposition of many puffs emitted at short time intervals.

spreading, (6.38*a*) describes the concentration distribution *relative to a fixed axis* and *not* relative to the center of mass of the puff. This distinction between diffusion relative to a fixed coordinate system and relative to one or more other particles is a subtle one but nevertheless an important one in turbulent diffusion. For the time being, we will restrict our attention to diffusion relative to a fixed axis.

Next we consider a continuously emitting point source at $(x_{1_0}, x_{2_0}, x_{3_0})$ into an unbounded atmosphere with a mean velocity $\bar{u}_1 = U$ in the x_1 direction. The source emits at a rate of S g/sec of pollutant. The continuous plume from this source can be considered the superposition of an infinite number of overlapping puffs carried along the x_1 axis by the mean wind U (see Fig. 6.1). The mean concentration from one of these puffs of strength $S\,dt'$ (g) emitted at $t = t'$ over a time interval dt' is given by

$$\langle c(\mathbf{x},t)\rangle = \frac{S\,dt'}{(2\pi)^{3/2}\sigma_1(t-t')\sigma_2(t-t')\sigma_3(t-t')}$$

$$\times \exp\left\{-\frac{[x_1 - x_{1_0} - U(t-t')]^2}{2\sigma_1^2(t-t')} - \frac{(x_2 - x_{2_0})^2}{2\sigma_2^2(t-t')} - \frac{(x_3 - x_{3_0})^2}{2\sigma_3^2(t-t')}\right\} \qquad (6.38b)$$

The mean concentration at $\mathbf{x}$ at time t from an infinite number of these puffs is a result of emission from $-\infty$ to t. Thus, for the continuous source

$$\langle c(\mathbf{x},t)\rangle = \int_{-\infty}^{t} \frac{S}{(2\pi)^{3/2}\,\sigma_1(t-t')\sigma_2(t-t')\sigma_3(t-t')}$$

$$\times \exp\left\{-\frac{[x_1 - x_{1_0} - U(t-t')]^2}{2\sigma_1^2(t-t')} - \frac{(x_2 - x_{2_0})^2}{2\sigma_2^2(t-t')} - \frac{(x_3 - x_{3_0})^2}{2\sigma_3^2(t-t')}\right\} dt'$$

We let $\beta = t - t'$ and this equation becomes

$$\langle c(\mathbf{x},t)\rangle = \int_{0}^{\infty} \frac{S}{(2\pi)^{3/2}\sigma_1(\beta)\sigma_2(\beta)\sigma_3(\beta)}$$

$$\times \exp\left[-\frac{(x_1 - x_{1_0} - U\beta)^2}{2\sigma_1^2(\beta)} - \frac{(x_2 - x_{2_0})^2}{2\sigma_2^2(\beta)} - \frac{(x_3 - x_{3_0})^2}{2\sigma_3^2(\beta)}\right] d\beta \qquad (6.38c)$$

We note that the right-hand side does not depend on t. Therefore, the mean concentration on the left-hand side must be independent of t and merely equal to $\langle c(\mathbf{x}) \rangle$.

Equation (6.38c) cannot, in principle, be simplified further. However, when the turbulent velocities are small when compared with the mean velocity, it is often possible to neglect turbulent diffusion in the direction of the mean velocity. If we make such an assumption, we can evaluate (6.38c). If turbulent diffusion in the x_1 direction is negligible, all the material emitted during the time interval $(t', t' + dt')$ will be found within a disk bounded by two planes perpendicular to the direction of the mean velocity and lying at positions $U(t - t')$ and $U(t - t' - dt')$. The continuous plume is then assumed to consist of an infinite number of such disks placed one against another downstream of the source. Equation (6.38b) was obtained by assuming that the continuous plume consisted of an infinite number of puffs. The mean concentration in a disk can be found by integrating the puff relation (6.38b) over x_1 from $-\infty$ to ∞, that is

$$\langle c(\mathbf{x},t) \rangle = \int_{-\infty}^{\infty} \frac{S}{(2\pi)^{3/2}\sigma_1(t-t')\sigma_2(t-t')\sigma_3(t-t')}$$

$$\times \exp\left\{-\frac{[x_1 - x_{1_0} - U(t-t')]^2}{2\sigma_1^2(t-t')} - \frac{(x_2 - x_{2_0})^2}{2\sigma_2^2(t-t')} - \frac{(x_3 - x_{3_0})^2}{2\sigma_3^2(t-t')}\right\} \frac{dx_1}{U}$$

In carrying out this integration, we express σ_2 and σ_3 as functions of the *distance* downwind, $x_1 - x_{1_0}$, since $t - t'$ and $x_1 - x_{1_0}$ are related for each disk. Then, the final result for the mean concentration is entry 2 of Table 6.1,

$$\langle c(\mathbf{x}) \rangle = \frac{S}{2\pi\sigma_2(x_1 - x_{1_0})\sigma_3(x_1 - x_{1_0})U} \exp\left[-\frac{(x_2 - x_{2_0})^2}{2\sigma_2^2(x_1 - x_{1_0})} - \frac{(x_3 - x_{3_0})^2}{2\sigma_3^2(x_1 - x_{1_0})}\right] \tag{6.39}$$

where $\sigma_2(x_1 - x_{1_0})$ and $\sigma_3(x_1 - x_{1_0})$ are the standard deviations of the mean concentration in the x_2 and x_3 planes at distance $x_1 - x_{1_0}$ from the source. Figure 6.2 shows typical distributions arising from (6.39).

In most problems of atmospheric diffusion we must take into account the ground at $x_3 = 0$ as an impenetrable barrier to diffusion (assuming no deposition of pollutants). To include the $x_3 = 0$ barrier in the gaussian puff and plume formulas we assume that there is an identical source at $(x_{1_0}, x_{2_0}, -x_{3_0})$ the contribution from which for $x_3 > 0$ is equivalent to reflection of the actual source effluent at $x_3 = 0$. Doing so with entries 1 and 2 of Table 6.1 yields entries 3 and 4, respectively.

So far we have not indicated how one determines the values of the $\sigma_i(t)$ for an instantaneous source or $\sigma_2(x_1 - x_1')$ and $\sigma_3(x_1 - x_1')$ for a continuous source. In principle, these would be measured by determining the concentration distributions

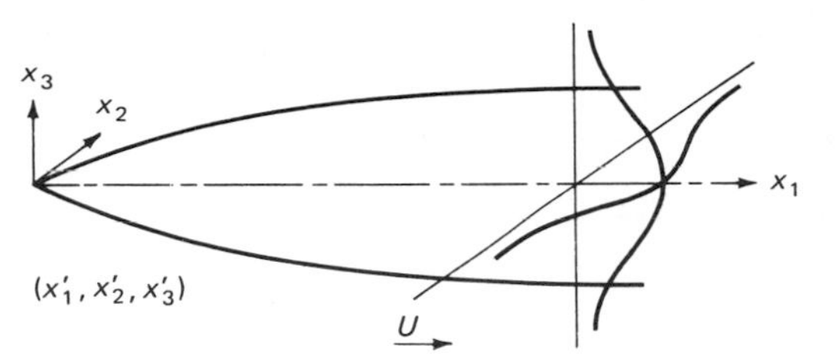

FIGURE 6.2
Concentration distributions from a continuous point source located at (x'_1,x'_2,x'_3) with a uniform wind U in the x_1 direction.

under varying conditions of turbulence for both instantaneous and continuous releases. There are, in fact, theoretical results available based on the diffusion of a single particle relative to a fixed axis to tell us how the σ_i's should vary with travel time from the source. We will not present these results at this point; we will do so later in this chapter.

The continuous-source formulas in Table 6.1 have been used extensively for predicting pollutant dispersion from point and line sources. They have been fairly successful in predicting long-time average of concentrations near the source ($x_i - x'_i < 10$ km) under steady meteorological conditions when used in conjunction with empirically determined values of the σ_i^2. Used in this way, the gaussian plume formulas are empirical expressions which can be applied with reasonable accuracy only to situations in which conditions are similar to those of the empirical studies from which the σ_i^2 have been obtained. However, since atmospheric turbulence is seldom homogeneous and almost never stationary, the gaussian plume formulas provide, at best, order-of-magnitude estimates of concentrations downwind of point and line sources under actual atmospheric conditions.

Mean Concentration from a Crosswind Line Source

A crosswind line source may be considered an infinite number of point sources along a line perpendicular to the direction of the mean wind. The mean concentration resulting from a segment of length dx_2 of the line source, emitting at a rate $S_l\, dx_2$ g/sec (the units of S_l are grams per meter-second) is given by (6.39) with S replaced by $S_l\, dx_2$. For an infinite, crosswind line source in an unbounded atmosphere, the mean concentration is obtained by integrating (6.39) over x_2 from $-\infty$ to ∞ (with S replaced by $S_l\, dx_2$). The result is

$$\langle c(x_1,x_3)\rangle = \frac{S_l}{\sqrt{2\pi}\,\sigma_3(x_1 - x_{1_0})U}\, e^{-(x_3 - x_{3_0})^2/2\sigma_3{}^2(x_1 - x_{1_0})} \qquad (6.40)$$

Estimation of Concentration Distributions Downwind of a Continuous Point Source by the Gaussian Plume Formula

Let us illustrate the use of the gaussian plume formula for estimating pollutant concentrations due to continuous point sources. As we have noted, this formula provides only rough estimates of concentrations since the complexities of true plume behavior—unsteady emissions rates, varying meteorological conditions, uncertainty of plume trajectories—are not accounted for.

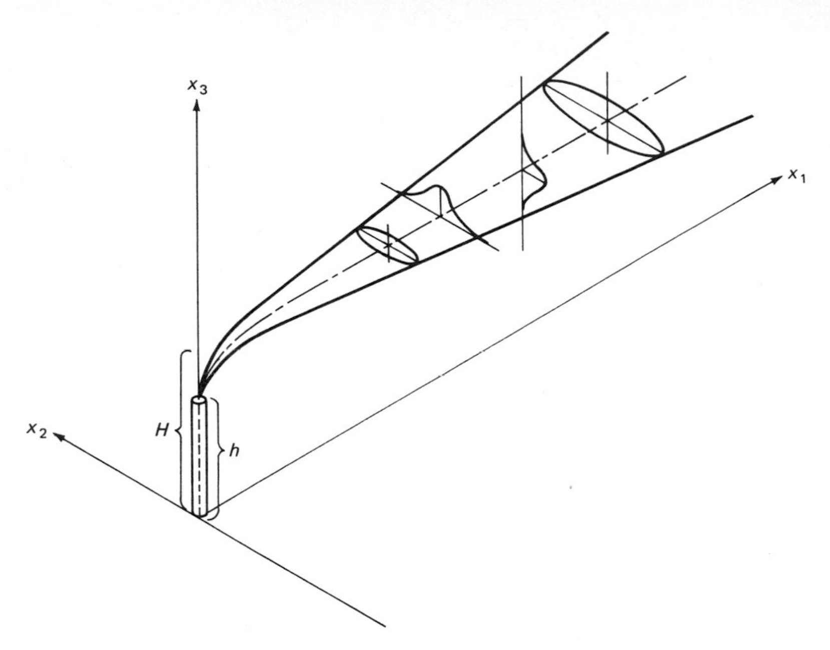

FIGURE 6.3
Concentration distributions from a continuous point source of effective height H (stack height h plus plume rise ΔH) as predicted by the gaussian plume equation (entry 4 of Table 6.1).

The problem of modeling a continuous, nonreacting plume by means of the gaussian plume formula consists of two parts:

1 Estimation of the height at which the buoyant plume becomes essentially horizontal (*plume rise*)
2 Estimation of the concentration distribution in the plume

In order to estimate plume rise, it is necessary to choose a suitable formula from the many suggested for this purpose. The "best" formula is probably one derived from tests on stacks having geometries, and operated under conditions, similar to those being considered. It is also desirable that the tests be made under meteorological conditions resembling those of the locale of interest. Reference should be made to the literature for guidance in making a choice; we have summarized several of the available theories on plume rise in Appendix B.

The steady concentration distribution at any point x, y, z (using x, y, z notation), resulting from a continuous source of strength S with an effective emission height H (the sum of stack height h and plume rise ΔH) is given by entry 4 of Table 6.1:

$$\langle c(x,y,z)\rangle = \frac{S}{2\pi\sigma_y\sigma_z U}\exp\left[-\frac{1}{2}\left(\frac{y}{\sigma_y}\right)^2\right]\left\{\exp\left[-\frac{1}{2}\left(\frac{z-H}{\sigma_z}\right)^2\right]+\exp\left[-\frac{1}{2}\left(\frac{z+H}{\sigma_z}\right)^2\right]\right\}$$

Figure 6.3 shows the form of the plume predicted with this equation.

This equation assumes that diffusion is limited in the vertical direction only by the ground. Often, however, an elevated inversion layer is present which may be considered to act as a lid to prevent

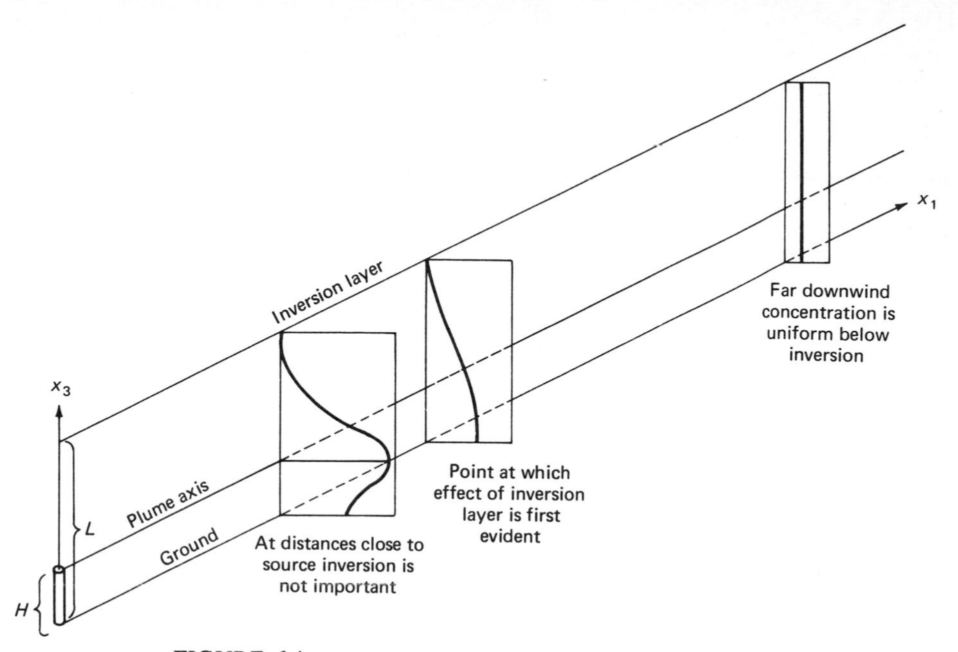

FIGURE 6.4
Concentration distributions from a continuous point source of effective height H below an inversion layer of height L.

further vertical mixing. When an inversion layer at $z = L$ is imposed, the steady concentration distribution along the plume center line ($y = 0$) is given by (for $H < L$)

$$\langle c(x,0,z)\rangle = \frac{S}{2\pi\sigma_y\sigma_z U}\left(\exp\left[-\frac{1}{2}\left(\frac{z-H}{\sigma_z}\right)^2\right] + \exp\left[-\frac{1}{2}\left(\frac{z+H}{\sigma_z}\right)^2\right]\right.$$
$$+\sum_{n=1}^{\infty}\left\{\exp\left[-\frac{1}{2}\left(\frac{z-H-2nL}{\sigma_z}\right)^2\right] + \exp\left[-\frac{1}{2}\left(\frac{z+H-2nL}{\sigma_z}\right)^2\right]\right.$$
$$\left.\left.+\exp\left[-\frac{1}{2}\left(\frac{z-H+2nL}{\sigma_z}\right)^2\right] + \exp\left[-\frac{1}{2}\left(\frac{z+H+2nL}{\sigma_z}\right)^2\right]\right\}\right)$$

In the use of this equation, four terms are usually sufficient to approximate the summation. The vertical concentration profiles from a point source below an inversion layer are illustrated in Fig. 6.4. As an alternative to this equation, Turner (1969) suggested that one allow σ_z to increase with distance downwind until it becomes equal to $0.47(L - H)$, where L is the height of the inversion base. At this distance, x_L, the plume is assumed to have a gaussian distribution in the vertical. Let us assume that by the time the plume travels twice this far, $2x_L$, it is uniformly distributed between the earth's surface and the inversion height L and its concentration distribution is given by

$$\langle c(x,y,z)\rangle = \frac{S}{\sqrt{2\pi}\,\sigma_y UL}\, e^{-1/2(y/\sigma_y)^2}$$

For $x_L < x < 2x_L$, the best approximation of the concentration is that read from a straight line drawn between the calculated concentrations at points x_L and $2x_L$ on a log-log plot of concentration as a function of distance.

In the case in which the effective emission height H is greater than the inversion height L, we can make the following assumptions:

1 If the physical stack height h is greater than L, an infinite value for L is assumed, with plume rise and dispersion estimated for stable atmospheric conditions.

2 If the stack height h is less than L, a minimum plume rise $\Delta H_{\min}$ is calculated, based on the coefficients for stably stratified air.

> If the minimum effective emission height, $H = h + \Delta H_{\min}$, is greater than the lid height L, H is used as the effective emission height. The plume is then analyzed as described in assumption 1.
>
> If $(h + \Delta H_{\min}) < L$ and if the plume rise ΔH based on the actual stability class yields an effective emission height greater than L, that is if $h + \Delta H_{\min} < L$ and $h + \Delta H \geq L$, the effective emission height is restricted to the lid height L.

Finally, we note that we have not yet specified how to determine the parameters σ_y and σ_z. We shall defer a discussion of these variables until later in the chapter.

We have noted the deficiencies in the gaussian formulas in describing pollutant concentrations under real atmospheric conditions. We have already derived the eulerian-based equation (6.14) which is capable, in principle, of treating unsteady, inhomogeneous conditions and chemical reactions among pollutants. We now ask: Can we determine $Q(\mathbf{x},t|\mathbf{x}',t')$ in nonstationary, inhomogeneous turbulence? The only ways this could be done are by extensive experimental measurements (which would be valid for only one particular situation) or numerical simulation of the turbulence. Unfortunately, either such measurements or such simulations are rare even for simplified situations. Thus, we ask: Is there any way of obtaining a "usable" equation for $\langle c(\mathbf{x},t)\rangle$ from (6.7) or (6.10), perhaps, as an alternative to (6.14)? In particular, can we derive a differential equation for $\langle c(\mathbf{x},t)\rangle$ which can be solved as a boundary-value problem and which is capable of use when chemical reactions are taking place? Fortunately, the answer is yes, and we now present the derivation of this equation. The primary reason for doing so is to gain further insight into the assumptions which must be made in order to obtain equations which can be used to compute species concentrations under actual atmospheric conditions. The derivation which we now present relies on some basic concepts of probability theory. Readers without the requisite background may wish to skip the derivation, although it is essential that the conditions under which the final equation is valid be noted carefully.

6.2.4 Derivation of a Differential Equation for $\langle c(\mathbf{x},t)\rangle$ from the Basic Lagrangian Equation (6.10)

Our object in this subsection is to derive a differential equation for $\langle c(\mathbf{x},t)\rangle$ from the lagrangian equation (6.10). Let us first consider the basic equation (6.5) which relates the probability density of the position of a single particle at time t to the probability

FIGURE 6.5
Trajectory of a fluid particle in a turbulent flow.

density of its position at some earlier time t' through the transition probability density $Q(\mathbf{x},t|\mathbf{x}',t')$. Let us envision, as in Fig. 6.5, the particle's trajectory from $\mathbf{x}'$ at t' to $\mathbf{x}$ at t to be composed of n time segments each of length Δt. Thus, we let $\mathbf{x}' = \mathbf{x}_1$ and $t' = t_1$ and $\mathbf{x} = x_n$ and $t = t_n$. At the end of each time interval $t_i, i = 2, \ldots, n$, the location of the particle is a continuous random vector $\mathbf{x}_i$. In the terminology of probability theory, the particle's locations at the end of each time step constitute a continuous-state stochastic process.

Let us analyze the process depicted in Fig. 6.5. The complete information on the process is embodied in the joint probability density function $p(\mathbf{x}_1,t_1;\ldots;\mathbf{x}_n,t_n)$, where $p(\mathbf{x}_1,t_1;\ldots;\mathbf{x}_n,t_n)\,d\mathbf{x}_1 \cdots d\mathbf{x}_n$ is the probability that at t_1 the particle lies in the volume element $\mathbf{x}_1$ to $\mathbf{x}_1 + d\mathbf{x}_1, \ldots,$ and at t_n in the volume element $\mathbf{x}_n$ to $\mathbf{x}_n + d\mathbf{x}_n$. This joint density can be written in terms of conditional densities as follows:

$$\begin{aligned} p(\mathbf{x}_1,t_1;\ldots;\mathbf{x}_n,t_n) &= p(\mathbf{x}_2,t_2;\ldots;\mathbf{x}_n,t_n|\mathbf{x}_1,t_1)p(\mathbf{x}_1,t_1) \\ &= p(\mathbf{x}_3,t_3;\ldots;\mathbf{x}_n,t_n|\mathbf{x}_1,t_1;\mathbf{x}_2,t_2)p(\mathbf{x}_1,t_1;\mathbf{x}_2,t_2) \\ &= p(\mathbf{x}_n,t_n|\mathbf{x}_{n-1},t_{n-1};\ldots;\mathbf{x}_1,t_1)p(\mathbf{x}_{n-1},t_{n-1};\ldots;\mathbf{x}_1,t_1) \end{aligned}$$

where $p(\mathbf{x}_1,t_1)\,d\mathbf{x}_1$ is the probability that at t_1 the particle lies in the volume element $\mathbf{x}_1$ to $\mathbf{x}_1 + d\mathbf{x}_1$; $p(\mathbf{x}_1,t_1;\mathbf{x}_2,t_2)\,d\mathbf{x}_1\,d\mathbf{x}_2$ is the joint probability that at t_1 the particle lies in $\mathbf{x}_1$ to $\mathbf{x}_1 + d\mathbf{x}_1$ and at t_2 in $\mathbf{x}_2$ to $\mathbf{x}_2 + d\mathbf{x}_2$.

Using the fundamental relations between marginal and joint probability densities, we can write

$$\begin{aligned} p(\mathbf{x}_n,t_n) &= \int_{-\infty}^{\infty} \cdots \int_{-\infty}^{\infty} p(\mathbf{x}_n,t_n;\mathbf{x}_{n-1},t_{n-1};\ldots;\mathbf{x}_1,t_1)\,d\mathbf{x}_{n-1}\cdots d\mathbf{x}_1 \\ &= \int_{-\infty}^{\infty} \cdots \int_{-\infty}^{\infty} p(\mathbf{x}_n,t_n|\mathbf{x}_{n-1},t_{n-1};\ldots;\mathbf{x}_1,t_1) \\ &\qquad \times p(\mathbf{x}_{n-1},t_{n-1};\ldots;\mathbf{x}_1,t_1)d\mathbf{x}_{n-1}\cdots d\mathbf{x}_1 \\ &= \int_{-\infty}^{\infty} \cdots \int_{-\infty}^{\infty} p(\mathbf{x}_n,t_n|\mathbf{x}_{n-1},t_{n-1};\ldots;\mathbf{x}_1,t_1) \\ &\qquad \times p(\mathbf{x}_{n-1},t_{n-1};\ldots;\mathbf{x}_2,t_2|\mathbf{x}_1,t_1)p(\mathbf{x}_1,t_1)\,d\mathbf{x}_{n-1}\cdots d\mathbf{x}_1 \end{aligned} \tag{6.41}$$

Therefore returning to (6.5), in the notation of (6.5), $\psi(\mathbf{x}_n,t_n) = p(\mathbf{x}_n,t_n)$ and

$$Q(\mathbf{x}_n,t_n|\mathbf{x}_1,t_1) = p(\mathbf{x}_n,t_n|\mathbf{x}_{n-1},t_{n-1};\ldots;\mathbf{x}_1,t_1)p(\mathbf{x}_{n-1},t_{n-1};\ldots;\mathbf{x}_2,t_2|\mathbf{x}_1,t_1) \tag{6.42}$$

We see that the computation of Q in general involves knowledge of the entire past history of the particle from t_1 to t_n. However, if the particle positions at each time increment are a *Markov process*, then an enormous simplification results, namely that

$$p(\mathbf{x}_n,t_n|\mathbf{x}_{n-1},t_{n-1};\ldots;\mathbf{x}_1,t_1) = p(\mathbf{x}_n,t_n|\mathbf{x}_{n-1},t_{n-1}) \tag{6.43}$$

In a Markov process, the current state depends only on the immediately prior state and not on any earlier states. Using (6.43) repeatedly, in the case in which the particle positions are a Markov process, (6.42) becomes

$$Q(\mathbf{x}_n,t_n|\mathbf{x}_1,t_1) = p(\mathbf{x}_n,t_n|\mathbf{x}_{n-1},t_{n-1})p(\mathbf{x}_{n-1},t_{n-1}|\mathbf{x}_{n-2},t_{n-2}) \cdots p(\mathbf{x}_2,t_2|\mathbf{x}_1,t_1) \tag{6.44}$$

We can therefore write (6.41) in a recursive fashion:

$$p(\mathbf{x}_n,t_n) = \int_{-\infty}^{\infty}\int_{-\infty}^{\infty}\int_{-\infty}^{\infty} p(\mathbf{x}_n,t_n|\mathbf{x}_{n-1},t_{n-1})p(\mathbf{x}_{n-1},t_{n-1})\, d\mathbf{x}_{n-1} \qquad n = 2, 3, \ldots \tag{6.45}$$

Returning to our original notation, we have

$$\psi(\mathbf{x}_n,t_n) = \int_{-\infty}^{\infty}\int_{-\infty}^{\infty}\int_{-\infty}^{\infty} Q(\mathbf{x}_n,t_n|\mathbf{x}_{n-1},t_{n-1})\psi(\mathbf{x}_{n-1},t_{n-1})\, d\mathbf{x}_{n-1} \tag{6.46}$$

where we have replaced $p(\mathbf{x}_n,t_n|\mathbf{x}_{n-1},t_{n-1})$ by $Q(\mathbf{x}_n,t_n|\mathbf{x}_{n-1},t_{n-1})$.

The key question is: Under what conditions might the diffusion of a fluid particle in turbulence be considered a Markov process? If the random component of the velocity of any particle, $v_i'(t)$, has a correlation function $\mathscr{R}_{ij}'(t;\tau) = \langle v_i'(t)v_j'(t+\tau)\rangle$ which vanishes sufficiently rapidly with increasing τ that a time scale

$$\mathscr{T}_{ij} = \langle v_i'(t)v_j'(t)\rangle^{-1}\int_0^{\infty} \mathscr{R}_{ij}'(t;\tau)\, d\tau$$

exists for all possible values of t and all possible points of release of the particle, it is to be expected that the motion of any particle at any time t will be statistically independent of its motions prior to the time $t - \Delta t$ as long as

$$\Delta t \gg \max_{i,j} \mathscr{T}_{ij} \tag{6.47}$$

Thus, if we choose a Δt that satisfies (6.47) we may assume that the positions of a particle at $t, t+\Delta t, t+2\Delta t, \ldots$, form a Markov process. Not only does this assumption greatly simplify (6.5) but also it enables the derivation of a differential equation for $\psi(\mathbf{x},t)$ [and, therefore, equivalently $\langle c(\mathbf{x},t)\rangle$].

The form of (6.46) corresponding to (6.10) is

$$\langle c(\mathbf{x}_n,t_n)\rangle = \int_{-\infty}^{\infty}\int_{-\infty}^{\infty}\int_{-\infty}^{\infty} Q(\mathbf{x}_n,t_n|\mathbf{x}_{n-1},t_{n-1})\langle c(\mathbf{x}_{n-1},t_{n-1})\rangle \times \exp\left[-\int_{t_{n-1}}^{t_n} k(\tau)\,d\tau\right] d\mathbf{x}_{n-1}$$
$$+\int_{-\infty}^{\infty}\int_{-\infty}^{\infty}\int_{-\infty}^{\infty}\int_{t_{n-1}}^{t_n} Q(\mathbf{x}_n,t_n|\mathbf{x}_p,t_p)S(\mathbf{x}_p,t_p)\exp\left[-\int_{t_p}^{t_n} k(\tau)\,d\tau\right] dt_p\,d\mathbf{x}_p \tag{6.48}$$

Equation (6.48) is a general relation for the mean concentration of exponentially decaying particles under the restriction that Δt is significantly greater than the lagrangian time scale of the turbulence. We would like to reduce (6.48) to a differential equation for $\langle c(\mathbf{x},t)\rangle$. To do this we must make additional assumptions. First, it is necessary to assume that temporal variations in $S(\mathbf{x},t)$ and $k(t)$ are small compared with Δt and that spatial variations in $S(\mathbf{x},t)$ are gradual compared with the mean distance traveled by a particle in Δt,

$$\frac{1}{S(\mathbf{x},t)}\frac{\partial S}{\partial t} \ll \frac{1}{\Delta t} \tag{6.49}$$

$$k(t) \ll \frac{1}{\Delta t} \tag{6.50}$$

$$\frac{1}{S(\mathbf{x},t)}\frac{\partial S}{\partial x_i} \ll [\Delta t^2(\langle v_i'^2\rangle + \bar{u}_i^2)]^{-1/2} \tag{6.51}$$

Secondly, we must assume that the spatial and temporal inhomogeneities in the turbulence are of such scales that the transition density Q of a particle released anywhere in the fluid is the gaussian density for travel times t in the range $0 \le t \le \Delta t$. Actually, this condition is probably automatically satisfied when (6.47) is true. Thus, over each interval of Δt a particle behaves as though it were in a field of stationary, homogeneous turbulence. We might refer to such a situation as locally stationary and homogeneous turbulence. Quantitatively, the fluid velocity components should satisfy the following conditions:

$$\frac{1}{\langle u_i'(\mathbf{x},t)u_j'(\mathbf{x},t)\rangle}\frac{\partial}{\partial t}\langle u_i'(\mathbf{x},t)u_j'(\mathbf{x},t)\rangle \ll \frac{1}{\Delta t} \tag{6.52}$$

$$\frac{1}{\bar{u}_i(\mathbf{x},t)}\frac{\partial \bar{u}_i}{\partial t} \ll \frac{1}{\Delta t} \tag{6.53}$$

$$\frac{1}{\langle u_i'(\mathbf{x},t)u_j'(\mathbf{x},t)\rangle}\frac{\partial}{\partial x_k}\langle u_i'(\mathbf{x},t)u_j'(\mathbf{x},t)\rangle \ll [(\mathscr{R}_{kk}' + \bar{u}_k^2\,\Delta t)\,\Delta t]^{-1/2} \tag{6.54}$$

$$\frac{1}{\bar{u}_i(\mathbf{x},t)}\frac{\partial \bar{u}_i}{\partial x_k} \ll [(\mathscr{R}_{kk}' + \bar{u}_k^2\,\Delta t)\,\Delta t]^{-1/2} \tag{6.55}$$

Under these conditions the transition density $Q(\mathbf{x}_n, t_n | \mathbf{x}_{n-1}, t_{n-1})$ can be written $Q(\Delta\mathbf{x} | \mathbf{x})$, that is, the probability density that the particle will be displaced a distance $\Delta\mathbf{x}$ from $\mathbf{x}$ during Δt. In addition, $Q(\Delta\mathbf{x} | \mathbf{x})$ will be gaussian:

$$Q(\Delta\mathbf{x} | \mathbf{x}) = \frac{1}{(2\pi)^{3/2} |P(\mathbf{x})|} \exp\left[-\frac{1}{2}(\Delta\mathbf{x} - \langle\Delta\mathbf{x}\rangle)^T \mathbf{P}^{-1}(\mathbf{x})(\Delta\mathbf{x} - \langle\Delta\mathbf{x}\rangle)\right]$$

Under all the above conditions, (6.48) takes the form

$$\langle c(\mathbf{x}, t + \Delta t)\rangle = \int_{-\infty}^{\infty}\int_{-\infty}^{\infty}\int_{-\infty}^{\infty} Q(\Delta\mathbf{x} | \mathbf{x} - \Delta\mathbf{x}) e^{-k(t)\,\Delta t} \times [\langle c(\mathbf{x} - \Delta\mathbf{x}, t)\rangle + S(\mathbf{x} - \Delta\mathbf{x}, t)\,\Delta t]\, d(\Delta\mathbf{x}) \tag{6.56}$$

The object is to derive a differential equation for $\langle c(\mathbf{x},t)\rangle$ from (6.56).

In order to illustrate the derivation of the differential equation we will use the simpler form of (6.56),

$$\langle c(\mathbf{x}, t + \Delta t)\rangle = \int_{-\infty}^{\infty}\int_{-\infty}^{\infty}\int_{-\infty}^{\infty} Q(\mathbf{x} | \mathbf{x} - \Delta x)\langle c(\mathbf{x} - \Delta\mathbf{x}, t)\rangle\, d(\Delta\mathbf{x}) \tag{6.57}$$

in which there is no decay or sources. We expand $\langle c(\mathbf{x}, t + \Delta t)\rangle$, $Q(\Delta\mathbf{x} | \mathbf{x} - \Delta\mathbf{x})$ and $\langle c(\mathbf{x} - \Delta\mathbf{x}, t)\rangle$ in terms of Taylor series about the point $(\mathbf{x},t)$ to obtain

$$\langle c(\mathbf{x}, t + \Delta t)\rangle = \langle c(\mathbf{x},t)\rangle + \frac{\partial\langle c\rangle}{\partial t}\Delta t + \cdots \tag{6.58}$$

$$Q(\Delta\mathbf{x} | \mathbf{x} - \Delta\mathbf{x}) = Q(\Delta\mathbf{x} | \mathbf{x}) - \frac{\partial Q}{\partial x_j}\Delta x_j + \frac{1}{2}\frac{\partial^2 Q}{\partial x_j\,\partial x_k}\Delta x_j\,\Delta x_k + \cdots \tag{6.59}$$

$$\langle c(\mathbf{x} - \Delta\mathbf{x},t)\rangle = \langle c(\mathbf{x},t)\rangle - \frac{\partial\langle c\rangle}{\partial x_j}\Delta x_j + \frac{1}{2}\frac{\partial^2\langle c\rangle}{\partial x_j\,\partial x_k}\Delta x_j\,\Delta x_k + \cdots \tag{6.60}$$

where the second-order terms are, for example,

$$\frac{1}{2}\frac{\partial^2 Q}{\partial x_j\,\partial x_k}\Delta x_j\,\Delta x_k = \frac{1}{2}\sum_j \frac{\partial^2 Q}{\partial {x_j}^2}\Delta {x_j}^2 + \sum_{j<k}\frac{\partial^2 Q}{\partial x_j\,\partial x_k}\Delta x_j\,\Delta x_k \tag{6.61}$$

Substituting (6.58) to (6.61) into (6.57) we obtain

$$\frac{\partial\langle c(\mathbf{x},t)\rangle}{\partial t}\Delta t = -\sum_j \frac{\partial}{\partial x_j}[\langle c(\mathbf{x},t)\rangle\langle\Delta x_j\rangle] + \frac{1}{2}\sum_j \frac{\partial^2}{\partial {x_j}^2}[\langle c(\mathbf{x},t)\rangle\langle\Delta {x_j}^2\rangle] + \sum_{j<k}\frac{\partial^2}{\partial x_j\,\partial x_k}[\langle c(\mathbf{x},t)\rangle\langle\Delta x_j\,\Delta x_k\rangle] \tag{6.62}$$

where $\langle\Delta x_j\rangle$ is the mean displacement in the jth direction,

$$\langle\Delta x_j\rangle = \int_{-\infty}^{\infty}\int_{-\infty}^{\infty}\int_{-\infty}^{\infty} \Delta x_j\, Q(\Delta\mathbf{x} | \mathbf{x})\, d(\Delta\mathbf{x}) \tag{6.63}$$

and $\langle \Delta x_j \, \Delta x_k \rangle$ is the mean-square displacement

$$\langle \Delta x_j \, \Delta x_k \rangle = \int_{-\infty}^{\infty} \int_{-\infty}^{\infty} \int_{-\infty}^{\infty} \Delta x_j \, \Delta x_k \, Q(\Delta \mathbf{x} \,|\, \mathbf{x}) \, d(\Delta \mathbf{x}) \qquad (6.64)$$

The mean displacement is given simply by

$$\langle \Delta x_j \rangle = \bar{u}_j(\mathbf{x},t) \, \Delta t \qquad (6.65)$$

In order to compute $\langle \Delta x_j \, \Delta x_k \rangle$ we divide Δx_j as $\langle \Delta x_j \rangle + \Delta x_j'$, where the displacement Δx_j has a mean and fluctuating component. Thus $\langle \Delta x_j \, \Delta x_k \rangle = \langle \Delta x_j \rangle \langle \Delta x_k \rangle + \langle \Delta x_j' \, \Delta x_k' \rangle$, the first term of which is $\bar{u}_j \bar{u}_k \, \Delta t^2$, and the second term of which is

$$\begin{aligned} \langle \Delta x_j' \, \Delta x_k' \rangle = P_{jk} &= \int_t^{t+\Delta t} \int_t^{t+\Delta t} \langle v_j'(\mathbf{x},t_1) v_k'(\mathbf{x},t_2) \rangle \, dt_1 \, dt_2 \\ &= \bar{K}_{jk}(\mathbf{x},t) \, \Delta t \end{aligned} \qquad (6.66)$$

where

$$\bar{K}_{jk} = \langle v_j' v_k' \rangle \mathscr{T}_{jk} + \langle v_k' v_j' \rangle \mathscr{T}_{kj} \qquad (6.67)$$

Substituting (6.65) and (6.66) into (6.62), dividing by Δt, and letting $\Delta t \to 0$, we obtain

$$\frac{\partial \langle c(\mathbf{x},t) \rangle}{\partial t} + \frac{\partial}{\partial x_j} [\bar{u}_j(\mathbf{x},t) \langle c(\mathbf{x},t) \rangle] = \frac{1}{2} \frac{\partial^2}{\partial x_j \, \partial x_k} [\bar{K}_{jk}(\mathbf{x},t) \langle c(\mathbf{x},t) \rangle] \qquad (6.68)$$

The form of (6.68) corresponding to the more general form (6.56) is[1]

$$\frac{\partial \langle c_i \rangle}{\partial t} + \frac{\partial}{\partial x_j} (\bar{u}_j \langle c_i \rangle) = \frac{1}{2} \frac{\partial^2}{\partial x_j \, \partial x_k} (\bar{K}_{jk} \langle c_i \rangle) - k(t) \langle c_i \rangle + S_i(\mathbf{x},t) \qquad (6.69)$$

If the coordinate axes coincide with the principal axes of the tensor $\{\bar{K}_{jk}\}$, (6.69) reduces to

$$\frac{\partial \langle c_i \rangle}{\partial t} + \frac{\partial}{\partial x_j} (\bar{u}_j \langle c_i \rangle) = \frac{1}{2} \frac{\partial^2}{\partial x_j \, \partial x_j} (\bar{K}_{jj} \langle c_i \rangle) - k(t) \langle c_i \rangle + S_i(\mathbf{x},t) \qquad (6.70)$$

[1] This equation is a parabolic partial differential equation of the diffusion type. Parabolic partial differential equations characterize molecular diffusion processes (such as of heat or mass); the classic heat conduction equation is a well-known example. Solutions of parabolic equations have the characteristic that some effect, be it heat or mass, is experienced everywhere except at $t = 0$. The implication of this property is that diffusion proceeds with an infinite velocity. Clearly, the speed of a real diffusion event, such as the dispersion of a smoke puff, must have some finite value. If we were to extend the Markov process model for successive particle displacements in such a way that prior locations other than just the immediately prior location influenced the next step, we could obtain a form of the "telegrapher's equation" for the mean concentration. The telegrapher's equation is of the hyperbolic type, like a wave equation, and therefore describes diffusion as occurring at a finite velocity. For most practical purposes the differences between the predictions of the two types of equations are negligible. The interested reader may consult Goldstein (1951) for more on this point.

From a strictly mathematical point of view, (6.69) and (6.70) are valid only for infinitely large time and space scales since we allowed $\Delta t \to 0$ in the derivation. From a practical standpoint, however, we can prescribe the relative time scales for use of these equations. In general, if T is a characteristic time for changes in $\langle c \rangle$, it is necessary that $T \gg \Delta t$. Also we have already prescribed that $\Delta t \gg \max \mathscr{T}_{jk}$, and so

$$\max_{j,k} \mathscr{T}_{jk} \ll \Delta t \ll T$$

The basic lagrangian equation (6.10) applies only to species which are inert or decay according to a first-order law, since it is assumed that the species reaction occurs independently of the particle motions. This assumption is clearly not always valid for species which react through collisions with themselves or other species. To handle this case by using lagrangian statistics is very difficult, and it appears that a lagrangian theory for nonlinear chemical reactions in turbulence is at present too complex to provide a viable basis for describing the behavior of reactive species in turbulence. Nevertheless, under certain conditions, (6.70) may be used for modeling turbulent diffusion and nonlinear chemical reaction because of the assumptions (6.49) to (6.51) on which (6.70) rests. By virtue of these assumptions, the temporal and spatial variations in the mean concentration field are so much larger in scale than the lagrangian time scale $\mathscr{T}_L = \max_{j,k} \mathscr{T}_{jk}$ and the average distance $[(\bar{u}_j^{\,2} + \langle v_j'^2 \rangle)\mathscr{T}_L^{\,2}]^{1/2}$ a particle travels in time $\mathscr{T}_L$ that the rate of reaction over a time $\mathscr{T}_L$ will not be affected greatly by changes in concentration and is the rate based on the local mean concentration. Thus, the terms $k\langle c_i \rangle$ and S_i may be used to represent not only the linear reaction rate and strength of sources, respectively, of the ith species but also the sources and sinks resulting from nonlinear chemical reaction among the N species. Specifically, the term $k\langle c_i \rangle$ can represent all those reactions which deplete i, and S_i can represent the formation of species i by reactions among the other species present.

6.2.5 Comparison of the Differential Equations for the Mean Concentrations from the Eulerian and Lagrangian Points of View

The basic eulerian equation for $\langle c(\mathbf{x},t) \rangle$ is (6.14), whereas that from the lagrangian approach is (6.70). Aside from the form of the terms accounting for chemical reactions, we note a fundamental discrepancy in the diffusion terms. First of all, the assumptions leading to (6.14) and hence (6.14) are probably valid provided that, in addition to (6.49), (6.51), and (6.52) to (6.55) the following condition is met: that the time scale of the fastest reaction described by R_i is much larger than the lagrangian time scale $\mathscr{T}_L$ of the turbulence. This list of restrictions is essentially identical to that placed on (6.70), and so as far as applicability to turbulent diffusion is concerned, (6.14) and (6.70) are virtually identical. The only significant difference therefore between these two equations lies in their diffusion terms. In (6.70) $\bar{K}_{jj}$ is dependent strictly on the lagrangian statistics of the turbulence whereas in (6.14) K_{jj} is an eulerian parameter, i.e., a function of each point in space, which must be determined by fitting data to the solution of the equation. However, since the lagrangian properties on which the $\bar{K}_{jj}$ are dependent are difficult to measure directly, the $\bar{K}_{jj}$ are usually determined by using them as parameters to fit the solution of (6.70) to experimental concentration data. Since the same procedure is used to determine the K_{jj}, there is little significant difference between (6.14) and (6.70). Because of the empirical nature of the "diffusivities," the accuracy of both equations is also dependent on the degree to which the conditions of the situation to which these equations are applied correspond to the conditions under which the diffusivities were measured.

Finally we point out that (6.14) and (6.70) apply only to time and space scales which are much larger than the corresponding scales $\mathscr{T}$ and L of the turbulence. Thus, to regard these as point equations is to imply that a "point" has spatial dimensions much larger than L_i ($i = 1,2,3$) and that an "instant" is a period long compared with $\mathscr{T}$.

In this section we have presented equations governing simultaneous diffusion and chemical reaction at several levels of generality. In order to summarize the distinguishing features of each model we itemize in Table 6.2 the assumptions which were employed in deriving each of the model's governing equations. Table 6.2 has been prepared from a similar table presented by Lamb (1971). The sequence in which the models are presented in Table 6.2 is indicative of the number of assumptions needed to derive the model from the basic starting point of (6.10).

In summary, in this section we have derived differential equations for the mean concentration of reactive species in turbulence from both the eulerian and lagrangian points of view. Although the two basic equations (6.14) and (6.70) look similar and have similar conditions of validity, the basic paths of their derivations were different. The study of turbulent diffusion is generally organized along the two lines of eulerian and lagrangian theories. The eulerian treatments often culminate in (6.14) (or more complex representations based on higher-order closure methods) whereas the lagrangian developments are generally concerned with the statistics of the motion of marked particles and do not usually attempt to obtain a differential equation for the ensemble mean concentration of a swarm of particles.

A major aspect of applying (6.14) to atmospheric diffusion problems is the solution of the boundary-value problems for $\langle c \rangle$ that arise for different source configurations, boundary conditions, and wind and eddy diffusivity profiles. Section 6.3 is, therefore, devoted to the solution of (6.14).

The lagrangian development showed that under certain idealized conditions the mean concentration from instantaneous or continuous sources would have gaussian distributions in which the variances of the distributions depend on the time of travel

Table 6.2 RESTRICTIONS ON THE APPLICABILITY OF VARIOUS TURBULENT DIFFUSION MODELS

	Model				
Parameter	(6.10)	(6.48)	(6.14) or (6.70)	Gaussian puff model (6.38c)†	Gaussian plume model (6.39)†
1 Coordinate system	None	None	Principal axes of $\overline{K}_{jk}$	Principal axes of the covariance **P**	Principal axes of the covariance **P**
2 Source-strength function	None	None	(6.49), (6.51)	$S(\mathbf{x})\,\delta(t-t_0)$	$S\,\delta(\mathbf{x}-\mathbf{x}_0)$, $S \neq f(t)$
3 Initial concentration distribution $\langle c(\mathbf{x},t_0)\rangle$	None	None	(6.51)	Arbitrary (although usually equal to zero)	$\langle c(\mathbf{x},t_0)\rangle = 0$
4 Chemical reactions	$k \neq f(c)$	$k \neq f(c)$	$\frac{1}{\Delta t} \gg \begin{cases} k(t) \\ \frac{\partial \ln k(t)}{\partial t} \end{cases}$	First-order decay or ground absorption	First-order decay or ground absorption
5 Turbulence u_i'	None	(6.52), (6.54)	(6.52), (6.54)	Stationary and homogeneous	Stationary and homogeneous
6 Deterministic velocity $\bar{u}_i$	None	(6.53), (6.55)	(1) (6.53), (6.55) (2) $\bar{u}_j \ll \left(\frac{\max \overline{K}_{jj}}{\Delta t}\right)^{1/2}$	$\frac{\partial \ln \bar{u}_i}{\partial x_k} \ll \frac{1}{D_k}$ D_k = puff diameter	$\bar{u}_i = \text{const}$
7 Resolvable time and space scales T and L_i	Arbitrary	$T > 2\,\Delta t$ arbitrary L_i	$T \gg \Delta t$ $L_j \gg (\overline{K}_{jj}\,\Delta t)^{1/2}$	Arbitrary	$T = \infty$ (i.e. steady state) L_i = arbitrary

SOURCE: Lamb (1971).

† Restrictions also apply to the form which includes reflection of the $x_3 = 0$ plane.

from the source. We have not as yet discussed how these variances depend on travel time, although clearly these dependences are of importance in estimating the rate of spreading of a plume. Information on the rate of spreading of particles in turbulence can be obtained from the classic statistical theory of turbulent diffusion, which is the subject of Sec. 6.4.

Comments on the General Form of the K Theory

The general form of the K theory for turbulent diffusion is

$$\begin{aligned} q_x &= -K_{xx}\frac{\partial\langle c\rangle}{\partial x} - K_{xy}\frac{\partial\langle c\rangle}{\partial y} - K_{xz}\frac{\partial\langle c\rangle}{\partial z} \\ q_y &= -K_{yx}\frac{\partial\langle c\rangle}{\partial x} - K_{yy}\frac{\partial\langle c\rangle}{\partial y} - K_{yz}\frac{\partial\langle c\rangle}{\partial z} \\ q_z &= -K_{zx}\frac{\partial\langle c\rangle}{\partial x} - K_{zy}\frac{\partial\langle c\rangle}{\partial y} - K_{zz}\frac{\partial\langle c\rangle}{\partial z} \end{aligned} \tag{6.71}$$

where q_x, q_y, q_z are shorthand notations for the turbulent mass fluxes $\langle u'c'\rangle$, $\langle v'c'\rangle$, $\langle w'c'\rangle$ relative to a coordinate system (x,y,z). The eddy diffusivities K_{xx}, etc. are the elements of a symmetric tensor. We note that the above relations represent simply a linear transformation of the gradient vector

$$\nabla\langle c\rangle = \left[\frac{\partial\langle c\rangle}{\partial x}\ \frac{\partial\langle c\rangle}{\partial y}\ \frac{\partial\langle c\rangle}{\partial z}\right]$$

to the flux vector $\mathbf{q} = [q_x q_y q_z]$. However the direction of the vector $\nabla\langle c\rangle$ differs in general from that of the mass flux vector. Since $\mathbf{K}$ is symmetric, there must exist some other orthogonal set of axes (ξ,η,ζ) according to which

$$\begin{aligned} q_\xi &= -K_{\xi\xi}\frac{\partial\langle c\rangle}{\partial \xi} \\ q_\eta &= -K_{\eta\eta}\frac{\partial\langle c\rangle}{\partial \eta} \\ q_\zeta &= -K_{\zeta\zeta}\frac{\partial\langle c\rangle}{\partial \zeta} \end{aligned} \tag{6.72}$$

These axes are called the *principal axes* of the tensor $\mathbf{K}$. Essentially all this means is that the coordinate axes can be rotated to a point where the fluxes in each of the new coordinate directions depend only on the gradients in each of the directions. Later we shall show how to determine these axes, given $\mathbf{K}$.

Here our objective is to determine K_{xx} etc., in terms of $K_{\xi\xi}$, $K_{\eta\eta}$, and $K_{\zeta\zeta}$. We can relate the fluxes in the two coordinate systems by

$$\begin{aligned} q_x &= l_1 q_\xi + l_2 q_\eta + l_3 q_\zeta \\ q_y &= m_1 q_\xi + m_2 q_\eta + m_3 q_\zeta \\ q_z &= n_1 q_\xi + n_2 q_\eta + n_3 q_\zeta \end{aligned} \tag{6.73}$$

where the l_i, m_i, and n_i are the direction cosines relating the two coordinate systems. Thus, the gradients relative to the two systems are related by

$$\begin{aligned}
\frac{\partial\langle c\rangle}{\partial x} &= l_1 \frac{\partial\langle c\rangle}{\partial \xi} + l_2 \frac{\partial\langle c\rangle}{\partial \eta} + l_3 \frac{\partial\langle c\rangle}{\partial \zeta} \\
\frac{\partial\langle c\rangle}{\partial y} &= m_1 \frac{\partial\langle c\rangle}{\partial \xi} + m_2 \frac{\partial\langle c\rangle}{\partial \eta} + m_3 \frac{\partial\langle c\rangle}{\partial \zeta} \\
\frac{\partial\langle c\rangle}{\partial z} &= n_1 \frac{\partial\langle c\rangle}{\partial \xi} + n_2 \frac{\partial\langle c\rangle}{\partial \eta} + n_3 \frac{\partial\langle c\rangle}{\partial \zeta}
\end{aligned} \tag{6.74}$$

or, equivalently,

$$\begin{aligned}
\frac{\partial\langle c\rangle}{\partial \xi} &= l_1 \frac{\partial\langle c\rangle}{\partial x} + m_1 \frac{\partial\langle c\rangle}{\partial y} + n_1 \frac{\partial\langle c\rangle}{\partial z} \\
\frac{\partial\langle c\rangle}{\partial \eta} &= l_2 \frac{\partial\langle c\rangle}{\partial x} + m_2 \frac{\partial\langle c\rangle}{\partial y} + n_2 \frac{\partial\langle c\rangle}{\partial z} \\
\frac{\partial\langle c\rangle}{\partial \zeta} &= l_3 \frac{\partial\langle c\rangle}{\partial x} + m_3 \frac{\partial\langle c\rangle}{\partial y} + n_3 \frac{\partial\langle c\rangle}{\partial z}
\end{aligned} \tag{6.75}$$

We can combine (6.72), (6.73), and (6.75) and obtain

$$\begin{aligned}
-q_x &= (K_{\xi\xi} l_1{}^2 + K_{\eta\eta} l_2{}^2 + K_{\zeta\zeta} l_3{}^2) \frac{\partial\langle c\rangle}{\partial x} + (K_{\xi\xi} l_1 m_1 + K_{\eta\eta} l_2 m_2 + K_{\zeta\zeta} l_3 m_3) \frac{\partial\langle c\rangle}{\partial y} \\
&\quad + (K_{\xi\xi} l_1 n_1 + K_{\eta\eta} l_2 n_2 + K_{\zeta\zeta} l_3 n_3) \frac{\partial\langle c\rangle}{\partial z} \\
-q_y &= (K_{\xi\xi} l_1 m_1 + K_{\eta\eta} l_2 m_2 + K_{\zeta\zeta} l_3 m_3) \frac{\partial\langle c\rangle}{\partial x} + (K_{\xi\xi} m_1{}^2 + K_{\eta\eta} m_2{}^2 + K_{\zeta\zeta} m_3{}^2) \frac{\partial\langle c\rangle}{\partial y} \\
&\quad + (K_{\xi\xi} m_1 n_1 + K_{\eta\eta} m_2 n_2 + K_{\zeta\zeta} m_3 n_3) \frac{\partial\langle c\rangle}{\partial z} \\
-q_z &= (K_{\xi\xi} l_1 n_1 + K_{\eta\eta} l_2 n_2 + K_{\zeta\zeta} l_3 n_3) \frac{\partial\langle c\rangle}{\partial x} + (K_{\xi\xi} m_1 n_1 + K_{\eta\eta} m_2 n_2 + K_{\zeta\zeta} m_3 n_3) \frac{\partial\langle c\rangle}{\partial y} \\
&\quad + (K_{\xi\xi} n_1{}^2 + K_{\eta\eta} n_2{}^2 + K_{\zeta\zeta} n_3{}^2) \frac{\partial\langle c\rangle}{\partial z}
\end{aligned} \tag{6.76}$$

Therefore, only if the axes (x,y,z) coincide with the principal axes (ξ,η,ζ) is the proper form of the atmospheric diffusion equation

$$\begin{aligned}
\frac{\partial\langle c\rangle}{\partial t} + \bar{u} \frac{\partial\langle c\rangle}{\partial \xi} + \bar{v} \frac{\partial\langle c\rangle}{\partial \eta} + \bar{w} \frac{\partial\langle c\rangle}{\partial \zeta} &= \frac{\partial}{\partial \xi}\left(K_{\xi\xi} \frac{\partial\langle c\rangle}{\partial \xi}\right) + \frac{\partial}{\partial \eta}\left(K_{\eta\eta} \frac{\partial\langle c\rangle}{\partial \eta}\right) \\
&\quad + \frac{\partial}{\partial \zeta}\left(K_{\xi\xi} \frac{\partial\langle c\rangle}{\partial \zeta}\right)
\end{aligned} \tag{6.77}$$

As an example, we consider diffusion from an infinite ground-level crosswind line source as shown in Fig. 6.6. The coordinate axes (x,z) are taken parallel and perpendicular to the ground as usual. However, let us assume that turbulence measurements have established that the principal axes of $\mathbf{K}$ are ξ and ζ, as shown in Fig. 6.6, where the (ξ,ζ) coordinate system is rotated an angle β

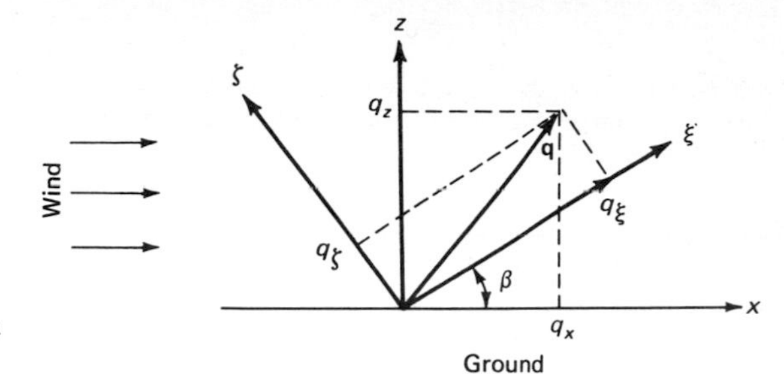

FIGURE 6.6
Relationship of turbulent flux vectors in (x,z) and (ξ,ζ) coordinate systems.

from the (x,z) system. The fluxes in the x and z directions are related to those in the ξ and ζ directions by

$$q_x = q_\xi \cos\beta - q_\zeta \sin\beta$$
$$q_z = q_\xi \sin\beta + q_\zeta \cos\beta$$

Thus,

$$q_x = -K_{\xi\xi}\cos\beta \frac{\partial\langle c\rangle}{\partial\xi} + K_{\zeta\zeta}\sin\beta \frac{\partial\langle c\rangle}{\partial\zeta}$$

$$q_z = -K_{\xi\xi}\sin\beta \frac{\partial\langle c\rangle}{\partial\xi} - K_{\zeta\zeta}\cos\beta \frac{\partial\langle c\rangle}{\partial\zeta}$$

Since

$$z = \xi \sin\beta = \zeta\cos\beta$$
$$x = \xi\cos\beta = -\zeta\sin\beta$$

we obtain

$$\frac{\partial\langle c\rangle}{\partial\xi} = \frac{\partial\langle c\rangle}{\partial x}\cos\beta + \frac{\partial\langle c\rangle}{\partial z}\sin\beta$$

$$\frac{\partial\langle c\rangle}{\partial\zeta} = -\frac{\partial\langle c\rangle}{\partial x}\sin\beta + \frac{\partial\langle c\rangle}{\partial z}\cos\beta$$

Thus, the turbulent fluxes in the x and z directions are

$$q_x = -(K_{\xi\xi}\cos^2\beta + K_{\zeta\zeta}\sin^2\beta)\frac{\partial\langle c\rangle}{\partial x} - (K_{\xi\xi} - K_{\zeta\zeta})\cos\beta\sin\beta\frac{\partial\langle c\rangle}{\partial z}$$

$$q_z = -(K_{\xi\xi} - K_{\zeta\zeta})\sin\beta\cos\beta\frac{\partial\langle c\rangle}{\partial x} - (K_{\xi\xi}\sin^2\beta + K_{\zeta\zeta}\cos^2\beta)\frac{\partial\langle c\rangle}{\partial z}$$

and the proper form of the turbulent diffusion equation is

$$\bar{u}\frac{\partial\langle c\rangle}{\partial x} = \frac{\partial}{\partial x}\left[(K_{\xi\xi}\cos^2\beta + K_{\zeta\zeta}\sin^2\beta)\frac{\partial\langle c\rangle}{\partial x} + (K_{\xi\xi} - K_{\zeta\zeta})\cos\beta\sin\beta\frac{\partial\langle c\rangle}{\partial z}\right]$$
$$+ \frac{\partial}{\partial z}\left[(K_{\xi\xi} - K_{\zeta\zeta})\sin\beta\cos\beta\frac{\partial\langle c\rangle}{\partial x} + (K_{\xi\xi}\sin^2\beta + K_{\zeta\zeta}\cos^2\beta)\frac{\partial\langle c\rangle}{\partial z}\right]$$

Clearly, only if $\beta = 0$ does this form reduce to that commonly used.

The final issue relates to the means of determining the principal axes of a given eddy diffusivity tensor that has been determined from, say, turbulence measurements. To illustrate this we require some basic concepts of matrix theory. Let us consider a two-dimensional case where, from measurements, we have established that the eddy diffusivity tensor with respect to a conventional (x,z) coordinate system is

$$\begin{bmatrix} K_{xx} & K_{xz} \\ K_{xz} & K_{zz} \end{bmatrix} = \begin{bmatrix} K_{11} & K_{12} \\ K_{12} & K_{22} \end{bmatrix}$$

Given a symmetric matrix $\mathbf{K}$, we can always represent $\mathbf{K}$ as

$$\mathbf{K} = \mathbf{V}\boldsymbol{\Lambda}\mathbf{V}^{-1}$$

where $\mathbf{V}$ is the 2×2 matrix whose columns are the eigenvectors of $\mathbf{K}$,

$$\mathbf{V} = \begin{bmatrix} v_1^{(1)} & v_1^{(2)} \\ v_2^{(1)} & v_2^{(2)} \end{bmatrix}$$

and $\boldsymbol{\Lambda}$ is the 2×2 diagonal matrix of the eigenvalues of $\mathbf{K}$:

$$\boldsymbol{\Lambda} = \begin{bmatrix} \lambda_1 & 0 \\ 0 & \lambda_2 \end{bmatrix}$$

The eigenvalues of $\mathbf{K}$ are found from the characteristic equation of $\mathbf{K}$:

$$\det\,[\mathbf{K} - \lambda\mathbf{I}] = 0$$

In this case the two eigenvalues of $\mathbf{K}$ are

$$\lambda_{1,2} = \tfrac{1}{2}\{(K_{11} + K_{22}) \pm [(K_{11} + K_{22})^2 - 4(K_{11}K_{22} - K_{12}^2)]^{1/2}\} \qquad (6.78)$$

The eigenvectors $\mathbf{v}^{(1)}$ and $\mathbf{v}^{(2)}$ are found from

$$\mathbf{K}\mathbf{v}^{(i)} = \lambda_i \mathbf{v}^{(i)} \qquad i = 1,2 \qquad (6.79)$$

Writing (6.79) in component form, we have

$$K_{11}v_1^{(i)} + K_{12}v_2^{(i)} = \lambda_i v_1^{(i)} \qquad (6.80)$$

$$K_{12}v_1^{(i)} + K_{22}v_2^{(i)} = \lambda_i v_2^{(i)} \qquad i = 1,2 \qquad (6.81)$$

It is easy to see from these two equations that the eigenvectors are not unique, since we can obtain only the ratio of $v_2^{(i)}$ to $v_1^{(i)}$ from each equation and not the individual values $v_1^{(i)}$ and $v_2^{(i)}$. Using (6.80) we obtain

$$\frac{v_2^{(i)}}{v_1^{(i)}} = \frac{\lambda_i - K_{11}}{K_{12}} \qquad (6.82)$$

and using (6.81) we obtain

$$\frac{v_2^{(i)}}{v_1^{(i)}} = \frac{K_{12}}{\lambda_i - K_{22}} \qquad (6.83)$$

By substituting either λ_i from (6.78) we will find that the right-hand sides of these two equations are identical. Thus, let us use only the first equation (6.82). Since we know only the ratio $v_2^{(i)}/v_1^{(i)}$ it is customary to set $v_1^{(i)} = 1$. Thus,

$$\mathbf{V} = \begin{bmatrix} 1 & 1 \\ \dfrac{\lambda_1 - K_{11}}{K_{12}} & \dfrac{\lambda_2 - K_{11}}{K_{12}} \end{bmatrix}$$

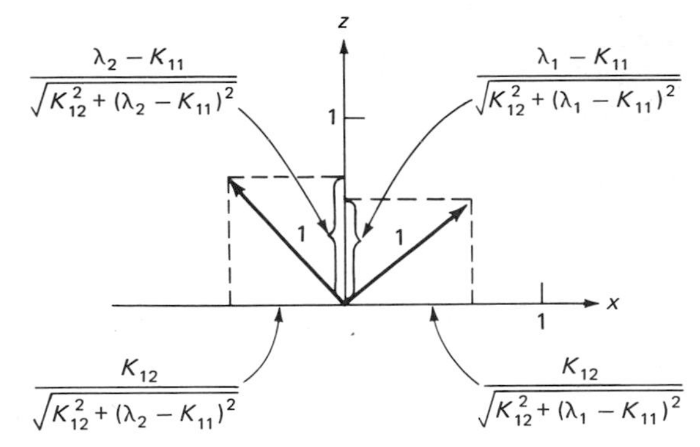

FIGURE 6.7
Principal axes of two-dimensional eddy diffusivity tensor.

Since $\mathbf{V}^{-1}\mathbf{K}\mathbf{V} = \mathbf{\Lambda}$, a diagonal matrix, it is now clear that $\lambda_1 = K_{\xi\xi}$ and $\lambda_2 = K_{\zeta\zeta}$. In addition, the two eigenvectors $\mathbf{v}^{(1)}$ and $\mathbf{v}^{(2)}$ are the principal axes of $\mathbf{K}$. (They are guaranteed to be orthogonal since $\mathbf{K}$ is symmetric.) In computing the principal axes for this two-dimensional example, it is customary to normalize the two eigenvectors so that their magnitude is unity. Doing so we obtain

$$\mathbf{V} = \begin{bmatrix} \dfrac{K_{12}}{\sqrt{K_{12}{}^2 + (\lambda_1 - K_{11})^2}} & \dfrac{K_{12}}{\sqrt{K_{12} + (\lambda_2 - K_{11})^2}} \\ \dfrac{\lambda_1 - K_{11}}{\sqrt{K_{12}{}^2 + (\lambda_1 - K_{11})^2}} & \dfrac{\lambda_2 - K_{11}}{\sqrt{K_{12}{}^2 + (\lambda_2 - K_{11})^2}} \end{bmatrix}$$

Therefore, the two columns of $\mathbf{V}$ are the two principal axes of $\mathbf{K}$; that is, the first and second elements of each eigenvector are the x and z components of the principal axes as shown in Fig. 6.7.

From a practical point of view the basic question is: What are the errors committed by neglecting the off-diagonal terms in (6.11), that is, by assuming that the coordinate axes Ox, Oy, and Oz are the principal axes of the eddy diffusivity tensor $\mathbf{K}$? In general, it is unlikely in an actual atmospheric situation with spatially and temporally varying wind speeds and directions that this condition will be fulfilled. In fact, because of the repeated changes in wind direction, in such a situation one cannot distinguish between K_{xx} and K_{yy}. Thus, we must set $K_{xx} = K_{yy}$ due to lack of well-defined downwind and crosswind directions.

Let us consider a two-dimensional turbulent shear flow. The turbulent mass flux of species in the direction of the mean flow (the x direction) is given by

$$\langle u'c' \rangle = -K_{xx}\frac{\partial\langle c \rangle}{\partial x} - K_{xz}\frac{\partial\langle c \rangle}{\partial z}$$

We can obtain some guidance as to the importance of the off-diagonal terms from the analogous heat transfer situation, for which more experimental data are available (Monin and Yaglom, 1971). Turbulent heat transfer in the direction of the mean flow is given by $\langle u'\theta' \rangle$. It has been found that $\langle u'\theta' \rangle$ depends on the vertical temperature gradient and that

$$\langle u'\theta' \rangle = \begin{cases} >0 & \partial\bar{\theta}/\partial z > 0 \\ <0 & \partial\bar{\theta}/\partial z < 0 \end{cases}$$

Not only does $\langle u'\theta' \rangle$ depend on $\partial\bar{\theta}/\partial z$, but $\langle u'\theta' \rangle$ is large (in near neutral conditions, $\langle u'\theta' \rangle$ is three times as great as $\langle w'\theta' \rangle$). Assuming that heat and material are transported by the same processes, we would expect K_{xz} to be about three times as large as K_{zz}. However, the fact that K_{xz} is larger in magnitude than K_{zz} does not necessarily mean that the term

$$\frac{\partial}{\partial x}\left(K_{xz}\frac{\partial\langle c\rangle}{\partial z}\right)$$

will be larger than the term

$$\frac{\partial}{\partial z}\left(K_{zz}\frac{\partial\langle c\rangle}{\partial z}\right)$$

in the atmospheric diffusion equation. The former term describes the flux of species in the direction of the mean wind due to vertical gradients in the mean concentration; the latter term describes the vertical flux due to vertical gradients. Thus, the two terms

$$\frac{\partial}{\partial x}\left(K_{xx}\frac{\partial\langle c\rangle}{\partial x}\right)+\frac{\partial}{\partial x}\left(K_{xz}\frac{\partial\langle c\rangle}{\partial z}\right)$$

may, in fact, *both* be small when compared with the convective term $\bar{u}\,\partial\langle c\rangle/\partial x$. Walters (1969) considered the importance of retaining the K_{xx} term in computing the concentration distribution downwind of a continuous, ground-level crosswind line source, and he found that the term may generally be neglected relative to $\bar{u}\,\partial\langle c\rangle/\partial x$, except under conditions of very light winds. However, if conditions are such that the term involving K_{xx} is to be retained, there is no reason why the term containing K_{xz} should not also be retained. In addition, if $K_{xz}\neq 0$, then $K_{zx}\neq 0$.

On the basis of heat transfer measurements and on physical grounds, we expect K_{xz} (and K_{zx}) to be negative. It is usually assumed that $K_{zx}=-K_{xx}/3$. Thus, in two dimensions with a mean flow in the x direction the general form of the atmospheric diffusion equation is

$$\frac{\partial\langle c\rangle}{\partial t}+\bar{u}\frac{\partial\langle c\rangle}{\partial x}=\frac{\partial}{\partial x}\left(K_{xx}\frac{\partial\langle c\rangle}{\partial x}\right)+\frac{\partial}{\partial x}\left(K_{xz}\frac{\partial\langle c\rangle}{\partial z}\right)+\frac{\partial}{\partial z}\left(K_{zz}\frac{\partial\langle c\rangle}{\partial z}\right)+\frac{\partial}{\partial z}\left(K_{zx}\frac{\partial\langle c\rangle}{\partial x}\right)$$

The last term on the right-hand side, like the second, is almost never retained, since the qualitative nature of the solution of the diffusion equation is not substantially altered by its inclusion (see Gee and Davies, 1963).

6.3 SOLUTIONS OF THE ATMOSPHERIC DIFFUSION EQUATION

Equation (6.14), the so called atmospheric diffusion equation, has been the basis of numerous theoretical and experimental investigations of pollutant dispersion in the atmosphere. In this section we present several solutions of (6.14) for inert pollutants under conditions of various source types and boundary conditions. Thus, the basic equation of this section is

$$\begin{aligned}\frac{\partial\langle c\rangle}{\partial t}&+\bar{u}(x,y,z,t)\frac{\partial\langle c\rangle}{\partial x}+\bar{v}(x,y,z,t)\frac{\partial\langle c\rangle}{\partial y}+\bar{w}(x,y,z,t)\frac{\partial\langle c\rangle}{\partial z}\\&=\frac{\partial}{\partial x}\left[K_{xx}(x,y,z,t)\frac{\partial\langle c\rangle}{\partial x}\right]+\frac{\partial}{\partial y}\left[K_{yy}(x,y,z,t)\frac{\partial\langle c\rangle}{\partial y}\right]\\&\quad+\frac{\partial}{\partial z}\left[K_{zz}(x,y,z,t)\frac{\partial\langle c\rangle}{\partial z}\right]+S(x,y,z,t)\qquad(6.84)\end{aligned}$$

Table 6.3 SOLUTIONS OF $\frac{\partial\langle c\rangle}{\partial t} + U\frac{\partial\langle c\rangle}{\partial x} = K_{xx}\frac{\partial^2\langle c\rangle}{\partial x^2} + K_{yy}\frac{\partial^2\langle c\rangle}{\partial y^2} + K_{zz}\frac{\partial^2\langle c\rangle}{\partial z^2}$

Conditions	Solution
Arbitrary initial distribution $\langle c(x_0,y_0,z_0)\rangle$, unbounded atmosphere	$\langle c(x,y,z,t)\rangle = \frac{1}{(2\pi)^{3/2}[P_{xx}(t)P_{yy}(t)P_{zz}(t)]^{1/2}} \cdot \int_{-\infty}^{\infty}\int_{-\infty}^{\infty}\int_{-\infty}^{\infty} \exp\left[-\frac{(x-x_0-Ut)^2}{2P_{xx}(t)} - \frac{(y-y_0)^2}{2P_{yy}(t)} - \frac{(z-z_0)^2}{2P_{zz}(t)}\right]\langle c(x_0,y_0,z_0)\rangle\, dx_0\, dy_0\, dz_0$ $P_{xx}(t) = 2K_{xx}t \qquad P_{yy}(t) = 2K_{yy}t \qquad P_{zz}(t) = 2K_{zz}t$
Instantaneous point source $\langle c(x_0,y_0,z_0)\rangle = \delta(x_0)\,\delta(y_0)\,\delta(z_0)$, unbounded atmosphere	$\langle c(x,y,z,t)\rangle = \frac{1}{(2\pi)^{3/2}[P_{xx}(t)P_{yy}(t)P_{zz}(t)]^{1/2}} \exp\left[-\frac{(x-x_0-Ut)^2}{2P_{xx}(t)} - \frac{(y-y_0)^2}{2P_{yy}(t)} - \frac{(z-z_0)^2}{P_{zz}(t)}\right]$
Continuous point source at the origin, unbounded atmosphere	$\langle c(x,y,z)\rangle = \frac{1}{4\pi}(K_{yy}K_{zz}x^2 + K_{xx}K_{zz}y^2 + K_{xx}K_{yy}z^2)^{-1/2} \exp\left\{-\frac{U}{2K_{xx}}\left[K_{xx}{}^{1/2}\left(\frac{x^2}{K_{xx}} + \frac{y^2}{K_{yy}} + \frac{z^2}{K_{zz}}\right)^{1/2} - x\right]\right\}$ When $x \gg \sqrt{y^2+z^2}$, (6.85) $\langle c(x,y,z)\rangle = \frac{1}{4\pi(K_{yy}K_{zz})^{1/2}x} \exp\left[-\frac{U}{4x}\left(\frac{y^2}{K_{yy}} + \frac{z^2}{K_{zz}}\right)\right]$

(continued)

Table 6.3—*continued*

Conditions	Solution	
Continuous line source on the y axis, unbounded atmosphere	$\langle c(x,z)\rangle = \frac{1}{2\pi(K_{xx}K_{zz})^{1/2}} K_0 \left[\frac{U}{2}\left(\frac{x^2}{K_{xx}^2}+\frac{z^2}{K_{xx}K_{zz}}\right)^{1/2}\right] e^{Ux/2K_{xx}}$ $K_0(\cdot)$ = modified Bessel function of the third kind. When $Ux \gg K_{xx}$ and $x \gg z$, $\langle c(x,z)\rangle = \frac{1}{2(\pi K_{zz}Ux)^{1/2}} e^{-Uz^2/4K_{zz}x}$	(6.86)
Continuous point source at $(0,0,h)$ reflecting plane at $z=0 (K_{xx}=0)$	$\langle c(x,y,z)\rangle = \frac{1}{4\pi x(K_{yy}K_{zz})^{1/2}} e^{-Uy^2/4K_{yy}x}(e^{-U(z-h)^2/4K_{zz}x}+e^{-U(z+h)^2/4K_{zz}x})$	(6.87)
Continuous line source at $x=0$ and $z=h (K_{xx}=0)$, reflecting plane at $z=0$	$\langle c(x,z)\rangle = \frac{1}{2(\pi UK_{zz}x)^{1/2}}(e^{-U(z-h)^2/4K_{zz}x}+e^{-U(z+h)^2/4K_{zz}x})$	(6.88)

the solution of which we will investigate for different source and boundary conditions. (We use x,y,z notation in this section to conform with the standard literature notation on this subject.)

The simplest case of (6.84) is that in which $\bar{u} = U$, $\bar{v} = \bar{w} = 0$, and K_{xx}, K_{yy}, and K_{zz} are constants independent of position. A number of solutions of (6.84) corresponding to this case are presented in Table 6.3. The assumption of constant $\bar{u}$ and K_{xx}, K_{yy}, and K_{zz} is tantamount to specifying stationary, homogeneous turbulence. The solutions presented in Table 6.3 predict certain dependences of the concentration at $z = 0$ on downwind distance x. Table 6.4 presents a comparison of these predicted dependences and those actually observed. It is clear that the agreement between predicted and observed dependences of mean concentration on x from continuous sources is poor. One must conclude that the assumptions on which the solutions in Table 6.3 are based are too restrictive to permit accurate prediction of downwind concentrations in the atmosphere. This conclusion is not unexpected, since it is well known that atmospheric turbulence is decidedly inhomogeneous, especially in the vertical direction, and that the wind speed varies with height above the ground. Since the eddy diffusivities are merely empirical coefficients, it is natural to allow these parameters to vary with position in an effort to have the predictions of (6.84) more closely match observed behavior. Unfortunately, when the coefficients of (6.84) are allowed to vary with position, the analytical solution of (6.84) becomes difficult. Consequently, solutions of (6.84) for variable wind speeds and eddy diffusivities are available only in a few specialized cases.

Since relatively few solutions of (6.84) are available for instantaneous sources we shall devote most of our attention to continuous sources. The concentration distribution downwind of a continuous, crosswind line source at a height h emitting at a rate S (g/m-sec) is governed by

$$\begin{aligned} \bar{u}(z)\frac{\partial\langle c\rangle}{\partial x} &= \frac{\partial}{\partial z}\left[K_{zz}(z)\frac{\partial\langle c\rangle}{\partial z}\right] \\ \langle c(0,z)\rangle &= (S/\bar{u}(h))\delta(z-h) \\ -K_{zz}(0)\frac{\partial\langle c\rangle}{\partial z}\bigg|_{z=0} &= 0 \qquad z=0,\ x>0 \\ \langle c(x,z)\rangle &= 0 \qquad z\to\infty \\ \int_0^\infty \bar{u}(z)\langle c(x,z)\rangle\,dz &= S \end{aligned} \tag{6.89}$$

where diffusion in the x direction has been neglected relative to convection. Commonly used representations of the wind velocity and vertical eddy diffusivity [recall (5.76)] are

$$\bar{u}(z) = u_1 z^m \qquad K_{zz} = K_1 z^n$$

The solution of (6.89) for a ground-level source ($h = 0$) with these power-law forms was obtained by Roberts (see Calder, 1949):

$$\langle c(x,z)\rangle = \frac{(m-n+2)S}{u_1\Gamma(s)}\left[\frac{u_1}{(m-n+2)^2K_1x}\right]^s e^{-u_1z^{m-n+2}/(m-n+2)^2K_1x} \tag{6.90}$$

where $s = (m+1)/(m-n+2)$, $\Gamma(s)$ is the gamma function, and $m - n + 2 > 0$. The solution of (6.89) for an elevated source was obtained by Smith (1957):

$$\langle c(x,z)\rangle = \frac{S(hz)^{(1-n)/2}}{(m-n+2)K_1x}\exp\left[-\frac{u_1(z^{m-n+2}+h^{m-n+2})}{(m-n+2)^2K_1x}\right] \times I_v\left[\frac{2u_1(hz)^{(m-n+2)/2}}{(m-n+2)^2K_1x}\right] \tag{6.91}$$

where I_v is the modified Bessel function of the first kind of order $v = -(1-n)/(m-n+2)$.

Solutions (6.90) and (6.91) have been obtained on the assumption that horizontal turbulent diffusion is negligible compared with advection. The validity of this common assumption was investigated by Walters (1969) for the case of a continuous, ground-level crosswind line source under the conditions

$$\bar{u}(z) = U = \text{const} \qquad K_{xx} = K_0 z \qquad K_{zz} = K_1 z$$

Thus, Walters solved the problem:

$$U\frac{\partial\langle c\rangle}{\partial x} = K_0 z\frac{\partial^2\langle c\rangle}{\partial x^2} + \frac{\partial}{\partial z}\left(K_1 z\frac{\partial\langle c\rangle}{\partial z}\right)$$

$$\langle c(0,z)\rangle = SU\,\delta(z)$$

$$-K_{zz}(0)\left.\frac{\partial\langle c\rangle}{\partial z}\right|_{z=0} = 0 \qquad z = 0,\ x > 0$$

$$\langle c(x,z)\rangle = 0 \qquad x \to \pm\infty, \qquad z \to \infty$$

$$\int_{-\infty}^{\infty}\left(-K_{zz}\frac{\partial\langle c\rangle}{\partial z}\right)dx = S$$

Table 6.4 COMPARISON OF DOWNWIND DECAY DEPENDENCE OF GROUND-LEVEL CONCENTRATION PREDICTED IN TABLE 6.3 AND OBSERVED

Source	Predicted in Table 6.3	Observed (neutral conditions)
Continuous point	$\sim x^{-1}$	$\sim x^{-2}$
Continuous line	$\sim x^{-1/2}$	$\sim x^{-1}$

The solution is

$$\langle c(x,z)\rangle = \frac{S}{K_1(1-e^{-\lambda\pi})}\frac{e^{-\lambda \tan^{-1}(\mu z/x)}}{(x^2+\mu^2 z^2)^{1/2}} \tag{6.92}$$

where $\lambda = U/(K_0K_1)^{1/2}$ and $\mu = (K_0/K_1)^{1/2}$. When horizontal diffusion is neglected in (6.91), the solution is [the $m = 0$, $n = 1$ case of (6.90)]

$$\langle c(x,z)\rangle = \frac{S}{K_1 x} e^{-Uz/K_1 x} \tag{6.93}$$

When $\mu z/x \ll 1$, that is, sufficiently close to the ground, the functional dependences of concentration on x and z are the same in the two cases. However, the ratio of the predicted magnitudes of the concentration varies from 1 (when $K_0 = 0$) to 2 (when $K_0 \to \infty$). In general, the addition of the horizontal diffusion term has little effect on the predicted ground-level concentrations except when the wind speed is very small, that is $\lambda \to 0$. Problem 6.4 further elaborates the effect of horizontal diffusion on concentration distributions downwind of a continuous, ground-level line source.

The solution of (6.84) for the mean concentration downwind of a continuous point source, i.e., of

$$\bar{u}(z)\frac{\partial\langle c\rangle}{\partial x} = K_{yy}\frac{\partial^2\langle c\rangle}{\partial y^2} + \frac{\partial}{\partial z}\left(K_{zz}\frac{\partial\langle c\rangle}{\partial z}\right) \tag{6.94}$$

for arbitrary source height and unrestricted values of m, n, and q in

$$\bar{u}(z) = u_1 z^m \qquad K_{zz} = K_1 z^n \qquad K_{yy} = K_2 z^q$$

has not yet been obtained. In fact, even in the surface layer case, in which $n = 1 - m$, and assuming isotropic turbulence so that $q = n$, a general solution of (6.93) has not been obtained. Smith (1957), however, solved for the ground-level concentration from a continuous, elevated source on the basis of the single assumption that the concentration profile in the y direction is gaussian, an assumption suggested by the way in which y variations appear in (6.94). (The restriction that the solution is for the ground level only is removed when the source itself is at ground level.) The ground-level solution has the form ($z = -h$ is taken as ground level and $z = 0$ as the source height)

$$\langle c(x,y,-h)\rangle = X(x)e^{-y^2/f(x)} \tag{6.95}$$

where $f(x)$ and $X(x)$ are related to the zeroth and second moments of the concentration distribution

$$c_0(x) = \int_{-\infty}^{\infty} \langle c(x,y,-h)\rangle\, dy$$

$$c_2(x) = \int_{-\infty}^{\infty} y^2\langle c(x,y,-h)\rangle\, dy$$

as follows:

$$f(x) - \frac{2c_2(x)}{c_0(x)} \qquad X(x) = c_0(x)\sqrt{\frac{c_0(x)}{2\pi c_2(x)}}$$

Smith shows that

$$c_0(x) = \frac{1}{(1+2m)^{1/(1+2m)}[-m/(1+2m)]!}\, x^{-(1+m)/(1+2m)} e^{-h^{1+2m}/(1+2m)^2 x}$$

$$c_2(x) = 2\sqrt{\frac{K_1}{K_0}\frac{{K_0}^{b-a}}{{u_0}^{b-a+1}}}\,(1+2m)^{(3b-4)/2}\frac{(b-1)!\,(b+a-2)!}{(a-1)!\,(2b-1)!}\,x^{b-1}$$

$$\times\, e^{-\eta}\left[\frac{(b-1)!}{(a-1)!}\,{}_1F_1(b;a;\eta) - \eta^b V(b;a;\eta)\right] \tag{6.96}$$

where $\qquad b = \dfrac{2+\mu}{1+2m} \qquad a = \dfrac{1+m}{1+2m} \qquad \eta = \dfrac{u_0 h^{1+2m}}{(1+2m)^2 K_0 x}$

$$V(b;a;\eta) = \sum_{r=0}^{\infty} \frac{(2b+r-1)!}{(b+r)!\,(b+a+r-1)!}\,\eta^r$$

and ${}_1F_1(b;a;\eta)$ is a confluent hypergeometric function (Abramowitz and Segun, 1965, p. 504). The solution (6.94) and (6.95) has the x dependence

$$\langle c \rangle \sim x^{-(m+5/2)/(2m+1)}$$

which for neutral conditions ($m = \frac{1}{7}$) gives $\langle c \rangle \sim x^{-5/3}$, agreeing reasonably well with observed data (see Table 6.4).

Mean Concentration from a Nonuniform Ground-level Area Source

Let us consider the prediction of the mean concentration as a function of time and location in an atmosphere above a spatially and temporally varying ground-level area source of finite extent. We assume that the transport and diffusion processes in the atmosphere may be adequately characterized by a constant mean wind $\bar{u} = U$ blowing in the x direction and constant horizontal and vertical eddy diffusivities K_H and K_V. The source strength at the ground is given by $S(x,y,t)$ g/m²-sec. We wish to compute $\langle c(x,y,z,t) \rangle$.

This solution is most easily developed by obtaining the solution for an instantaneous point source at, say, the origin and then using the principle of superposition to obtain the mean concentration resulting from the continuous distribution of sources. The problem to be solved for the instantaneous point source of strength S_p grams is given by

$$\frac{\partial\langle c\rangle}{\partial t} + U\frac{\partial\langle c\rangle}{\partial x} = K_H\left(\frac{\partial^2\langle c\rangle}{\partial x^2} + \frac{\partial^2\langle c\rangle}{\partial y^2}\right) + K_V\frac{\partial^2\langle c\rangle}{\partial z^2}$$

subject to

$$\langle c(x,y,z,0)\rangle = S_p\,\delta(x)\,\delta(y)\,\delta(z)$$

$$\langle c(x,y,z,t)\rangle = 0 \qquad x, y \to \pm\infty$$

$$-K_V\left.\frac{\partial\langle c\rangle}{\partial z}\right|_{z=0} = 0 \qquad z = 0$$

$$\langle c(x,y,z,t)\rangle = 0 \qquad z \to \infty$$

In solving this equation it is convenient to change the independent variables to $\xi = x - Ut$, $\eta = y$, $\rho = z$, and $\tau = t$, in which case it becomes

$$\frac{\partial\langle c\rangle}{\partial\tau} = K_H\left(\frac{\partial^2\langle c\rangle}{\partial\xi^2} + \frac{\partial^2\langle c\rangle}{\partial\eta^2}\right) + K_V\frac{\partial^2\langle c\rangle}{\partial\rho^2}$$

In ξ, η, ρ, τ coordinates this equation describes the dispersion of a puff of inert contaminant relative to its horizontal center of mass.

The solution of this equation may be obtained by standard methods and is

$$\langle c(\xi,\eta,\rho,\tau)\rangle = \frac{S_p}{4\pi^{3/2}K_H K_V{}^{1/2}\tau^{3/2}}\exp\left(-\frac{\xi^2+\eta^2}{4K_H\tau} - \frac{\rho^2}{4K_V\tau}\right)$$

As a first step in obtaining the solution for a spatially and temporally varying area source, we extend this result to the case of an instantaneous area source of strength $S_a(x,y)$. The instantaneous point source strength (S_p, in grams) can be replaced by $S_a\,dx\,dy$ (S_a, in grams per square meter), so that the mean concentration at (x,y,z) at time t resulting from the point source $S_a\,dx\,dy$ at location $x = \alpha$ and $y = \beta$ is given by

$$\langle c(x,y,z,t)\rangle = \frac{S_a\,dx\,dy}{4\pi^{3/2}K_H K_V{}^{1/2}t^{3/2}}\exp\left[-\frac{(x-Ut-\alpha)^2+(y-\beta)^2}{4K_H t} - \frac{z^2}{4K_V t}\right]$$

If the region over which the source is distributed is, say, $0 \le x \le X$, $0 \le y \le Y$, the mean concentration due to the instantaneous area source is

$$\langle c(x,y,z,t)\rangle = \frac{e^{-z^2/4K_V t}}{4\pi^{3/2}K_H K_V{}^{1/2}t^{3/2}}\int_0^X\int_0^Y \exp\left[-\frac{(x-Ut-\alpha)^2+(y-\beta)^2}{4K_H t}\right] S_a(\alpha,\beta)\,d\alpha\,d\beta$$

Finally, if we now consider the time-varying source $S(x,y,t)$, we may represent $S_a(x,y)$ as $S(x,y,t)\,dt$; that is, the instantaneous area source S_a is the contribution from the continuous area source S during a time interval dt. Therefore, the desired concentration at time t is given by

$$\langle c(x,y,z,t)\rangle = \frac{1}{4\pi^{3/2}K_H K_V{}^{1/2}}\int_0^t \frac{e^{-z^2/4K_V(t-\lambda)}}{(t-\lambda)^{3/2}}$$

$$\int_0^X\int_0^Y \exp\left[-\frac{(x-U(t-\lambda)-\alpha)^2+(y-\beta)^2}{4K_H(t-\lambda)}\right] S(\alpha,\beta,\lambda)\,d\alpha\,d\beta\,d\lambda$$

Although the solutions presented above are based on empirical power-law models for the turbulent eddy diffusivities, somewhat more universal forms of K_{zz} have been determined on the basis of the Monin and Obukhov similarity theory. Monin and Obukhov (1954) showed that the asymptotic dependence of K_{zz} in the surface layer should be

$$K_{zz}(z) \sim \begin{cases} z & \text{when } z \ll |L| \\ z^{4/3} & \text{when } z \gg -L \text{ and } L < 0 \\ \text{const} & \text{when } z \gg L \text{ and } L > 0 \end{cases} \tag{6.97}$$

where L is the Monin-Obukhov length. Monin and Yaglom (1971) indicate that the transition between the dynamic (shear-dominated) and thermal (buoyancy-dominated) sublayers in the case of unstable stratification ($L < 0$) is narrow and occurs at values of $-z/L$ between 0.03 and 0.05. In the case of stable stratification ($L > 0$), the transition probably occurs at values of z/L in the range 0.05 to 0.3.

On the basis of the above observations, Yordanov (1968) proposed the following models for K_{zz} in the surface layer:

$$\text{Unstable stratification:} \qquad K_{zz}(z) = \begin{cases} K_1 z^n & \text{when } z \leq aL \\ K_1 (aL)^{n-4/3} z^{4/3} & \text{when } z \geq aL \end{cases}$$

$$\text{Stable stratification:} \qquad K_{zz}(z) = \begin{cases} K_1 z^n & \text{when } z \leq aL \\ K_1 (aL)^n & \text{when } z \geq aL \end{cases} \tag{6.98}$$

For unstable stratification $a \cong -0.05$ and for stable stratification $a \cong 0.3$. The exponent n lies in the intervals:

Unstable: $1 < n \leq \frac{4}{3}$

Stable: $0 < n \leq 1$

For the entire planetary boundary layer Blackadar (1965) proposed that under adiabatic conditions

$$K_{zz}(z) = \begin{cases} \kappa u_* z & z \leq a_1 \lambda \\ \kappa u_* a_1 \lambda & z \geq a_1 \lambda \end{cases} \qquad a_1 \cong 0.1 \tag{6.99}$$

where $\lambda = \kappa u_*/f$, the height scale of the planetary boundary layer, and f is the Coriolis parameter.

The horizontal diffusivities K_{xx} and K_{yy} have been less frequently studied than K_{zz}. For most applied studies it is reasonable to replace these functions with constant values which reflect the average horizontal diffusivities in the layer.

6.4 STATISTICAL THEORIES OF TURBULENT DIFFUSION

Up to this point in this chapter we have developed the common theories of turbulent diffusion in a purely formal manner. We have done this so that the relationship of the approximate models for turbulent diffusion, such as the K theory and the gaussian puff and plume formulas, to the basic underlying theory is clearly evident. When such relationships are clear, the limitations inherent in each model can be appreciated. We have in some cases applied the models obtained to the prediction of the mean concentration resulting from an instantaneous or continuous source in idealized stationary, homogeneous turbulence or in the atmosphere. However, we have not discussed the physical processes responsible for the dispersion of a cloud or a plume other than to attribute the phenomenon to velocity fluctuations. A great deal of insight into the actual nature of turbulent diffusion can be gained by considering, in turn, the dispersion of an instantaneous puff and a continuous plume of pollutants. Such a consideration will also enable us to predict the statistical parameters of cloud and plume dispersion, such as the variances $\sigma_i^2(t)$, which are needed in the actual use of the gaussian dispersion formulas.

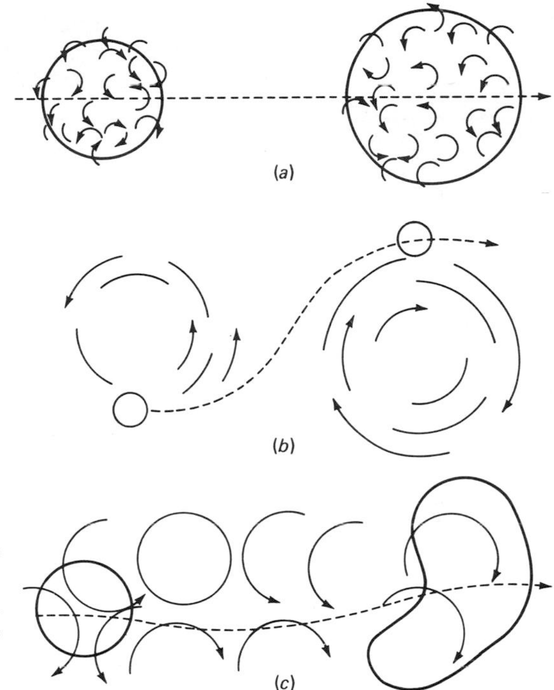

FIGURE 6.8
Dispersion of a puff under three different turbulence conditions: (*a*) turbulent eddies smaller than the puff; (*b*) turbulent eddies much larger than the puff; (*c*) turbulent eddies comparable in size to the puff (*Source: Slade*, 1968).

We begin with a brief discussion of the qualitative features of atmospheric diffusion. We then present analyses of the diffusion of one particle relative to a fixed axis and of two particles relative to each other. As we shall see, these two idealized situations provide the needed statistical information on the behavior of a plume or puff. We shall have occasion in this section to call upon some basic elements of the statistical description of turbulence, a short summary of which can be found in Appendix A.

6.4.1 Qualitative Features of Atmospheric Diffusion

The two idealized source types commonly used in atmospheric turbulent diffusion are the instantaneous point source and the continuous point source. An instantaneous point source is the conventional approximation to a rapid release of a quantity of material. Obviously, an "instantaneous point" is a mathematical idealization since any rapid release has finite spatial dimensions. As the puff is carried away from its source by the wind, it will disperse under the action of turbulent velocity fluctuations. Figure 6.8 shows the dispersion of a puff under three different turbulence conditions. Figure 6.8*a* shows a puff embedded in a turbulent field in which all the turbulent eddies

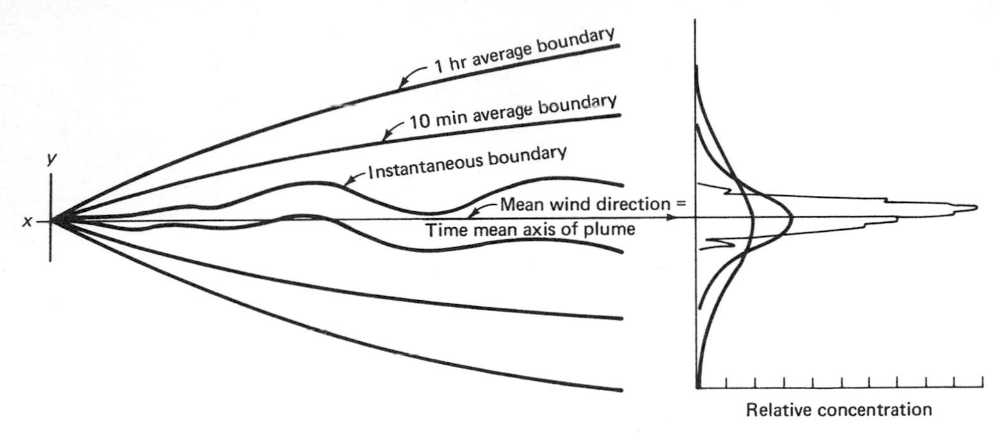

FIGURE 6.9
Plume boundaries as a function of averaging time (*Source: Slade*, 1968).

are smaller than the puff. The puff will disperse uniformly as the turbulent eddies at its boundary entrain fresh air. In Fig. 6.8*b*, a puff is embedded in a turbulent field all of whose eddies are considerably larger than the puff. In this case the puff will appear to the turbulent field as a small patch of fluid which will be convected through the field with little dilution. Ultimately, molecular diffusion will dissipate the puff. Figure 6.8*c* shows a puff in a turbulent field of eddies of size comparable to the puff. In this case the puff will be both dispersed and distorted. In the atmosphere, a cloud of material is always dispersed since there are almost always eddies of size smaller than the cloud. From Fig. 6.8 we can see that the dispersion of a puff relative to its center of mass depends on the initial size of the puff relative to the length scales of the turbulence. In order to describe such relative dispersion, we shall see that we must consider the statistics of the separation of *two* representative fluid particles in the puff. The analysis of the wandering of a single particle is insufficient to tell us about the dispersion of a cloud.

A continuous source emits a plume which might be envisioned, as we have noted, as an infinite number of puffs released sequentially with an infinitesimal time interval between puffs. The quantity of material released is expressed in terms of a rate, say grams per minute. The dimensions of a plume perpendicular to the plume axis are generally given in terms of the standard deviation of the mean concentration distribution (see Fig. 6.2) since the mean cross-sectional distributions are often nearly gaussian. Figure 6.9 shows the plume "boundaries" and concentration distributions as might be seen in an instantaneous snapshot and exposures of a few minutes and several hours. An instantaneous picture of a plume reveals a meandering behavior with the width of the plume gradually growing downwind of the source. Longer-time averages give a more regular appearance to the plume and a smoother concentration distribution.

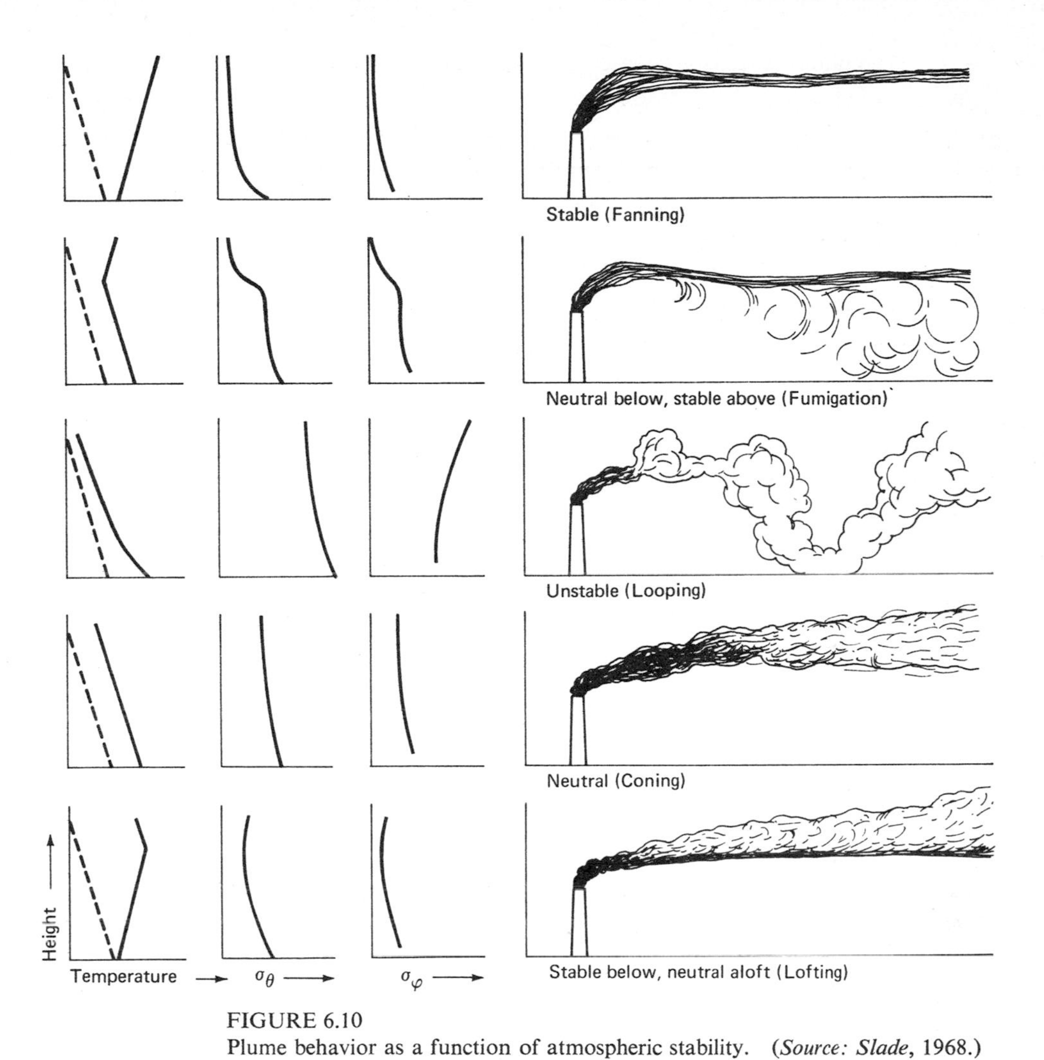

FIGURE 6.10
Plume behavior as a function of atmospheric stability. (*Source: Slade*, 1968.)

If we were to take a time exposure of the plume at large distances from the source, we would find that the boundaries of the time-averaged plume would begin to meander, because the plume would come under the influences of larger and larger eddies, and the averaging time, say several hours, would still be too brief to time-average adequately the effect of these larger eddies. Eddies larger in size than the plume dimension tend to transport the plume intact whereas those that are smaller tend to disperse it. As the plume becomes wider, larger and larger eddies become effective in dispersing the plume and the smaller eddies become increasingly ineffective.

As the plume widens, its centerline concentration decreases with travel distance. This rate of decrease depends on the wind profile and the stability condition of the atmosphere. Much of conventional gaussian plume theory has been concerned with estimating the proper functional form of the σ's with downwind distance so that predictions of ground-level concentrations downwind of a continuous source will match those measured experimentally. Various types of smoke plumes observed in the atmosphere are summarized in Fig. 6.10, together with temperature profiles and the standard deviations of the horizontal and vertical wind velocity components, σ_θ and σ_ϕ, respectively.

The theoretical analysis of the spread of a plume from a continuous point source can be achieved by considering the statistics of the diffusion of a *single* fluid particle relative to a fixed axis. The actual plume would then consist of a very large number of such identical particles, the average over the behavior of which yields the ensemble statistics of the plume.

From this discussion it has become apparent that in order to understand the laws of growth of puffs and plumes we must understand the diffusion of two particles relative to each other and one particle relative to a fixed axis, respectively. We present first the single-particle diffusion problem.

6.4.2 Motion of a Single Particle Relative to a Fixed Axis

Let us consider, as shown in Fig. 6.11, a single particle which is at position $\mathbf{x}_0$ at time t_0 and is at position $\mathbf{x}$ at some later time t in a turbulent fluid. The complete statistical properties of the particle's motion are embodied in the transition probability density $Q(\mathbf{x},t|\mathbf{x}',t')$. An analysis of this problem for stationary, homogeneous turbulence was presented by Taylor (1921) in one of the classic papers in the field of turbulence. If the turbulence is stationary and homogeneous, $Q(\mathbf{x},t|\mathbf{x}',t') = Q(\mathbf{x} - \mathbf{x}'; t - t')$; that is, Q depends only on the displacements in space and time and not on the initial position or time. The single particle may be envisioned as one of a very large number of particles which are emitted sequentially from a source located at $\mathbf{x}_0$. The distribution of the concentration of marked particles in the fluid is known once the statistical behavior of one representative marked particle is known. For convenience we assume that the particle is released at the origin at $t = 0$, so that its displacement corresponds to its coordinate location at time t.

The most important statistical quantity is the mean-square displacement of the particle from the source after a time t since the mean displacement from the axis parallel to the flow direction will be zero. If we envision this particle as one being emitted from a continuous source, we can see that the mean-square displacement of the particle from the axis of the plume will tell us the width of the plume and hence the variances $\sigma_i^2(t)$.

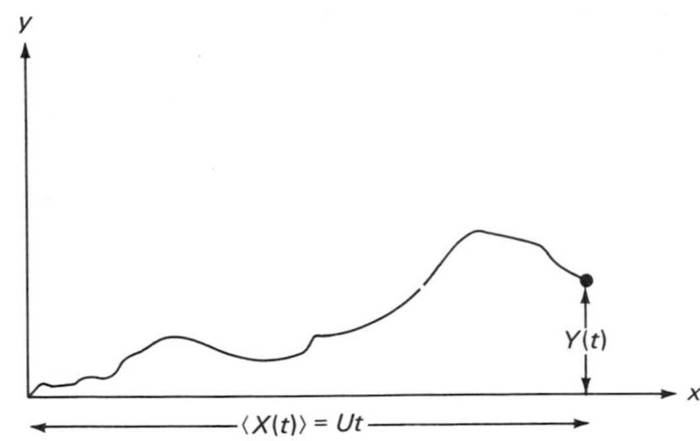

FIGURE 6.11
Diffusion of a single particle relative to a fixed axis.

The mean displacement of the particle along the ith coordinate is defined by

$$\langle X_i(t) \rangle = \int_{-\infty}^{\infty} x_i Q(\mathbf{x};t)\, d\mathbf{x} \qquad (6.100)$$

where the braces ($\langle\ \rangle$) indicate an ensemble average over an infinite number of identical marked particles. If the velocity of the particle in the ith direction at any time is $v_i(t)$, the position of the particle at time t is given by

$$X_i(t) = \int_0^t v_i(t')\, dt' \qquad (6.101)$$

where the velocity of the particle at any instant is equal, by definition, to the fluid velocity at the spot where the particle happens to be at that instant,

$$v_i(t) = u_i[\mathbf{X}(t),t] \qquad (6.102)$$

Let us consider a situation in which there is no mean velocity, so that $v_i(t) = u_i'[\mathbf{X}(t),t]$. If there is a mean velocity, say in the x_1 direction, we will be interested in the dispersion about a point moving with the mean velocity. Therefore, the influence of the fluctuating eulerian velocities on the wanderings of the marked particle from its axis is the key issue here, not its translation in the mean flow. We might also note that, in discussing the statistics of *particle motion*, all averages are conceptually ensemble averages, carried out over a very large number of similar particle releases. For this reason, we denote mean quantities by braces, as in (6.100), as opposed to overbars, which have been reserved for time averages. It is understood, however, that when discussing mean properties of the *velocity field* itself, due to the condition of stationarity, the ensemble average and the time average are identical. The mean displacement can also be computed by ensemble averaging of (6.101),

$$\langle X_i(t) \rangle = \int_0^t \langle v_i(t') \rangle\, dt' \qquad (6.103)$$

where the averaging can be taken inside the integral.

The variance of the displacements is given by a tensor the elements of which are

$$P_{ij}(t) = \langle X_i(t)X_j(t)\rangle = \int_{-\infty}^{\infty} x_i x_j Q(\mathbf{x};t)\, d\mathbf{x} \tag{6.104}$$

The diagonal elements $\langle X_i^2(t)\rangle$ are of principal interest since they describe the rate of spreading along each axis. Using (6.101) in (6.104), we obtain

$$\begin{aligned} P_{ij}(t) &= \left\langle \int_0^t v_i(t')\, dt' \int_0^t v_j(t'')\, dt'' \right\rangle \\ &= \int_0^t \int_0^t \langle v_i(t')v_j(t'')\rangle\, dt'\, dt'' \end{aligned} \tag{6.105}$$

The integrand of (6.105) is recognized as the lagrangian correlation function $\mathscr{R}_{ij}(t' - t'')$, so that (6.105) may be rewritten

$$\begin{aligned} P_{ij}(t) &= \int_0^t \int_0^t \mathscr{R}_{ij}(t' - t'')\, dt'\, dt'' \\ &= \int_0^t \int_0^{t-t'} \mathscr{R}_{ij}(\zeta)\, d\zeta\, dt' \end{aligned} \tag{6.106}$$

By definition $\mathscr{R}_{ij}(\zeta) = \mathscr{R}_{ji}(-\zeta)$, so that

$$P_{ij}(t) = \int_0^t (t - \zeta)[\mathscr{R}_{ij}(\zeta) + \mathscr{R}_{ji}(\zeta)]\, d\zeta \tag{6.107}$$

We now consider the form of $P_{ij}(t)$ in the two limiting situations, $t \to 0$ and $t \to \infty$. First, for $t \to 0$, $\mathscr{R}_{ij}(\xi) \cong \mathscr{R}_{ij}(0) = \langle u_i' u_j'\rangle \equiv \overline{u_i' u_j'}$. Thus,

$$P_{ij}(t) = \overline{u_i' u_j'}\, t^2 \qquad t \to 0 \tag{6.108}$$

and the dispersion increases as t^2. Next, as $t \to \infty$ we expect the lagrangian correlation function $\mathscr{R}_{ij}(t)$ to approach zero as the motion of the particle becomes uncorrelated with its original velocity. We expect convergence of the following two integrals,

$$\int_0^{\infty} [\mathscr{R}_{ij}(\zeta) + \mathscr{R}_{ji}(\zeta)]\, d\zeta = I_{ij}$$

$$\int_0^{\infty} \zeta[\mathscr{R}_{ij}(\zeta) + \mathscr{R}_{ji}(\zeta)]\, d\zeta = J_{ij}$$

where I_{ij} is proportional to the lagrangian time scale of the turbulence. Thus, as $t \to \infty$, $P_{ij}(t)$ becomes proportional to t:

$$P_{ij}(t) = I_{ij} t - J_{ij} \tag{6.109}$$

We can obtain these results by a slightly different route. Let us, for example, compute the rate of change of the dispersion $\langle X_i^2(t)\rangle$:

$$\begin{aligned}\frac{d}{dt}\langle X_i^2(t)\rangle &= 2\langle X_i(t)v_i(t)\rangle \\ &= 2\left\langle v_i(t)\int_0^t v_i(t')\,dt'\right\rangle \\ &= 2\int_0^t \langle v_i(t)v_i(t')\rangle\,dt' \\ &= 2\int_0^t \mathscr{R}_{ii}(t-t')\,dt' \qquad (6.110)\end{aligned}$$

Integration of (6.110) with respect to t gives (6.106).

In summary, we have found that the mean-square dispersion of a particle in stationary, homogeneous turbulence has the following dependence on time:

$$P_{ii}(t) = \begin{cases}\overline{u_i'^2}\,t^2 & t \to 0 \\ 2K_{ii}t & t \to \infty\end{cases} \qquad (6.111)$$

where

$$K_{ii} = \lim_{t\to\infty}\int_0^t \mathscr{R}_{ii}(t-t')\,dt' \qquad (6.112)$$

Since $\overline{u_i'^2}$ is proportional to the total turbulent kinetic energy, the total energy of the turbulence is important in the early dispersion. After long times the largest eddies will contribute to $\mathscr{R}_{ii}$, and $\mathscr{R}_{ii}$ will not go to zero until the particle can escape the influence of the largest eddies. From its definition, K_{ii} has the dimensions of a diffusivity, since as $t \to \infty$

$$\frac{1}{2}\frac{d\langle X_i^2(t)\rangle}{dt} = K_{ii} \qquad (6.113)$$

Now, if we return to Sec. 6.3 we see that this K_{ii} is precisely the constant eddy diffusivity used in the atmospheric diffusion equation for stationary, homogeneous turbulence. Since the lagrangian time scale is defined by

$$\mathscr{T}_L = \max_i\left[\frac{1}{\overline{u_i'^2}}\int_0^\infty \mathscr{R}_{ii}(\tau)\,d\tau\right]$$

it becomes clear that the Fickian diffusion theory, with constant K's, should apply only when the diffusion time t is much greater than $\mathscr{T}_L$. Because of large atmospheric eddies, $\mathscr{T}_L$ might be quite large, indicating that Fickian diffusion theory will not be very useful in the atmosphere. In general, large eddies dominate atmospheric diffusion when diffusion is measured relative to a fixed coordinate system.

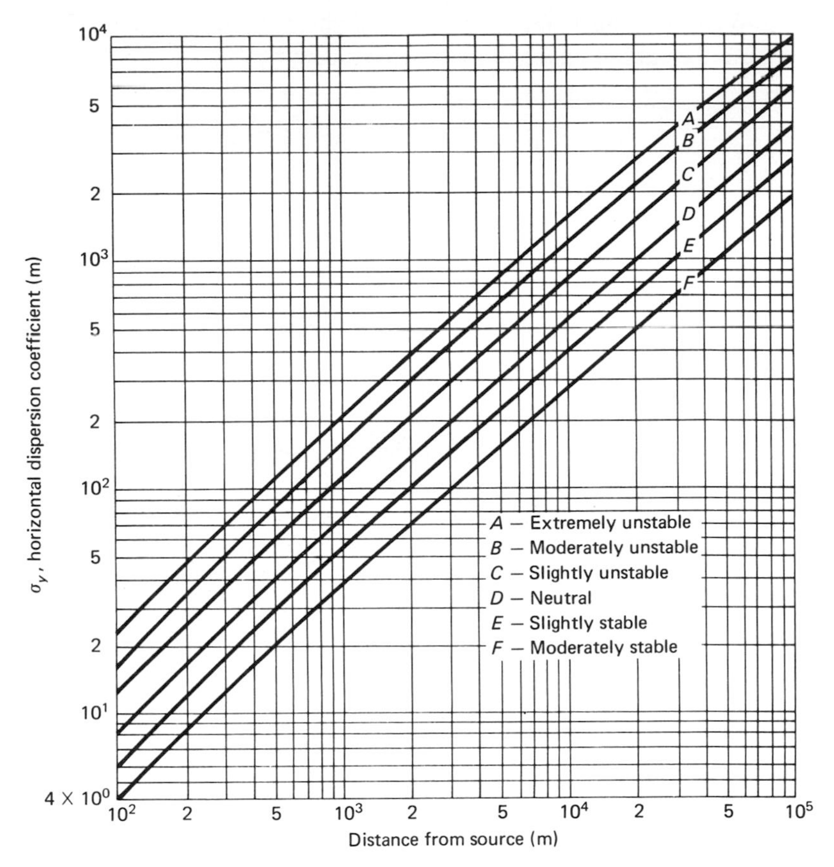

FIGURE 6.12
σ_y as a function of downwind distance for the Pasquill-Gifford stability categories.

There is no precise way of evaluating the times when the two limiting cases in (6.111) will apply in the atmosphere. If it were possible to measure $\mathscr{R}_{ii}(\tau)$ accurately, these times could be evaluated, although such measurements are difficult to make. The small time limit is probably good for diffusion times up to at least a few minutes. whereas the long time limit may not be valid until scales of several hundred kilometers are reached. As an interpolation formula, Sutton (1953) suggested

$$P_{ii}(t) = \tfrac{1}{2}C_i^2(\bar{u}_i t)^{2-n} \qquad 0 < n < 1 \qquad (6.114)$$

Most practical studies have employed a set of empirical correlations of Pasquill (1961, 1962) and Gifford (1961) for the P_{ii} (or the σ_i^2) when the gaussian puff or plume

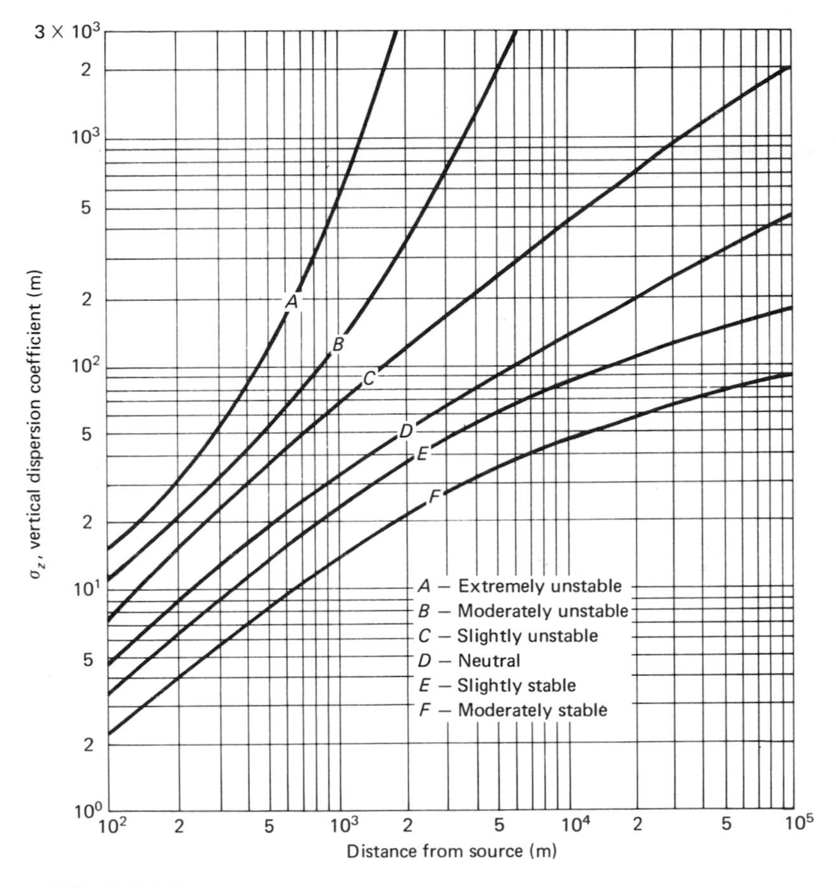

FIGURE 6.13
σ_z as a function of downwind distance for the Pasquill-Gifford stability categories.

equations are used to estimate atmospheric dispersion. Based on experimental observations of the dispersion of real plumes, Pasquill suggested six categories of stability, summarized in Table 6.5, which could be used to determine the values of the lateral and vertical standard deviations σ_y and σ_z as a function of downwind distance for a continuous plume. The correlations were presented as two figures, reproduced here as Figs. 6.12 and 6.13.

In an effort to express stability in terms of more fundamental physical properties, Pasquill and Smith (1971) estimated the values of the Richardson number and Monin-Obukhov length associated with the stability classes defined in Table 6.5. These values are given in Table 6.6.

Table 6.5 DEFINITION OF SIX STABILITY CLASSES FOR USE WITH THE PASQUILL-GIFFORD CURVES IN FIGS. 6.12 AND 6.13

Surface wind speed, m/sec	Daytime insolation			Nighttime conditions	
	Strong	Moderate	Slight	Thin overcast or $\geq 4/8$ cloudiness‡	$\leq 3/8$ Cloudiness
<2	A	A-B	B		
2	A-B	B	C	E	F
4	B	B-C	C	D	E
6	C	C-D	D	D	D
>6	C	D	D	D	D

A: extremely unstable conditions
B: moderately unstable conditions
C: slightly unstable conditions
D: neutral conditions†
E: slightly stable conditions
F: moderately stable conditions

SOURCE: Gifford (1968).
† Applicable to heavy overcast, day or night.
‡ The degree of cloudiness is defined as that fraction of the sky above the local apparent horizon which is covered by clouds.

Table 6.6 ESTIMATES OF THE CORRESPONDENCE BETWEEN PASQUILL'S STABILITY CATEGORIES, Ri AND L, FOR SHORT GRASS

Stability category	Ri (at $z_1 = 2$ m)	L, m
A	−1.0 to 0.7	−2 to −3
B	−0.5 to −0.4	−4 to −5
C	−0.17 to −0.13	−12 to −15
D	0	∞
E	0.03 to 0.05	35 to 75
F	0.05 to 0.11	8 to 35

SOURCE: Pasquill and Smith (1971).

The stability class can be obtained from σ_θ, the standard deviation of the horizontal wind direction, as follows (Gifford, 1968):

Category	σ_θ, deg
A	25
B	20
C	15
D	10
E	5
F	2.5

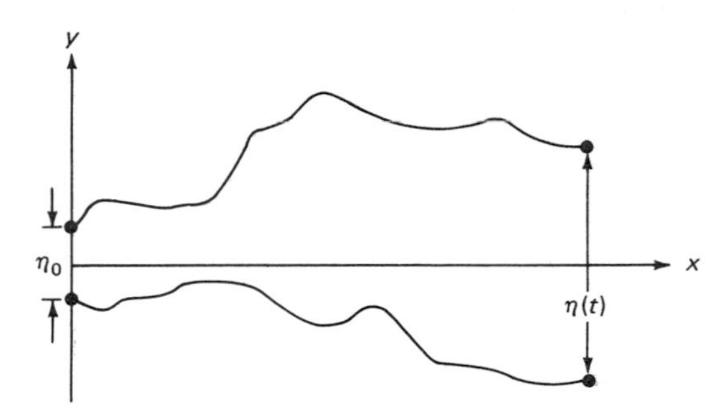

FIGURE 6.14
Relative diffusion of two particles.

Measurements of σ_θ can be carried out easily with a wind vane and a recorder.

The horizontal standard deviation σ_y has been studied most, and most data indicate that $\sigma_y \sim x^{0.8-0.9}$. The vertical standard deviation σ_z is difficult to measure directly since elevated measurements are necessary. Consequently, σ_z is usually inferred from ground-level data. No one model for the σ's emerges as clearly superior, and the curves in Figs. 6.12 and 6.13 provide as good an estimate of these parameters as any. The basic problem is that, although the gaussian dispersion theories are based on stationary and homogeneous turbulence, the atmosphere is not stationary, is never vertically homogeneous, and only seldom is horizontally homogeneous. Thus, the use of the gaussian plume formulas provides at best a rough estimate of true atmospheric plume dispersion.

6.4.3. Relative Diffusion of Two Particles

In the preceding development, the diffusion of particles relative to a fixed coordinate system is completely characterized by the statistics of the motion of a single fluid particle. From the viewpoint of diffusion with respect to a fixed axis, the motion of any two particles released, say, one after another should be completely independent. However, if we are interested in the rate of spreading of a cloud of fluid particles, that is, diffusion relative to the center of mass of the cloud, then from this viewpoint the motions of two particles are *not* independent. Because the particles all start out together in the puff, the motions of the particles are at first strongly correlated. In fact, if it is specified that all the particles start out infinitely close together, they will never separate since they are always acted on by the same velocity fluctuations. For this reason, the motion of a single particle does not provide information on the rate of dispersion of a cloud. Rather, we must consider the spreading of *two* particles with some given initial separation. Just as the fixed-source problem can be reduced to a study of the statistics of a single marked particle, the dispersion of a cloud can be effectively studied by considering the statistical behavior of two fluid particles.

All the information we seek is embodied in the joint pdf $Q(\mathbf{x},\mathbf{x}',t\,|\,\mathbf{x}_0,\mathbf{x}'_0,t_0)$ which expresses the probability that two particles located at $\mathbf{x}_0$ and $\mathbf{x}'_0$ at t_0 are subsequently at $\mathbf{x}$ and $\mathbf{x}'$ at t. Since this density is, in general, unknown, we shall resort to an analysis based on similarity (Batchelor, 1950).

Let us consider two marked particles separated by a distance $\mathbf{l}_0$ at $t = 0$ and by $\mathbf{l}(t)$ at some later time t, as shown in Fig. 6.14. Stationary and homogeneous turbulence is again assumed. The elements of $\mathbf{l}(t)$ will be denoted by $\xi(t)$, $\eta(t)$, and $\zeta(t)$ and the magnitude of $\mathbf{l}(t)$ by $l(t)$. Let $\delta u(t)$ be the x_1 component of the relative velocity of the two particles at time t, that is

$$\delta u(t) = u_1(\mathbf{x} + \mathbf{l}(t),t) - u_1(\mathbf{x},t) \tag{6.115}$$

If a large number of particle pairs with initial separation l_0 are released, the average rate of increase of the separation along the x_1 axis is zero,

$$\left\langle \frac{d\xi(t)}{dt} \right\rangle = \frac{d\langle \xi(t) \rangle}{dt} = \langle \delta u(t) \rangle = 0 \tag{6.116}$$

since there is no net preference for any relative velocity. The average rate of increase of the dispersion of the two particles (the mean-square separation) is

$$\frac{d\langle \xi^2(t) \rangle}{dt} = 2\,\langle \xi(t)\,\delta u(t) \rangle \tag{6.117}$$

but

$$\xi(t) = \xi_0 + \int_0^t \delta u(t')\,dt' \tag{6.118}$$

Combining (6.117) and (6.118), we have

$$\begin{aligned} \frac{d\langle \xi^2(t) \rangle}{dt} &= 2\left\langle \left[\xi_0 + \int_0^t \delta u(t')\,dt' \right] \delta u(t) \right\rangle \\ &= 2 \int_0^t \langle \delta u(t)\,\delta u(t') \rangle\,dt' \end{aligned} \tag{6.119}$$

We consider first the behavior of the dispersion at small times. For $t \to 0$, $\delta u(t')$ is approximately equal to $\delta u(0)$ over the range of 0 to t, and (6.119) becomes

$$\frac{d\langle \xi^2(t) \rangle}{dt} \cong 2t\,\langle \delta u^2(0) \rangle = 2t \langle [u_1(\mathbf{x} + \mathbf{l}_0) - u_1(\mathbf{x})]^2 \rangle \tag{6.120}$$

The quantity on the right-hand side of (6.120) is the eulerian spatial correlation function $R_{11}(\mathbf{l}_0)$ for stationary, homogeneous turbulence. In order to obtain the dependence of this correlation on the turbulence we can resort to the similarity theory of Kolmogorov. When l_0 (the magnitude of $\mathbf{l}_0$) is small, we expect this correlation to be governed chiefly by the small eddies. However, as $t \to \infty$, the two particles will travel independently since their separation will be greater than the sizes of any eddies in the field. At this extreme we expect

$$\frac{d\langle \xi^2(t) \rangle}{dt} = 2\,\frac{d\langle X_1{}^2(t) \rangle}{dt} \tag{6.121}$$

where $\langle X_1{}^2(t) \rangle$ is the mean-square displacement of a single particle along the x_1 axis. The relative velocities in the general expression (6.119) depend on the small-scale structure of the turbulence as long as $l(t)$ is comparable to sizes of the small eddies. Within this range of l_0 and t, we may apply the similarity hypotheses.

Let us consider the correlation $\langle [u(\mathbf{x} + \mathbf{l}) - u(\mathbf{x})]^2 \rangle \equiv \overline{[u(\mathbf{x} + \mathbf{l}) - u(\mathbf{x})]^2}$, where u is any one of the three velocity components. There will be little or no contribution to this correlation from eddies whose sizes are substantially smaller than l, the magnitude of $\mathbf{l}$, whereas eddies larger than l will contribute to the correlation. Based on Kolmogorov's hypothesis, we expect this correlation to be a function of ε, ν, and $\mathbf{l}$ only. The scaling parameters are the Kolmogorov length $\eta = (\nu^3/\varepsilon)^{1/4}$ and velocity $v = (\varepsilon\nu)^{1/2}$. Thus,

$$\overline{[u(\mathbf{x} + \mathbf{l}) - u(\mathbf{x})]^2} = (\varepsilon\nu)^{1/2} F\left(\mathbf{l} \left(\frac{\varepsilon}{\nu^3} \right)^{1/4} \right) \tag{6.122}$$

In the inertial subrange, the entire right-hand side of (6.122) should be independent of ν. To achieve this, F should assume the form

$$F\left(\mathbf{l}\left(\frac{\varepsilon}{\nu^3}\right)^{1/4}\right) = c\left[l\left(\frac{\varepsilon}{\nu^3}\right)^{1/4}\right]^{2/3} \tag{6.123}$$

where, owing to the isotropy of the small eddies, we have replaced the vector separation $\mathbf{l}$ by its magnitude l. Combining (6.122) and (6.123), we have

$$\overline{[u(\mathbf{x}+\mathbf{l}) - u(\mathbf{x})]^2} = c\varepsilon^{2/3} l^{2/3} \tag{6.124}$$

We can apply the similarity hypothesis to $d\langle\xi^2(t)\rangle/dt$ over a limited range of l_0 and t. Within this range we assume that $d\langle\xi^2(t)\rangle/dt$ is a function only of the parameters ν, ε, t, and l_0. Analogous to (6.122), we can write

$$\frac{d\langle\xi^2(t)\rangle}{dt} = \nu G\left(l_0\left(\frac{\varepsilon}{\nu^3}\right)^{1/4}, t\left(\frac{\varepsilon}{\nu}\right)^{1/2}\right) \tag{6.125}$$

We can make (6.125) more specific by considering some special cases.

CASE 1 $t \to 0$ As we saw in (6.120), when $t \to 0$, $d\langle\xi^2(t)\rangle/dt$ is linear in t. Thus, (6.125) should reduce to

$$\frac{d\langle\xi^2(t)\rangle}{dt} = (\varepsilon\nu)^{1/2} t G'\left(l_0\left(\frac{\varepsilon}{\nu^3}\right)^{1/4}\right) \tag{6.126}$$

This is the general expression for the rate of spreading. If, in addition, l_0 is in the inertial subrange, the rate of spreading should not depend on ν. In this case Batchelor has shown that

$$\frac{d\langle\xi^2(t)\rangle}{dt} = 2c_1 t(\varepsilon l_0)^{2/3}\left(1 + \frac{1}{3}\frac{\xi_0^2}{l_0^2}\right) \tag{6.127}$$

CASE 2 t Intermediate When t is large enough so that $\langle\xi^2(t)\rangle$ is independent of l_0 but not so large that $\langle\xi^2(t)\rangle$ is of the order of the energy-containing eddies, the function G in (6.125) should be a function only of $t(\varepsilon/\nu)^{1/2}$:

$$\frac{d\langle\xi^2(t)\rangle}{dt} = \nu G''\left(t\left(\frac{\varepsilon}{\nu}\right)^{1/2}\right) \tag{6.128}$$

If, in addition, the separation is still in the inertial subrange,

$$\frac{d\langle\xi^2(t)\rangle}{dt} = c_2 \varepsilon t^2 \tag{6.129}$$

The time t_1 which marks the division between short and intermediate times depends on the similarity range of eddies and is governed solely by l_0 and ε, that is, $t_1 \sim l_0^{2/3}\varepsilon^{-1/3}$. Thus, case 1 above is valid for $t \ll l_0^{2/3}\varepsilon^{-1/3}$, and case 2 is valid for $t \gg l_0^{2/3}\varepsilon^{-1/3}$, provided in both cases that the separation is much less than the size of the energy-containing eddies.

The results for small and intermediate t can be combined:

$$\frac{d\langle\xi^2(t)\rangle}{dt} = 2c_1 t(\varepsilon l_0)^{2/3}\left(1 + \frac{1}{3}\frac{\xi_0^2}{l_0^2}\right) + c_2\varepsilon t^2 \tag{6.130}$$

Integrating (6.130), we obtain

$$\frac{\langle l^2(t)\rangle - \xi_0^2}{l_0^2} = c_1\left(\frac{t\varepsilon^{1/3}}{l_0^{2/3}}\right)^2\left(1 + \frac{1}{3}\frac{\xi_0^2}{l_0^2}\right) + \frac{1}{3}c_2\left(\frac{t\varepsilon^{1/3}}{l_0^{2/3}}\right)^3 \tag{6.131}$$

Identical formulas will apply for $\langle\eta^2(t)\rangle$ and $\langle\zeta^2(t)\rangle$. Therefore, adding the three gives the result for the mean-square displacement between the two particles at any time t:

$$\frac{\langle l^2(t)\rangle - l_0^2}{l_0^2} = \frac{10}{6} c_1 \left(\frac{t\varepsilon^{1/3}}{l_0^{2/3}}\right)^2 + c_2 \left(\frac{t\varepsilon^{1/3}}{l_0^{2/3}}\right)^3 \tag{6.132}$$

If $\mathbf{l}_0$ is parallel to the x_1 axis, then when t is small

$$\langle\xi^2(t)\rangle - \xi_0^2 = \frac{4}{3}\langle\eta^2(t)\rangle = \frac{4}{3}\langle\zeta^2(t)\rangle \tag{6.133}$$

The question of small versus intermediate values of t is readily resolved by the magnitude of the left-hand side of (6.131). If $\ll 1$ or $\gg 1$ the time scales are short and intermediate, respectively.

When t is large

$$\frac{\langle\xi^2(t)\rangle - \xi_0^2}{l_0^2} = \frac{1}{3} c_2 \left(\frac{t\varepsilon^{1/3}}{l_0^{2/3}}\right)^3 \tag{6.134}$$

Using (6.129) in conjunction with (6.134), we obtain

$$\frac{d}{dt}\frac{\langle\xi^2(t)\rangle - \xi_0^2}{l_0^2} = (9c_2\varepsilon)^{1/3} l_0^{-2/3} \left[\frac{\langle\xi^2(t)\rangle - \xi_0^2}{l_0^2}\right]^{2/3} \tag{6.135}$$

so that the rate of spreading is proportional to the two-thirds power of the current mean-square separation;

$$\frac{d\langle\xi^2(t)\rangle}{dt} \sim \langle\xi^2(t)\rangle^{2/3} \tag{6.136}$$

In summary, relative diffusion, the spreading of a cloud of marked particles or the spreading of a plume from its center line, is described by the joint lagrangian statistics of two particles. On the other hand, the mean spreading of a plume about a fixed axis is described by single-particle lagrangian statistics. An instantaneous picture of plume with its meandering and irregularity would be described by relative diffusion, whereas a long time exposure of the same plume would be described by single-particle diffusion.

Relative diffusion was seen to depend on eddies of approximately the same size as the puff or the width of the plume. Average plume dispersion, described by single-particle statistics, was seen, however, to depend on the large energy-containing eddies. In comparison, the temporal dependences of single- and two-particle diffusion are

$$\langle X_i^2(t)\rangle \sim \begin{cases} t^2 & t \to 0 \\ t & t \to \infty \end{cases} \qquad \text{single-particle}$$

$$\langle\xi^2(t)\rangle - \xi_0^2 \sim \begin{cases} t^2 & t < t_1 \\ t^3 & t > t_1 \end{cases} \qquad \text{two particle}$$

6.5 SIMILARITY THEORY FOR DIFFUSION IN THE ATMOSPHERIC SURFACE LAYER

The treatments of the previous section are based on the assumption of stationary, homogeneous turbulence. The atmospheric boundary layer, and in particular the surface layer, is characterized by marked inhomogeneity of turbulence. This inhomogeneity is especially important in the vertical direction as a result of wind shear and stability. The K theories of Sec. 6.3 handle this problem by the assumption of some functional form for $\bar{u}(z)$ and $K_{zz}(z)$, usually a power law. The problem is not totally solved since there exist in $\bar{u}(z)$ and $K_{zz}(z)$ parameters which can be determined only from experimental data. Ellison (1959) and Batchelor (1959, 1964) proposed a dimensional method for the determination of the statistics of a particle released into a logarithmic (adiabatic) surface layer. Further results were presented by Gifford (1962), Cermak (1963), Pasquill (1966), and Chatwin (1968).

The statistical behavior of a marked particle in the surface layer is a function of the lagrangian fluid velocities. The eulerian characteristics of neutral surface-layer turbulent flow are, as we have seen in Chap. 5, completely characterized by u_*, the friction velocity. If it can be assumed that lagrangian statistical properties of the surface layer are also governed by u_* we can make two hypotheses (so-called lagrangian similarity hypotheses):

1 The statistical properties of the velocity of a marked fluid particle after time t after release from the ground depend only on u_* and t.

2 For a marked particle released at height h at $t = 0$, the height h undoubtedly affects the statistical properties of the motion in the neighborhood of the release. However, after a sufficiently long time the statistical properties of the particle velocity will have lost any dependence on h and will approximate those of a particle released at the ground at $-t_1$, provided $t \gg t_1$, where t_1 is of order h/u_*.

Let us denote the mean position at time t of a representative particle released into a mean wind blowing in the x direction by $[\langle X(t)\rangle, \langle Y(t)\rangle, \langle Z(t)\rangle]$. On the basis of hypotheses 1 and 2, on purely dimensional grounds, we can write

$$\frac{d^2\langle X\rangle}{dt^2} \sim \frac{u_*}{t - t_1}$$

$$\frac{d^2\langle Y\rangle}{dt^2} \sim \frac{u_*}{t - t_1}$$

$$\frac{d^2\langle Z\rangle}{dt^2} \sim \frac{u_*}{t - t_1}$$

By symmetry, $\langle Y\rangle = d\langle Y\rangle/dt = d^2\langle Y\rangle/dt^2 = 0$. Also, in order for $d\langle Z\rangle/dt$ to be finite for all finite time, $d^2\langle Z\rangle/dt^2 = 0$ and $d\langle Z\rangle/dt = \text{const}$. On dimensional grounds, $d\langle Z\rangle/dt \sim u_*$, and so we set

$$\frac{d\langle Z\rangle}{dt} = bu_* \qquad t \gg \frac{h}{u_*}$$

where b is a constant. Integrating subject to $\langle Z\rangle = h$ at $t = 0$ gives

$$\langle Z\rangle = bu_*t + h \tag{6.137}$$

Neglecting diffusion in the x direction, $d\langle X\rangle/dt$ can be set equal to the mean fluid velocity at $z = \langle Z\rangle$ corresponding to $x = \langle X\rangle$:

$$\frac{d\langle X\rangle}{dt} = u_* f\left(\frac{\langle Z\rangle}{z_0}\right) \tag{6.138}$$

Using (6.137), we have

$$\frac{d\langle X\rangle}{d\langle Z\rangle} = \frac{1}{b} f\left(\frac{\langle Z\rangle}{z_0}\right) \tag{6.139}$$

To develop an initial condition for (6.139) we note that particles released at $z = h$ at $t = 0$ do not acquire a motion identical to the fluid motion until a time t_1 of order h/u_*. During this time of relaxation the particle travels a distance x_1 of order $\bar{u}(h)h/u_*$. Thus, as an approximation of this distance we can set x_1 equal to $\bar{u}(h)h/u_*$. The initial condition for (6.139) is therefore taken to be

$$\langle X\rangle = \frac{\bar{u}(h)h}{u_*} \qquad \text{at } \langle Z\rangle = h \tag{6.140}$$

Fortunately, the uncertainty in this distance is a defect only at distances close to the source.

What we would like to compute is the probability that on any one realization a particle will reach location (x,y,z). This is identical in principle to considering the concentration distribution $\langle c(x,y,z,t)\rangle$ of M particles released at the origin at $t = 0$.[1] From dimensional reasoning, the following relationship can be postulated:

$$F\left(\frac{\langle c\rangle\langle Z\rangle^3}{M}, \frac{x - \langle X\rangle}{\langle Z\rangle}, \frac{y}{\langle Z\rangle}, \frac{z - \langle Z\rangle}{\langle Z\rangle}\right) = 0$$

Solving for the ratio containing $\langle c\rangle$, we obtain

$$\langle c(x,y,z,t)\rangle = \frac{M}{\langle Z\rangle^3}\Psi\left(\frac{x - \langle X\rangle}{\langle Z\rangle}, \frac{y}{\langle Z\rangle}, \frac{z - \langle Z\rangle}{\langle Z\rangle}\right) \tag{6.141}$$

[1] The M particles would be released in an initial volume whose dimensions are small compared with the downwind distances.

For a continuous source of M particles/time,

$$\langle c(x,y,z)\rangle = M\int_0^\infty \frac{1}{\langle Z\rangle^3}\Psi\left(\frac{x-\langle X\rangle}{\langle Z\rangle}, \frac{y}{\langle Z\rangle}, \frac{z-\langle Z\rangle}{\langle Z\rangle}\right) dt \tag{6.142}$$

Equation (6.141) specifies that the concentration distribution about the point $(\langle X\rangle, 0, \langle Z\rangle)$ always has the same shape as the cloud expands. The concept of similarity is an old and useful one. For example, turbulent jets have distributions of mean velocity and mean-square velocity fluctuations over cross-sectional areas that have the same shape (Batchelor, 1957).

The integration in (6.142) is difficult in general because each of the dimensionless ratios is a function of time. However, we can change the integration variable by letting

$$\eta_x = \frac{x-\langle X\rangle}{\langle Z\rangle} \qquad \eta_y = \frac{y}{\langle Z\rangle} \qquad \eta_z = \frac{z-\langle Z\rangle}{\langle Z\rangle}$$

and noting that

$$\frac{d\eta_x}{dt} = \frac{bu_*}{\langle Z\rangle}\left[-\frac{1}{b}f\left(\frac{\langle Z\rangle}{z_0}\right) - \frac{x-\langle X\rangle}{\langle Z\rangle}\right]$$

Then, (6.142) can be written

$$\langle c(x,y,z)\rangle = \frac{M}{bu_*}\int_0^\infty \frac{1}{\langle Z\rangle^2}\Psi\left(\frac{x-\langle X\rangle}{\langle Z\rangle}, \frac{y}{\langle Z\rangle}, \frac{z-\langle Z\rangle}{\langle Z\rangle}\right)$$
$$\times \left[-\frac{x-\langle X\rangle}{\langle Z\rangle} - \frac{1}{b}f\left(\frac{\langle Z\rangle}{z_0}\right)\right]^{-1} d\left(\frac{x-\langle X\rangle}{\langle Z\rangle}\right) \tag{6.143}$$

Consider the centerline ground-level concentration, $y=0$, $z=0$. Also, at a sufficient distance downwind, the particles will be swept past any point x rather quickly compared with the time taken to reach that point, so that we may assume Ψ is sharply peaked around $x = \langle X\rangle$. Thus, we assume $x \cong \langle X\rangle$, in which case (6.143) becomes

$$\langle c(x,0,0)\rangle = \frac{M}{u_*}\frac{1}{\langle Z^2(\langle X\rangle)\rangle f(\langle Z\rangle/z_0)} \tag{6.144}$$

This relationship can be used together with the integrated form of (6.143) to predict the x dependence of the ground-level concentration from a continuous source, depending on the particular form of the velocity profile f. We will now present the application of the lagrangian similarity theory to neutral and diabatic surface layers.

6.5.1 Neutral Surface Layer (Cermak, 1963; Batchelor, 1964)

In a neutral surface layer the mean velocity obeys a logarithmic distribution and so $f(z/z_0) = (1/\kappa) \ln z/z_0$. Using this form in conjunction with (6.139) and (6.140) yields

$$b\kappa \frac{\langle X \rangle}{z_0} = \frac{\langle Z \rangle}{z_0} \ln \frac{\langle Z \rangle}{z_0} - \frac{\langle Z \rangle - h}{z_0} + (b-1)\frac{h}{z_0} \ln \frac{h}{z_0}$$

For a ground-level source, $h = 0$ and

$$b\kappa \langle X \rangle = \langle Z \rangle \ln \frac{\langle X \rangle}{z_0} - \langle Z \rangle$$

As $\langle X \rangle \to \infty$, $b\kappa\langle X \rangle \cong \langle Z \rangle \ln(\langle Z \rangle / z_0)$, and so (6.144) becomes (letting $x = \langle X \rangle$)

$$\langle c(x,0,0) \rangle \sim \frac{M}{u_* x^2} \left(\ln \frac{\langle Z \rangle}{z_0} \right)_{x = \langle X \rangle}$$

which is in basic accord with the observed dependence of ground-level concentration on x in Table 6.4

6.5.2 Diabatic Surface Layer (Gifford, 1962)

We now assume that the lagrangian particle statistics are determined entirely by u_* and L, the Monin-Obukhov length. The relation for $d\langle Z \rangle/dt$ in a diabatic layer is assumed to be

$$\frac{d\langle Z \rangle}{dt} = b u_* \phi(\zeta) \qquad \zeta = \frac{\langle Z \rangle}{L}$$

where $\phi(0) = 1$. Similarly, the more general form of (6.138) is

$$\frac{d\langle X \rangle}{dt} = \frac{u_*}{\kappa} [f(\zeta) - f(\zeta_0)] \qquad \zeta_0 = \frac{z_0}{L}$$

The extension of the similarity theory using these forms is outlined by Gifford. Table 6.7 shows a comparison of the ground-level concentration dependence on x from a continuous point source predicted by the similarity theory and observed at Project Prarie Grass. Additional experimental evaluation of the predictions of the lagrangian similarity theory is presented by Klug (1968).

Table 6.7 COMPARISON OF THE DEPENDENCE OF THE GROUND-LEVEL CONCENTRATION FROM A CONTINUOUS POINT SOURCE ON DOWNWIND DISTANCE AS PREDICTED BY THE LAGRANGIAN SIMILARITY THEORY AND OBSERVED ($c \sim x^m$)

Stability	ζ_0	$\langle X \rangle$, m	Observed m	Calculated m
Very stable	+0.1	100–200	−1.3	−1.3
Moderately stable	+0.01	400–800	−1.4	−1.6
Neutral	±0.001	400–800	−1.8	−2.0
Moderately unstable	−0.01	400–800	−2.6	−2.8
Very unstable	−0.1	400–800	−3.3	−3.0

SOURCE: Gifford (1962).

6.6 SUMMARY OF ATMOSPHERIC DIFFUSION THEORIES

Turbulent diffusion is concerned with the behavior of individual particles which are supposed to follow faithfully the airflow or, in principle, are simply marked minute elements of the air itself. Because of the inherently random character of winds in the atmosphere, one can never predict with certainty the distribution of concentration of marked particles emitted from a source. Although the basic equations describing turbulent diffusion are available, there does not exist a single mathematical model that can be used as a practical means of computing atmospheric concentrations over all ranges of conditions.

There are two basic ways of considering the problem of turbulent diffusion, the so-called eulerian and lagrangian approaches. The eulerian method is based on carrying out a material balance over an infinitesimal region fixed in space, whereas the lagrangian approach is based on considering the meandering of marked fluid particles in the flow. Each approach can be shown to have certain inherent difficulties which render impossible an exact solution for the mean concentration of particles in turbulent flow. For the purposes of practical computation, several approximate theories have been used for calculating mean concentrations of species in turbulence. These theories are the *K theory*, based on the semiempirical equation of atmospheric diffusion; the *statistical theory*, based on the behavior of individual particles in stationary, homogeneous turbulence; and the *similarity theory*, based on dimensional analysis of the surface layer. The salient features of these three theories are summarized in Table 6.8.

The K theory is perhaps the best known. The basic issues of interest with respect to the K theory are (1) under what conditions on the source configuration and the turbulent field can this theory be applied and (2) to what extent can the eddy diffusivities be specified in an a priori manner from measured properties of the turbulence. The first question has been rather thoroughly answered in Subsec. 6.2.2. In summary, the spatial and temporal scales of the turbulence should be small in comparison with the corresponding scales of the concentration field.

The statistical theory, as developed by Taylor and Batchelor, is concerned with the actual velocities of individual particles in stationary, homogeneous turbulence. Under this assumption the statistics of the motion of one typical particle provides a statistical estimate of the behavior of all particles, and that of two particles an estimate of the behavior of a cluster of particles. In the atmosphere one may expect the crosswind component (v) of turbulence to be nearly homogeneous since the variations in the scale and intensity of v with height are often small. On the other hand, the vertical velocity component (w) is decidedly inhomogeneous, since characteristically w increases with height above the ground. Thus, the statistical theory should be suitable for describing the spread of a plume in the crosswind direction regardless of

Table 6.8 THEORIES OF ATMOSPHERIC DIFFUSION

	K theory	Statistical theory	Similarity theory
Key assumptions	Spatial and temporal variations in the source and velocity fields large with respect to the resolution of the equation; chemical reactions slow compared with temporal scale of resolution	Stationary, homogeneous turbulence	Surface layer. Lagrangian particle statistics depend on same variables as the eulerian velocity statistics.
Basic parameters	K_{xx}, K_{yy}, K_{zz}	$\mathscr{R}_{ii}(\tau)$	Friction velocity u_* Vertical heat flux $\bar{q}$
Resulting form	Parabolic partial differential equation for $\langle c(x,y,z,t)\rangle$	$\sigma_i^2(t)$ for single particle dispersion from a fixed axis; and $\langle l^2(t)\rangle$ for two particles	$\frac{d\langle Z\rangle}{dt}$, $\frac{d\langle X\rangle}{dt}$ for single particle in surface layer
Required measurements	Wind and temperature profiles; intensity and scale of turbulence	Spectrum of turbulence	Wind and temperature profiles
Basic references	Roberts (1923) Richardson (1926) Sutton (1932) Calder (1949, 1965)	Taylor (1921) Batchelor (1949, 1950, 1952)	Monin (1959) Batchelor (1959, 1964) Gifford (1962) Cermak (1963) Pasquill (1966) Chatwin (1968)

the height but for vertical spread only in the early stages of travel from a source considerably elevated above the ground.

The basic parameter of the statistical theory is the lagrangian time scale of the turbulence, i.e. the time integral of the lagrangian autocorrelation function. Unfortunately, the lagrangian time scale is difficult to measure directly, and therefore it would be desirable to determine $\mathcal{T}_L$ from the usual fixed-point (eulerian) turbulence data, a problem which has yet to be solved theoretically. Thus, a rigorous relationship between the lagrangian and eulerian time scales of the turbulence, $\mathcal{T}_L$ and $\mathcal{T}_E$, is unknown. Several approximate theories lead to the conclusion that $\mathcal{T}_L/\mathcal{T}_E$ is inversely proportional to the intensity of the turbulence. The relationship $\mathcal{T}_L = \beta\mathcal{T}_E$, where β is about 4, is often used as a rough guide to $\mathcal{T}_L$ (Hay and Pasquill, 1959). In the statistical theory the crosswind and vertical standard deviations for single-particle diffusion relative to a fixed axis can be written in the general form (Pasquill, 1971)

$$\sigma_2 = \sigma_{u_2} t f_1\left(\frac{t}{\mathcal{T}_L}\right)$$

$$\sigma_3 = \sigma_{u_3} t f_2\left(\frac{t}{\mathcal{T}_L}\right)$$

where t is the time of travel of the particle.

The similarity theory originally developed by Monin (1959) and Batchelor (1959, 1964) predicts the dependence of the mean downwind and vertical position of a particle released in the atmospheric surface layer. The prediction of the similarity theory for vertical spread from the ground under neutral conditions is identical to that following from K theory with $K_{zz} = \kappa u_* z$, provided that the constant b in the similarity theory is set equal to κ. It is only for vertical spread that the theory is justifiable. In general, similarity theory gives predictions for downwind concentration dependence on distance that are in substantial agreement with observations. However, the similarity theory does not yield the functional form of the probability density Ψ and so cannot be used for absolute predictions of concentration.

Several features of real atmospheric diffusion have been disregarded in this chapter. For example, actual source effluents are often emitted with appreciable exit velocities and at temperatures considerably above ambient to ensure that the plume will travel upward to a certain height before it is carried downwind and dispersed. In order to predict downwind concentrations from a continuous elevated source, it is usually necessary to add the "plume rise" to the stack height to obtain an "effective" stack height from which it can be assumed the plume is emitted with zero net buoyancy. The theory and some practical formulas for plume rise estimates are summarized in Appendix B.

We have also assumed that the particles being diffused behave as ideal fluid particles following precisely the motion of the air. This assumption is a safe one for gas molecules and very small particles; however, many atmospheric particles are in a size range for which the ideal-fluid-particle assumption is not a good one. Moreover, for heavier particles, processes other than diffusion, such as settling and deposition, may significantly affect atmospheric concentrations. The detailed description of simultaneous processes of turbulent diffusion, settling, and agglomeration is difficult and beyond our scope.

The deciding factor in judging the validity of a theory for atmospheric diffusion is the comparison of its predictions with experimental data. It must be kept in mind, however, that all the theories we have discussed are based on predicting the ensemble mean concentration $\langle c \rangle$, whereas a single experimental observation constitutes only one sample from the hypothetically infinite ensemble of observations from that identical experiment. Thus, it is not to be expected that any one realization should agree precisely with the predicted mean concentration even if the theory used is applicable to the set of conditions under which the experiment has been carried out. Nevertheless, because it is practically impossible to repeat an experiment more than a few times under identical conditions in the atmosphere, one must be content with at most a few experimental realizations when testing any available theory.

A lengthy review of comparisons of theory and experiment in atmospheric turbulent diffusion would be inappropriate here. Several rather nice comparisons can be found elsewhere; e.g. see Pasquill (1962, 1971), Slade (1968), and Klug (1968). We describe here two recent studies in which diffusion data have been analyzed theoretically. The first deals with the dispersion of an instantaneous line source of tracer from an urban highway; the second is concerned with concentration distributions from high sources.

Dispersion of a Crosswind Line Source of Tracer Released from an Urban Highway (Drivas and Shair, 1974)

In this study sulfur hexafluoride (SF_6) was used as a tracer in a quasi-instantaneous line source released by an automobile moving along an urban highway in Los Angeles. Ground-level concentrations at locations from 0.4 to 3.2 km downwind of the highway were recorded as a function of time. The tracer data were used to test the validity of both the instantaneous gaussian puff model and the atmospheric diffusion equation with empirical-power-law velocity and diffusivity profiles.

Rates of SF_6 release and speeds of the releasing automobile are shown in Table 6.9. The section of the highway used, Interstate 405, runs parallel to the Pacific Ocean coastline about 6 km inland. During the afternoon a brisk sea breeze normally blows inland perpendicularly across the highway. The general weather conditions and heights of temperature inversions (if any) are given in Table 6.9. The horizontal wind direction was measured during each run by means of an anemometer wind vane 2 m above ground level. From the recorded wind direction data, the standard deviation (σ_θ, as shown in Table 6.9) was calculated, using a 3-sec averaging period and a 10-min sampling time.

Typical SF_6 concentration versus time curves at various distances downwind of the line-source release are shown in Fig. 6.15 and 6.16; these data represent runs 3 and 4, respectively. As expected,

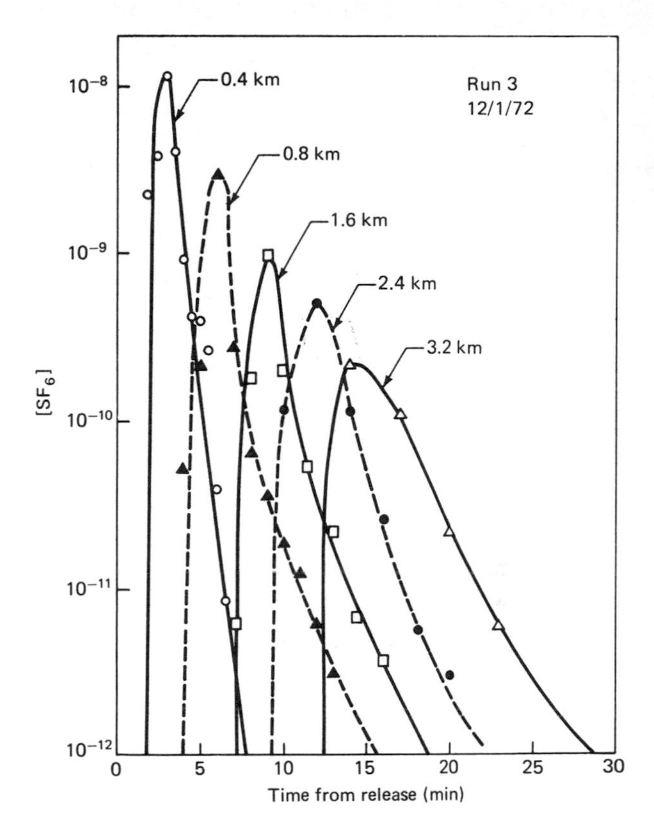

FIGURE 6.15
Tracer concentration at the five downwind distances tested in run 3 as a function of time from the automobile line-source release. (*Source: Drivas and Shair, 1974. With the permission of Microfirm International Marketing Corporation, exclusive copyright licensee of Pergamon Press journal back files.*)

Table 6.9 DESCRIPTION OF EXPERIMENTAL RUNS BY DRIVAS AND SHAIR (1974)

Run no.	Rate of SF_6 release, 1/min	Car speed, km/hr	Weather conditions	Temperature inversion height, m†	σ_θ deg
1	31	80	Sunny, smoggy	271	5.6
2	33	80	Sunny, smoggy	230	21.2
3	44	88	Partly cloudy	none	8.6
4	31	80	Sunny, clear	none	10.1
5	31	80	Overcast	510	9.4
6	36	96	Sunny, clear	none	9.8

† Data obtained from the United States Weather Bureau station at Los Angeles International Airport. The airport is about 8 km from the testing site, and the temperature inversion readings were taken about 2 hr before each experimental run.

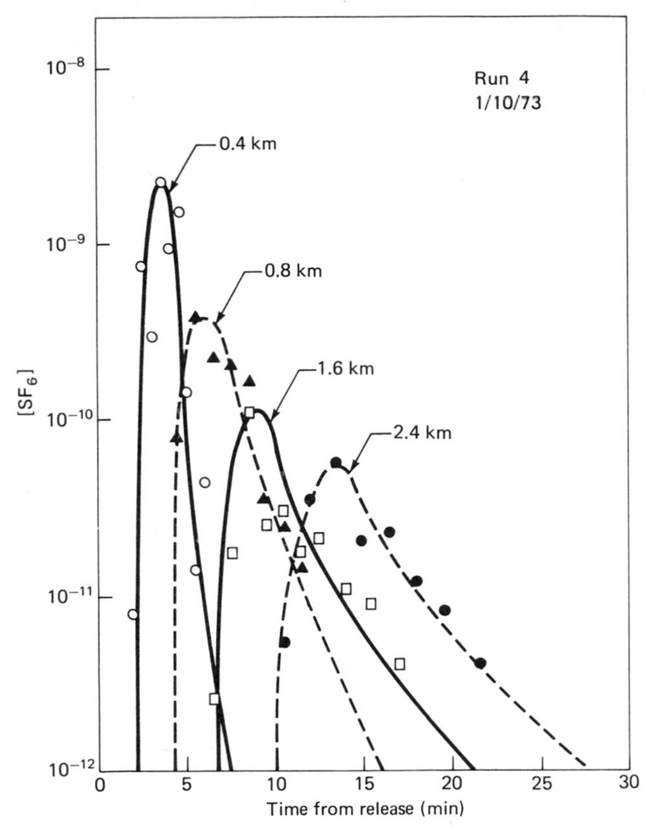

FIGURE 6.16
Tracer concentration at the four downwind distances tested in run 4 as a function of time from the automobile line-source release. (*Source: Drivas and Shair,* 1974. *With the permission of Microfirm International Marketing Corporation, exclusive copyright licensee of Pergamon Press journal back files.*)

the peak concentrations decrease and the curves spread with increasing distance downwind. It is clear, except for the shortest downwind distance at 0.4 km, the curves are decidedly nongaussian; they exhibit a skewness to the right. This skewness increases with increasing downwind distance. The average time of travel ($\langle t \rangle$) was calculated at each location by

$$\langle t \rangle = \frac{\int_0^\infty t \langle c \rangle \, dt}{\int_0^\infty \langle c \rangle \, dt}$$

The corresponding apparent average velocity ($\bar{u}$) was calculated by dividing the downwind distance by $\langle t \rangle$. Except for one point in run 2, the apparent average velocity increases with travel time.

The experimental results of nongaussian concentration profiles and an apparent velocity which increases with time can be interpreted by the effect of wind shear, i.e. a horizontal wind velocity which increases with height, coupled with vertical turbulent diffusion. A cloud of pollutant released near the ground spreads out in the direction of flow as well as vertically. Vertical spreading is, of course, due to vertical turbulent mixing, as a result of which the center of mass of the cloud will rise with time. Horizontal spreading, on the other hand, results from turbulent diffusion vertically into layers moving at different velocities as well as from turbulent mixing in the direction of flow. To envision the former process, let us suppose many trains are traveling on parallel tracks, with the trains on the outer tracks moving progressively faster. Starting with many mailbags on the slowest moving train, we begin throwing them back and forth between trains. The throwing corresponds to vertical diffusion. Since bags are thrown onto faster moving trains, their distribution will spread out in the direction of travel, resulting in what appears to be a horizontal diffusion but what is actually the combined effect of diffusion normal to the direction of flow and the increasing velocity of different layers of the flow. Thus, as the cloud grows vertically with time, the effective mean wind velocity which is transporting the cloud increases. The gaussian puff model cannot account for this effect. The atmospheric diffusion equation can do so through proper choice of $\bar{u}(z)$ and $K_{zz}(z)$. Unfortunately, analytical solutions of (6.84) for arbitrary forms of $\bar{u}(z)$ and $K_{zz}(z)$ and instantaneous sources are not available. The theoretical description of the dispersion of a cloud of pollutants in the atmosphere has been considered by Saffman (1962), Smith (1965), Tyldesley and Wallington (1965), and Chatwin (1968). The approach usually followed is to compute certain integral properties, called *moments*, of the concentration distribution $\langle c(x,y,z,t)\rangle$.

We now consider the mean concentration distribution from an instantaneous point source. If the total release consisted of M grams then

$$\iiint_{\text{All space}} \langle c(x,y,z,t)\rangle \, dx\, dy\, dz = M$$

The position of the x component of the center of mass of the cloud at time t is given by

$$\langle X(t)\rangle = \iiint x\langle c(x,y,z,t)\rangle \, dx\, dy\, dz$$

with similar relations for $\langle Y(t)\rangle$ and $\langle Z(t)\rangle$. The variance of spread of the distribution at time t, $\sigma_x^2(t)$, is then defined by

$$\sigma_x^2(t) = \iiint [x - \langle X(t)\rangle]^2 \langle c(x,y,z,t)\rangle \, dx\, dy\, dz$$

with similar expressions for $\sigma_y^2(t)$ and $\sigma_z^2(t)$. The distribution $\langle c\rangle$ need not be gaussian in order to define these properties.

The mean of a distribution is its first *noncentral moment*, whereas the variance is its *second central moment* (taken with respect to the mean). The values of the moments of a distribution give information about the shape of the distribution. Central moments of even order give information on the width of the distribution, whereas those of odd order describe the degree of symmetry of the distribution. Thus, the third central moment,

$$\gamma_x^3(t) = \iiint [x - \langle X(t)\rangle]^3 \langle c(x,y,z,t)\rangle \, dx\, dy\, dz$$

is called the *skewness* of $\langle c\rangle$.

Chatwin (1968) used the method of moments to analyze (6.84) for the case of an instantaneous, ground-level line source of tracer in neutrally stable conditions, i.e.,

$$\bar{u}(z) = \frac{u_*}{\kappa} \ln \frac{z}{z_0}$$

$$K_{zz}(z) = \kappa u_* z$$

when turbulent diffusion in the downwind direction could be neglected. Chatwin predicted that the standard deviation of the concentration distribution in the downwind direction, σ_x, should be proportional to t under adiabatic conditions. A linear least-squares fit of ln σ_x versus ln $\langle t \rangle$ from the data obtained by Drivas and Shair yielded slopes (which represent a power-law exponent, $\sigma_x \sim \langle t \rangle^b$) of 1.11 to 1.47, as shown in Table 6.10. Thus, the experimental curves spread slightly faster than Chatwin's theoretical prediction applicable to adiabatic conditions.

Saffman (1962) used the method of moments for the more general case of an instantaneous source and an unbounded upper atmosphere with power-law velocity and vertical eddy diffusivity profiles, namely

$$\bar{u} = u_0 z^m \qquad K_{zz}(z) = K_0 z^n$$

His basic results for the asymptotic dependence of the apparent ground-level velocity and horizontal standard deviation were

$$\bar{u} \sim t^{m/(2-n)} \qquad \sigma_x \sim t^{1+m/(2-n)}$$

A linear least-squares fit of ln $\bar{u}$ versus ln $\langle t \rangle$ from the data yielded values of $m/(2-n)$ of 0.13 to 0.55 as shown in Table 6.10. A comparison of the $\bar{u}$ and σ_x time exponents for each run with the theoretical predictions of Saffman exhibits excellent agreement. Also, Saffman's analysis of the third-order moment yields the theoretical prediction that the skewness of the tracer curves should increase with time, a prediction which is verified by the experimental curves.

Use of the constant-shear-stress assumption ($n = 1 - m$) and the above values of $m/(2-n)$ yielded average values of m and n as shown in Table 6.10. Run 1, which occurred under the most stable conditions tested (the lowest value of the wind-direction standard deviation), resulted in essentially a linear velocity profile and a constant vertical eddy diffusivity. Run 2, which occurred under the largest turbulent intensity tested, resulted in a velocity exponent of about one-seventh. The relationship between the constant-shear-stress average velocity exponent and the wind-direction standard deviation is shown in Fig. 6.17.

Concentration Distributions from High Sources (Yordanov, 1972)

A problem of considerable practical importance is the dispersion of effluents from high stacks under different meteorological conditions, since a frequent policy aimed at reducing ground-level concentrations is to release pollutants from tall chimneys. We consider here the process of vertical turbulent mixing in the planetary boundary layer. Beginning with (6.84) as the basic describing equation, we can define $\theta(z,t)$ as the total quantity of pollutant at height z and time t following a release. Thus,

$$\theta(z,t) = \int_{-\infty}^{\infty}\int_{-\infty}^{\infty} \langle c(x,y,z,t) \rangle \, dx \, dy$$

Table 6.10 COMPARISON OF EXPERIMENTAL RESULTS WITH THEORY

Run no.	Exponent $\bar{u} \sim \langle t \rangle^b$	Exponent $\sigma_z \sim \langle t \rangle^b$	Constant stress region m	n	σ_θ, deg
1	0.55	1.47	1.04	−0.04	5.6
2	0.13	1.11	0.14	0.86	21.2
3	0.33	1.22	0.38	0.62	8.6
4	0.30	1.33	0.46	0.54	10.1
5	0.28	1.22	0.33	0.67	9.4
6	0.33	1.29	0.45	0.55	9.8

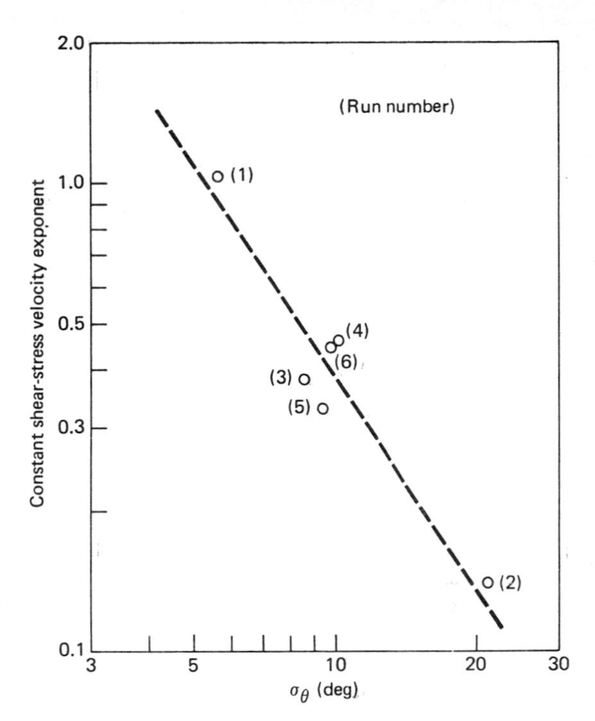

FIGURE 6.17
Average constant shear-stress velocity exponent as a function of the horizontal wind-direction standard deviation σ_θ. (*Source: Drivas and Shair*, 1974. *With the permission of Microfirm International Marketing Corporation, exclusive copyright licensee of Pergamon Press journal back files.*)

Integrating (6.84) term by term over x and y, subject to conditions that $\langle c\rangle \to 0$ as $x, y \to \pm\infty$, we obtain the equation governing $\theta(z,t)$ as

$$\frac{\partial \theta}{\partial t} = \frac{\partial}{\partial z}\left[K_{zz}(z)\frac{\partial \theta}{\partial z}\right] \tag{6.145}$$

with the following boundary and initial conditions for an instantaneous horizontal source of intensity S at height $z = h$:

$$\begin{aligned} K_{zz}\frac{\partial \theta}{\partial z} &= 0 & z &= 0 \\ \theta &= 0 & z &\to \infty \\ \theta(h,t) &= S\,\delta(z-h) & t &= 0 \end{aligned} \tag{6.146}$$

As a result of the integration over x and y, the boundary value problem (6.145) and (6.146) describes the process of vertical diffusion only. Thus, $\theta(0,t)$ is the total amount of pollutant at ground level at any time t.

Yordanov (1968) investigated the solution of (6.145) and (6.146) for K_{zz} given by (6.98) in the surface layer. For high sources, expressions for K_{zz} for the planetary boundary layer are needed. Under neutral conditions, (6.99) has been proposed for K_{zz} in the planetary boundary layer. In addition, Yordanov (1972) used the following expressions for unstable and stable stratification:

Unstable stratification:

$$K_{zz}(z) = \begin{cases} \kappa u_* z & z \le a_2 L \\ \kappa u_*(a_2 L)^{-1/3} z^{4/3} & z > a_2 L \end{cases} \qquad -0.16 \le a_2 \le -0.04 \tag{6.147}$$

Stable stratification:

$$K_{zz}(z) = \begin{cases} \kappa u_* z & z \leq a_3 L \\ \kappa u_* a_3 L & z > a_3 L \end{cases} \qquad a_3 \cong 1 \tag{6.148}$$

The general solution of (6.145) and (6.146) subject to either (6.99), (6.147), or (6.148) is shown by Yordanov to be

$$\theta(z,t) = \frac{S}{\pi a L} \int_0^\infty A(x) I(\zeta,x) I(\eta,x) e^{x^2 \tau/4}\, dx \tag{6.149}$$

where $A(x) = \dfrac{1}{\{[(1/r_1 x)J_0(x)]^{r_2} - J_1(x)\}^2 + J_0{}^2(x)}$

$$I(\zeta,x) = \begin{cases} J_0(x\zeta^{1/2}) & \zeta \leq 1 \\ \zeta^{-r_3}\left\{\left[\dfrac{1}{r_1 x} J_0(x)\right]^{r_2} - J_1(x)\right\} \sin r_1 x(\zeta^{r_4} - 1) + \zeta^{-r_3} J_0(x) \\ \qquad \cos r_1 x(\zeta^{r_4} - 1) & \zeta \geq 1 \end{cases}$$

and $\zeta = z/aL$, $\tau = u_* t/aL$, $\eta = h/aL$, and a denotes a_1, a_2, or a_3 depending on which expression of (6.99), (6.147), or (6.148) is used. For unstable stratification, $r_1 = 1.5$, $r_2 = 1$, $r_3 = 1/3$, and for stable stratification, $r_1 = 0.5$, $r_2 = r_3 = 0$, $r_4 = 1$. $J_\nu(x)$ is the Bessel function of the first kind of order ν.

In the case of a continuous line source with a wind velocity $\bar{u}$ independent of height, the concentration at any level z and downwind position x is found from

$$\theta(z,x) = \frac{S_l}{\bar{u} S}\, \theta\left(z, \frac{x}{\bar{u}}\right) \tag{6.150}$$

where S_l is the strength of the line source per unit length.

For unstable conditions (6.147) is most valid for small values of $a_2 L$ (say $a_2 L \leq 10$ m). Under these conditions, Yordanov showed that the ground-level concentration from a continuous line source can be approximated by

$$\theta(0,x) = \frac{9 S_l \sqrt{\bar{u} a_2 L}}{2\sqrt{\pi}(\kappa u_*)^{3/2} x^{3/2}} \exp\left[-\frac{9 h^{2/3} \bar{u}}{4 \kappa u_* (a_2 L)^{-1/3} x}\right] \tag{6.151}$$

which is valid for $x \gg \bar{u} a_2 L/\kappa u_*$ and $z \gg a_2 L$ or $h \gg a_2 L$. For sources for which $h > 10 a_2 L$ and distances from the source for which $x > 10 \bar{u} a_2 L/\kappa u_*$ these requirements are met. The physical significance of these requirements is that at low heights of the dynamic sublayer ($a_2 L < 10$ m) the process of diffusion from a high source takes place mainly in the upper thermal sublayer.

For stable stratification the ground-level concentration is given by

$$\theta(0,x) = \frac{S_l}{(\pi \kappa \bar{u} u_* a_3 L x)^{1/2}}\, e^{-h^2 \bar{u}/4 \kappa u_* a_3 L x} \tag{6.152}$$

which is valid for the same conditions as (6.151) if a_2 is replaced by a_3.

Equations (6.151) and (6.152) can be expressed as

$$\theta(0,x) = \frac{S_l}{M \bar{u} x^m}\, {}^{-}e^{h^n/Nx} \tag{6.153}$$

where

Unstable stratification:

$$M = \frac{2\sqrt{\pi}(\kappa u_*)^{3/2}}{9\sqrt{a_2 L}\,\bar{u}^{3/2}} \qquad m = \frac{3}{2} \qquad N = \frac{4 \kappa u_*}{9 (a_2 L)^{1/3} \bar{u}} \qquad n = \frac{2}{3}$$

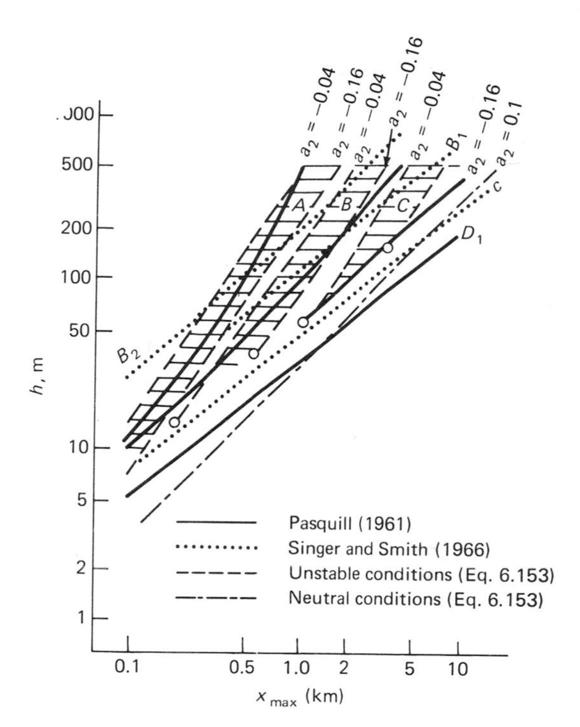

FIGURE 6.18
Source height h as a function of location of maximum ground-level concentration as predicted by (6.153) and as measured by Pasquill (1961) and Singer and Smith (1966). Equation (6.153) is not valid below the circles. (*Source:* Yordanov, 1972. *With the permission of Microfirm International Marketing Corporation, exclusive copyright licensee of Pergamon Press journal back files.*)

Stable stratification:

$$M = \left(\frac{\pi \kappa u_* a_3 L}{\bar{u}}\right)^{1/2} \qquad m = \frac{1}{2} \qquad N = \frac{4\kappa u_* a_3 L}{\bar{u}} \qquad n = 2$$

We can consider the six stability categories given in Table 6.5 in determining L, $u_*/\bar{u}$, and the other parameters above. This information is given in Table 6.11. In this table $\bar{u}$ is the wind velocity at a height of 2 m. The values of M and N are given in Table 6.11 along with the limits of validity of (6.153). The inequalities $x \gg uaL/u_*$ and $h \gg aL$ are considered to be satisfied when the left-hand side is at least 10 times greater than the right-hand side.

The source height is related to the distance where the ground-level concentration is maximum, $x_{\max}$, for unstable stratification by

$$h = (\tfrac{3}{2} N x_{\max})^{3/2} \tag{6.154}$$

Figure 6.18 shows comparisons between data of Pasquill (1961) and Singer and Smith (1966) and (6.154) for $x_{\max}$. For stability categories A and B the data agree reasonably well with the theoretical prediction; for stability class C the region of validity of (6.153) is quite small.

For stable conditions a direct comparison of (6.153) with the gaussian plume equation can be made, in which case the vertical standard deviation $\sigma_z(x)$ is found to be given by

$$\sigma_z(x) = \sqrt{\frac{Nx}{2}} \tag{6.155}$$

The comparison of $\sigma_z(x)$ measured by Pasquill and Singer and Smith and that predicted by (6.155) is shown in Fig. 6.19.

Table 6.11 PARAMETER VALUES OF m, M, AND n, N FOR THE CALCULATION OF THE CONCENTRATION DISTRIBUTION OF A LINE SOURCE (6.153) OR A POINT SOURCE (6.155) AND OF THE LIMITS OF VALIDITY OF THESE EQUATIONS IN TERMS OF THE DIFFUSION CATEGORY

			Unstable stratification $m = 3/2;\ n = 2/3$							
	$L(m)$	$\frac{u_*}{\bar{u}}$	$M = \frac{2\sqrt{\pi}(\kappa u_*)^{3/2}}{9\sqrt{a_2 L}\bar{u}^{3/2}}\ (m^{-1/2})$		$N = \frac{4\kappa u_*}{9\bar{u}(a_2 L)^{1/3}}\ (m^{-1/3})$		$x > \frac{10\bar{u}a_2 L}{\kappa u_*}\ (m)$		$h > 10a_2 L(m)$	
			$a_2 = -0.04$	$a_2 = -0.16$	$a_2 = -0.04$	$a_2 = -0.16$	$a_2 = -0.04$	$a_2 = -0.16$	$a_2 = -0.04$	$a_2 = -0.16$
A	−2	0.10	1.1×10^{-3}	5.5×10^{-3}	4.1×10^{-2}	2.6×10^{-2}	20	80	1	3
B	−10	0.09	4.3×10^{-3}	2.2×10^{-3}	2.2×10^{-2}	1.4×10^{-2}	110	440	4	16
C	−100	0.08	11.1×10^{-3}	0.57×10^{-3}	0.90×10^{-2}	0.57×10^{-2}	1250	5000	40	160
			Stable stratification $m = 1/2;\ n = 2$							
	$L(m)$	$\frac{u_*}{\bar{u}}$	$M = \sqrt{\frac{\pi\kappa u_* a_3 L}{\bar{u}}}\ (m^{1/2})$		$N = \frac{4\kappa u_* a_3 L}{\bar{u}}\ (m)$		$x > \frac{10\bar{u}a_3 L}{\kappa u_*}\ (m)$		$h > 10a_3 L(m)$	
			$a_3 = 1$	$a_3 = 2$	$a_3 = 1$	$a_3 = 2$	$a_3 = 1$	$a_3 = 2$	$a_3 = 1$	$a_3 = 2$
D	20	0.08	1.4	2.0	2.5	5.1	6250	1250	200	400
E	4	0.07	0.59	0.84	0.45	0.90	1430	2860	40	80
F	1.6	0.06	0.35	0.49	0.15	0.31	670	1340	16	30
			Adiabatic stratification $m = n = 1$							
	$L(m)$	$\frac{u_*}{\bar{u}}$	$M = N = \kappa u_*/u$				$x < \frac{a_1\lambda\bar{u}}{\kappa u_*} = \frac{a_1\bar{u}}{f_1}\ (m)$		$h < 10a_1\lambda(m)$	
C							$a_1 = 0.1 f_1 = 10^{-4}$ (sec^{-1})		$a_1 = 0.1 f_1 = 10^{-4}$ (sec^{-1})	
D	$\pm\infty$	0.08	3.2×10^{-2}				$x < 1000\ \bar{u}$ (m/sec^{-1})		$h < 32\ \bar{u}$ (m/sec^{-1})	

SOURCE: Yordanov (1972).

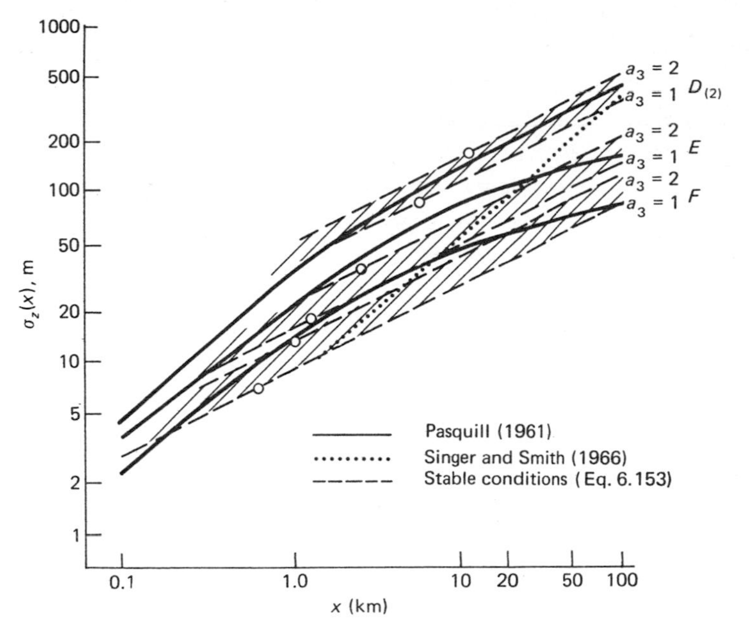

FIGURE 6.19
Standard deviation of the vertical spread σ_z as a function of downwind distance as predicted by (6.153) and as measured by Pasquill (1961) and Singer and Smith (1966). The predictions are not valid to the left of the circles, if they are approximated to ≤ 3 percent. (*Source: Yordanov*, 1972. *With the permission of Microfirm International Marketing Corporation, exclusive copyright licensee of Pergamon Press journal back files.*)

The form of (6.153) can be used under adiabatic conditions if $h < a_1\lambda$ and $x \ll a_1\lambda\bar{u}/u_* = \bar{u}a_1/f$ [see (6.99)], where $M = N = \kappa u_*/\bar{u}$, and $n = m = 1$. Taking $a_1 = 0.1$ and $f = 10^{-4}$ sec^{-1}, we obtain the limits of validity as $h < 32u(m)$ and $x < 100\bar{u}(m)$.

From Figs. 6.18 and 6.19 we see that at short distances, $x \ll \bar{u}aL/u_*$, Pasquill's $\sigma_z(x)$ data approach $\sigma_z(x) \sim x^{1/2}$. These results support the general model in which $\sigma_z(x) \sim x^p$ and where p changes with distance. Yordanov's study leads to the conclusion that if a gaussian distribution is assumed in the vertical direction for the concentration then the data must be classified not only in accordance with Pasquill's diffusion categories but also according to the height of the source, a conclusion supported by Singer and Smith's data.

6.7 EMISSION LEVEL–AIR QUALITY MODELS

In Section 1.6 two classes of models relating emissions to air quality were delineated: (1) physical models, and (2) mathematical models. Of the latter class, we identified two subclasses: (1) statistical models based on air quality data, and (2) models based on the fundamental description of atmospheric transport and chemistry. Physical simulation involves constructing actual physical models of air pollution problems. This approach may be useful for very special problems, for example, for simulating dispersion in a street canyon and atmospheric chemical reactions, but is unable to incorporate aspects of meteorology and air dynamics on the scale relevant to regional air pollution problems. Because of the inherent limitations of physical models for simulation of atmospheric processes of transport, dispersion, and chemical reaction, mathematical models must serve as the primary tools in emission level–air quality relationships.

The first category of mathematical models involves statistical modeling based on past air quality data. An assumption inherent in nearly all statistical models is that, for given meteorological conditions, primary contaminant concentrations are proportional to total regionwide emission levels, with appropriate allowances for natural background concentrations. This assumption would hold exactly for an inert species, by the linearity of the equation of species continuity, if emission level changes occurred homogeneously in space and time. Thus, statistical models for relating changes in air quality to those in emission levels are not applicable to situations in which significant alterations in the spatial and temporal distribution of emissions occur. They are intended primarily for estimating the gross effects of overall emission level changes. In addition, the development of statistical models for secondary pollutants, such as ozone, has proven to be a very difficult task. The basic problem in dealing with reactive pollutants is that the statistical model is usually based on contaminant concentrations measured at one monitoring station, even though the maximum concentration of secondary pollutants may occur many kilometers downwind several hours later. Thus, effects of pollutant transport and continuous addition of emissions usually cannot be included in statistical models. In summary, while statistical models have an advantage in realism afforded by a dependence on actual aerometric data, they are based on simplifying assumptions which limit their applicability to prediction of gross effects.

In the remainder of this section we consider mathematical models of pollutant behavior in the atmosphere. To be generally useful in urban planning studies, a mathematical model must meet several requirements:

1 It must be able to deal with completely arbitrary distributions of sources with emission rates variable in space and time.

2 It must allow spatial and temporal variations of wind and temperature structure (stability), including variations of an inversion height if an inversion is present.

3 It must be able to predict accurately the ground-level concentrations of inert pollutants, as well as those formed in the atmosphere by chemical reactions.

The model should have a spatial and temporal resolution appropriate for the uses to which the model will be put. The resolution of the model will, of course, be influenced by the availability of data of similar resolution.

It is useful to divide mathematical models according to the categories: (1) models for inert species, and (2) models for reactive species. In addition, for each of inert and reactive species, a model may be descriptive of dynamic behavior (say, over the course of a day), or it may predict the steady-state concentration of pollutants under conditions of unchanging emissions and meteorology. One of the primary aims of this chapter has been to develop the equations upon which the various mathematical emission level–air quality models are based.

6.7.1 Inert and Linearly Reactive Contaminants

We consider first species which are inert or decay by first-order reaction (so-called linearly reactive contaminants), since the linearity of their behavior affords us a wider variety of modeling approaches than are available for nonlinearly reactive pollutants. We will discuss only inert species here, since any formulas applicable for inert species can be modified for first-order decay by an appropriate exponential decay.

There are essentially two ways by which the development of a dynamic model may proceed, the lagrangian and eulerian approaches. The fundamental relation on which the lagrangian approaches are based is (6.7) [or (6.10) in the case of a linearly reactive species]

$$\langle c(\mathbf{x},t)\rangle = \int_{-\infty}^{\infty}\int_{-\infty}^{\infty}\int_{-\infty}^{\infty} Q(\mathbf{x},t\,|\,\mathbf{x}_0,t_0)\langle c(\mathbf{x}_0,t_0)\rangle\, d\mathbf{x}_0 + \int_{-\infty}^{\infty}\int_{-\infty}^{\infty}\int_{-\infty}^{\infty}\int_{t_0}^{t} Q(\mathbf{x},t\,|\,\mathbf{x}',t')S(\mathbf{x}',t')dt'\, d\mathbf{x}' \tag{6.156}$$

The basic problem in the use of such models is the proper specification of the transition probability Q. Q is generally determined in either of two ways. First, Q can be ascribed a gaussian form directly, in which case

$$Q(\mathbf{x},t\,|\,x',t') = Q(\mathbf{x}-\mathbf{x}';t-t')$$

Then (6.156) is written

$$\langle c(\mathbf{x},t)\rangle = \int_{-\infty}^{\infty}\int_{-\infty}^{\infty}\int_{-\infty}^{\infty} Q(\mathbf{x}-\mathbf{x}_0;t-t_0)\langle c(\mathbf{x}_0,t_0)\rangle\, d\mathbf{x}_0 + \int_{-\infty}^{\infty}\int_{-\infty}^{\infty}\int_{-\infty}^{\infty}\int_{t_0}^{t} Q(\mathbf{x}-\mathbf{x}';t-t')S(\mathbf{x}',t')\, dt'\, d\mathbf{x}' \tag{6.157}$$

Second, Q can be determined such that $\langle c\rangle$ is the solution of an appropriate form of the atmospheric diffusion equation. In fact, (6.157) is just a general way of expressing the solution of a linear partial differential equation containing a source term $S(\mathbf{x},t)$ and an initial condition $\langle c(\mathbf{x}_0\, t_0)\rangle$. In such a context, Q is called a Green's function. The transition probability, or Green's function, Q is simply the concentration at $(\mathbf{x},t)$ resulting from an instantaneous point source of unit strength at $(\mathbf{x}',t')$. To illustrate this second way of obtaining Q, we note that if conditions are such that $\langle c\rangle$ satisfies

$$\frac{\partial\langle c\rangle}{\partial t} + U\frac{\partial\langle c\rangle}{\partial x_1} = K_{11}\frac{\partial^2\langle c\rangle}{\partial {x_1}^2} + K_{22}\frac{\partial^2\langle c\rangle}{\partial {x_2}^2} + K_{33}\frac{\partial^2\langle c\rangle}{\partial {x_3}^2} + S(\mathbf{x},t) \tag{6.158}$$

subject to

$$\langle c(\mathbf{x},t_0)\rangle = 0$$

$$\langle c(\mathbf{x},t)\rangle \to 0 \qquad x_1,\, x_2 \to \pm\infty,\, x_3 \to +\infty$$

$$\frac{\partial\langle c\rangle}{\partial x_3} = 0 \qquad x_3 = 0$$

then Q is found to be given by

$$Q(\mathbf{x}-\mathbf{x}';\tau) = \frac{1}{(2\pi\tau)^{3/2}(K_{11}K_{22}K_{33})^{1/2}} \exp\left[-\frac{(x_1 - x_1' - U\tau)^2}{4K_{11}\tau} - \frac{(x_2 - x_2')^2}{4K_{22}\tau}\right] \times \left\{\exp\left[-\frac{(x_3 - x_3')^2}{4K_{33}\tau}\right] + \exp\left[-\frac{(x_3 + x_3')^2}{4K_{33}\tau}\right]\right\} \tag{6.159}$$

where $\tau = t - t'$.

The source function is specified on the basis of the strengths and locations of all significant sources in the airshed. For the implementation of (6.157), it is convenient to divide sources according to point, line, and area sources. Doing so, the source function can be written in the form

$$S(\mathbf{x},t) = \sum_{i=1}^{M_p} S_i^{\,p}(t)\,\delta(\mathbf{x}-\mathbf{x}_i) + \sum_{i=1}^{M_l} S_i^{\,l}(x_1,x_2,t)\,\delta(x_3)\,\delta(x_2 - s_i(x_1)) + \sum_{i=1}^{M_a} S_i^{\,a}(x_1,x_2,t)\,\delta(x_3) \tag{6.160}$$

where M_p, M_l, and M_a are the number of point, line, and area sources, respectively; $\mathbf{x}_i$, $i = 1, 2, \ldots, M_p$, are the location of the point sources; and $s_i(x_1)$ is the equation of the ith line source in the $x_1\, x_2$ plane. The line and area sources are assumed to be at ground level in (6.160). $S_i^{\,p}$, $S_i^{\,l}$, and $S_i^{\,a}$ are the strengths of the three types of sources.

The airshed model is given by (6.157) and (6.160) together with an appropriate form of Q. This model can be referred to as a lagrangian puff model, since the kernel Q represents the concentration resulting from a puff of unit strength. A detailed application of the lagrangian puff model is given by Lamb and Neiburger (1971).

The representation of the dynamic behavior of inert contaminants under conditions of spatially and temporally varying wind speeds and directions such that an analytical solution for Q cannot be obtained requires the numerical solution of (6.14). The solution of (6.14) involves the choice of an array of points (a grid) covering the volume of the airshed with a spacing fine enough for the desired spatial resolution. The spatial derivatives in (6.14) are then approximated by difference formulas involving the values of the concentrations at one or more neighboring grid points. Time derivatives are similarly expressed in terms of the values at two or more successive time intervals. The approximations yield a set of difference equations which enable a step-by-step numerical calculation of the concentrations at each grid point at each time interval. The major problems associated with the use of (6.14) are that computing times for the solution of several coupled equations (for reactive species) of the form (6.14) can be excessive, and that difference methods used to approximate the spatial derivatives may introduce spurious effects, such as an artificial diffusion of pollutants.

Up to this point we have been discussing dynamic models for inert contaminants. When air quality standards are expressed in terms of an annual average, it is necessary to devise a model that predicts this quantity as a function of location in a region. The mean concentration at any location and time is given by (6.156). We wish to predict the concentration averaged over a time period T,

$$\overline{\langle c(\mathbf{x})\rangle} = \frac{1}{T}\int_0^T \langle c(\mathbf{x},t)\rangle\, dt \qquad (6.161)$$

where T might normally be 1 yr. Thus, $\overline{\langle c(x)\rangle}$ is the predicted annual average concentration at location $\mathbf{x}$.

Upon substituting (6.156), with $\langle c(\mathbf{x},t_0)\rangle = 0$ and $t_0 = 0$ for convenience, we obtain

$$\overline{\langle c(\mathbf{x})\rangle} = \frac{1}{T}\int_0^T \int_{-\infty}^{\infty}\int_{-\infty}^{\infty}\int_{-\infty}^{\infty}\int_0^t Q(\mathbf{x},t\,|\,\mathbf{x}',t')S(\mathbf{x}',t')\, dt'\, d\mathbf{x}'\, dt \qquad (6.162)$$

Reversing the order of integration we obtain

$$\overline{\langle c(\mathbf{x})\rangle} = \int_{-\infty}^{\infty}\int_{-\infty}^{\infty}\int_{-\infty}^{\infty}\int_0^T \bar{Q}(\mathbf{x},\mathbf{x}',t';\,T)S(\mathbf{x}',t')\, dt'\, d\mathbf{x}' \qquad (6.163)$$

where

$$Q(\mathbf{x},\,\mathbf{x}',\,t';\,T) = \frac{1}{T}\int_{t'}^T Q(\mathbf{x},t\,|\,\mathbf{x}',t')\, dt \qquad (6.164)$$

The critical aspect in using (6.163) and (6.164) is, of course, the proper specification of Q. If $T = 1$ yr, it is clear that over the period $(0,T)$ there are a very large number of different meteorological conditions, each of which, in theory, demands a different form for Q. It is not possible to consider individually each of the myriad of these conditions in evaluating (6.163). Therefore, for the purpose of evaluating (6.163) we can divide the time interval $(0,T)$ into two regimes:[1]

Ω_s = domain of $(0,T)$ in which the wind speed exceeds a defined threshold value and the meteorology is steady

Ω_u = domain of $(0,T)$ in which either the wind speed does not exceed a defined threshold value (stagnation conditions) or the meteorology is unsteady

Ω_s is then the collection of all separate time periods during $(0,T)$ in which the wind velocity exceeds a defined threshold value (the lower limit of the anemometers employed), and in which the wind speed and direction are steady. Ω_u is the collection of all separate time periods during $(0,T)$ which do not fall into Ω_s, that is those in which the wind speed is below the threshold measurement level (stagnation conditions) or in which wind speed and direction are changing with time.

On the basis of the definition of Ω_s and Ω_u we can rewrite the inner integral of (6.163) as

$$\int_0^T \bar{Q}(\mathbf{x},\mathbf{x}',t';\, T)S(\mathbf{x}',t')\, dt' = \int_{\Omega s} \bar{Q}(\mathbf{x},\mathbf{x}',t';\, T)S(\mathbf{x}',t')\, dt' + \int_{\Omega u} \bar{Q}(\mathbf{x},\mathbf{x}',t';\, T)S(\mathbf{x}',t')\, dt' \qquad (6.165)$$

Note that the time periods contributing to Ω_s and Ω_u need not all be coincident.

Let us consider the evaluation of the first of the integrals on the right-hand side of (6.165). If t_m and Δt_m denote the beginning and length of the mth interval contained in Ω_s, then that integral can be written as

$$\int_{\Omega s} \bar{Q}(\mathbf{x},\mathbf{x}',t';\, T)S(\mathbf{x}',t')\, dt' = \sum_{m=1}^{M} \int_{t_m}^{t_m+\Delta t_m} \bar{Q}(\mathbf{x},\mathbf{x}',t';\, T)S(\mathbf{x}',t')\, dt' \qquad (6.166)$$

where it is assumed that there is a total of M time periods contributing to Ω_s.[2] From (6.156) and (6.164), we recognize that $T\bar{Q}(\mathbf{x},\mathbf{x}',t';T)$ is equivalent to the mean concentration at location $\mathbf{x}$ at time T, due to a point source at $\mathbf{x}'$ which began emitting at t' (with unchanging meteorological conditions). If T is large, $T\bar{Q}$ is just the steady-state concentration from the point source at $\mathbf{x}'$.

[1] The author is indebted to Dr. Robert G. Lamb for pointing out this development.

[2] Steady meteorology implies that each time period Δt_m is long compared with the characteristic time for advection across the region and that variations in wind speed and direction during Δt_m are sufficiently small. A reasonable value of Δt_m, if T is 1 yr, is 1 day.

For example, for a point source at $(0,0,x_3')$, a common form for $\bar{Q}(\mathbf{x},\mathbf{x}',t';T)$ for large T is

$$\bar{Q}(\mathbf{x},\mathbf{x}',t';T) = \frac{1}{T}\left(\frac{1}{2\pi U\sigma_2\sigma_3}\right)\exp\left(-\frac{x_2^2}{2\sigma_2^2}\right)\left[\exp\left(-\frac{(x_3-x_3')^2}{2\sigma_3^2}\right) + \exp\left(-\frac{(x_3+x_3')^2}{2\sigma_3^2}\right)\right] \tag{6.167}$$

Let us denote the right-hand side of (6.167) by $(1/T)g(\mathbf{x},\mathbf{x}',U,\boldsymbol{\sigma},\theta)$, where θ denotes the wind direction and $\boldsymbol{\sigma}$ denotes (σ_2,σ_3). The parameters U, θ, and $\boldsymbol{\sigma}$ are determined by the meteorological conditions prevailing over the interval, and are the conditions which began at t'. Thus, we write (6.167) as

$$\bar{Q}(\mathbf{x},\mathbf{x}',t';T) = \frac{1}{T}g(\mathbf{x},\mathbf{x}',U(t'),\theta(t'),\boldsymbol{\sigma}(t')) \tag{6.168}$$

Although the wind speed and direction have a continuous range of values, it is convenient in the evaluation of (6.166) to consider wind speed and direction as lying in a certain number of discrete ranges. Thus, wind speed can be divided into N_1 categories, for example

$$n = 1 \qquad 4 \le U < 6 \text{ km/hr}$$

$$n = 2 \qquad 6 \le U < 8 \text{ km/hr, etc.}$$

Wind direction can be divided into N_2 categories, for example

$$n = 1 \qquad 0^\circ \le \theta < 45^\circ$$

$$n = 2 \qquad 45^\circ \le \theta < 90^\circ\text{, etc.}$$

Let Δt_{ijm} be the length of time that the mean speed U is within the ith speed-class interval and θ is within the jth direction-class interval during the mth time period of Ω_s. Assuming that the source strength is constant over each interval Δt_{ijm}, we can rewrite (6.166) as

$$\int_{\Omega_s} \bar{Q}(\mathbf{x},\mathbf{x}',t';T)S(\mathbf{x}',t')\,dt'$$

$$= \frac{1}{T}\sum_{m=1}^{M}\sum_{i=1}^{N_1}\sum_{j=1}^{N_2} g(\mathbf{x},\mathbf{x}',U_i,\theta_j,\boldsymbol{\sigma}(t_{ijm}))S(\mathbf{x}',t_{ijm})\,\Delta t_{ijm} \tag{6.169}$$

where t_{ijm} is the time at which the interval commences.

Then, if we define

$$f_{ijm} = \frac{\Delta t_{ijm}}{T} \tag{6.170}$$

as the fraction of the total time period T during which (U,θ,σ) lie in the class interval (U_i,θ_j,σ_m), (6.169) becomes

$$\int_{\Omega_s} \bar{Q}(\mathbf{x},\mathbf{x}',t';T)S(\mathbf{x}',t')\,dt' = \sum_{m=1}^{M}\sum_{i=1}^{N_1}\sum_{j=1}^{N_2} f_{ijm}\,g(\mathbf{x},\mathbf{x}',U_i,\theta_j,\boldsymbol{\sigma}(t_{ijm}))S(\mathbf{x}',t_{ijm}) \tag{6.171}$$

This equation essentially states that the contribution to the long-time average concentration at point **x**, due to sources at **x**′ during periods in which the meteorology is steady, is a sum of contributions from a number of periods over which meteorological conditions fall into discrete categories.

The total long-time average concentration is obtained from (6.163) and (6.165) as

$$\begin{aligned}\overline{\langle c(\mathbf{x})\rangle} &= \int_{-\infty}^{\infty}\int_{-\infty}^{\infty}\int_{-\infty}^{\infty}\sum_{m=1}^{M}\sum_{i=1}^{N_1}\sum_{j=1}^{N_2} f_{ijm}\,g(\mathbf{x},\mathbf{x}',U_i,\theta_j,\boldsymbol{\sigma}(t_{ijm}))S(\mathbf{x}',t_{ijm})\,\Delta t_{ijm}\,d\mathbf{x}' \\ &\quad + \int_{-\infty}^{\infty}\int_{-\infty}^{\infty}\int_{-\infty}^{\infty}\int_{\Omega_u}\bar{Q}(\mathbf{x},\mathbf{x}',t';T)S(\mathbf{x}',t')\,dt'\,d\mathbf{x}'\end{aligned} \tag{6.172}$$

Because the form of $\bar{Q}$ under dynamic or stagnant conditions is very difficult to determine, most steady-state models based on (6.172) neglect the contribution to $\overline{\langle c(x)\rangle}$ from Ω_u. Unfortunately, during stagnation periods the contribution to the overall long-term average concentration $\overline{\langle c(\mathbf{x})\rangle}$ is the most important. Thus, steady-state models based only on Ω_s may underestimate the long-term average concentration.

6.7.2 Reactive Contaminants

A dynamic model for reactive contaminants must be based on (6.14). In addition to requiring a source description, giving the spatial and temporal distribution of emissions from all significant sources in the region, and a meteorological description, including wind speed and direction at each location in the region as a function of time, the vertical atmospheric temperature profile, and radiation intensity, a kinetic mechanism describing the rates of atmospheric chemical reactions as a function of the concentrations of the various species present is required.

Steady-state models for reactive contaminants do not exist because conditions under which reactive pollutant concentrations are not changing with time are virtually nonexistent. Therefore, to estimate the annual average concentration at a particular location, one would have to perform a number of dynamic simulations representing typical days in the spectrum of meteorological conditions, and then average the results according to the expected frequency of occurrence of each condition.

Table 6.12 URBAN AIRSHED MODELING STUDIES

Model used	Investigator	Region applied to	Pollutants
	Steady-State Models		
Box model†	Reiquam (1970a,b)	Willamette Valley	SO_2
		Northern Europe	SO_2
Gaussian plume model	Pooler (1961)	Nashville, Tenn.	SO_2
	Turner (1964)	Nashville, Tenn.	SO_2
	Clarke (1964)	Cincinnati, Ohio	SO_2, NO_X
	Fortak (1966)	Bremen, Germany	SO_2
	Hilst and Bowne (1966)	Ft. Wayne, Ind.	Particles
	Pooler (1966)	St. Louis, Mo.	Particles
	Miller and Holzworth (1967)	Washington, D.C.	NO_X
		Los Angeles, Calif.	NO_X
		Nashville, Tenn.	SO_2
	Davidson (1967)	New York, N.Y.	SO_2
	Koogler et al. (1967)	Jacksonville, Fla.	SO_2
	Panofsky and Prasad (1967)	Johnstown, Pa.	Particles
	Hilst el al. (1967)	Connecticut	Hydrocarbons CO, NO_X, SO_2, particles
	Martin (1971)	St. Louis	SO_2
	Johnson et al. (1972, 1973)	Chicago, San Jose	CO
	Dynamic Models		
Lagrangian puff model	Croke et al. (1968a,b,c), Roberts et al. (1971)	Chicago, New York	SO_2
	Shieh et al. (1970)	New York	SO_2
	Lamb and Neiburger (1971)	Los Angeles	CO
K theory	Randerson (1970)	Nashville	SO_2
	Eschenroeder and Martinez (1972)‡	Los Angeles	CO, NO, NO_2, O_3 hydrocarbons
	Reynolds et al. (1973a,b, 1974), Roth et al. (1974)	Los Angeles	CO, NO, NO_2, O_3 hydrocarbons
	Shir and Shieh (1973)	St. Louis	SO_2
	MacCracken et al. (1972)	San Francisco	CO

† In the box model pollutant concentrations are assumed to be uniform in each of a number of well-mixed cells that comprise the airshed. Within each cell it is assumed that sources are distributed uniformly, emitted pollutants are instantaneously and uniformly mixed, a uniform wind characterizes transport, and a constant cell size is typical of time-averaged meteorology. Unless a large number of cells are used, the box model lacks the spatial resolution needed to represent spatial concentration variations in the airshed.

‡ The model used by Eschenroeder and Martinez is based on computing the concentration changes that occur in a hypothetical vertical column of air which is advected through the airshed by the mean ground-level wind. Thus, this model, a so-called trajectory model, predicts concentration changes along a chosen wind trajectory.

6.7.3 Model Validation

Models are evaluated or "validated" by comparing their predictions with actual air monitoring data under conditions in which all model inputs (emissions, meteorology, etc.) are the same as those during the period the measurements were made. Clearly, for comparisons to be meaningful, predictions and observations must be commensurate. In the case of models based on the numerical solution of species continuity equations, the predicted concentrations represent average concentrations over a region of grid dimensions (say 2 × 2 km), whereas observations are made at a point and are representative of only the local region, say 10 or 50 m^2. As these quantities are not commensurate, formal comparisons can be misleading. Either subgrid scale models must be employed or area-wide observations (as from airborne platforms) carried out.

The models that have been developed to date based on numerical solution of the species continuity equations do not produce predictions that are commensurate with observation. Hence, statistical techniques or other formal modes of comparison have not been needed; comparisons of a qualitative nature have been the appropriate form of representation for evaluation of models. In the future, however, as subgrid scale models are incorporated into airshed models, formal means of comparison will be required.

Table 6.12 summarizes a number of the reported urban air pollution modeling studies available.

REFERENCES

ABRAMOWITZ, M., and I. A. SEGUN (eds.): "Handbook of Mathematical Functions," Dover, New York, 1965.

BATCHELOR, G. K.: Diffusion in a Field of Homogeneous Turbulence. I. Eulerian Analysis, *Australian J. Sci. Res.*, **2**:437 (1949).

BATCHELOR, G. K.: Application of the Similarity Theory of Turbulence to Atmospheric Diffusion, *Q. J. Roy. Meteorol. Soc.*, **76**:133 (1950).

BATCHELOR, G. K.: Diffusion in a Field of Homogeneous Turbulence. II. The Relative Motion of Particles, *Proc. Cambridge Phil. Soc.*, **48**:345 (1952).

BATCHELOR, G. K.: Diffusion in Free Turbulent Shear Flows, *J. Fluid Mech.*, **3**:67 (1957).

BATCHELOR, G. K.: Note on the Diffusion from Sources in a Turbulent Boundary Layer, unpublished, 1959.

BATCHELOR, G. K.: Diffusion from Sources in a Turbulent Boundary Layer, *Arch. Mech. Stosowanej*, **3**:661 (1964).

BLACKADAR, A. K.: A Simplified Two-layer Model of the Baroclinic Neutral Atmospheric Boundary Layer, *Pennsylvania State Univ. AF CRL–65–531*, *Final rept. Contract AF* 604–6641, 49, 1965.

CALDER, K. L.: Eddy Diffusion and Evaporation in Flow Over Aerodynamically Smooth and Rough Surfaces: A Treatment Based on Laboratory Laws of Turbulent Flow with Special Reference to Conditions in the Lower Atmosphere, *Q. J. Mech. Appl. Math.*, **2**:153 (1949).

CALDER, K. L.: On the Equation of Atmospheric Diffusion, *Q. J. Roy. Meteorol. Soc.*, **91**: 514 (1965).

CERMAK, J. E.: Lagrangian Similarity Hypothesis Applied to Diffusion in Turbulent Shear Flow, *J. Fluid Mech.*, **15**:49 (1963).

CHATWIN, P. D.: The Dispersion of a Puff of Passive Contaminant in the Constant Stress Region, *Q. J. Roy. Meteorol. Soc.*, **94**:350 (1968).

CLARKE, J. F.: A Simple Diffusion Model for Calculating Point Concentrations from Multiple Sources, *J. Air Pollut. Control Assoc.*, **14**:347 (1964).

CROKE, E., J. CARSON, F. CLARK, A. KENNEDY, and J. ROBERTS: Chicago Air Pollution System Model, *Argonne Natl. Lab.*, *First Q. Prog. Rept. ANL/ES–CC*–001, 1968a.

CROKE, E. J., J. E. CARSON, D. F. GATZ, H. MOSES, F. L. CLARK, A. S. KENNEDY, J. A. GREGORY, J. J. ROBERTS, R. P. CARTER, and D. B. TURNER: Chicago Air Pollution System Model, *Argonne Natl. Lab. Second Q. Prog. Rept. ANL/ES–CC*–002, 1968b.

CROKE, E. J., J. E. CARSON, D. F. GATZ, H. MOSES, A. S. KENNEDY, J. A. GREGORY, J. J. ROBERTS, K. CROKE, J. ANDERSON, D. PARSONS, J. ASH, J. NORSO, and R. P. CARTER: Chicago Air Pollution System Model, *Argonne Natl. Lab.*, *Third Q. Prog. Rept. ANL/ES–CC*–003, 1968c.

CSANADY, G. T.: Diffusion in an Ekman Layer, *J. Atmos. Sci.*, **26**:414 (1969).

CSANADY, G. T.: Crosswind Shear Effects on Atmospheric Diffusion, *Atmos. Environ.*, **6**:221 (1972).

DAVIDSON, B.: A Summary of the New York Urban Air Pollution Dynamics Research Program, *J. Air Pollut. Control Assoc.*, **17**:154 (1967).

DEACON, E. L.: Vertical Diffusion in the Lowest Layers of the Atmosphere, *Q. J. Roy. Meteorol. Soc.*, **75**:89 (1949).

DRIVAS, P. J., and F. H. SHAIR: Dispersion of a Crosswind Line Source of Tracer Released from an Urban Highway, *Atmos. Environ.*, **8**:475 (1974).

ELLISON, T. H.: Meteorology, *Sci. Prog.* **47**:187, 495 (1959).

ESCHENROEDER, A. Q., and J. R. MARTINEZ: Concepts and Applications of Photochemical Smog Models, *Adv. Chem.*, **113**:101 (1972).

FORTAK, H.: Rechnerische Ermittlung der SO_2-Grundbelastung aus Emissiondaten, Anwendung auf die Verhaltnisse des Stadtgebietes von Bremen mit Abbildungsteil, Institut fur theoretische Meteorologie der freien Universitat Berlin, 1966.

GEE, J. H., and D. R. DAVIES: A Note on Horizontal Dispersion from an Instantaneous Ground Source, *Q. J. Roy. Meteorol. Soc.*, **89**:542 (1963).

GIFFORD, F. A., JR.: Uses of Routine Meterological Observations for Estimating Atmospheric Dispersion, *Nuclear Safety*, **2**:47 (1961).

GIFFORD, F. A., JR.: Diffusion in the Diabatic Surface Layer, *J. Geophys. Res.*, **67**:3207 (1962).

GIFFORD, F. A., JR.: An Outline of Theories of Diffusion in Lower Layers of the Atmosphere, in D. H. SLADE (ed.), "Meteorology and Atomic Energy, 1968," U.S. Atomic Energy Commission, Oak Ridge, Tenn., 1968.

GOLDSTEIN, S.: On Diffusion by Discontinuous Movements and on the Telegraph Equation, *Q. J. Mech. Appl. Math.*, **4**:129 (1951).

HAY, J. S., and F. PASQUILL: Diffusion from a Continuous Source in Relation to the Spectrum and Scale of Turbulence in F. N. FRENKIEL and P. A. SHEPPARD (eds.), "Advances in Geophysics," vol. 6, Academic Press, New York, 1959.

HILST, G. R., and N. E. BOWNE: A Study of Diffusion of Aerosols Released from Aerial Line Sources Upwind of an Urban Complex, A Final Report to the U.S. Army Dugway Proving Ground, vol. I, The Travelers Research Center, Hartford, Conn., 1966.

HILST, G. R., F. I. BADGLEY, J. B. YOCUM, and N. E. BOWNE: The Development of a Simulation Model for Air Pollution Over Connecticut, A Final Report to the Connecticut Research Commission, vols. I and II, The Travelers Research Center, Hartford, Conn., 1967.

JOHNSON, W. B., F. L. LUDWIG, W. F. DABBERDT, and R. J. ALLEN: An Urban Diffusion Simulation Model for Carbon Monoxide, *J. Air. Pollut. Control. Assoc.*, **23**:490 (1973).

KLUG, W.: Diffusion in the Atmospheric Surface Layer: Comparison of Similarity Theory with Observations, *Q. J. Roy. Meteorol. Soc.*, **94**:555 (1968).

KOOGLER, J. B., R. S. SHOLTES, A. L. DAVIS, and C. I. HARDING: A Multivariable Model for Atmospheric Dispersion Prediction, *J. Air Pollut. Control Assoc.*, **17**:211 (1967).

LAMB, R, G.: Numerical Modeling of Urban Air Pollution, Ph. D. thesis, University of California, Los Angeles, 1971.

LAMB, R. G.: Note on Application of K-Theory to Turbulent Diffusion Problems Involving Chemical Reaction, *Atmos. Environ.*, **7**:235 (1973).

LAMB, R. G., and M. NEIBURGER: An Interim Version of a Generalized Urban Air Pollution Model, *Atmos. Environ.*, **5**:239 (1971).

MACCRACKEN, M. C., T. V. CRAWFORD, K. R. PETERSON, and J. B. KNOX: Initial Application of a Multi-box Air Pollution Model to the San Francisco Bay Area, 1972 Joint Automatic Control Conference, Stanford Univ., Stanford, Calif., 1972.

MARTIN, D. O.: An Urban Diffusion Model for Estimating Long Term Average Values of Air Quality, *J. Air Pollut. Control Assoc.*, **21**:16 (1971).

MILLER, M. E., and G. C. HOLZWORTH: An Atmospheric Model for Metropolitan Areas, *J. Air Pollut. Control Assoc.*, **17**:46 (1967).

MONIN, A. S.: Smoke Propagation in the Surface Layer of the Atmosphere, in F. N. FRENKIEL and P. A. SHEPPARD (eds.), "Advances in Geophysics," vol. 6, Academic Press, New York, 1959.

MONIN, A. S., and A. M. OBUKHOV: Basic Laws of Turbulent Mixing in the Ground Layer of the Atmosphere, *Trudy Geofiz. Inst., Akad. Nauk, SSSR*, **24**:151, 163 (1954).

MONIN, A. S., and A. M. YAGLOM: "Statistical Fluid Mechanics," M.I.T., Cambridge, Mass., 1971.

PANOFSKY, H. A., and B. PRASAD: The Effect of Meteorological Factors on Air Pollution in a Narrow Valley, *J. Appl. Meteorol.* **6**:493 (1967).

PASQUILL, F.: The Estimation of the Dispersion of Windborne Material, *Meteorol. Mag.*, **90**:33 (1961).

PASQUILL, F.: "Atmospheric Diffusion," Van Nostrand, London, 1962.

PASQUILL, F.: Lagrangian Similarity and Vertical Diffusion from a Source at Ground Level, *Q. J. Roy. Meteorol. Soc.*, **92**:185 (1966).

PASQUILL, F.: Atmospheric Dispersion of Pollution, *Q. J. Roy. Meteorol. Soc.*, **97**:369 (1971).

PASQUILL, F. and F. B. SMITH: The Physical and Meteorological Basis for the Estimation of the Dispersion, in H. M. ENGLUND and W. T. BEERY, (eds.), "Proceedings of the Second International Clean Air Congress," Academic Press, New York, 1971.

POOLER, F.: A Prediction Model of Mean Urban Pollution for Use With Standard Wind Roses, *Int. J. Air Water Pollut.*, **4**:199 (1961).

POOLER, F.: A Tracer Study of Dispersion Over a City, *J. Air Pollut. Control Assoc.*, **16**:677 (1966).

RANDERSON, D.: A Numerical Experiment in Simulating the Transport of Sulfur Dioxide Through the Atmosphere, *Atmos. Environ.*, **4**:615 (1970).

REIQUAM, H.: An Atmospheric Transport and Accumulation Model for Airsheds, *Atmos. Environ.*, **4**:233 (1970a).

REIQUAM, H.: Sulfur: Simulated Long-range Transport in the Atmosphere, *Science*, **170**:3955, 318 (1970b).

REYNOLDS, S. D., P. M. ROTH, and J. H. SEINFELD: Mathematical Modeling of Photochemical Air Pollution. I. Formulation of the Model, *Atmos. Environ.*, **7**:1033 (1973a).

REYNOLDS, S. D., M. LIU, T. A. HECHT, P. M. ROTH, and J. H. SEINFELD: Mathematical Modeling of Photochemical Air Pollution. III. Evaluation of the Model, *Atmos. Environ.*, **8**: (1974).

REYNOLDS, S. D., M. LIU, T. A. HECHT, P. M. ROTH, and J. H. SEINFELD: Further Development and Validation of a Simulation Model for Estimating Ground Level Concentrations of Photochemical Pollutants, Systems Applications, Inc., Rept. R73-19, San Rafael, Calif., 1973b.

RICHARDSON, L. F.: Atmospheric Diffusion Shown on a Distance-Neighbor Graph, *Proc. Roy. Soc. (London), Ser, A*, **110**:709 (1926).

ROBERTS, O. F. T.: The Theoretical Scattering of Smoke in a Turbulent Atmosphere, *Proc. Roy. Soc. (London), Ser, A*, **104**:640 (1923).

ROTH, P. M., P. J. W. ROBERTS, M. LIU, S. D. REYNOLDS, and J. H. SEINFELD: Mathematical Modeling of Photochemical Air Pollution. II. A Model and Inventory of Pollutant Emissions, *Atmos. Environ.*, **8**:97 (1974).

SAFFMAN, P. G., The Effect of Wind Shear on Horizontal Spread from an Instantaneous Ground Source, *Q. J. Roy. Meteorol. Soc.*, **88**:382 (1962).

SHIEH, L. J., B. DAVIDSON, and J. P. FRIEND: A Model of Diffusion in Urban Atmospheres: SO_2 in Greater New York, Proceedings of Symposium on Multiple Source Urban Diffusion Models, *U.S. Environ. Prot. Agency Publ. AP*–86, 1970.

SHIR, C. C., and L. J. SHIEH: A Generalized Urban Air Pollution Model and Its Application to the Study of SO_2 Distributions in the St. Louis Metropolitan Area, *IBM Res. Lab. Rept. RJ* 1227, San Jose, Calif., 1973.

SINGER, I. A., and N. E. SMITH: Atmospheric Dispersion at Brookhaven National Laboratory, *Int. J. Air Water Pollut.*, **10**:125 (1966).

SKLAREW, R. C., A. J. FABRIK, and J. E. PRAGER: A Particle-in-cell Method for Numerical Solution of the Atmospheric Diffusion Equation, and Applications to Air Pollution Problems, Systems, Science and Software, LaJolla, Calif., 1971.

SLADE, D. H. (ed.): "Meteorology and Atomic Energy, 1968," U.S. Atomic Energy Commission, Oak Ridge, Tenn., 1968.

SMITH, F. B.: The Diffusion of Smoke from a Continuous Elevated Point Source into a Turbulent Atmosphere, *J. Fluid Mech.*, **2**:49 (1957).

SMITH, F. B.: The Role of Wind Shear in Horizontal Diffusion of Ambient Particles, *Q.J. Roy Meteorol. Soc.*, **91**:318 (1965).

SUTTON, O. G.: A Theory of Eddy Diffusion in the Atmosphere, *Proc. Roy. Soc.* (*London*), *Ser. A*, **135**:143 (1932).

SUTTON, O. G.: "Micrometeorology," McGraw-Hill, New York, 1953.

TAYLOR, G. I.: Diffusion by Continuous Movements, *Proc. London Math. Soc.*, *Ser.* 2, **20**:196 (1921).

TURNER, D. B.: A Diffusion Model for an Urban Area, *J. Appl. Meteorol.*, **3**:85 (1964).

TURNER, D. B.: Workbook of Atmospheric Dispersion Estimates, *Public Health Service Publ.* 999–AP–26, 1967.

TYLDESLEY, J. B., and C. E. WALLINGTON: The Effect of Wind Shear and Vertical Diffusion on Horizontal Dispersion, *Q. J. Roy. Meteorol. Soc.*, **91**:158 (1965).

WALTERS, T. S.: The Importance of Diffusion Along the Mean Wind Direction for a Ground-level Crosswind Line Source, *Atmos. Environ.*, **3**:461 (1969).

YORDANOV, D.: On Some Asymptotic Formulae Describing Diffusion in the Surface Layer of the Atmosphere, *Atmos. Environ.*, **2**:167 (1968).

YORDANOV, D.: Simple Approximation Formulae for Determining the Concentration Distribution of High Sources, *Atmos. Environ.*, **6**:389 (1972).

PROBLEMS

6.1 A power plant burns 12 tons per hour of coal containing 2.5 percent sulfur. The effluent is released from a single stack of height 70 m. The plume rise is normally about 30 m, so that the effective height of emission is 100 m. The wind on a particular day at the 10-m level is blowing at 4 m/sec. An elevated inversion at 700 m will limit vertical mixing aloft.

(*a*) Assuming that the wind is uniform with altitude, a rough estimate of downwind SO_2 concentrations can be obtained by using the gaussian plume formulas together with an empirical evaluation of the σ's. Assume that the day in question is a sunny summer day. Using the Pasquill-Gifford curves to estimate the σ's, determine the distance to the point of maximum ground-level SO_2 concentration and the SO_2 concentration at this point. Repeat the calculation for an overcast day with the same wind speed.

(*b*) Based on the gaussian plume formula used in (*a*), construct a graph of ground-level centerline SO_2 concentration over distances of 100 m to 20 km for the conditions (sunny) given above. Use log-log coordinates. For the same conditions, draw a graph of ground-level SO_2 concentration versus crosswind distance at downwind distances of 200 m and 1 km.

(*c*) For the conditions above, construct a plot of the vertical SO_2 centerline concentration profile from ground level to the inversion height at distances of 200 m, 1 km, and 5 km.

6.2 An eight-lane freeway is oriented so that the prevailing wind direction is usually normal to the freeway. During a typical day the average traffic flow rate per lane is 30 cars per minute, and the average speed of vehicles in both directions is 80 km/hr. The emission rate of CO from an average vehicle is 90 g/km traveled.

(*a*) Assuming a 5 km/hr wind and conditions of neutral stability, with no elevated inversion layers present, determine the average ground-level CO concentration as a function of downwind distance, using the appropriate gaussian plume formula.

(*b*) A more accurate estimate of downwind concentrations can be obtained by taking into account the variation of wind velocity and turbulent mixing with height. For neutral conditions we have seen that

$$\bar{u}(z) = u_0 z^{1/7}$$

and

$$K_{zz} = \kappa u_* z = 0.4 u_* z$$

should be used. Assume that the 5 km/hr wind reading was taken at a 10-m height and that for the surface downwind of the freeway $u_* = 0.6$ km/hr (grassy field). Repeat the calculation of (*a*). Discuss your result.

6.3 You have been asked to formulate the problem of determining the best location for a new power plant from the standpoint of minimizing the average yearly exposure of the population to its emissions of SO_2. Assume that the long-term average concentration of SO_2 downwind of the plant can be fairly accurately represented by the gaussian plume equation,

$$\langle c(x',y',z') \rangle = \frac{S}{2\pi \bar{u} \sigma_y \sigma_z} e^{-(y'-Y')^2/2\sigma_y^2} \left(e^{-(z'-Z')^2/2\sigma_z^2} + e^{-(z'+Z')^2/2\sigma_z^2} \right) \qquad x' > X' \qquad \text{(A)}$$

where S is the average emission rate of SO_2 in grams per minute, $\bar{u}$ is the wind velocity (assume constant) in the x'-direction, and (X',Y',Z') is the location of the source in the (x',y',z') coordinate system. Assume that

$$\begin{aligned} \sigma_y^2 &= a\bar{u}^\alpha (x' - X')^\beta \\ \sigma_z^2 &= b\bar{u}^\alpha (x' - X')^\beta \end{aligned} \qquad \text{(B)}$$

where a, b, α, and β depend on the atmospheric stability. The wind $\bar{u}$ is assumed to be in the x' direction in (A). A conventional fixed (x,y,z) coordinate system with x and y pointing in the east and north directions, respectively, will not necessarily coincide with the (x',y',z') system which is always chosen so that the wind direction is parallel to x'. The vertical coordinate is the same in each system, that is, $z = z'$. If the (x',y') coordinate system is at an angle θ to the fixed (x,y) system as shown below, then

$$\begin{aligned} x' &= x\cos\theta + y\sin\theta \\ y' &= -x\sin\theta + y\cos\theta \end{aligned} \tag{C}$$

You want to convert the ground-level concentration $\langle c(x',y',0)\rangle$ to $\langle c(x,y,0)\rangle$, the average concentration at $(x,y,0)$ from a point source at (X,Y,Z) with the wind in a direction at an angle of θ to the x axis. Show that the result is

$$\langle c(x,y,0)\rangle = \frac{S}{\pi\sqrt{ab}\,\bar{u}^{1+\alpha}[(x-X)\cos\theta + (y-Y)\sin\theta]^{\beta}} \times \exp\left\{-\frac{b[(X-x)\sin\theta + (y-Y)\cos\theta]^2 + aZ^2}{2\bar{u}^{\alpha}ab[(x-X)\cos\theta + (y-Y)\sin\theta]^{\beta}}\right\} \tag{D}$$

and that this equation is valid as long as

$$(x-X)\cos\theta + (y-Y)\sin\theta > 0 \tag{E}$$

Now let $f(\bar{u}_j,\theta_k)$ be the fraction of the time over a year that the wind blows with speed $\bar{u}_j$ in direction θ_k. Thus, if you consider J discrete wind speed classes and K directions,

$$\sum_{j=1}^{J}\sum_{k=1}^{K} f(\bar{u}_j,\theta_k) = 1 \tag{F}$$

The yearly average concentration at location $(x,y,0)$ from the source at (X,Y,Z) is given by

$$\langle c_{av}(x,y,0)\rangle = \sum_{j=1}^{J}\sum_{k=1}^{K} c(x,y,0) f(\bar{u}_j,\theta_k) \tag{G}$$

where $\langle c(x,y,0)\rangle$ is given by (D). The total exposure of the region Ω is defined by

$$E = \iint_{\Omega} \langle c_{av}(x,y,0)\rangle \, dx\, dy \tag{H}$$

The optimal source location problem is then: Choose (X,Y) (assuming the stack height Z is fixed) to minimize E subject to the constraint that (X,Y) lies in Ω.

Carry through the solution for the optimal location X of a ground-level crosswind line source (parallel to the y axis) on a region $0 \le x \le L$. Let $f_0(\bar{u}_j)$ and $f_1(\bar{u}_j)$ be the fractions of the time that the wind blows in the $+x$ and $-x$ directions, respectively. Thus, determine X to minimize E subject to $0 \le X \le L$. Discuss your result.

Ans. $$X_{opt} = \begin{cases} 0 & \alpha_0 < \alpha_1 \\ L & \alpha_0 > \alpha_1 \end{cases}$$

$$\alpha_i = \frac{S}{\sqrt{\pi}\,a}\frac{L^{1-\beta/2}}{1-\beta/2}\sum_{j=1}^{J} \bar{u}_j^{-\alpha/2} f_i(\bar{u}_j) \qquad i = 0, 1$$

6.4 We have noted that in computing the mean concentration distribution downwind of a continuous line source it is often reasonable to neglect the effect of turbulent diffusion in the direction of the mean wind relative to that of convection. Walters (1969) considered the errors induced by neglecting horizontal diffusion in a situation in which the mean velocity $\bar{u} = U = \text{const}$, $K_{xx} = K_0 z$, and $K_{zz} = K_1 z$. Consider the simpler case of $\bar{u} = U$, $K_{xx} = K$, and $K_{zz} = K$ as one which might be applicable in the vicinity of a heavily traveled roadway. Thus, it is necessary to solve

$$U \frac{\partial \langle c \rangle}{\partial x} = K\left(\frac{\partial^2 \langle c \rangle}{\partial x^2} + \frac{\partial^2 \langle c \rangle}{\partial z^2}\right)$$

subject to

$$\langle c(0,z) \rangle = \delta(z)$$

$$\left.\frac{\partial \langle c \rangle}{\partial z}\right|_{z=0} = 0 \qquad z = 0,\ x > 0$$

$$\langle c(x,z) \rangle = 0 \qquad x \to \pm\infty,\ z \to \infty$$

$$\int_{-\infty}^{\infty} \left(K \frac{\partial \langle c \rangle}{\partial z}\right) dx = S$$

(*a*) Assume that the solution has the form

$$\langle c(x,z) \rangle = \phi(r) e^{Ux/2K}$$

where $r = \sqrt{x^2 + z^2}$. Show that the solution for ϕ is

$$\phi = \frac{S}{\pi K} K_0 \left[\frac{U}{2K}(x^2 + z^2)^{1/2}\right]$$

where $K_0(\cdot)$ is a modified Bessel function of the second kind of order zero. (A related solution appears in Table 6.3.)

(*b*) Compare the concentration distributions predicted by this solution and that predicted by (6.93) for $K_0 = K_1 = KH/2$, where H is the depth of a layer near the surface, up to $z = H$. What general conclusions can you draw about the effect of linearly increasing eddy diffusivities as opposed to uniform diffusivities?

6.5 To account properly for terrain variations in a region over which the transport and diffusion of pollutants are to be predicted, the following dimensionless coordinate transformation is used:

$$\rho = \frac{z - h(x,y)}{Z} \qquad Z = H - h(x,y)$$

where $h(x,y)$ is the ground elevation at point (x,y) and H is the assumed extent of vertical mixing. Likewise, a similar change of variables for the horizontal coordinates may be performed:

$$\xi = \frac{x - x_S}{X} \qquad X = x_N - x_S$$

$$\eta = \frac{y - y_W}{Y} \qquad Y = y_E - y_W$$

where x_N, x_S, y_E, and y_W are the coordinates of the horizontal boundaries of the region. Show that the form of (6.84) in ξ, η, ρ, t coordinates is

$$\frac{\partial\langle c_i\rangle}{\partial t}+\frac{\bar{u}}{X}\frac{\partial\langle c_i\rangle}{\partial\xi}+\frac{\bar{v}}{Y}\frac{\partial\langle c_i\rangle}{\partial\eta}+\frac{W}{Z}\frac{\partial\langle c_i\rangle}{\partial\rho}=$$

$$\frac{1}{X^2}\frac{\partial}{\partial\xi}\left[K_{xx}\left(\frac{\partial\langle c\rangle}{\partial\rho}-\Lambda_\xi\frac{\partial\langle c\rangle}{\partial\rho}\right)\right]-\frac{\Lambda_\xi}{X}\frac{\partial}{\partial\rho}\left[\frac{K_{xx}}{X}\left(\frac{\partial\langle c\rangle}{\partial\xi}-\Lambda_\xi\frac{\partial\langle c\rangle}{\partial\rho}\right)\right]$$

$$+\frac{1}{Y^2}\frac{\partial}{\partial\eta}\left[K_{yy}\left(\frac{\partial\langle c\rangle}{\partial\eta}-\Lambda_\eta\frac{\partial\langle c\rangle}{\partial\rho}\right)\right]-\frac{\Lambda_\eta}{Y}\frac{\partial}{\partial\rho}\left[\frac{K_{yy}}{Y}\left(\frac{\partial\langle c\rangle}{\partial\eta}-\Lambda_\eta\frac{\partial\langle c\rangle}{\partial\rho}\right)\right]$$

$$+\frac{1}{Z^2}\frac{\partial}{\partial\rho}\left(K_{zz}\frac{\partial\langle c\rangle}{\partial\rho}\right)$$

where $W=\bar{w}-\frac{\bar{u}}{X}\Lambda_\xi Z-\frac{\bar{v}}{Y}\Lambda_\eta Z$ and where

$$\Lambda_\xi=\frac{1}{Z}\left(\frac{\partial h}{\partial\xi}+\rho\frac{\partial Z}{\partial\xi}\right)\qquad\Lambda_\eta=\frac{1}{Z}\left(\frac{\partial h}{\partial\eta}+\rho\frac{\partial Z}{\partial\eta}\right)$$

6.6 When a cloud of pollutant is deep enough to occupy a substantial fraction of the Ekman layer, its lateral spread will be influenced or possibly dominated by variations in the direction of the mean velocity with height. The mean vertical position of a cloud released at ground level increases at a velocity proportional to the friction velocity u_*. If the thickness of the Ekman layer can be estimated as $0.2u_*/f$, where f is the Coriolis parameter, estimate the distance that a cloud must travel from its source in mid-latitudes for crosswind shear effects to become important. (For experimental data relating to this question see Csanady, 1969, 1972.)

6.7 Consider a continuous, ground-level crosswind line source of finite length b and strength S (g km^{-1} sec^{-1}). Assume that conditions are such that a gaussian plume model is applicable.

(*a*) Taking the origin of the coordinate system as the center of the line, show that the mean ground-level concentration of pollutant at any point downwind of the source is given by

$$\langle c(x,y,0)\rangle=\frac{S}{\sqrt{2\pi}\,\sigma_z U}\left[\operatorname{erf}\left(\frac{b/2-y}{\sqrt{2}\,\sigma_y}\right)+\operatorname{erf}\left(\frac{b/2+y}{\sqrt{2}\,\sigma_y}\right)\right]$$

(*b*) For large distances from the source, show that the ground-level concentration along the axis of the plume ($y=0$) may be approximated by

$$\langle c(x,0,0)\rangle=\frac{Sb}{\pi\sigma_y\sigma_z U}$$

(*c*) The width w of a diffusing plume is often defined as the distance between the two points where the concentration drops to 10 percent of the axial value. For the finite line source, at large enough distances from the source, the line source can be considered a point source. Show that under these conditions σ_y may be determined from a measurement of w from

$$\sigma_y = \frac{w}{4.3}$$

6.8 A 1000-MW power plant continuously releases from a 70-m stack a plume into the ambient atmosphere of temperature 298°K.

(*a*) Consider a neutral environment and an ambient wind of 20 km/hr. How high above the stack is the plume 200 m downwind?

(*b*) Consider a ground-based inversion 150 m thick through which the temperature changes 0.3°C. If the wind is blowing at 10 km/hr will the effluent plume penetrate the ground-based inversion? What if no wind is blowing?

(*c*) Consider an elevated inversion between 200 and 1200 m in height through which the temperature changes 2°C. Will the plume penetrate the elevated inversion?

(NOTE: Appendix B is required to solve this problem.)

.9 The presence of a fog in an urban area may result in a removal mechanism for certain gaseous pollutants through absorption by the pollutant in the fog droplets. Let us assume that the rate of loss of gaseous pollutant to the fog can be described as a first-order decay. Thus, if the mean concentration of the gaseous pollutant is governed by the atmospheric diffusion equation, a term, $-\alpha\langle c\rangle$, is added to the R.H.S. to account for loss by absorption. This term can be related to the nature of the fog by

$$\alpha\langle c\rangle = \int_0^\infty m(r)g(r)\,dr$$

where $m(r)$ is the mass of contaminant absorbed by a fog droplet of radius r per unit time, and $g(r)$ is the size distribution function of the fog aerosol.

(*a*) Assuming that the absorption can be described by steady-state molecular diffusion to a drop with rapid absorption at the drop surface, show that $m(r) = 4\pi D\langle c\rangle r$, where D is the molecular diffusivity of the species in air.

(*b*) For the size distribution $g(r) = ar^2e^{-br}$, determine α.

(*c*) For a continuous, ground-level line source, determine the ratio of the downwind ground-level gas phase concentration in the presence of the fog to that in its absence.

7

COMBUSTION PROCESSES AND THE FORMATION OF GASEOUS AND PARTICULATE POLLUTANTS

The final two chapters of this book are devoted to sources of air pollutants and their control. Whereas in Chap. 2 we considered global sources of air pollutants, we now focus on those specific processes and equipment which represent the major man-made sources of air pollutants. The present chapter is devoted to an explanation of the processes through which pollutants are formed, since an understanding of these processes is vital to the appreciation of the methods employed to control formation at the source. Chapter 8 presents a discussion of the principles underlying common control methods.

7.1 INTERNAL COMBUSTION ENGINE

7.1.1 Operation of Internal Combustion Engines

There are three common types of internal combustion engines in wide use in the world. The most common is the four-stroke-cycle, spark-ignited internal combustion engine, which is used primarily for passenger cars and light-duty trucks. The second most common is the four- and two-stroke-cycle, compression-ignition internal combustion engine, commonly referred to as a diesel engine. This engine is used for large trucks,

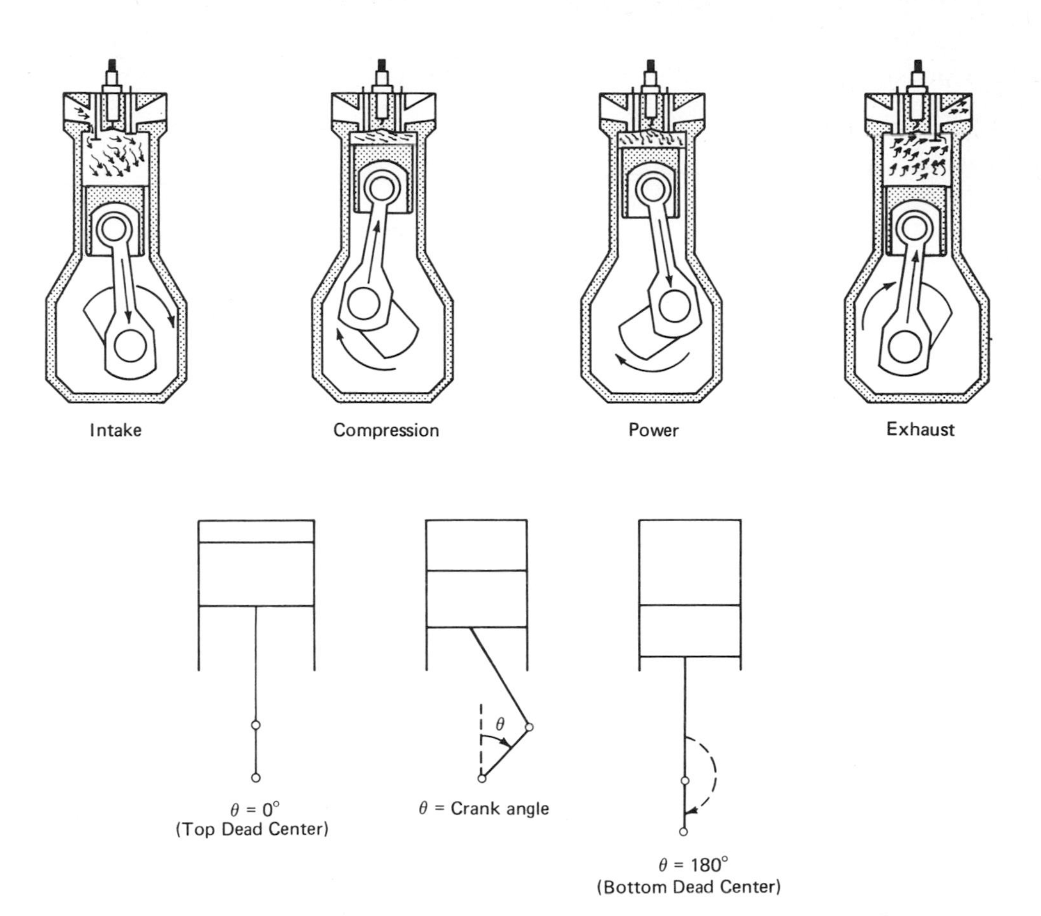

FIGURE 7.1
Four-stroke-cycle internal combustion engine: stroke 1, intake; stroke 2, compression; stroke 3, power; stroke 4, exhaust.

buses, locomotives, and ships. Finally, the third type of internal combustion engine is the aircraft gas-turbine engine. We will not discuss the operation of the gas-turbine engine here.

The operating cycle of the spark-ignited internal combustion engine is shown in Fig. 7.1. The basic principle of operation is that a piston moves up and down within a cylinder, transmitting its motion through a connecting rod to the crankshaft, which drives the vehicle. The four strokes of the spark-ignited internal combustion engine are:

1 Intake: The descending piston draws a mixture of gasoline and air in through the open intake valve.

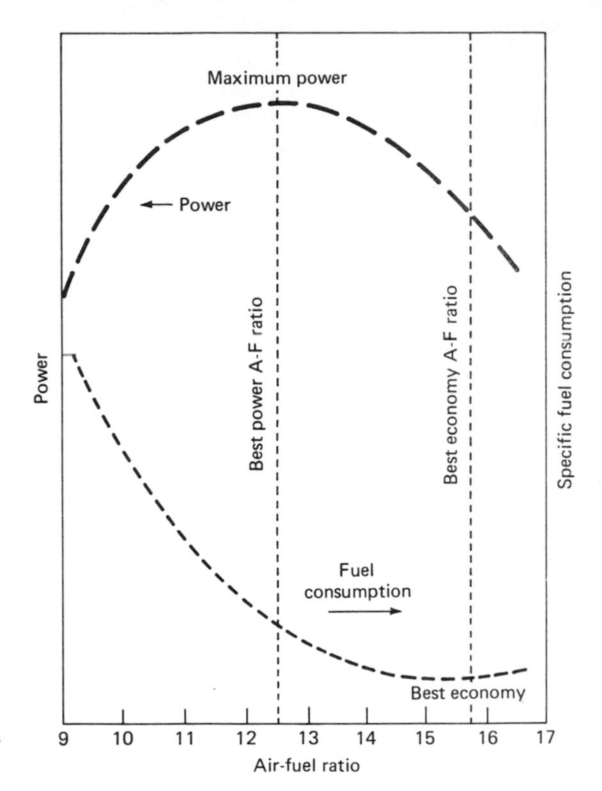

FIGURE 7.2
Effect of air-fuel ratio on power and economy.

2 Compression: The rising piston compresses the fuel-air mixture. Near or at the top of the stroke the spark plug fires, igniting the mixture.

3 Expansion: The burning mixture expands, driving the piston down and delivering power.

4 Exhaust: The exhaust valve opens as the piston rises, expelling the burned gases from the cylinder.

The fuel-air mixture is prepared in the carburetor. This mixture is characterized by the air-fuel ratio, the weight of air per weight of fuel. Figure 7.2 shows power and fuel consumption versus air-fuel ratios commonly used in internal combustion engines. Ratios below 9 and above 20 are generally not combustible. As seen from Fig. 7.2, maximum power is obtained at a lower ratio than for minimum fuel consumption. Mixtures with low air-fuel ratios are referred to as *rich*, whereas those with high ratios are called *lean*. During acceleration, when power is needed, a richer mixture is required than during cruising. We will return to the question of the air-fuel ratio when we consider pollutant formation, since this ratio is one of the key factors governing the type and quantity of pollutants formed in the cylinder.

The ignition system is designed to ignite the air-fuel mixture at the instant when conditions favor optimum power. As engine speed increases, it is most favorable to advance the time of ignition to a point on the compression stroke before the piston reaches top dead center (TDC). This is because the firing of the mixture takes a certain amount of time, and optimum power is developed if the completion of the combustion coincides with the piston arriving at top dead center. The distributor shaft has centrifugal weights attached that advance the spark as engine speed increases. Also, a pressure diaphragm senses airflow through the carburetor and advances the spark as airflow increases. However, as a compromise to reduce the octane requirement of the gasoline, the spark is usually retarded slightly in the compression stroke from the point of optimum power.

In a diesel engine air and fuel are not mixed prior to being passed into the cylinder. Air is drawn in through the intake valve, and while it is being compressed to a high temperature, fuel is injected into the chamber as a spray under high pressure in precise quantities. As the piston nears the top position, the high temperature and pressure of compression cause ignition of the fuel without the aid of a spark. Ignition timing is governed by timing the injection of the fuel, and the power delivered is controlled by the amount of fuel injected in each cycle. The air-fuel mixture in a diesel engine is generally much leaner than that in a spark-ignition engine.

The advantage of good fuel economy in the diesel engine is offset by several disadvantages which have discouraged its use in passenger cars. These disadvantages include (Control Techniques for CO, NO and Hydrocarbon Emissions from Mobile Sources, 1970):

1 High weight-to-power ratio: The high compression ratio necessary to achieve compression ignition requires more rugged construction for the diesel than for the spark-ignition engine.

2 Noise: High rates of pressure rise during combustion make the diesel noisier than a spark-ignition engine.

3 Large size for comparable power: The maximum speed of diesel engines is generally lower than that of gasoline engines because of the difficulty in getting efficient combustion at high speeds. Thus, a larger diesel engine is required to produce the same amount of power that can be obtained from a smaller, higher-speed, spark-ignition engine.

4 Cost: The requirement for rugged engine construction and precision fuel-injection equipment makes the diesel more expensive to manufacture than the spark-ignition engine.

5 Odor and smoke: Exhaust odor and smoke from diesel engines are generally considered more objectionable than those from spark-ignition engines.

7.1.2 Crankcase Emissions

Crankcase emissions are caused by the escape of gases from the cylinder during the compression and power strokes. The gases escape between the sealing surfaces of the piston and cylinder wall into the crankcase. This leakage around the piston rings is commonly called *blowby*. Emissions increase with increasing engine airflow, that is, under heavy load conditions. The resulting gases emitted from the crankcase consist of a mixture of approximately 85 percent unburned fuel-air charge and 15 percent exhaust products. Because these gases are primarily the carbureted fuel-air mixture, hydrocarbons are the main pollutants. Hydrocarbon concentrations in blowby gases range from 6000 to 15,000 ppm. Blowby emissions increase with engine wear as the seal between the piston and cylinder wall becomes less effective. On uncontrolled cars (i.e. a car manufactured before any emission controls were required), blowby gases are vented to the atmosphere by a draft tube and account for about 25 percent of the hydrocarbon emissions.

Blowby was the first source of automotive emissions to be controlled. Beginning with 1963 model year cars, this category of vehicular emissions has been totally controlled in cars made in the United States. The control is accomplished by recycling the blowby gas from the crankcase into the engine air intake to be burned in the cylinders. Control methods for crankcase blowby will be discussed in a little more detail in Chap. 8.

7.1.3 Evaporative Emissions

Evaporative emissions issue from the fuel tank and the carburetor. Fuel-tank losses result from the evaporation of fuel and the displacement of vapors when fuel is added to the tank. The amount of evaporation depends on the composition of the fuel and its temperature. Obviously, evaporative losses will be high if the fuel tank is exposed to high ambient temperatures for a prolonged period of time. The quantity of vapor expelled when fuel is added to the tank is equal to the volume of the fuel added.

Evaporation of fuel from the carburetor occurs primarily during the period just after the engine is turned off. During operation the carburetor and the fuel in the carburetor remain at about the temperature of the air under the hood. But the airflow ceases when the engine is stopped, and the carburetor bowl absorbs heat from the hot engine, causing fuel temperatures to reach 60 to 70°F above ambient. The vaporized gasoline leaves through the carburetor vents to the atmosphere. This condition is called a *hot soak*. The amount and composition of the vapors depend on the fuel volatility, volume of the bowl, and temperature of the engine prior to shutdown. Roughly 10 g of hydrocarbons may be emitted during a hot soak.

Fuel evaporation from both the fuel tank and the carburetor accounts for approximately 20 percent of the hydrocarbon emissions from an uncontrolled automobile. Table 7.1 shows the mass emissions of various hydrocarbons from fuel-tank and carburetor evaporative losses.

It is clear that gasoline volatility is a primary factor in evaporative losses. The measure of fuel volatility is the empirically determined *Reid vapor pressure*, which is a composite value reflecting the cumulative effect of the individual vapor pressures of the different gasoline constituents. It provides both a measure of how readily a fuel can be vaporized to provide a combustible mixture at low temperatures and an indicator of the tendency of the fuel to vaporize. In a complex mixture of hydrocarbons, such as gasoline, the lowest-molecular-weight molecules have the greatest tendency to vaporize and thus contribute more to the overall vapor pressure than do the higher-molecular-weight constituents. As the fuel is depleted of low-molecular-weight constituents by evaporation, the fuel vapor pressure decreases. The measured vapor pressure of gasoline, therefore, depends on the extent of vaporization during the test. The Reid vapor pressure determination is a standard test at 100°F in which the final ratio of vapor volume to liquid volume is constant (4 : 1) so that the extent of vaporization is always the same. Therefore, the Reid vapor pressure for various fuels can be used as a comparative measure of fuel volatility.

Figure 7.3 shows carburetor evaporative loss as a function of temperature and Reid vapor pressure. The volatility and thus the evaporative loss increase with Reid vapor pressure.

Table 7.1 MASS EMISSIONS OF HYDROCARBON CLASSES FROM EVAPORATIVE LOSSES

	Hydrocarbons, g/day	
Components	Tank loss	Carburetor-soak loss
Paraffins		
C_1 to C_5	17.70	13.10
C_6 and heavier	4.50	18.10
Olefins		
C_2 to C_4	0.45	0.45
C_5 and heavier	5.90	7.26
Aromatics		
Total less benzene	1.36	1.81
Benzene	0.90	0.45
Total	30.81	41.17

SOURCE: Hurn (1968).

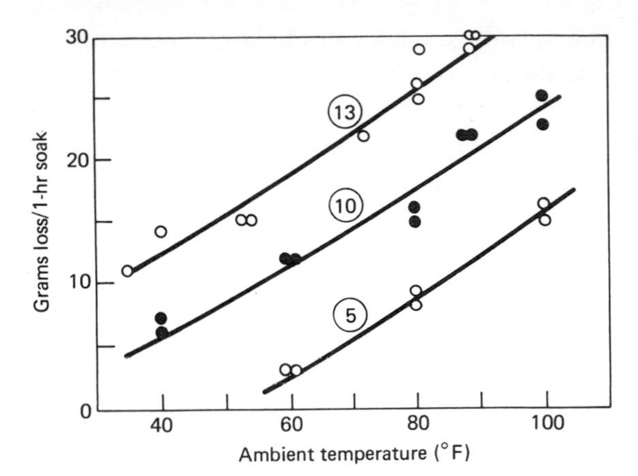

FIGURE 7.3
Variation of carburetor evaporative loss with fuel vapor pressure and ambient temperature. Numbers in large circles are Reid vapor pressure. (*Source: Muller et al.*, 1967.)

7.1.4 Exhaust Hydrocarbon Emissions

Ideal complete oxidation of hydrocarbon fuel yields only CO_2 and H_2O as combustion products. Unfortunately, under the conditions of combustion in an internal combustion engine, other products are formed, including CO, H_2, and partially oxidized hydrocarbons, such as aldehydes. In addition, some of the gasoline remains unburned, and some is thermally cracked to smaller hydrocarbon molecules. When air is used as the source of oxygen, some of the N_2 and O_2 combine to form NO. Finally, particulate matter, particularly involving lead compounds, accompanies the gaseous emissions.

The type and quantity of exhaust contaminants depend on a number of factors including the following:

1 Air-fuel ratio
2 Ignition timing
3 Compression ratio
4 Combustion chamber geometry
5 Engine speed
6 Type of fuel

It is our objective in this section to examine the effect of these variables on exhaust hydrocarbon, CO, and particulate emissions. In the next section we will do the same for exhaust NO emissions.

For a given quantity of fuel, a precise amount of oxygen is required for complete combustion according to the fundamental relationship

$$C_x H_y + (x + \tfrac{1}{4}y)O_2 \rightarrow x\, CO_2 + \tfrac{1}{2}y\, H_2O$$

For example, most hydrocarbon fuels are accurately represented as consisting of 1.85 hydrogen atoms per carbon atoms. Thus, for $x = 1$, $y = 1.85$,

$$CH_{1.85} + 1.460O_2 \rightarrow CO_2 + 0.925H_2O$$

An air-fuel mixture which is theoretically of the precise ratio to obtain complete combustion, with no excess of O_2, is termed a *stoichiometric mixture*. A stoichiometric mixture for a typical gasoline requires about 14.6 parts by weight of air per part by weight of fuel. If the mixture contains less than the stoichiometric amount of air it is said to be rich, whereas with excess air the mixture is termed lean. A common measure, then, of the combustion conditions is the air-fuel ratio, the weight of air divided by the weight of fuel, delivered from the carburetor to the cylinders. Another related measure is the so-called equivalence ratio, the ratio of the actual fuel-air to the stoichiometric fuel-air ratio (note fuel-air and not air-fuel). The straight air-fuel ratio is more commonly used, although the equivalence ratio has the advantage of indicating exactly the deviation from stoichiometric conditions. In Fig. 7.2, we see that maximum power requires a mixture richer than stoichiometric, whereas best fuel economy is achieved with a mixture leaner than stoichiometric.

Figure 7.4 shows the relationship of combustion products to air-fuel ratio, the single most important factor in determining emissions. Combustion of rich mixtures leads to CO formation as well as to the presence of residual fuel in the exhaust, either unburned or partially burned. Lean mixtures, on the other hand, produce considerably less CO and unburned hydrocarbons. However, if the mixture is too lean, above an air-fuel ratio of about 17, the mixture may not ignite properly, leading to misfiring and large amounts of fuel passing through unburned.

The air-fuel ratio is a function of the driving speed, as we described previously. Thus, we expect exhaust emissions to vary depending on the driving mode. Table 7.2 gives typical exhaust gas constituents from an uncontrolled vehicle as a function of driving mode. (Oxides of nitrogen will be discussed in Sec. 7.2.)

During idling, most engines require rich mixtures to compensate for residual combustion products in the cylinder. Thus, CO emissions are high during idling.

Table 7.2 TYPICAL EXHAUST GAS CONSTITUENTS AS A FUNCTION OF DRIVING MODE

Pollutant	Idling	Acceleration	Cruising	Deceleration
Carbon monoxide, %	4–9	0–8	1–7	2–9
Hydrocarbons (as hexane), ppm	500–1000	50–800	200–800	3000–12,000
Oxides of nitrogen, ppm	10–50	1000–4000	1000–3000	5–50

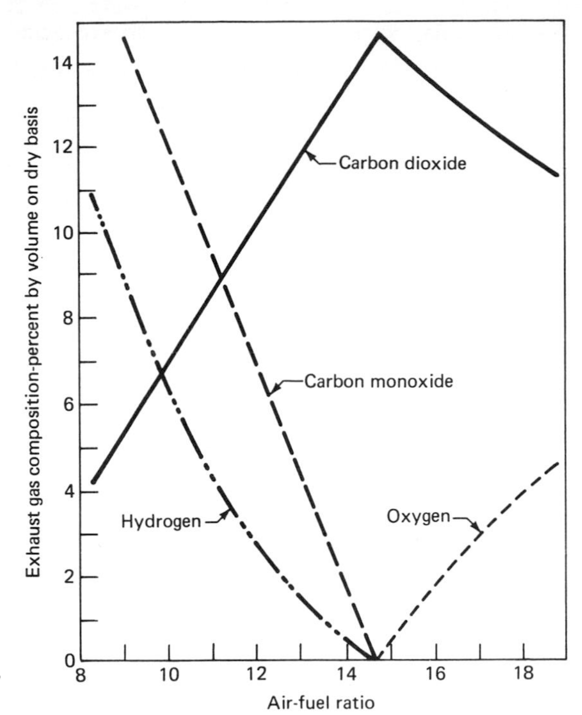

FIGURE 7.4
Relationship of hydrocarbon combustion products to air-fuel ratio.

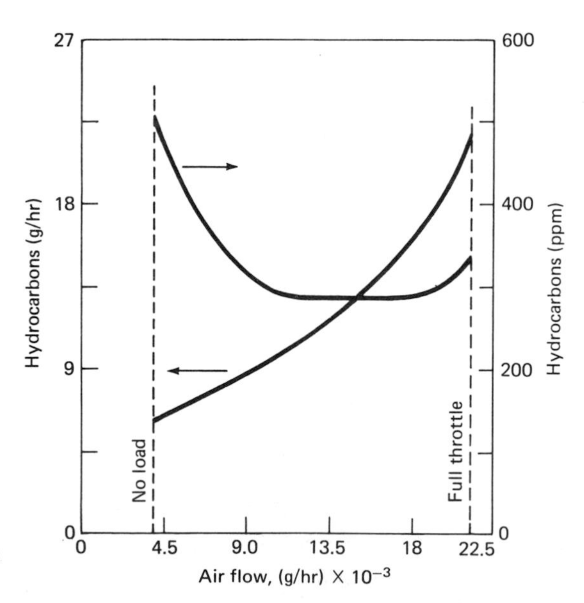

FIGURE 7.5
Effect of engine airflow on hydrocarbon emissions. (*Source: Temple*, 1971.)

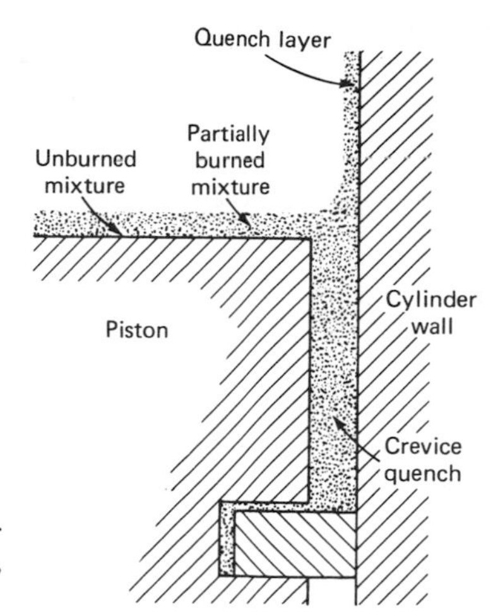

FIGURE 7.6
View of piston and cylinder wall, showing quench layers on piston head, cylinder walls, and piston ring crevice.

At deceleration, residual fuel is present in the cylinders, leading to high levels of unburned hydrocarbons. The concentration of CO increases as the air-fuel ratio decreases, so that the concentration of CO is at a maximum during idle and deceleration. High power requirements, such as maximum acceleration, also produce higher CO concentrations than moderate-power cruising where CO emission levels are at a minimum.

It is important at this point to stress that a clear distinction must be made between emissions expressed as a mass rate (g/hr) or as a fraction of exhaust volume in ppm. For example, Fig. 7.5 shows the effect of engine airflow on hydrocarbon emissions, expressed on both a mass and volume concentration basis. Hydrocarbon emissions are seen to increase nearly linearly when expressed on a weight basis (due to the greater volume of exhaust and fuel consumed) but decrease when expressed on a concentration basis. Clearly, a mass basis is the more relevant measure of exhaust emissions, since it is a direct indication of the quantity of pollutants being emitted. For this reason current automotive emission standards are expressed on a mass basis.

The key question with respect to hydrocarbon emissions is: Why does some of the fuel apparently remain uncombusted in the cylinder when flame propagation appears to be complete? The answer is that flame propagation is actually not complete since the flame fails to touch the cylinder walls. The phenomenon is commonly called *wall quenching* and was first proposed by Daniel and Wentworth (1964) as the mechanism responsible for unburned hydrocarbons in exhaust from internal combustion engines.

In this early investigation it was found by means of magnified photographs that the flame failed to propagate through the mixture located within 0.005 to 0.03 cm of the combustion chamber wall. Figure 7.6 shows the nature of these wall-quench

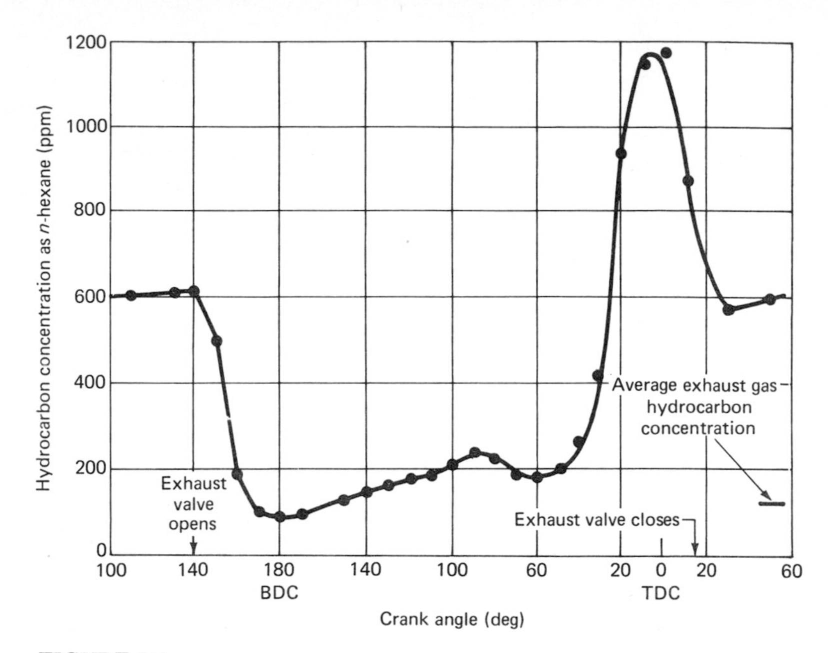

FIGURE 7.7
Variation in hydrocarbon concentration with the crank angle measured 5 cm downstream of the exhaust valve. (*Source: Daniel and Wentworth*, 1964.)

regions. Experiments were performed in which gases in the quench zone of an operating engine were sampled. It was found that the proportion of quench-zone hydrocarbons exhausted from the cylinder is less than the proportion of total gases exhausted, because of trapping in the boundary layer on the walls. Although the residual gas fraction in a normally operating engine may be low, the residual gas hydrocarbon concentration tends to be very high, so that the amount of hydrocarbon recycled may be a significant proportion of the total amount of hydrocarbon left unburned at the wall. The trapping effect can be explained as follows: Gases adjacent to the wall opposite the exhaust valve are farthest from the exit and least likely to be exhausted. Gases along the walls near the exhaust valve have a better chance to be exhausted, but viscous drag slows their movement. Some quenched gases do escape, but on the whole the more completely burned gases at the center of the chamber are preferentially exhausted first with the result that the residual gas has a higher concentration of hydrocarbons than the exhaust gas. The actual fate of that fraction of the quench-zone hydrocarbons remaining in the chamber on a given cycle is unknown; however, it is likely that they are burned in the succeeding cycle. In the experiments reported by Daniel and Wentworth, about one-third of the total hydrocarbons were recycled and probably burned in succeeding cycles.

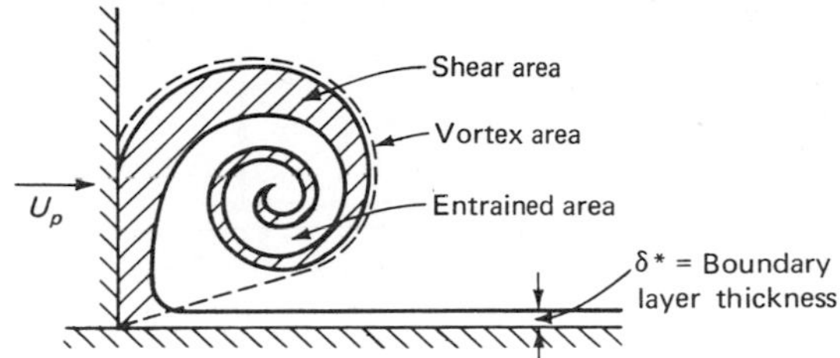

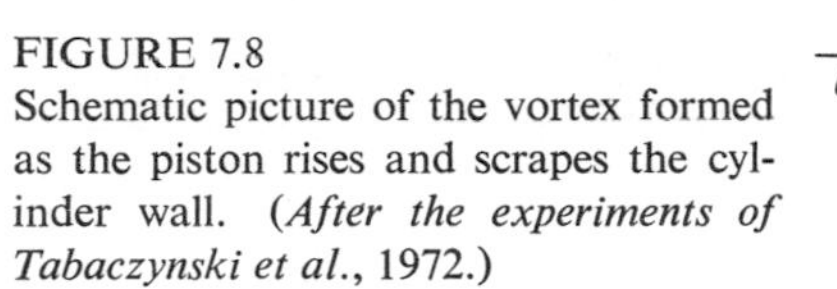
FIGURE 7.8
Schematic picture of the vortex formed as the piston rises and scrapes the cylinder wall. (*After the experiments of Tabaczynski et al.*, 1972.)

Figure 7.7 shows the variation in exhaust hydrocarbon concentration with crank angle experimentally determined by Daniel and Wentworth. As the exhaust valve opened and the emptying of the combustion chamber started, the hydrocarbon concentration decreased rapidly to 100 ppm. As the piston rose, the hydrocarbon concentration rose to 1200 ppm when the piston was near top dead center. At this point the hydrocarbon concentration dropped back to 600 ppm. The early low hydrocarbon concentration is most probably due to gases which were selectively exhausted from the center of the combustion chamber. As the piston rose, the proportion of gases exhausted from near the walls rose, with the hydrocarbon concentration increasing to 1200 ppm, slightly higher than the residual hydrocarbon concentration of 1085 ppm found for the same cylinder. The decrease in hydrocarbon concentration from 1200 to 600 ppm just as the exhaust valve closed was found to be attributable to a flow reversal near the end of the exhaust stroke. Exhausted low-hydrocarbon gases flowed back past the sampling valve which was placed 5 cm from the cylinder exhaust valve.

An attempt to explain the behavior shown in Fig. 7.7 involves a consideration of the fate of the hydrocarbon-rich quench layers adjacent to the walls during the expansion and exhaust strokes. Tabaczynski et al. (1972) have shown what happens to these quench layers by simulating in the laboratory the aerodynamics of the cylinder gases. They found that as the piston moves down on the expansion stroke the gases in the crevices are laid along the cylinder wall. As the piston moves up during the exhaust stroke the layer on the cylinder wall is scraped off the wall and rolled up into a vortex, as depicted in Fig. 7.8. Measurements of hydrocarbon concentrations and mass flow rates as a function of crank angle by Tabaczynski et al. show that, as suggested by the data of Daniel and Wentworth in Fig. 7.7, the unburned hydrocarbons exit the cylinder in two distinct peaks: one at the beginning of the exhaust stroke and one at the end of the stroke. About half the total mass of hydrocarbon emissions is found in each peak. The first peak probably results from entrainment of the quench layer on the top of the cylinder in the first gases to leave the cylinder. The second peak is attributable to the exit of the side wall vortices shown in Fig. 7.8.

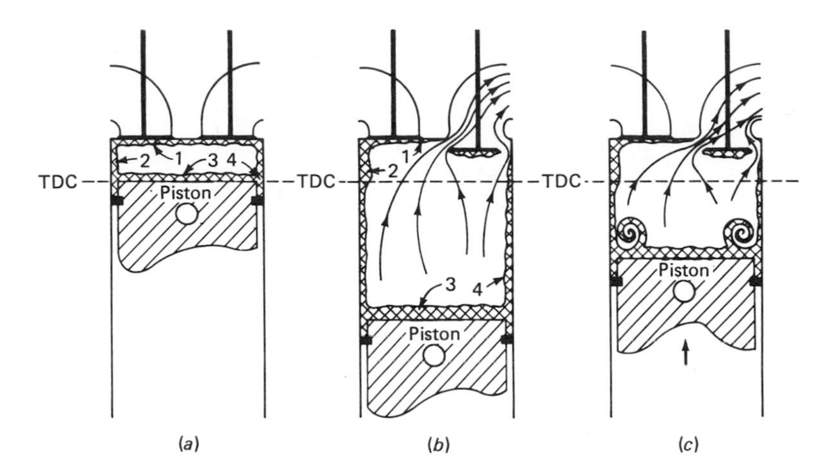

FIGURE 7.9
Summary of processes important in exhaust hydrocarbon emissions in the internal combustion engine. (*a*) Formation of quench layers 1, 2, 3, and crevice quench 4 as flame is extinguished at cool walls. (*b*) Gas in quench volume between piston crown and cylinder wall, 4, expands as cylinder pressure falls and is laid along cylinder walls. When exhaust valve opens, quench layers 1 and 2 exit cylinder. (*c*) Roll-up of hydrocarbon-rich cylinder wall boundary layer into a vortex as piston moves up cylinder during exhaust stroke. [*Reprinted from J. B. Heywood and J. C. Keck*, (1973). *Copyright* 1973 *by the American Chemical Society, Reprinted by permission of the copyright owner.*]

Figure 7.9 summarizes the processes important in the exhaust hydrocarbon emissions we have just discussed. Daniel (1970) formulated an analytical model which relates the effect of engine variables such as air-fuel ratio, airflow rate, compression ratio, engine speed, and ignition timing on the quantity of exhaust hydrocarbon emissions. Using his model and experimental data as a basis, Daniel summarized the following effects of engine variables on hydrocarbon exhaust emissions:

1 Increasing the air-fuel ratio decreased the exhaust hydrocarbon concentration as a result of decreases in the unburned fuel from the crevices, in the fractions of the quench and crevice gases not reacted in the chamber, and in the fraction not reacted in the exhaust system (the fraction of unburned fuel which entered the exhaust system and did not oxidize in the exhaust pipe).
2 Decreasing the airflow rate decreased the exhaust hydrocarbon concentration as a result of decreases of unburned fuel from the quench zone and the crevices and in the fraction of unburned hydrocarbons exhausted from the chamber, even though the fractions of quench and crevice gases not oxidized increased.

3 Lowering the compression ratio decreased the exhaust hydrocarbon concentration as a result of decreases in all factors except the fraction of the quench hydrocarbons not oxidized.

4 Increasing the engine speed decreased the exhaust hydrocarbon concentration as a result of decreases of all factors.

5 Retarding the ignition timing decreased the exhaust hydrocarbon concentration as a result of decreases in the unburned fuel from the crevices and primarily because of decreases in the fraction not oxidized in the exhaust system.

In summary, Daniel found that the unburned fuel from the crevices and, with one exception when it did not change, the fraction not oxidized in the exhaust system always changed in the same direction as the changes in exhaust hydrocarbon concentration. The change in unburned fuel from the crevices consistently caused a large part of the total change in exhaust hydrocarbon concentration. In addition, there appears to be significant oxidation of hydrocarbons in the exhaust system.

Whereas up to now we have concentrated on studies of *total* hydrocarbon concentration, Jackson (1966) and Daniel (1967) studied the effect of engine variables on the composition of exhaust hydrocarbons. As we saw in Chap. 4, total hydrocarbon measures can be misleading unless the breakdown into individual species is given, particularly when smog-forming potential is involved. Because of the tremendous reactivity differences among hydrocarbons, it is conceivable that substantial reductions in total hydrocarbons could lead to no significant decrease in smog-forming potential of the mixture if the high-reactivity components were not selectively removed. Table 7.3, based on data of Jackson (1966), lists the 10 hydrocarbons generally having the highest concentration in the exhaust together with a reactivity rating for each (a product of concentration and reactivity index), using the reactivity scale of Glasson and Tuesday (see Chap. 4). Six of the ten hydrocarbons were low-boiling compounds not present in the original fuel: ethylene, methane, propylene, acetylene, 1-butene (and *i*-butene and 1,3-butadiene), and ethane. The hydrocarbons produced by the engine accounted for 64 percent of the total hydrocarbons. Only four of the ten most prevalent were present originally in the fuel; these were toluene, *p*-, *m*-, and *o*-xylene, *i*-pentene, and *n*-butane. Thus, of the ten most reactive hydrocarbons, five were produced in the engine.

In general, the C_5 and heavier hydrocarbons in the exhaust reflect the composition of the gasoline, whereas the C_4 and lighter components are formed primarily from fuel cracking and do not necessarily reflect the fuel composition. The more reactive of the hydrocarbon species are, in fact, these lighter hydrocarbons. Table 7.3 shows that about two-thirds of the reactive hydrocarbons are formed in the engine and also that this two-thirds accounts for three-quarters of the total reactivity. The exhaust also consists of partial oxidation products, such as aldehydes, alcohols, esters, and ketones. Typical exhaust concentrations of this class are between 50 and 100 ppm. Formaldehyde and acetaldehyde are the most common of the oxygenated species emitted.

Figure 7.10 illustrates the effect of air-fuel ratio on reactivity index and total hydrocarbon concentration as measured by Jackson. Figure 7.11 shows a breakdown by hydrocarbon class. Because of the increase of olefins and aromatics as the air-fuel ratio is increased above 16, the decrease in the reactivity index for lean mixtures was only about half as much as the decrease in total hydrocarbons. Jackson also examined the effect of retarded spark timing and combined lean air-fuel mixtures and retarded spark timing. The key point to be made is that for realistic evaluations of

Table 7.3 PRINCIPAL HYDROCARBONS IN EXHAUST—ORIGIN AND CONTRIBUTION TO PHOTOCHEMICAL REACTIVITY

Most prominent on basis of concentration	Percent of total hydrocarbons
Ethylene	19.0
Methane	13.8
Propylene	9.1
Toluene†	7.9
Acetylene	7.8
1-Butene, *i*-butene, and 1,3-butadiene	6.0
p-, *m*-, and *o*-Xylene†	2.5
i-Pentane†	2.4
n-Butane†	2.3
Ethane	2.3
Total	73.1

Most prominent on basis of reactivity	Percent of total reactivity
Ethylene	17.6
Propylene	16.9
1-Butene, *i*-butene, and 1,3-butadiene	11.3
2-Methyl-2-butene†	8.5
t- and *c*-2-Butene	7.9
Toluene†	5.4
p-, *m*-, and *o*-Xylene†	3.7
Propadiene and methylacetylene	2.2
t- and *c*-2-Pentene†	2.2
2-Methyl-2-pentene†	1.9
Total	77.6

SOURCE: Jackson (1966).

† Fuel components; others are products of fuel cracking or rearrangement in the engine.

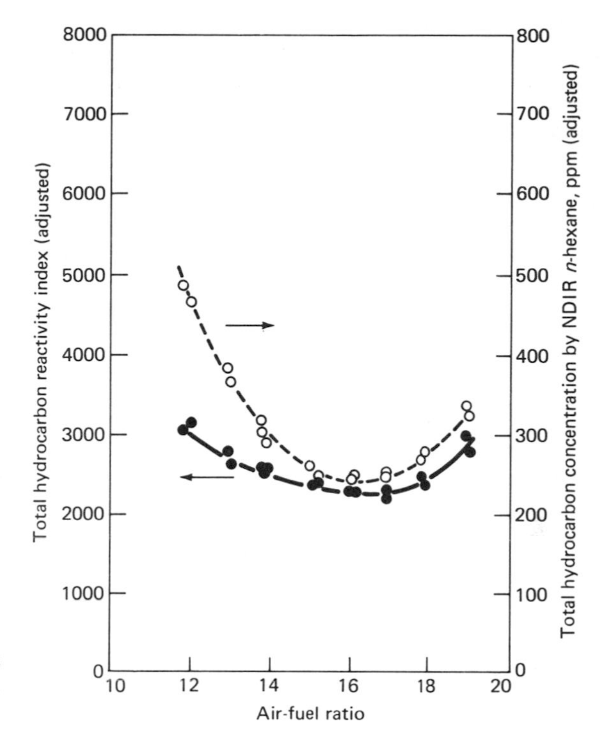

FIGURE 7.10
Effect of air-fuel ratio on reactivity index and total hydrocarbon concentration. (*Source: Jackson*, 1966.)

changes in engine design variables and exhaust control systems it is necessary that individual hydrocarbon exhaust components rather than simply the total hydrocarbon concentration be measured.

An important question is: What will be the effect on emissions of fuel composition changes necessary with lead removal to maintain the same fuel antiknock properties as before the lead was removed? To provide information necessary to assess the effect of changing from typical leaded fuels to unleaded fuels of comparable octane quality, Eccleston and Hurn (1970) experimentally studied the emissions resulting from leaded and unleaded fuels used in a small number of typical vehicles for a typical driving cycle. The tests were designed to determine the characteristics, including photochemical reactivity, of both the exhaust and the evaporative emissions. The principal guideline used in preparation of the lead-free fuel was to match the octane and volatility characteristics of leaded fuels marketed in the United States in 1966 (octane ratings of 95.5 for premium fuel and 89.5 for regular grade fuel with a Reid vapor pressure of 10 lb). Photochemical reactivity of the emissions was experimentally determined by reacting, separately, representative samples of exhaust and evaporative emissions in a smog chamber. Rate of NO_2 formation was used as

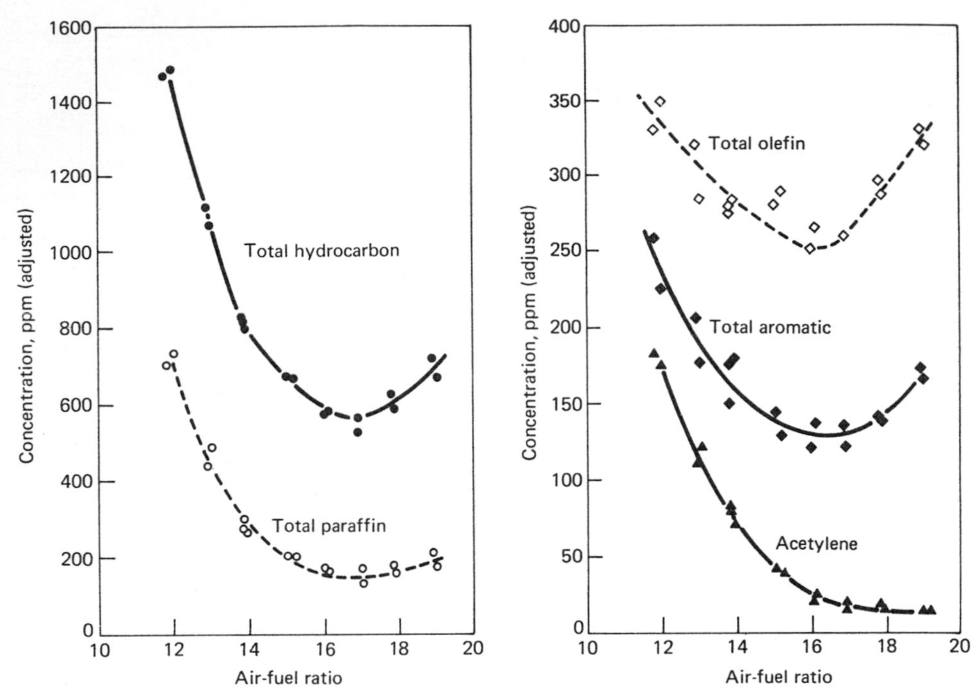

FIGURE 7.11
Effect of air-fuel ratio on hydrocarbon composition by class. (*Source: Jackson*, 1966.)

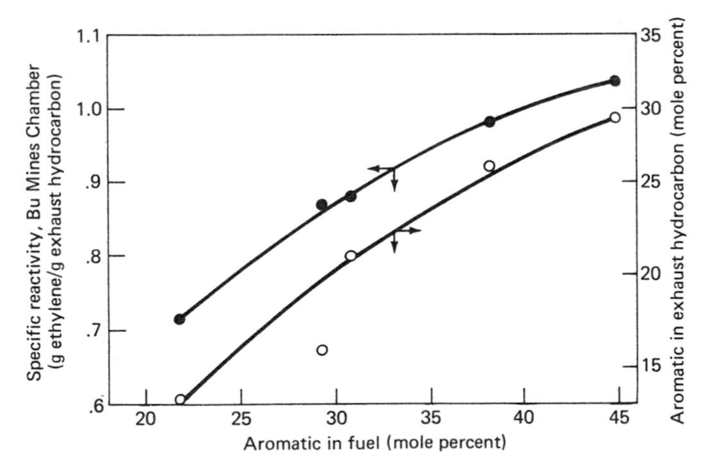

FIGURE 7.12
Influence of fuel composition on exhaust composition and reactivity. (*Source: Eccleston and Hurn*, 1970. *Courtesy of the U.S. Department of the Interior, Bureau of Mines. Reproduced from Report of Investigations* 7390.)

the indicator of reactivity, and the reactivity was expressed as the weight of ethylene which would produce the same rate of NO_2 formation as the sample.

Eccleston and Hurn reached the following conclusions:

1 The gross weights of exhaust hydrocarbons were very nearly equal for the leaded and unleaded fuels. The gross weight of evaporative emissions varied somewhat among all the fuels tested, but the differences could not be attributed to any fuel characteristics related to maintaining octane quality. Thus, the evaporative losses were not influenced by the presence or absence of lead in the fuel.

2 The photochemical reactivity of the exhaust emissions increased about 29 percent in changing from leaded to lead-free gasoline. Figure 7.12 shows that the higher reactivity of exhaust from lead-free fuels is attributable to the higher aromatic content of the fuel.

7.1.5 Exhaust Particulate Emissions

Particulate matter, consisting of carbon, metallic ash, and hydrocarbons, is emitted in the exhaust and crankcase blowby[1] gases of internal combustion engines. Metal-based particles result from lead antiknock compounds in the fuel, metallic lubricating oil additives, and engine wear particles. Carbonaceous and hydrocarbon aerosol results from incomplete combustion of fuel and leakage of crankcase oil past the piston rings into the combustion chamber.

Some of the particulate matter in the exhaust is generated during the combustion process and subsequently nucleated in the exhaust system prior to leaving the automobile. On the other hand, a fraction of the particulate matter deposits on the surfaces of the exhaust system to later flake off and become entrained in the exhaust gas. Therefore, the quantity and nature of the exhaust particulate emissions from an automobile at any time are influenced by several different physical and chemical processes, making the complete description of the character of these emissions a very difficult undertaking. Many factors, such as the mode of vehicle operation, the age and mileage of the car, and the type of fuel being burned, influence the composition and total mass emission rate of exhaust particulate matter.

Under certain driving conditions, the most significant component of exhaust particles is lead compounds resulting from the use of tetraethyl lead as an antiknock

[1] Particles in blowby gases consist basically of lubricating oil, although this loss is now fully controlled.

fuel additive.[1] The total amount of lead in the exhaust is directly proportional to the amount of tetraethyl lead in the gasoline. Most of the major studies of exhaust particulate matter have focused on the lead component. The reasons for this attention are threefold: (1) Lead is a toxic substance (as we noted in Chap. 1), (2) lead deposits in the engine can lead to increased emissions of unburned hydrocarbons, and (3) lead compounds in the exhaust add considerable difficulties to the catalytic treatment of exhaust gases (see Chap. 8). Early studies showed that the sizes of exhausted lead particles vary from 0.1 μm to several millimeters in diameter and that the ratio of fine to coarse particles emitted changes with vehicle speed.

Habibi (1970, 1973) studied the effect of driving conditions and vehicle age on both particulate lead and total particulate emissions. He made the following observations with respect to lead emissions:

1 The percentage of the total lead in the gasoline that is emitted in the exhaust increases as the vehicle speed increases. Percentages measured varied from 10 percent at 20 mi/hr to 90 percent at 70 mi/hr.
2 After 3000 to 5000 mi of operation, a base point emission value of lead salts is reached for a new car. As mileage is accumulated beyond this point, there is wide fluctuation in the rate of lead emission. (Emissions, reported as grams per mile of PbBrCl, varied from 0.05 g/mi, brand new, to 0.1 to 0.15 g/mi, after 3000 to 5000 mi, to 0.2 to 0.4 g/mi after 20,000 mi.)

Both of these observations were explained on the basis of the flake-off and reentrainment of deposited lead during high-speed operation and after aging of the car. The entrainment hypothesis is further confirmed by measurements of the effect of mileage accumulation on the size distribution of lead particles. Habibi showed that over the course of 28,000 miles on a test vehicle the percentage of emitted lead in the coarse-particle fraction increased, whereas there was an associated reduction in the percentage of lead emitted in particles of diameter less than 0.3 μm. For an average car which has accumulated over 20,000 miles, roughly 50 percent of the exhausted lead is associated with particles $<5\mu$m/ in diameter, and about 25 percent is associated with particles $<1\ \mu$m.

[1] Antiknock additives, particularly organometallic compounds, limit the tendency of the fuel-air mixture to autoignite. The most widely used, and effective, of such compounds is tetraethyl lead, $(C_2H_5)_4Pb$. The tendency of a fuel to detonate upon compression is measured by its *octane number*. The octane number of a fuel is defined as the percentage (by volume) of isooctane (2,3,4-trimethyl-pentane) in a mixture of isooctane and *n*-heptane which will just autoignite under the same conditions as the fuel under test. Thus, the addition of tetraethyl lead to a fuel increases its octane number. Tetraethyl lead produces nonvolatile combustion products which accumulate on the spark plugs. However, when ethylene dibromide and ethylene dichloride are also added to the gasoline, the lead compounds formed during combustion are sufficiently volatile to leave with the exhaust gases.

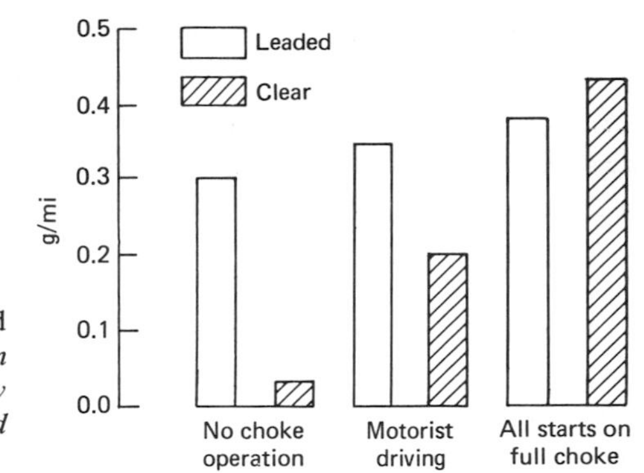

FIGURE 7.13
Total particulate emissions with leaded and unleaded fuels. [*Reprinted from K. Habibi,* (1973). *Copyright* 1973 *by the American Chemical Society. Reprinted by permission of the copyright owner.*]

Studies of the composition of exhaust particles have also been carried out (Lee et al., 1971; Habibi, 1973). Particles larger than 200 μm have a composition similar to exhaust system deposits, confirming that they represent reentrained material. These particles contain about 60 to 65 percent lead salts, 30 to 35 percent Fe_2O_3, and 2 to 3 percent carbonaceous material. The major lead salt is PbBrCl (in particles of sizes greater than about 2μm). Submicron lead particles are primarily $2PbBrCl \cdot NH_4Cl$. Also, there is considerably more carbonaceous material associated with the small particles than with the large ones.

As we have noted, most of the studies on exhaust particulate matter have been concerned with lead compounds. The characterization of the total particulate matter is difficult for several reasons. First, it has been observed that during the first few miles of vehicle operation after a cold start the amount of particulate matter emitted is considerably greater than the amount emitted during the subsequent hot operation (Habibi, 1970). Also, it has been observed that cooling of exhaust gases results in a substantial increase in the quantity of particulate matter. Apparently, as the temperature of the exhaust decreases, components initially in vapor form condense.

Total particulate emissions (in grams per mile) of a test car with both leaded and unleaded fuel are compared in Fig. 7.13 for no choke operation, average motorist driving (one-third starts with full choke action, two-thirds with no choke action), and all starts on full choke. The total particulate emissions with leaded fuel seem to be less sensitive to the degree of choking. Under no choke operation (or hot-running conditions) the unleaded fuel gave substantially lower emissions, with this difference diminishing as the percentage of cold starts increased. In general, cold operation produces from two to eight times more particulate matter than hot operation. The carbon content of the aerosol is about 35 and 70 percent for leaded and unleaded fuel, respectively.

The very strong dependence of total particulate emissions on driving mode points out the need to determine a representative "driving cycle," if one wishes to estimate the total mass emissions from a typical automobile over a typical trip in an urban area. We shall discuss this problem in the next chapter.

7.2 FORMATION OF OXIDES OF NITROGEN DURING COMBUSTION PROCESSES

Oxides of nitrogen are emitted by both stationary and mobile sources. The stationary sources of NO_X fall into two categories: combustion processes and noncombustion processes. Since oxides of nitrogen are formed in virtually all fossil fuel combustion processes, these constitute the major man-made source of NO_X emissions. At the temperatures of a flame, atmospheric nitrogen and oxygen combine to form NO. As the combustion gases cool, the NO formed at the high temperatures becomes thermodynamically unstable. However, at these lower temperatures the rate of decomposition of NO is very slow, so that the NO formed at high temperatures is essentially "frozen." Subsequently, at lower temperatures, a small fraction of the NO produced may be oxidized to NO_2 by the excess O_2 in the combustion gases. Although this reaction rate increases with decreasing temperature, it depends on the square of the NO concentration, which rapidly diminishes as the combustion gases mix with the air. Consequently, over 95 percent of the NO_X emissions from combustion processes are probably NO, with the remainder being NO_2.

In this section we shall discuss the various factors important in the formation of NO in combustion processes. We begin by considering the chemistry of NO formation through the reaction of N_2 and O_2 during combustion.

7.2.1 Formation of NO from N_2-O_2 Mixtures

Although the formation of NO from N_2 and O_2 proceeds by several steps, the overall reaction can be written

$$N_2 + O_2 \rightleftharpoons 2NO$$

The equilibrium constant for this reaction is

$$K_{eq} = \frac{[NO]^2}{[N_2][O_2]} = 21.9e^{-43,200/RT}$$

where $R = 1.987$ cal g mole^{-1} °K^{-1}, T is in °K, and [NO], [N_2], and [O_2] are mole fractions. Using this relation, we can compute the mole fraction of NO formed at equilibrium as a function of temperature from a mixture with the initial composition

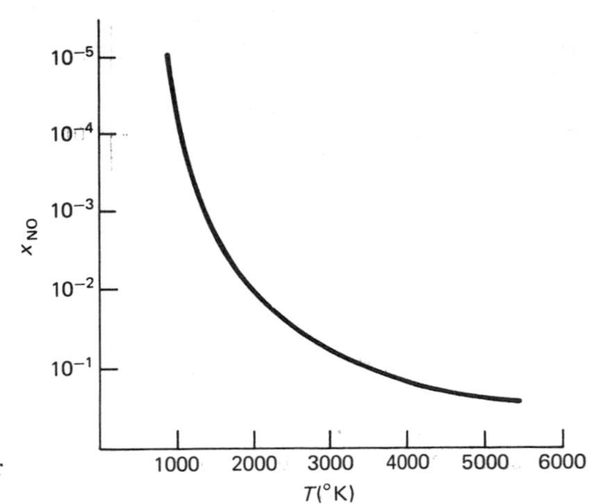

FIGURE 7.14
Equilibrium mole fraction of NO in air as a function of temperature.

of air: $[N_2] = 0.781$, $[O_2] = 0.209$. Figure 7.14 shows [NO] at equilibrium as a function of T°K. We note the very strong temperature dependence of the equilibrium mole fraction of NO.

In 1946 Zeldovich proposed a free-radical chain mechanism for NO formation from air at high temperature. The mechanism is

$$O_2 + M \rightleftharpoons 2O + M$$
$$O + N_2 \rightleftharpoons NO + N$$
$$N + O_2 \rightleftharpoons NO + O$$

A simple rate expression can be derived for NO formation from the Zeldovich mechanism, assuming that O is at equilibrium from the first reaction, N is at pseudo-steady state, and that there is excess air present. The expression is

$$\frac{d[NO]}{dt} = k_f[N_2][O_2]^{1/2} - k_b[NO]^2[O_2]^{-1/2}$$

where $k_f = 9 \times 10^{14}e^{-135{,}000/RT}$, $k_b = 4.1 \times 10^{13}e^{-91{,}600/RT}$, concentrations are in g mole/cm^3, and T is in °K. The extremely large activation energy for NO formation (135,000 cal) indicates that the reaction rate is very temperature-dependent. Thus, modest reduction in flame temperatures can lead to substantial reductions in the amount of NO formed.

Other reactions in the N_2-O_2 system besides those in the Zeldovich mechanism have been identified, and all are summarized in Table 7.4. The first two reactions are simply the dissociation-recombination reactions for O_2 and N_2. The next two are exchange reactions and are quite rapid. Finally, the last two reactions represent the formation of NO directly from molecular and atomic oxygen and nitrogen.

We are interested in NO formation in both the internal combustion engine and stationary combustion equipment such as coal, oil, and gas burners. In each case, fuel is present in addition to air. Although the fuel itself does not play a role in NO formation from atmospheric N_2 and O_2, certain free radicals such as OH and H which result from the fuel combustion may.

7.2.2 Nitric Oxide Formation in the Internal Combustion Engine

Let us first review the experimental evidence related to NO formation in internal combustion engines. Since NO formation is favored by high temperatures, it is clear that NO is formed primarily in the bulk gases in the cylinder as opposed to within a quench zone near the relatively cold chamber walls.

Figure 7.15 compares actual exhaust NO, CO, and hydrocarbon concentrations as a function of air-fuel ratio for a typical automobile. Actually, as we saw with hydrocarbons and CO, the quantity of NO formed depends markedly on the mode of operation of the vehicle. Emissions of NO under varying driving conditions were shown roughly in Table 7.2 and are elaborated in Table 7.5. Concentrations are highest during acceleration and cruising, and mass emissions are highest during acceleration, due to the high volume of exhaust gases produced. At low air-fuel ratios both the amount of available O_2 and the flame temperatures are low, resulting in low NO. As the ratio increases, so do the available O_2, flame temperature, and NO

Table 7.4 REACTIONS AND RATE CONSTANTS FOR THE OXYGEN-NITROGEN SYSTEM

Reaction	k_f (forward rate)	k_r (reverse rate)
$O_2 + M = 2O + M$	$1.13 \times 10^{18} T^{-1/2} \times e^{-118,000/RT}$ cm^3 mole^{-1} sec^{-1}	0.94×10^{15} cm^3 mole^{-2} sec^{-1}
$N_2 + M = 2N + M$	$1.93 \times 10^{17} T^{-1/2} \times e^{-224,900/RT}$ cm^3 mole^{-1} sec^{-1}	$1.09 \times 10^{7} T^{-1/2}$ (cm^3)^2 mole^{-2} sec^{-1}
$O + N_2 = NO + N$	$7 \times 10^{13} \times e^{-75,500/RT}$ cm^3 mole^{-1}sec^{-1}	1.55×10^{13} cm^3 mole^{-1} sec^{-1}
$N + O_2 = NO + O$	$1.33 \times 10^{10} \times e^{-7,080/RT}$ cm^3 mole^{-1} sec^{-1}	6.3×10^{14} cm^3 mole^{-1} sec^{-1}
$N_2 + O_2 = 2NO$	$9.1 \times 10^{24} T^{-5/2} \times e^{-128,500/RT}$ cm^3 mole^{-1} sec^{-1}	$4.8 \times 10^{23} T^{-5/2} \times e^{-85,520/RT}$ cm^3 mole^{-1} sec^{-1}
$NO + M = N + O + M$	$5.3 \times 10^{15} T^{-1/2} \times e^{-150,000/RT}$ cm^3 mole^{-1} sec^{-1}	$1.33 \times 10^{16} T^{-1/2}$ (cm^3)^2 mole^{-2} sec^{-1}

SOURCE: Ammann and Timmins (1966).

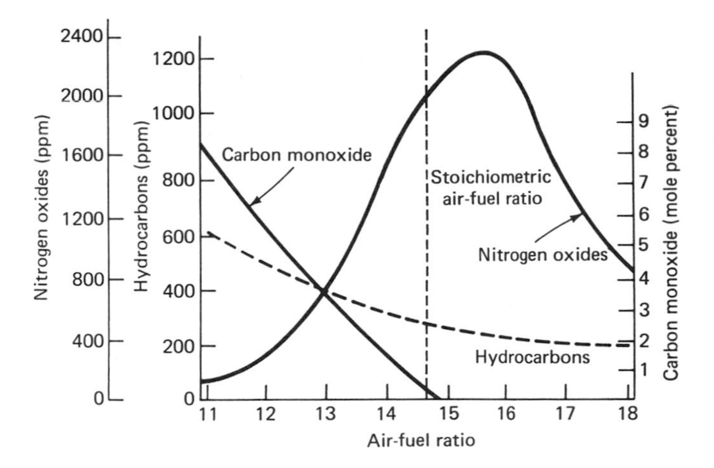

FIGURE 7.15
Exhaust hydrocarbons, carbon monoxide, and nitric oxide as a function of air-fuel ratio.

concentrations. However, as the air-fuel ratio is increased beyond about 16, the flame temperature and thus the NO begin to decrease owing to dilution of the combustion mixture with excess air. Measurements of NO concentrations in exhaust gases as a function of engine operating parameters, such as air-fuel ratio, spark advance, engine speed, etc., have been reported by Wimmer and McReynolds (1961), Hazen and Holliday (1962), Alperstein and Bradow (1966), and Huls and Nichol (1967).

In short, experimental data show that modifications in conditions which increase the peak temperature or the oxygen concentration in the combustion gases increase the NO concentration in the exhaust. In addition, NO levels are found to be nearer the equilibrium concentration corresponding to the peak cycle temperature and pressure than to the equilibrium concentration at exhaust conditions.

Table 7.5 NITRIC OXIDE EMISSIONS UNDER VARYING DRIVING CONDITIONS FOR A TYPICAL VEHICLE

Mode of operation	Air-fuel ratio	Exhaust gas flow, l/min	NO conc., ppm	Mass emission, g/min
Idle	11.9	192	30	0.0075
Cruise 30	13.3	695	1057	1.44
50	13.9	1370	1450	3.78
Acceleration 20–45	12.7	2570	940	4.84
Deceleration 50–0	11.9	192	60	0.0226

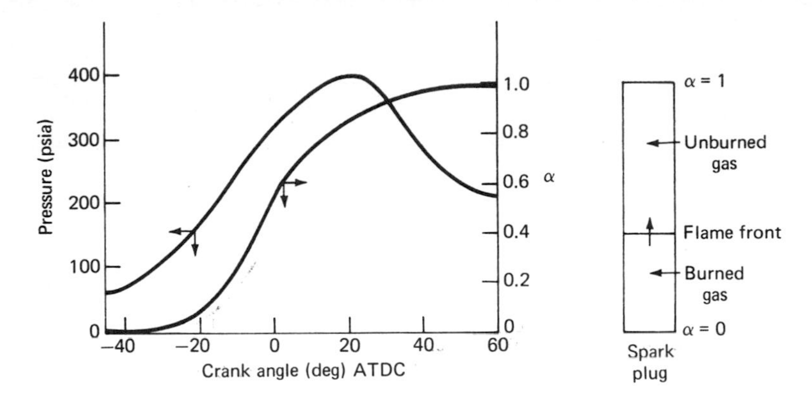

FIGURE 7.16
Pressure and mass fraction of fuel burned as a function of crank angle. Model of the cylinder as a tube with a spark plug at one end. Spark ignites a flame which travels the length of the tube.

Having presented the experimental observations relative to NO emissions as a function of various operating conditions, we now pose the question: Can we account for the levels and behavior of NO concentrations observed? To do this will increase our understanding of the combustion process as well as enable us to assess various methods of control of NO emissions in the internal combustion engine.

The combustion process in an internal combustion engine is shown in Fig. 7.16. Both the pressure p and the mass fraction of the charge, α, are shown as a function of crank angle θ. In this particular case, combustion begins at $\theta = -40°$ with the spark and continues until $\theta = 30°$ (recall $\theta = 0°$ is top dead center, TDC). Because the fuel-air mixture burns over a finite time, during the end of the compression stroke through the beginning of the expansion stroke, different elements of the fuel-air mixture burn at different temperatures and pressures. As shown in Fig. 7.16, we can envision the cylinder as a long tube with spark ignition at one end. The reaction zone, of negligible volume, propagates through the tube, so that at any time the gas within the cylinder consists of a burned fraction and an unburned fraction at the original composition of the charge.

The analysis of NO formation consists of two parts. In the first, the temperature-pressure versus crank angle history of each element of gas in the charge is computed based on the thermodynamics of the fuel combustion. In the second, the reaction rate equations for NO formation are solved. We consider, in turn, these two aspects of the NO formation problem.

7.2.2.1 Thermodynamics of internal combustion The basic model, now widely accepted, for the thermodynamics of internal combustion was presented by Lavoie et al. (1970). The model developed permits calculation of the thermodynamic state of the burned and unburned gases from the pressure and volume of the system as a function of time. The following assumptions relative to the nature of the combustion process are made:

1 The hydrocarbon constituents of the burned gases are at equilibrium; i.e. in the burning zone the combustion reactions are so fast that equilibrium is established.
2 The original charge is homogeneous.
3 The pressure throughout the cylinder is uniform at any time.
4 The volume of gas in the burning zone is negligible.
5 The unburned gas is "frozen" at its original composition and undergoes an adiabatic compression as other elements burn and the pressure rises.

Thus, it is assumed that the energy-producing reactions in the flame, that is, those of the carbon-oxygen-hydrogen system, are sufficiently fast to be in chemical equilibrium, whereas, as we shall see, those associated with NO formation are not. In addition, the charge is considered unmixed in that each element is assumed to be independent of the others and follows its own temperature-pressure history.

Using assumptions 1 to 5, we can determine the mass fraction burned, α, from the laws of conservation of mass and energy and the equations of state for the burned and unburned gases.

The first stage consists of compression of the uniformly mixed unburned charge from bottom dead center (BDC) to θ_0, the crank angle at which combustion begins. The compression is taken to be adiabatic and reversible so that

$$\frac{p(\theta)}{p_m} = \left(\frac{V_i}{V(\theta)}\right)^{\gamma_u} \qquad -180^\circ \leq \theta \leq \theta_0 \tag{7.1}$$

$$\frac{T_u(\theta)}{T_i} = \left[\frac{p(\theta)}{p_m}\right]^{(\gamma_u - 1)/\gamma_u} \qquad -180^\circ \leq \theta \leq \theta_0 \tag{7.2}$$

where γ_u is the ratio of specific heats, p_m is the pressure at BDC (the manifold pressure), T_i is the temperature of the charge at BDC, V_i is the total cylinder volume, and $p(\theta)$, $V(\theta)$, and $T(\theta)$ are the pressure, volume, and temperature at a crank angle θ during the compression stroke. An analytical form of $V(\theta)$ is given by Blumberg and Kummer (1971). At crank angle θ_0 combustion begins. During the combustion process the mass fraction burned, α, rises from 0 to 1.0 according to the relationship

$$\alpha = \frac{1}{2}\left[1 - \cos\left(\frac{\theta - \theta_0}{\Delta\theta_c}\pi\right)\right] \tag{7.3}$$

where $\Delta\theta_c$ is the duration of the combustion in crank angle degrees.

As combustion proceeds, part of the gas will be in the burned state and part will be unburned, as shown in Fig. 7.16. A temperature gradient will exist in the burned gas. Elements which burn late in the charge compress the earlier burned elements adiabatically. Thus, we let $T(\alpha,\alpha')$ be the temperature of an element of burned gas at position α' in the charge when the mass fraction burned is at a value α. The average temperature of the burned gas is defined by

$$\overline{T}_b(\alpha) = \frac{1}{\alpha}\int_0^{\alpha} T_b(\alpha,\alpha')\, d\alpha' \tag{7.4}$$

The objective now is to compute the pressure-temperature history of each element of gas from the beginning to the end of combustion. To do so requires that we simply solve the equations of conservation of mass and energy for the gas in the cylinder. We shall follow the treatment of Blumberg and Kummer (1971).

The internal energies of the burned and unburned gases may be computed if the chemical composition of each fraction is known. Over the temperature ranges of interest it is a reasonable approximation to assume constant specific heats for both burned and unburned gases. Thus, the internal energies are represented by

$$e_u = c_{vu} T_u + a_u \qquad e_b = c_{vb} T_b + a_b \tag{7.5}$$

where a_u and a_b are constants. The average internal energy of the burned fraction is therefore

$$\bar{e}_b(\alpha) = c_{vb}\overline{T}_b(\alpha) + a_b \tag{7.6}$$

An energy balance on the total charge (the first law of thermodynamics) is

$$M\frac{de_T}{d\theta} = \frac{dQ}{d\theta} - p\frac{dV}{d\theta} \tag{7.7}$$

where e_T is the total specific internal energy, $-Q$ is the positive heat transferred out of the charge, M is the total mass of the charge, and V is the cylinder volume at any crank angle θ. We can express e_T as

$$e_T = \bar{e}_b + (1-\alpha)e_u \tag{7.8}$$

Then (7.7) becomes

$$M\left[\alpha\frac{d\bar{e}_b}{d\theta} + (1-\alpha)\frac{de_u}{d\theta} + (\bar{e}_b - e_u)\frac{d\alpha}{d\theta}\right] = \frac{dQ}{d\theta} - p\frac{dV}{d\theta} \tag{7.9}$$

The object is to convert (7.9) into a differential equation for p as a function of θ. To do so we employ

$$T_u(\theta) = T_{u_0}\left[\frac{p(\theta)}{p_0}\right]^{(\gamma_u-1)/\gamma_u} \tag{7.10}$$

which describes the adiabatic compression of the unburned gas during the combustion pressure rise from an initial state at θ_0 of T_{uo}, p_0. Conservation of mass for ideal gases can be written

$$\frac{M\alpha R_b T_b(\theta)}{p(\theta)} + \frac{M(1-\alpha)R_u T_u(\theta)}{p(\theta)} = V(\theta) \tag{7.11}$$

where $R_b = N_b R/M$ and $R_u = N_u R/M$, and N_b and N_u are the number of moles of burned and unburned gas, respectively. Substituting (7.6), (7.10), and (7.11) into (7.9) yields the following differential equation for $p(\theta)$:

$$M\frac{d\alpha(\theta)}{d\theta}\left\{c_{vu}T_{uo}\left[\frac{p(\theta)}{p_0}\right]^{(\gamma_u-1)/\gamma_u}(\gamma_b-\gamma_u)-(a_b-a_u)(\gamma_b-1)\right\}\frac{dp(\theta)}{d\theta}$$

$$= \left[-\gamma_b p(\theta)\frac{dV}{d\theta}+(\gamma_b-1)\frac{dQ}{d\theta}\right]\left(V(\theta)+V_0\left\{\frac{\gamma_b-\gamma_u}{\gamma_u}\left[\frac{p(\theta)}{p_0}\right]^{-1/\gamma_u}[1-\alpha(\theta)]\right\}\right)^{-1} \tag{7.12}$$

In order to specify the temperature of each element of gas we use the relations

$$T_b(\alpha,\alpha) = \frac{c_{pu}}{c_{pb}}T_{uo}\left[\frac{p(\alpha)}{p_0}\right]^{(\gamma_u-1)/\gamma_u} + \frac{a_u-a_b}{c_{pb}} \tag{7.13}$$

and

$$T_b(\alpha,\alpha') = T_b(\alpha',\alpha')\left[\frac{p(\alpha)}{p(\alpha')}\right]^{(\gamma_b-1)/\gamma_b} \tag{7.14}$$

the first of which gives the temperature of the burned gas at the instant of its combustion (each element burns isenthalpically), and the second of which expresses the continuing adiabatic compression (or expansion) after the flame front passes position α'.

From (7.12) to (7.14) we can determine the pressure-temperature history of each element of the charge from the beginning to the end of combustion. Figure 7.17 shows the temperature as a function of crank angle at three points in the charge as computed by Blumberg and Kummer (1971) for a specified set of operating conditions. We see that the first element initially burns at 4450°R and then reaches a peak temperature of 5030°R owing to subsequent compression. The middle element also rises in temperature before the work extracted results in a temperature decline. The last element shows no temperature rise as there is no increase in pressure after it burns. The fuel used in the calculations shown in Fig. 7.17 is 70 percent isooctane and 30 percent xylene. In the calculations $dQ/d\theta$ was set equal to zero on the basis that the flame front movement is very rapid and heat losses should be negligible.

7.2.2.2 Kinetics of NO formation The next step in the analysis is to compute the rate of NO formation in the charge during combustion. We have already seen that NO formation is highly temperature-dependent, so that we expect that NO will form at different rates in the charge, depending on the temperature history of each

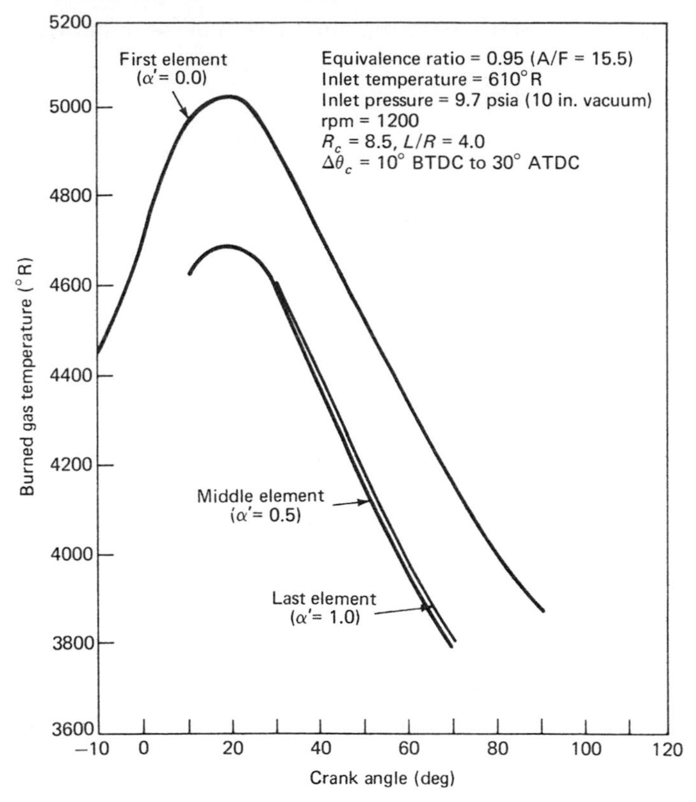

FIGURE 7.17
Burned-gas temperature as a function of crank angle for the first element to burn, the middle element to burn, and the last element to burn. (*Source: Blumberg and Kummer*, 1971.)

element. Since the NO reactions require the thermal energy released by the combustion process, NO formation will take place in the charge only after the flame front has passed through it.

The reactions believed to be most influential in NO formation are those of the Zeldovich mechanism with the addition of the reaction between N and OH:

$$N + NO \underset{}{\overset{k_1}{\rightleftharpoons}} N_2 + O$$

$$N + O_2 \underset{}{\overset{k_2}{\rightleftharpoons}} NO + O$$

$$N + OH \underset{}{\overset{k_3}{\rightleftharpoons}} NO + H$$

where the forward rate constants are

$$k_1 = 1.32 \times 10^{13} \text{ cm}^3 \text{ mole}^{-1} \text{ sec}^{-1}$$
$$k_2 = 1.81 \times 10^8 T^{1.5} e^{-3000/T} \text{ cm}^3 \text{ mole}^{-1} \text{ sec}^{-1} (T\,{}^\circ\text{K})$$
$$k_3 = 4.2 \times 10^{13} \text{ cm}^3 \text{ mole}^{-1} \text{ sec}^{-1}$$

The reverse rate constants can be obtained from the equilibrium constants (Campbell and Thrush, 1968; Bortner, 1969). It is assumed that O, OH, and H are present at their equilibrium concentrations. Thus, O, OH, and H result from the hydrocarbon combustion. Assuming that O, OH, and H are at their equilibrium concentrations and that N atoms are at pseudo steady state, we obtain the following rate equation for NO formation and decomposition (Blumberg and Kummer, 1971):

$$\frac{dy_{NO}}{d\theta} = \left(\frac{RT}{p}\,\frac{\text{cm}^3}{\text{mole}}\right)\left(\frac{60}{360 \text{ rpm}}\,\frac{\text{sec}}{\text{deg}}\right)\left[\frac{2R_1(1-\beta^2)}{\beta R_1/(R_2+R_3)+1}\,\frac{\text{mole}}{\text{cm}^3 \text{ sec}}\right] \quad (7.15)$$

where
y_{NO} = mole fraction of NO
$\beta = y_{NO}/y_{NO_e}$, fractional attainment of equilibrium
y_{NO_e} = equilibrium mole fraction of NO
R_i = forward reaction rate of reaction i evaluated at equilibrium conditions, $i = 1, 2, 3$

When $\beta < 1$, $dy_{NO}/d\theta > 0$ and NO tends to form; when $\beta > 1$, $dy_{NO}/d\theta < 0$ and NO tends to decompose. Equation (7.15) is integrated at each point α' in the charge from the crank angle at which that element initially burns to a crank angle at which the reaction rates are negligible. At this point the quenched value of the NO concentration, $y_{NO}{}^q$, is achieved. The overall mole fraction of NO in the entire charge is given by

$$\bar{y}_{NO} = \int_0^1 y_{NO}{}^q(\alpha')\,d\alpha' \quad (7.16)$$

Nitric oxide concentrations versus crank angle were computed by Blumberg and Kummer at three equivalence ratios, as shown in Figs. 7.18 to 7.20. For each air-fuel ratio, (A/F) both rate calculated and equilibrium NO are shown at three positions in the charge, $\alpha' = 0, 0.5, 1.0$. All other parameters are identical for the three cases. In each case, the major contribution to the total NO formed results from the elements that burn first. They experience the highest temperatures and have the longest time in which to react. Under rich conditions considerable decomposition of NO occurs in the first element because of the high temperatures. However, as the first element cools during expansion under rich conditions, the rate of NO decomposition decreases rapidly with temperature so that after about 40 crank angle degrees the equilibrium

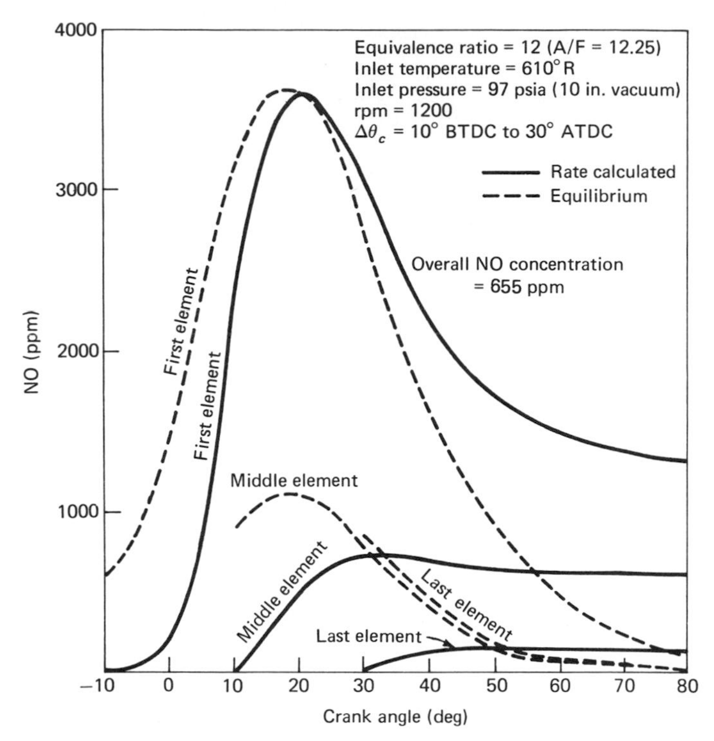

FIGURE 7.18
Nitric oxide concentration in the burned gas as a function of crank angle for the first, middle, and last element of the charge to burn for a rich charge (A/F = 12.25). (*Source: Blumberg and Kummer*, 1971.)

NO level can no longer be obtained. At this point the NO formed is effectively frozen. Under lean operation the temperature of the first element is not as high as under rich operation. Thus, the rate of NO formation is not rapid enough to achieve equilibrium as in the rich-burning case. However, in lean operation there is significantly more oxygen and nitrogen available so that even though equilibrium NO levels are not achieved, the final frozen NO levels are higher than those under rich conditions. Near stoichiometric operation conditions are such that *both* high temperatures and reasonable availability of O_2 and N_2 occur, so that NO levels are largest for this case.

We can now summarize the processes responsible for the production of hydrocarbons and nitric oxide in the internal combustion engine. First, the compressed air-fuel mixture is ignited by the spark plug, and a flame front propagates across the cylinder. As the flame approaches the walls, which are relatively cooler, the flame is

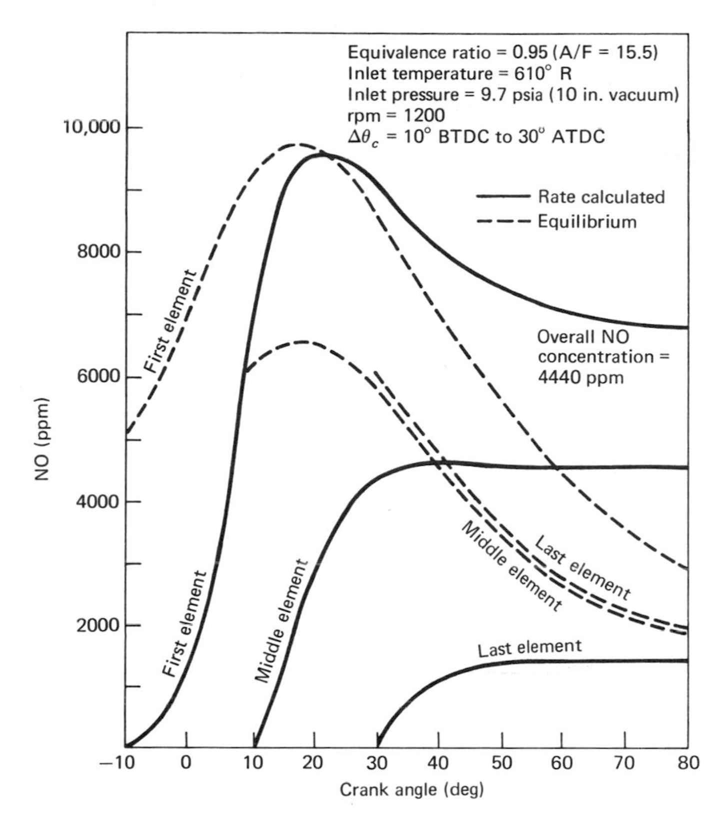

FIGURE 7.19
Nitric oxide concentration in the burned gas as a function of crank angle for the first, middle, and last element of the charge to burn for a charge near stoichiometric (A/F = 15.5). (*Source: Blumberg and Kummer*, 1971.)

quenched, leaving a very thin layer of unburned fuel on the walls and in the crevice between the piston and cylinder wall above the piston ring. During the flame propagation, NO is formed by chemical reactions in the hot just-burned gases. As the piston recedes, the temperatures of the different burned elements drop sharply, "freezing" the NO (i.e. the chemical reactions which would remove the NO become much slower) at the levels formed during combustion, levels well above those corresponding to equilibrium at exhaust temperatures. As the valve opens on the exhaust stroke, the bulk gases containing the NO exit, entraining the unburned hydrocarbons in the wall layers. It is to these processes that we must devote our attention if we wish to reduce both hydrocarbon and NO formation in the cylinder. Control methods for exhaust hydrocarbons and NO, including those directed at the processes in the cylinder, will be studied in the next chapter.

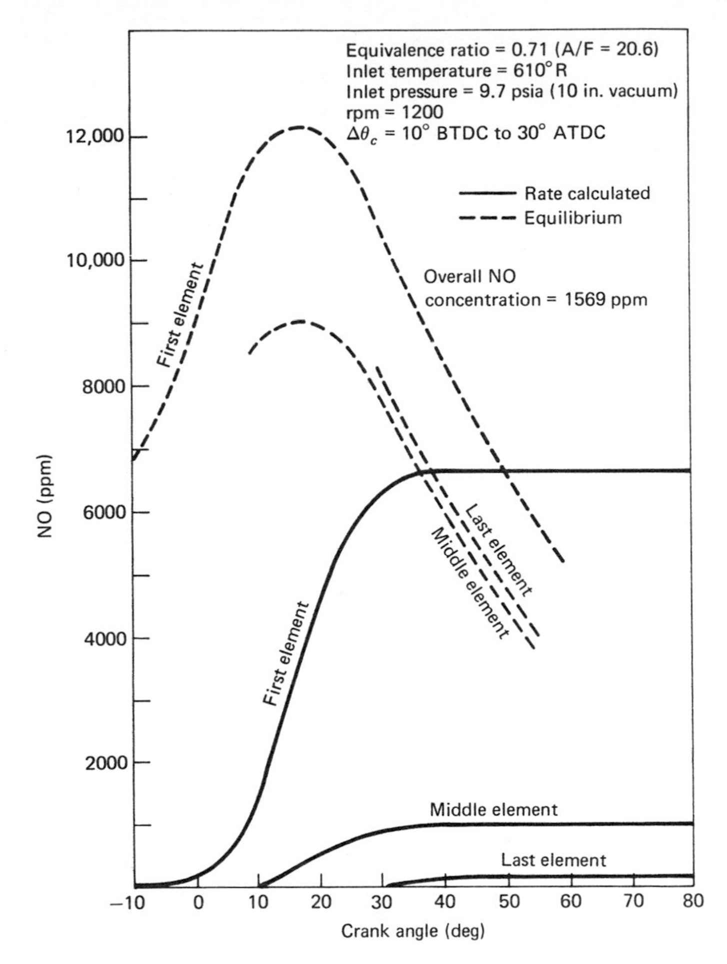

FIGURE 7.20
Nitric oxide concentration in the burned gas as a function of crank angle for the first, middle, and last element of the charge to burn for a lean charge (A/F = 20.6). (*Source: Blumberg and Kummer*, 1971.)

7.2.3 Nitric Oxide Formation in Stationary Combustion

We have already discussed in detail the formation of NO in the internal combustion engine. In this subsection our focus will be on the formation of NO during fuel combustion in burners. Nitric oxide is formed by either of two avenues: (1) high-temperature reaction of atmospheric nitrogen with oxygen and (2) oxidation of chemically bound nitrogen in the fuel. In the case of the internal combustion engine we considered only the former route because nitrogen is virtually nonexistent in gasoline. In coal, however, chemically bound nitrogen is common, resulting in NO formation by both avenues. Because atmospheric nitrogen is extremely stable (bond dissociation energy of 225 kcal/mole) relative to carbon-nitrogen bonds in fuel molecules (bond dissociation energies from 60 to 150 kcal/mole), the activation energy required for oxygen to react with nitrogen in the fuel is considerably lower than that required for reacting with molecular nitrogen. Thus, NO formation by oxidation of nitrogen in fuel occurs rapidly and is generally unaffected by changes in combustion conditions. Henceforth, we will consider NO formation only by fixation of atmospheric molecular nitrogen.

As we have previously noted, the only oxide of nitrogen of appreciable importance formed during combustion is NO. Thermodynamic equilibrium highly favors

Table 7.6 EMISSION FACTORS FOR NITROGEN OXIDES DURING COMBUSTION OF FUELS AND OTHER MATERIALS

Source	Average emission factor (as NO_2)
Fuels	
Coal	
Household and commercial	4 g/kg
Industry and utility	10 g/kg
Fuel oil	
Household and commercial	5.45–32.7 g/gal
Industry	32.7 g/gal
Utility	43.3 g/gal
Natural gas	
Household and commercial	1680 kg/10^6 m^3
Industry	3100 kg/10^6 m^3
Utility	5650 kg/10^6 m^3
Wood	5.5 g/kg
Combustion sources	
Gas engines	
Oil and gas production	11,150 kg/10^6 m^3
Gas plant	62,400 kg/10^6 m^3
Pipeline	106,000 kg/10^6 m^3
Refinery	64,000 kg/10^6 m^3
Gas turbines	2900 kg/10^6 m^3
Waste disposal	
Open burning	5.5 g/kg
Conical incinerator	0.33 g/kg
Municipal incinerator	1 g/kg
Chemical industries	
Nitric acid manufacture	28.5 g/kg HNO_3 product
Adipic acid	6 g/kg product
Terephthalic acid	6.5 g/kg product
Nitrations	
Large operations	0.1–7 g/kg HNO_3 used
Small batches	1–130 g/kg HNO_3 used

SOURCE: Control Techniques for Nitrogen Oxide Emissions from Stationary Sources (1970).

the formation of NO over NO_2 at temperatures in a flame. Once NO is formed, it can be oxidized to NO_2, although residence times in conventional combustion equipment are too short for an appreciable fraction (more than 5 percent) of the NO to be oxidized to NO_2.

Our objective, then, is to predict the rate of NO formation in a combustion chamber of given geometry as a function of the operating variables available, such as amount of excess air, type of fuel, amount of premixing of fuel and air, flow rates of fuel and air, etc., The main factors influencing NO formation are flame temperature, amount of excess air present, and the length of time that combustion gases are maintained at flame temperature. The lower the excess air, the lower is the amount of NO formed. However, too much excess air can dilute the flame and decrease the flame temperature, decreasing NO. For very short residence times, NO will not reach equilibrium. In most burners, gases are cooled rapidly, "freezing" the NO at a level exceeding the equilibrium level at flue gas temperatures.

In reality, the physical and chemical nature of a flame is quite complex, varying with fuel type, fuel state (gaseous to fine droplets), and burner design. For this reason, NO concentrations vary widely in flames and are difficult to predict. The major factors influencing NO formation in combustion processes are the amount of excess air used, the rate of heat removal, fuel type and composition, and the transport rates of air and fuel as influenced by the burner geometry. We shall discuss these factors in more detail in Chap. 8.

Table 7.6 presents NO_X emission factors for combustion processes. (The values given are as NO_2.)

7.3 PARTICULATE EMISSIONS FROM STATIONARY SOURCES

The main stationary sources of particulate matter are combustion processes and industrial operations.

The principal particulate emission from combustion is fly ash, incombustible solid portions of fuels too small to settle out in the combustion chamber. Fly ash is usually composed of a large number of inorganic compounds. A typical fly ash from coal might have the following composition.

Silica, SiO_2	~40%
Alumina, Al_2O_3	~20%
Iron oxide, Fe_2O_3	~15%
Calcium oxide, CaO	~10%
Carbon Magnesium oxide, MgO Other	~15%

Table 7.7 presents a summary of particulate emissions for industrial processes. Discussion of the various processes can be found in Control Techniques for Particulate Air Pollutants (1969), the source of Table 7.7.

Table 7.8 presents a detailed summary of emission factors for particulate air pollutants. The factors are for uncontrolled sources except as noted in the table.

Table 7.7 INDUSTRIAL PROCESS SUMMARY FOR PARTICULATE EMISSIONS

	Particulate emissions	
Industry or process	Nature	Principal sources
Iron and steel mills	Iron oxide dust, smoke	Blast furnaces, steel making furnaces, sintering machines
Gray-iron foundries	Iron oxide dust, smoke, oil and grease, metal fumes	Cupolas, shakeout systems, core making
Nonferrous smelters	Smoke, metal fumes, oil and grease	Smelting and melting furnaces
Petroleum refineries and asphalt blowing	Catalyst dust, ash, sulfuric acid mist, liquid aerosols	Catalyst regenerator, sludge incineration, air blowing of asphalt
Portland cement	Alkali and product dusts	Kilns, coolers, dryers, material handling systems
Kraft pulp mills	Chemical dusts, mists	Chemical reclaiming furnaces, smelt tanks, lime kilns
Asphalt batch plants	Aggregate dusts	Dryers, material handling systems
Acid manufacture:		
Phosphoric	Acid mist, dust	Thermal processes—phosphate rock acidulating, grinding, and handling system
Sulfuric	Acid mist	
Coke manufacturing	Coal and coke dusts, coal tars	Charging and discharging oven cells, quenching, material handling
Glass furnaces and glass fiber manufacture	Sulfuric acid mist, raw material dusts, alkaline oxides, resin aerosols	Raw-material handling, glass furnaces, glass fiber forming and curing
Coffee processing	Chaff, oil aerosols, ash dehydrated coffee dusts	Roasters, spray dryers, waste heat boilers, coolers, stoners, conveying equipment, chaff burning
Cotton ginning	Cotton fiber, dust, and smoke	Gins, trash incineration
Carbon black	Carbon black	Carbon black generators
Soap and detergent manufacturing	Detergent dusts	Spray dryers, product and raw-material handling systems
Gypsum processing	Product dusts	Calciners, dryers, grinding and material handling systems
Coal cleaning	Coal dusts	Washed coal dryers

SOURCE: Control Techniques for Particulate Air Pollutants (1969).

Table 7.8 **PARTICULATE EMISSION FACTORS FOR STATIONARY SOURCES**

Source	Particulate emission rate†
Fuel combustion—stationary sources	
Coal (pulverized)	
General (anthracite and bituminous)	8A‡ g/kg of coal burned
Wet bottom	6.5A g/kg of coal burned
Dry bottom	8.5A g/kg of coal burned
Cyclone	1A g/kg of coal burned
Spreader stoker	6.5A g/kg of coal burned
Hand-fired units	10 g/kg of coal burned
Fuel oil	
Power plant	1 $kg/10^3$ l of oil burned
Residual unit	2.75 $kg/10^3$ l of oil burned
Distillate unit	1.8 $kg/10^3$ l of oil burned
Domestic unit	1.2 $kg/10^3$ l of oil burned
Natural gas	
Power plant	240 $kg/10^6$ m^3 of gas burned
Industrial process boiler	290 $kg/10^6$ m^3 of gas burned
Domestic and commercial heating	302 $kg/10^6$ m^3 of gas burned
Solid waste disposal	
Open burning of leaves and brush	8.5 g/kg of refuse burned
Open burning dump	8 g/kg of refuse burned
Municipal incinerator	8.5 g/kg of refuse burned
On-site commercial and industrial multiple-chamber incinerator	15 g/kg of refuse burned
Single-chamber incinerator	7.5 g/kg of refuse burned
Process industries	
Paint and varnish manufacture	
Varnish cooker	30–60 g/kg of feed
Alkyl resin production	40–60 g/kg of feed
Cooking and blowing of oils	10–30 g/kg of feed
Polymerization of acrylic resins	10 g/kg of feed
Phosphoric acid manufacture—thermal process, absorber tail gas with control	0.1–5.4 g/kg of phosphorus burned
Sulfuric acid manufacture—contact process	0.15–3.75 g/kg of acid produced
Iron and steel manufacture	
Sintering machine gases	10 g/kg of sinter
Open hearth furnace	
Oxygen lance	11 g/kg of steel
No oxygen lance	7 g/kg of steel
Basic oxygen furnace	23 g/kg of steel
Electric arc furnace	
Oxygen lance	5.5 g/kg of steel
No oxygen lance	3.5 g/kg of steel
Blast furnace	
Ore charging	55 g/kg of iron
Aluminum smelting	
Chlorination—lancing of chlorine gas into molten metal bath	500 g/kg of chlorine used
Crucible furnace	0.95 g/kg of metal processed
Reverberatory furnace	2.15 g/kg of metal processed
Sweating furnace	16.1 g/kg of metal charged
Brass and bronze smelting	
Crucible furnace	1.95 g/kg of metal charged

Table 7.8—*continued*

Source	Particulate emission rate
Electric furnace	1.5 g/kg of metal charged
Reverberatory furnace	13.2 g/kg of metal charged
Rotary furnace	10.5 g/kg of metal charged
Gray-iron foundry	
Cupola	8.7 g/kg of metal charged
Electric induction furnace	1.0 g/kg of metal charged
Reverberatory furnace	1.0 g/kg of metal charged
Lead smelting	
Cupola	150 g/kg of metal charged
Pot furnace	0.05 g/kg of metal charged
Reverberatory and sweating furnace	77 g/kg of metal charged
Magnesium smelting: pot furnace	2.2 g/kg of metal charged
Steel foundry	
Electric arc furnace	7.5 g/kg of metal charged
Electric induction furnace	0.05 g/kg of metal charged
Open hearth furnace	5.3 g/kg of metal charged
Zinc smelting	
Galvanizing kettles	2.7 g/kg of metal charged
Calcine kilns	44.4 g/kg of metal charged
Pot furnaces	0.05 g/kg of metal charged
Sweating furnace	5.4 g/kg of metal charged
Calcium carbide plant	
Coke dryer	0.1 g/kg of product
Electric furnace hood	0.85 g/kg of product
Furnace room vents	1.3 g/kg of product
Cement manufacture	
Dry process—kiln	20.8 kg/bbl of cement
Wet process—kiln	17.2 kg/bbl of cement
Line production	
Rotary kiln	100 g/kg of lime
Vertical kiln	10 g/kg of lime
Rock, gravel, and sand production	
Crushing	10 g/kg of product
Conveying, screening, shaking	0.35 g/kg of product
Storage piles—wind erosion	10 g/kg of product
Petroleum industry	
Fluid catalytic crackers	0.05–0.1 g/kg of catalyst circulated
Moving-bed catalytic crackers	
TCC-type unit	0.025–0.075 g/kg of catalyst circulated
HCC-type unit	0.075–0.125 g/kg of catalyst circulated
Kraft pulp industry	
Smelt tank	
Uncontrolled	10 g/kg of dry pulp produced
Water spray	2.5 g/kg of dry pulp produced
Mesh demister	0.5–1 g/kg of dry pulp produced

† Emission rates are those from uncontrolled sources, unless otherwise noted.

‡ Where letter A is shown, multiply the number given by the percentage of ash in the coal.

SOURCES: Control Techniques for Particulate Air Pollutants (1969); Compilation of Air Pollutant Emission Factors (1973).

Table 7.9 EMISSION FACTORS FOR SULFUR OXIDES FROM STATIONARY SOURCES

Source or process	Emission factor (SO_2)
1 Fuel combustion	
Coal	19S† g/kg (assumes 5% of sulfur remains in ash)
Natural gas	6.4 kg/10^6 m^3
Process gas	45.6C‡ kg/10^6 m^3
Fuel oil	19.8S† kg/10^3 l
Gasoline-powered engine	1.1 kg/10^3 l (assumed sulfur content of 0.07%)
Diesel-powered engine	5 kg/10^3 l (assumed sulfur content of 0.3 %)
2 Nonferrous primary smelters	
Several important metallic ores, such as copper, lead, and zinc, occur as sulfides. The natural metal ores are usually mixed with large amounts of worthless rock. The process of removing the worthless rock, concentrating the metallic ore, and finally driving off the sulfur (as SO_2) is called smelting. For example, for copper and lead the important reactions are $Cu_2S + O_2 \rightarrow 2Cu + SO_2$; $2PbS + 3O_2 \rightarrow 2PbO + 2SO_2$	
Exit SO_2 concentrations with moderate control are often as high as 8000 ppm.	
Copper smelting—primary	625 g/kg of ore
Lead smelting—primary	330 g/kg of ore
Lead smelting—secondary cupola	32 g/kg of metal charged
Lead smelting—secondary reverbatory and sweat furnaces	75 g/kg of metal charged
Zinc smelting—primary	265 g/kg of ore
3 Sulfuric acid plants	Range: 10–35 g/kg of 100% acid produced
Sulfuric acid is essentially made by burning elemental sulfur with a controlled amount of excess air, producing SO_2, and then catalytically oxidizing the SO_2 to SO_3, and finally absorbing SO_3 in water to yield H_2SO_4. The heart of the sulfuric acid plant is the converter in which SO_2 is catalytically converted to SO_3 in a fixed bed. The ultimate SO_2 remaining after the final absorption step (and thus emitted from the plant if uncontrolled) depends on the operation of the converter. Exit gas concentrations of SO_2 vary from 2000 to 3500 ppm.	
4 Pulp and paper mills	
In pulping, wood is reduced to fiber, bleached, and dried in preparation for making paper at the paper mill. Most pulp mill processes use some type of cooking liquor to dissolve lignins in the wood and free the wood fibers. To make this process economical, spent cooking liquor is recovered, usually by some process involving combustion. It is primarily in recovery processes that particulate matter, odorous sulfur compounds (H_2S and organic sulfides), and SO_2 are produced.	
Kraft type—recovery furnace	1.2–6.7 g/kg of air-dried pulp
Sulfite type—recovery furnace	20 g/kg of air-dried pulp (assumes 90% recovery of SO_2)

† S = percent sulfur by weight.
‡ C = grains of sulfur/100 m^3 of gas.

7.4 STATIONARY SOURCES OF SULFUR OXIDES

Sulfur oxides, primarily SO_2, are generated during combustion of any sulfur-containing fuel and are emitted by industrial processes that consume sulfur-containing raw materials. Because of the relatively high sulfur content of bituminous coals and fuel oils, which are burned in great quantities in the world, fuel combustion accounts for roughly 75 percent of all SO_2 emitted. The major industrial sources of SO_2 are smelting of metallic ores and refining of oil. Table 7.9 summarizes the type of processes, and their SO_2 emission factors, that are important sulfur-contributing emitters.

REFERENCES

ALPERSTEIN, M., and R. L. BRADOW: Exhaust Emissions Related to Engine Combustion Reactions, SAE Fuels and Lubricants Meeting, Paper 660871, 1966.

AMMANN, P. R., and R. S. TIMMINS: Chemical Reactions During Rapid Quenching of Oxygen-Nitrogen Mixtures from Very High Temperatures, *Am. Inst. Chem. Engrs. J.*, **12**:956 (1966).

BLUMBERG, P., and J. T. KUMMER: Prediction of NO Formation in Spark-ignited Engines—An Analysis of Methods of Control, *Combust. Sci. Technol.*, **4**:73 (1971).

BORTNER, H. M.: Review of Rate Constants of Selected Reactions of Interest in Re-entry Flow Fields in the Atmosphere, *NBS Tech. Note* 484, 1969.

CAMPBELL, I. M., and B. A. THRUSH: Reactivity of Hydrogen to Atomic Nitrogen and Atomic Oxygen, *Trans. Faraday Soc.*, **64**:1265 (1968).

Compilation of Air Pollutant Emission Factors, *U.S. Dept. Health, Education, and Welfare, Publ. AP*–42, 1973.

Control Techniques for CO, NO and Hydrocarbon Emissions from Mobile Sources, *U.S. Dept. Health, Education, and Welfare Publ. AP*–66, 1970.

Control Techniques for Nitrogen Oxide Emissions from Stationary Sources, *U.S. Dept. Health, Education, and Welfare Pulb. AP*–67, 1970.

Control Techniques for Particulate Air Pollutants, *U.S. Dept. Health, Education, and Welfare Publ. AP*–51, 1969.

Control Techniques for Sulfur Oxide Air Pollutants, *U.S. Dept. Health, Education, and Welfare Publ. AP*–52, 1969.

DANIEL, W. A.: Engine Variable Effects on Exhaust Hydrocarbon Composition (A Single Cylinder Engine Study with Propane as the Fuel) *SAE Trans.*, **76**:774 (1967).

DANIEL, W. A.: Why Engine Variables Affect Exhaust Hydrocarbon Emission, presented at Automotive Engineering Congress, *SAE Paper* 700108, 1970.

DANIEL, W. A., and J. T. WENTWORTH: Exhaust Gas Hydrocarbons—Genesis and Exodus, in "Vehicle Emissions," SAE Technical Progress Series, vol. 6, Society of Automotive Engineers, New York, 1964.

ECCLESTON, B. H., and R. W. HURN: Comparative Emissions from Some Leaded and Prototype Lead-free Automobile Fuels, *U.S. Bur. Mines, Rept.* 7390, 1970.

HABIBI, K.: Characterization of Particulate Lead in Vechicle Exhaust-Experimental Techniques, *Environ. Sci. Technol.*, **4**:239 (1970).

HABIBI, K.: Characterization of Particulate Matter in Vehicle Exhaust, *Environ. Sci. Technol.*, **7**:223 (1973).

HAZEN, D. E., and G. W. HOLLIDAY: The Effects of Engine Operating and Design Variables on Exhaust Emissions, SAE National Automobile Week, *SAE Paper* 486C, 1962.

HEYWOOD, J. B., and J. C. KECK: Formation of Hydrocarbons and Oxides of Nitrogen in Automobile Engines, *Environ. Sci. Technol.*, **7**: 216 (1973).

HULS, T. A., and H. A. NICHOL: Influence of Engine Variables on Exhaust Oxides of Nitrogen Concentrations from a Multi-cylinder Engine, SAE Mid-Year Meeting, *SAE Paper* 670482, 1967.

HURN, R. W.: Mobile Combustion Sources, in A. C. STERN (ed.), "Air Pollution," vol. III, Academic Press, New York, 1968.

JACKSON, M. W.: Effects of Some Engine Variables and Control Systems on Composition and Reactivity of Exhaust Hydrocarbons, presented at Mid-Year Meeting, *SAE Paper* 660404, 1966.

LAVOIE, G. A., J. B. HEYWOOD, and J. C. KECK: Experimental and Theoretical Study of Nitric Oxide Formation in Internal Combustion Engines, *Combust. Sci. Technol.*, **1**:313 (1970).

LEE, R. E., JR., R. K. PATTERSON, W. L. CRIDER, and J. WAGMAN: Concentration and Particle Size Distribution of Particulate Emissions in Automobile Exhaust, *Atmos. Environ.*, **5**:225 (1971).

MAGA, J. A.: Motor Vehicle Emissions in Air Pollution and Their Control, in J. N. PITTS, JR., and R. L. METCALF (ed.), "Advances in Environmental Science and Technology," vol. I, Wiley Interscience, New York, 1971.

MULLER, H. L., R. E. KAY, and T. O. WAGNER: Determining the Amount and Composition of Evaporation Losses from Automotive Fuel Systems, *SAE Trans.*, **75**:720 (1967).

TABACZYNSKI, R. J., J. B. HEYWOOD, and J. C. KECK: Time-resolved Measurements of Hydrocarbon Mass Flowrate in the Exhaust of a Spark-ignition Engine, *SAE Trans.*, **81**:379 (1972).

TEMPLE, R. G.: Control of Internal Combustion Engines, in W. STRAUSS (ed.), "Air Pollution Control," Wiley Interscience, New York, 1971.

WIMMER, D. B., and L. A. MCREYNOLDS: Nitrogen Oxides and Engine Combustion, SAE Summer Meeting, *SAE Paper* 380E, 1961.

ZELDOVICH, Y. B.: The Oxidation of Nitrogen in Combustion Explosions, *Acta Physicochimica* USSR, **21**:577 (1946).

PROBLEMS

7.1 The pressure-volume relationships in the four-cycle internal combustion engine of Fig. 7.1 are shown in Fig. 1. As an idealization of this cycle, one may use the so-called air standard cycle shown in Fig. 2, in which air is considered as the working fluid and combustion is neglected. In the air standard cycle, compression is followed by heating,

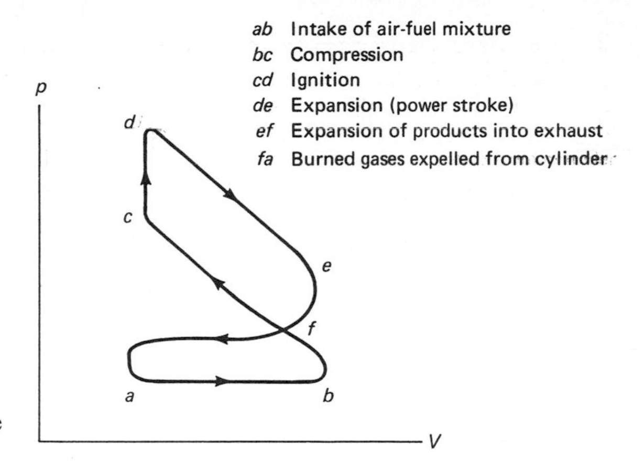

FIGURE 1
Pressure-volume relationships in the four-cycle internal combustion engine.

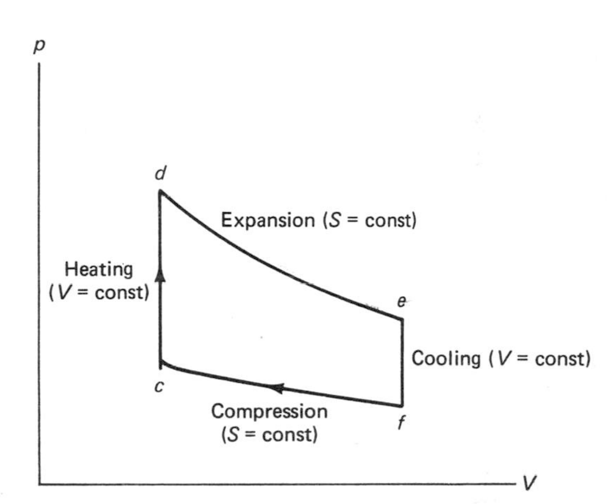

FIGURE 2
Air standard cycle.

then expansion, and cooling. The heat added along path *cd* is $q_1 = C_V(T_d - T_c)$, and the heat rejected along path *ef* is $q_2 = C_V(T_e - T_f)$. The thermodynamic efficiency of the cycle is defined by

$$\eta = \frac{q_1 - q_2}{q_1}$$

Show that the efficiency is related to the compression ratio $r = v_f/v_c$ by

$$\eta = 1 - r^{1-\gamma}$$

where ideal-gas behavior is assumed and $\gamma = C_p/C_V$, the ratio of specific heats.

7.2 The precise amount of air required for the complete combustion of a hydrocarbon fuel can be calculated by considering the stoichiometric conversion of the hydrocarbon to CO_2 and H_2O. Determine the stoichiometric air-fuel ratios for combustion of *n*-heptane and isooctane.

7.3 We have shown that the complete combustion of a hydrocarbon C_xH_y can be expressed chemically as

$$C_xH_y + (x + \tfrac{1}{4}y)O_2 + (x + \tfrac{1}{4}y)3.76N_2 \rightarrow xCO_2 + \tfrac{1}{2}yH_2O + (x + \tfrac{1}{4}y)3.76N_2$$

Thus, the combustion of 1 mole of hydrocarbon produces x moles of CO_2 and $\frac{1}{2}y$ moles of H_2O, in addition to the $(x + \frac{1}{4}y)$ 3.76 moles of N_2 assumed to be unaltered during the reaction. Show that if f is the fraction of incomplete combustion the ppm by volume, λ, of unburned fuel in the exhaust is given by

$$\lambda = \frac{10^6 f}{M + f(1 + F - M)}$$

where M is the total number of moles of products resulting from 1 mole of fuel and $1 + F$ is the total number of reacting moles (1 mole of fuel plus F moles of air). Evaluate λ for Indolene, $C_7H_{13.02}$, a standardized blended fuel, for $f = 0.001$, 0.01, and 0.10.

7.4 The quantities of various exhaust products of internal combustion can be estimated on the basis of chemical equilibrium considerations. In particular, one can estimate the mole fractions of CO, NO, and NO_2 at equilibrium under different conditions.

(*a*) Compute the equilibrium mole fraction of CO in a system of CO, CO_2, and O_2 at 1 atm pressure at 2000, 3000, and 4000°K according to

$$CO_2 \rightleftarrows CO + \tfrac{1}{2}O_2$$

under conditions in which the ratio of the number of oxygen atoms to the number of carbon atoms is 2, 3.125 (stoichiometric combustion of octane), and 5. The equilibrium constant is

$$K = 3 \times 10^4 e^{-67{,}000/RT}$$

(*b*) Compute the equilibrium mole fraction of NO in a system of N_2, O_2, and NO at 1 atm pressure at 2000, 3000, and 4000°K according to

$$\tfrac{1}{2}N_2 + \tfrac{1}{2}O_2 \rightleftarrows NO$$

under conditions in which the ratio of the number of nitrogen atoms to the number of oxygen atoms is 4 (air) and 40 (combustion flue gases at 10 percent excess air).

(*c*) It is of interest to estimate the quantity of NO_2 that can be formed during combustion. Determine the mole fraction of NO_2 that is attained at equilibrium in the system

$$\tfrac{1}{2}N_2 + \tfrac{1}{2}O_2 \rightleftarrows NO$$

$$NO + \tfrac{1}{2}O_2 \rightleftarrows NO_2$$

at 25, 1000, and 1600°C both in air and in flue gas of composition 3.3 percent O_2 and 76 percent N_2. The equilibrium constant for the second reaction is

$$K = 2.5 \times 10^{-4} e^{13{,}720/RT}$$

7.5 The Zeldovich mechanism for NO formation during combustion is

$$O_2 + M \rightleftharpoons O + O + M$$

$$O + N_2 \rightleftharpoons NO + N$$

$$N + O_2 \rightleftharpoons NO + O$$

Derive the rate expression for NO given in Subsec 7.2.1.

7.6 Small quantities of NO_2 can be formed in combustion exhaust gases by the third-order reaction

$$2NO + O_2 \xrightarrow{k} 2NO_2$$

You wish to estimate the amount of NO_2 that can be formed by this route under conditions typical of those in the exhaust of the automobile. Assume that the exhaust system of a car can be represented as a straight pipe through which the exhaust gases flow in so-called plug flow (each element travels through the pipe independently of the other elements). Assume that the concentration of NO at the beginning of the exhaust system is 2000 ppm. For initial O_2 concentrations of 10^2, 10^3, and 10^4 ppm, compute the concentration of NO_2 formed for a residence time of 2 sec if the temperature of the exhaust gases (*a*) is constant at 800°C; (*b*) is constant at 300°C; (*c*) decreases linearly from 800 to 300°C.

The following rate constant data are available:

T°C	k, l^2 mole^{-2} sec^{-1}
0	17.9×10^3
40	13.1×10^3
200	6.6×10^3
390	5.1×10^3

7.7 A typical coal particle burned in an industrial furnace may be considered a porous sphere of pure carbon. At time zero the particle is placed in an air stream of temperature T_0 and oxygen concentration c_0. The carbon pellet reacts with oxygen according to

$$C + O_2 \rightarrow CO_2$$

This reaction takes place not only at the outer surface of the particle but also internally because of the diffusion of O_2 into the pores of the particle. Assume that the process of oxygen diffusion in the particle can be characterized by an effective diffusivity D_E.

(*a*) Write the differential equations which describe the subsequent spatial and temporal variations of oxygen concentration and temperature in the pellet. Formulate a consistent set of boundary conditions. Place these equations in dimensionless form. Sketch the oxygen concentration profiles which might be expected for situations in which the rate of reaction is much faster than the rate of diffusion of oxygen and vice versa.

(*b*) In the situation in which the reaction rate $\gg$ diffusion rate, the pellet burns only at its outer surface. In such a case, the rate of burning depends on oxygen diffusion through the gas phase to the surface of the particle. Assume that the particle burns at a temperature T_0, equal to the air temperature. Based on these assumptions, show how you could estimate the lifetime of the particle, i.e. derive an expression for the time it takes to burn completely a particle initially of diameter D_p.

(*c*) Estimate the lifetime of a particle for which $D_p = 50\ \mu m$, the median size for pulverized coal burning in air.

(*d*) It was assumed above that the carbon is oxidized completely to CO_2. How would you estimate when CO formation becomes important in the combustion of a coal pellet?

7.8 (*a*) A source burns 10^7kg/yr of coal in a spreader stoker. The process is outfitted with 85 percent efficient multiple cyclone separators. The coal contains 10 percent ash. Determine the yearly mass of particulate emissions from this source.

(*b*) A secondary brass and bronze smelting operation is equipped with a fabric filter of 99 percent efficiency for removing particles. The furnace used is electric, with metal charging directly into the furnace. The charge is 2×10^7 kg/yr. What is the yearly mass of particulate emissions from this operation?

8

AIR POLLUTION CONTROL PRINCIPLES

Control techniques for limiting pollutant formation and emission at the source fall into three classes: (1) modification of the basic process to result in "cleaner" operation, (2) substitution of alternative cleaner-burning fuels in combustion processes, and (3) cleaning of the effluent gases before release to the atmosphere. The particular technique chosen for a problem depends on the pollutant involved, the process responsible for pollutant formation, and the required degree of control. Considering combustion processes, for example, we can identify three categories of pollutants:

1 Products of incomplete combustion, such as hydrocarbons, CO, and combustible particulate matter

2 Pollutants resulting from fuel contaminants, such as sulfur oxides and nitrogen oxides from fuel nitrogen

3 Oxides of nitrogen from high-temperature reactions involving N_2 and O_2

Modifications of the original process can play a key role in eliminating emissions in categories 1 and 3, for example, by making the combustion more efficient and altering the time-temperature history of the burning mixture. Substitution of alternative fuels is a prime strategy when dealing with pollutants of class 2. Cleaning of effluent gases is a basic strategy amenable, in principle, to pollutants of all three classes above.

Of the three classes of control techniques listed above, gas cleaning techniques have probably received the most attention and are probably the most widely used source control measure. The type of gas cleaning method adopted depends on the particular pollutant(s) that must be removed.

For gaseous contaminants, essentially two alternative classes of methods are available. The first class is absorption of the gas, usually achieved by passing the pollutant-bearing gas through a piece of equipment in which the gas intimately contacts a liquid solution capable of preferentially dissolving the pollutant gas. Gas absorption is a common means of removing SO_2, NH_3, NO and NO_2, and H_2S from gas streams. A process related to absorption of the gas by a liquid is adsorption on a solid, such as, for example, the removal of trace quantities of hydrocarbons by activated charcoal. The second basic method applicable to gaseous pollutants involves chemical alteration of the pollutant, usually through combustion or catalytic processing of the waste gases.

For particulate pollutants there are a number of methods which take special advantage of the nature and properties of particulate matter. The basic mechanisms that can be employed for removal of particulate matter from gas streams are (Strauss, 1966):

1. Gravity separation
2. Centrifugal separation
3. Inertial impaction
4. Direct interception
5. Brownian diffusion
6. Turbulent diffusion
7. Thermal precipitation
8. Electrostatic precipitation
9. Magnetic precipitation
10. Brownian agglomeration
11. Sonic agglomeration

Often more than one of the above mechanisms are used to cleanse a particular stream.

The remainder of this chapter is devoted to an exposition of the principles underlying the more common air pollution control measures.

8.1 INTERNAL COMBUSTION ENGINE

8.1.1 United States Motor Vehicle Emission Standards

In 1963 the United States government enacted the Clean Air Act, aimed at stimulating state and local air pollution control activity. Amendments to the Clean Air Act in 1965 and 1970 authorized the setting of national standards for emissions from all new motor vehicles commencing with 1968.

The magnitude of contaminant emissions from a motor vehicle is a variable in time and is a function of the percentage of time the vehicle is operated in each driving mode (accelerate, cruise, decelerate, idle). The modal split is in turn dependent on the habits of the driver, the type of street on which the vehicle is operated, and the degree of congestion on that street. Also affecting emissions are the presence or absence of an emission control device, the condition of the car, its size, and other factors.

Because of all these factors, measured automobile exhaust emissions depend on, in addition to the make and year of the car, the driving condition. The basic approach underlying the specification of exhaust emission rates has been determination of an "average trip," that is, one representative of the average driving habits of the population (usually in an urban area). The trip, usually termed a *driving cycle*, is composed of a series of driving modes (idle, accelerate, cruise, and decelerate) in which a predetermined length of time is spent in each mode. Such a cycle is formulated, in principle, by "tagging" a substantial number of vehicles on a particular day and analyzing their trips according to the sequencing of the different driving modes and the time spent in each. (For example, see the study of Smith and Manos, 1972.) Once the driving cycle has been determined, a standard emission rate for the cycle is determined by running a representative sample of vehicles of varying makes and ages through the cycle in a stationary test (on a chassis dynamometer) and measuring their emissions.

The first driving cycle used as a basis for exhaust emission tests was the so-called California Driving Cycle (CDC) or the Federal Test Procedure (FTP). The FTP, shown in Fig. 8.1*a*, is a seven-mode test procedure and was originally specified as a typical Los Angeles commuter trip. To determine emissions from the FTP, the vehicle is run through the seven cycles shown in Fig. 8.1*a* from an initial cold start. The first four cycles are designated "warm-up" cycles, and emissions from these are weighted 35 percent. The fifth cycle is run but not measured, and the sixth and seventh are classed as "hot cycles," and emissions from these are weighted 65 percent. In the FTP, emissions were measured on a concentration basis. The FTP was used to certify 1966 and newer vehicles by the State of California and to certify 1968 to 1971 vehicles by the federal government. However, it was recognized that the most relevant emission parameter is the mass of emissions rather than the volume concentration. Thus, the 1970 and 1971 federal standards, although based on the FTP, were computed as mass emissions by multiplying the exhaust concentrations by the vehicle exhaust volume (which varies with vehicle weight).

For 1972 and newer motor vehicles, a new test procedure has been adopted. In the procedure, a new driving cycle has been specified, and emissions are measured on a true mass basis. The new procedure has been given the designation the CVS (constant volume sampling) cycle. The test consists of a 12-hr wait at a temperature between 15.5 and 30°C, a cold-engine start-up, and a continuous sequence of driving

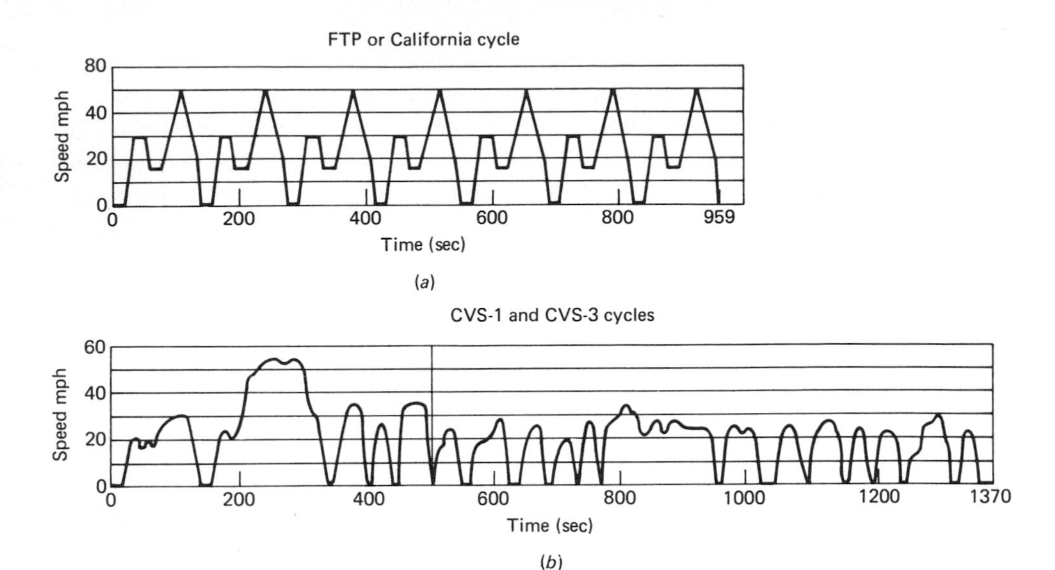

FIGURE 8.1
Automotive emission test cycles. (*a*) The Federal Test Procedure (the seven-mode California Driving Cycle.) (*b*) The CVS cycles. The dashed line connecting the two cycles indicates the end of the first cycle of the seven-mode FTP. The vertical line at 505 sec in the constant volume sampling (CVS) indicates the end of that part of the cycle which is repeated for the CVS-3 test.

modes (Fig. 8.1*b*) for 23 mins. For 1972 to 1974 vehicles, a single exhaust sampling bag is used, and the cycle is called CVS-1. To better assess the effect of cold versus hot starting conditions, standards for 1975 and 1976 vehicles are based on the so-called CVS-3 test, in which the CVS-1 test is followed by a 10-min shutdown and then a hot-engine restart and a repeat of the first 505 secs of the 23-min cycle. Exhaust-gas sampling begins immediately after the key is turned on. Diluted exhaust emissions are collected during the first 505 sec in one bag, those during the remainder of the 23-min cycle in a second bag, and those from the hot-restart phase in a third bag. Contents of the three bags are analyzed and then weighted to obtain the final mass emissions, in grams per mile, of hydrocarbons, CO, and NO_X. Table 8.1 summarizes United States federal automotive exhaust emssion standards.[1]

[1] Since the test procedures based on the FTP and the CVS cycles are different, the emissions rates determined for each on the same vehicle are different. Unfortunately, the grams per mile emissions over the two tests differ considerably, particularly for hydrocarbons and CO. This raises the question as to which test procedure more accurately reflects actual vehicle operation under typical urban driving conditions. The seriousness of this question is readily apparent when one realizes that current motor vehicle emissions standards are based on the CVS cycle, standards which must be met by all new cars manufactured after 1972.

Because of the length of time required for the CVS-3 test, it is clear that all assembly-line vehicles cannot be tested to ensure compliance with the standards. In addition, measurements of exhaust emissions often show poor repeatability, some of the reasons for which are the difficulty of following the speed-time curve of the driving cycle (Fig. 8.1*b*), lack of repeatability of engine functioning, and difficulties in measuring accurately the low concentrations involved. Since the real concern, in terms of air quality, is that the average emissions from a population of cars meet the required standards, assembly-line regulations should be expressed in such a way as to ensure that emissions on the average comply with the standards. Therefore, assembly-line tests involve selecting a sufficient sample of vehicles such that the full population of vehicles considered is properly represented. Each vehicle in the sample is then run through

Table 8.1 UNITED STATES AUTOMOTIVE EXHAUST EMISSION STANDARDS

Year	Hydrocarbons	CO	Oxides of nitrogen
Prior to controls	850 ppm (11 g/mi)†	3.4% (80 g/mi)	1000 ppm (4 g/mi)
FTP cycle			
1958–1969	275 ppm (3.2 g/mi)	1.5% (33 g/mi)	
1970	2.2 g/mi	23 g/mi	
1971	2.2 g/mi	23 g/mi	4.0 g/mi‡
CVS-1 cycle			
1971	4.6 g/mi	47 g/mi	
1972	3.4 g/mi	39 g/mi	4.0 g/mi
1973–1974	3.4 g/mi	39 g/mi	3.0 g/mi
CVS-3 cycle			
1975	0.41 g/mi	3.4 g/mi	3.0 g/mi
1976§	0.41 g/mi	3.4 g/mi	0.4 g/mi

† The values in parentheses are approximately equivalent to those above.

‡ State of California standard only.

§ On April 11, 1973 the Environmental Protection Agency delayed these standards for one year for all new cars except those to be sold in the state of California. The interim standards for 1975 are:

	California	Other states
HC	0.9	1.5
CO	9	15
NOx	3.1	2

NOTE: Evaporative losses are restricted to 6.0 g/test (24 hr) in 1970 in California, and 1971 nationwide, and 2.0 g/test (1 hr) beginning 1972 nationwide. Particulate emission standards of 0.1 g/mi have been proposed.

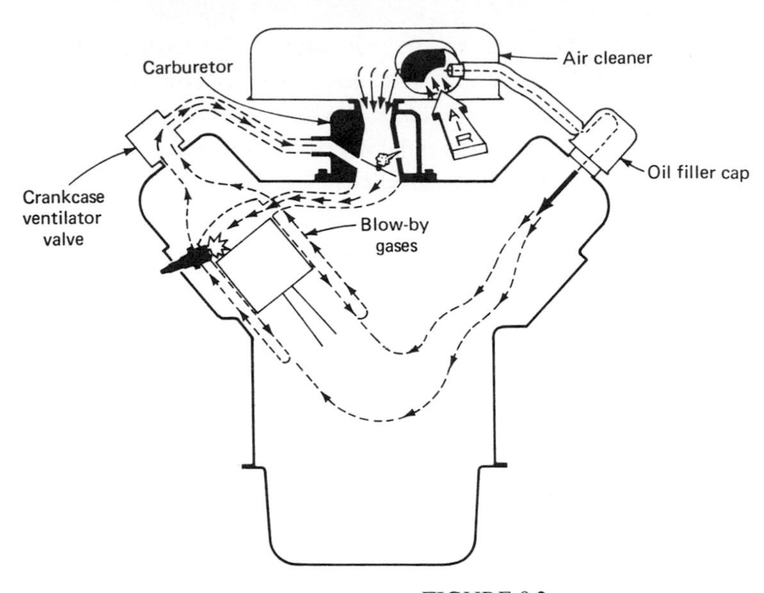

FIGURE 8.2
Crankcase emission control system.

the CVS-3 cycle, and the emissions from the sample are plotted as a frequency distribution. A reasonable criterion for compliance is that no more than a certain fraction of the tested vehicles exceed the emissions standards.

To obtain a certificate of conformity for a class of vehicles, the manufacturer must demonstrate the effectiveness of the emission controls over the probable life of the car. One fleet is driven 4000 mi, at which point emissions are measured by the CVS-3 procedure. A second fleet is driven for 50,000 mi and tested for emissions every 4000 mi. Emission deterioration factors are determined separately for HC, CO, and NO_X by fitting a least-squares straight line to the plots of emissions versus mileage for the fleet. Deterioration factors are determined as the ratio of emissions at 50,000 mi to those at 4000 mi. The emission test results, at 4000 mi, for each vehicle are then multiplied by the appropriate deterioration factor to give the adjusted emissions for each vehicle. These adjusted emissions for each vehicle must comply with the standards before the entire vehicle class can be certified.

8.1.2 Crankcase Emission Control

The primary purpose of crankcase emission control is to keep blowby gases from escaping into the atmosphere. All control systems use essentially the same approach, which involves recycling the blowby gases from the engine oil pump to the air intake system. A typical system is shown in Fig. 8.2. Ventilation air is drawn down into the crankcase and then up through a ventilator valve and hose and into the intake manifold. When airflow through the carburetor is high, additional air from the crankcase ventilating system has little effect on engine operation. However, during idling, airflow through the carburetor is so low that the returned blowby gases could alter the air-fuel ratio and cause rough idling. For this reason, the flow control valve restricts the ventilation flow at high intake manifold vacuum (low engine speed) and permits free flow at low manifold vacuum (high engine speed). Thus, high ventilation rates occur in conjunction with the large volume of blowby associated with high speeds; low ventilation rates occur with low-speed operation. Generally, this principle of controlling blowby emissions is called *positive crankcase ventilation* (PCV).

There are a wide variety of systems in actual use, all based on the general principle outlined above and shown in Fig. 8.2. As we noted in Subsec. 7.1.2, crankcase emissions have been controlled on all new cars manufactured in the United States since 1963.

8.1.3 Evaporative Emission Control

In principle, evaporative emissions can be reduced by reducing gasoline volatility. However, a decrease in fuel volatility below the 8 to 12 Reid vapor pressure range, commonly used in temperate climates, would necessitate modifications in carburetor and intake manifold design, required when low-vapor-pressure fuel is burned. In view of costly carburetion changes associated with reduction of fuel volatility, evaporative emission control techniques have been based on mechanical design changes. Two evaporative emission control methods are the *vapor-recovery system* and the *adsorption-regeneration system*.

In the vapor-recovery system, the crankcase is used as a storage tank for vapors from the fuel tank and carburetor. Figure 8.3*a* shows the routes of hydrocarbon vapors during shutdown and hot soak. During the hot soak period the declining temperature in the crankcase causes a reduction in crankcase pressure, sufficient to draw in vapors. During the hot soak, vapors from the carburetor are drawn into the crankcase. Vapor from the fuel tank is first carried to a condenser and vapor-liquid separator, with the vapor then being sent to the crankcase and the condensate to the fuel tank. When the engine is started, the vapors stored in the crankcase are sent to the air intake system by the positive crankcase ventilation system.

In the adsorption-regeneration system, a canister of activated charcoal collects the vapors and retains them until they can be fed back into the intake manifold to be burned. The system is shown in Fig. 8.3*b*. The essential elements of the system are the canister, a pressure-balancing valve, and a purge control valve. During the hot soak period, hydrocarbon vapors from the carburetor are routed by the pressure balance valve to the canister. Vapor from the fuel tank is sent to a condenser and

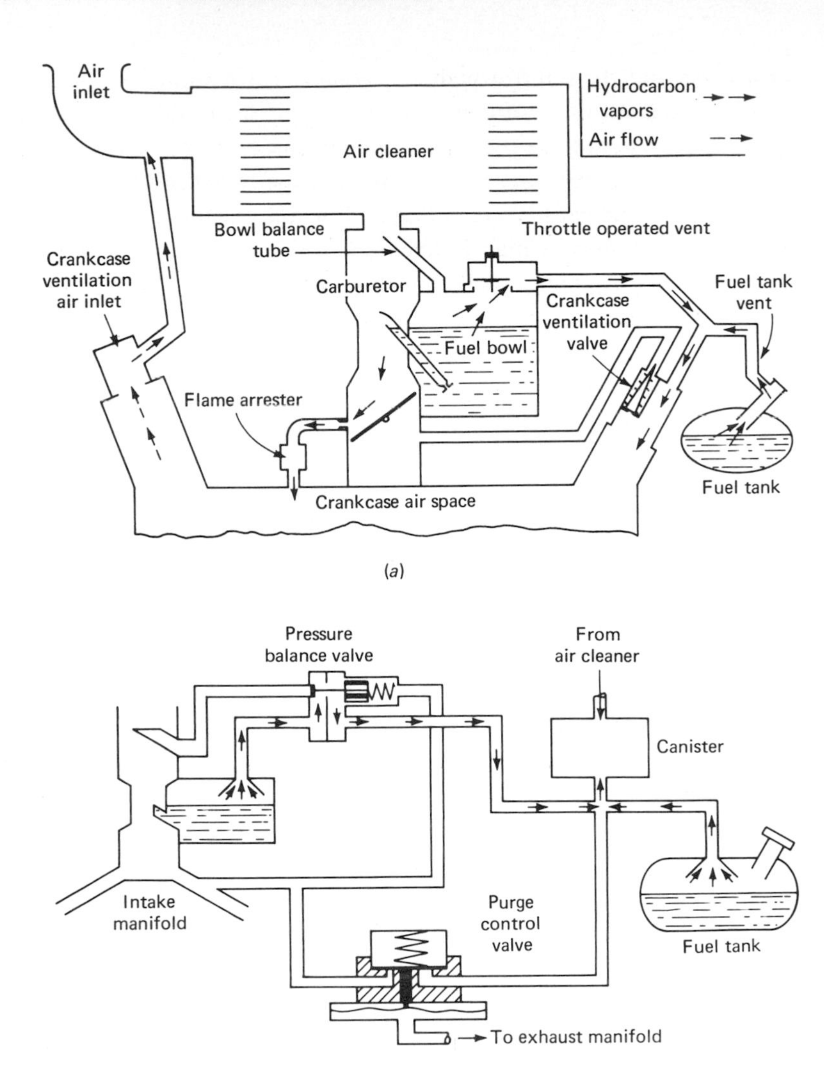

FIGURE 8.3
Evaporative emission control systems. (*a*) Use of crankcase air space. (*b*) Adsorption-regeneration system.

separator, with liquid fuel returned to the tank. When the engine is started, the pressure control valve causes air to be drawn through the canister, carrying the trapped hydrocarbons to the intake manifold to be burned.

8.1.4 Exhaust Emission Control

There are basically four ways in which exhaust emissions of HC, CO, and NO_X can be reduced in a motor vehicle:

1 Modifications in operation
2 Modifications in engine design
3 Modifications in the fuel burned
4 Exhaust gas treatment

Modifications in operation of the conventional internal combustion engine include those changes which can be instituted without the need for engine redesign. Changes in this category are exemplified by modified air-fuel ratios and ignition timing. Modifications in engine design are considerably more costly to implement than are changes in operating conditions since they may involve significant changes in parts and therefore in assembly-line equipment. Substitution of other hydrocarbon fuels for gasoline constitutes the third general category of control methods. If the new fuel can be used in an unmodified engine, the entire cost of this alternative is related to fuel use. Finally, the fourth class of control methods involves treatment of the exhaust gases (usually together with some modifications in operation) by means of reactors placed in the exhaust system of the automobile. In this subsection we will discuss each of these four approaches to automotive exhaust emission control.

8.1.4.1 Modifications in operation As we saw in Chap. 7, and as illustrated in Fig. 7.15, those operating changes, namely adjustment of air-fuel ratio, which result in lower levels of HC and CO generally lead to higher levels of NO. For example, slightly lean operation minimizes HC and CO formation but favors NO formation. As shown in Table 8.1, exhaust emission standards were enacted for HC and CO 5 years before those for NO_X. In order to meet the HC and CO standards, manufacturers employed several different systems on 1966 to 1970 motor vehicles. Features common to essentially all versions of the systems were (1) relatively lean air-fuel ratios for idle and cruise and (2) higher engine idle speeds. Although these modifications were successful in reducing HC and CO emissions, they did so at the expense of increasing NO emissions. Because of the opposite effect of air-fuel ratio on the three species, it is not possible to reduce all three pollutants solely through modifications in air-fuel ratio and ignition timing.

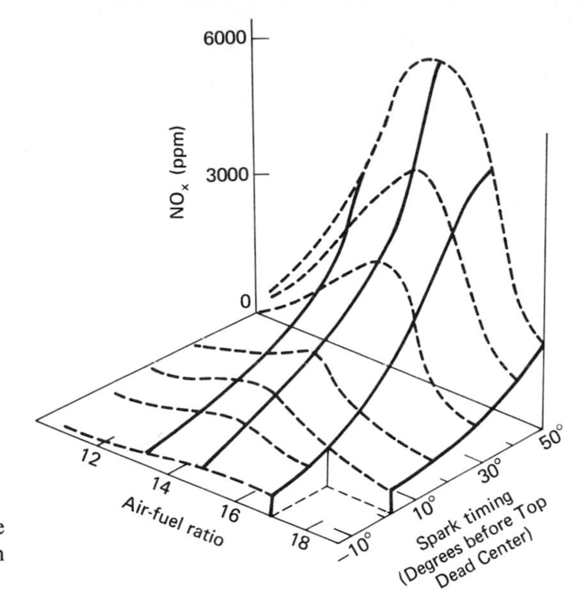

FIGURE 8.4
Effect of spark timing on nitric oxide formation in the internal combustion engine.

8.1.4.2 Modifications in engine design In order to reduce HC and CO emissions, the engine should be operated on a lean air-fuel mixture, whereas the optimum operation with respect to NO_X is either very rich or very lean combustion (assuming that no exhaust gas treatment is employed). Lean operation is effective as a means of reducing NO formation because the peak cycle temperature decreases as the air-fuel ratio increases. Even though the equilibrium NO levels at the reduced combustion temperatures of lean operation are much higher than desired, the high overall activation energy for NO formation results in a rate of NO formation that declines with temperature much more rapidly than does the equilibrium concentration of NO. As a result, the peak equilibrium NO levels are never attained, as seen in Fig. 7.20.

Nitric oxide can also be reduced by retarding the spark at a fixed air-fuel ratio, since this also has the effect of lowering the peak cycle temperature. The effect of spark timing on NO formation, illustrated in Fig. 8.4, can be explained on the basis of the lower pressures and temperatures attained when the combustion interval is moved farther into the expansion stroke. Blumberg and Kummer (1971) have shown, however, that to achieve a given NO level it is less costly in fuel consumption to operate at lean air-fuel ratios with efficient combustion timing than to retard the combustion interval into the expansion stroke.

One of the most advantageous means to achieve lean operation is exhaust gas recycle (EGR). The advantages of EGR result mainly from the fact that the air-fuel

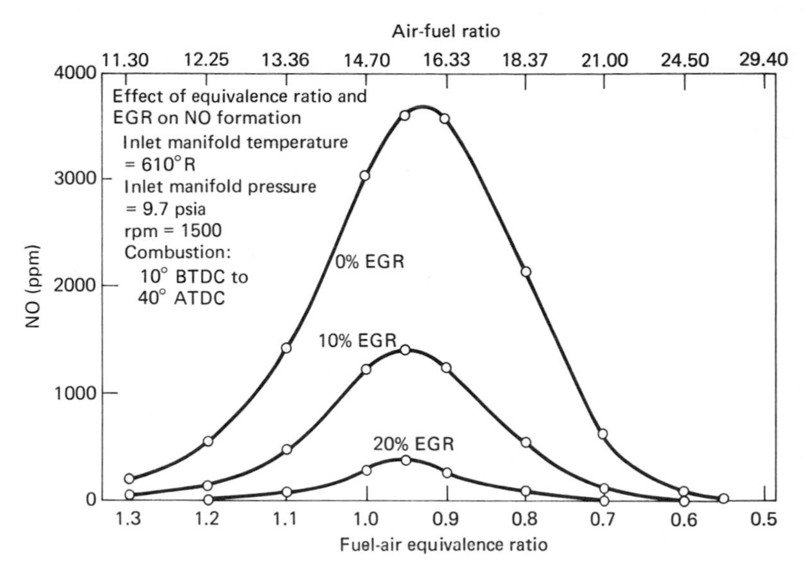

FIGURE 8.5
Calculated exhaust NO concentrations as a function of fuel-air equivalence ratio for three different exhaust-gas recirculation rates. (*Source: Blumberg and Kummer*, 1971.)

mixture can be diluted without the addition of excess O_2 (from which NO is formed) and that dilution with exhaust gas results in the introduction of species, such as CO_2 and H_2O, into the charge with higher heat capacities than N_2. Figure 8.5 shows NO concentrations as a function of equivalence ratio at different levels of recirculation, as computed by Blumberg and Kummer. The amount of exhaust gas recirculated is expressed as a percentage of the total mass of charge introduced into the cylinder, i.e. as a percentage of the sum of the exhaust gas and fresh air-fuel mixture fed. The exhaust gas is assumed to be cooled to the inlet mainfold temperature. Also apparent from Fig. 8.5 is that dilution to a given gas-fuel ratio gives greater reductions in NO when the diluent is exhaust gas rather than air. Since either lean operation or exhaust recycle will result in a loss in power at constant fuel economy, for the same level of NO control, higher power output per unit of diluent is achieved by exhaust gas recirulation than by purely lean operation.

Campau (1971) summarized the effects of using EGR; this summary is presented in Fig. 8.6. An EGR in excess of 20 percent results in a large fuel-economy penalty, severe drivability problems, and an increase in HC and CO emissions. It is not practical at this stage to achieve NO_X emission levels approaching 0.4 g/mi with EGR alone in a conventional engine.

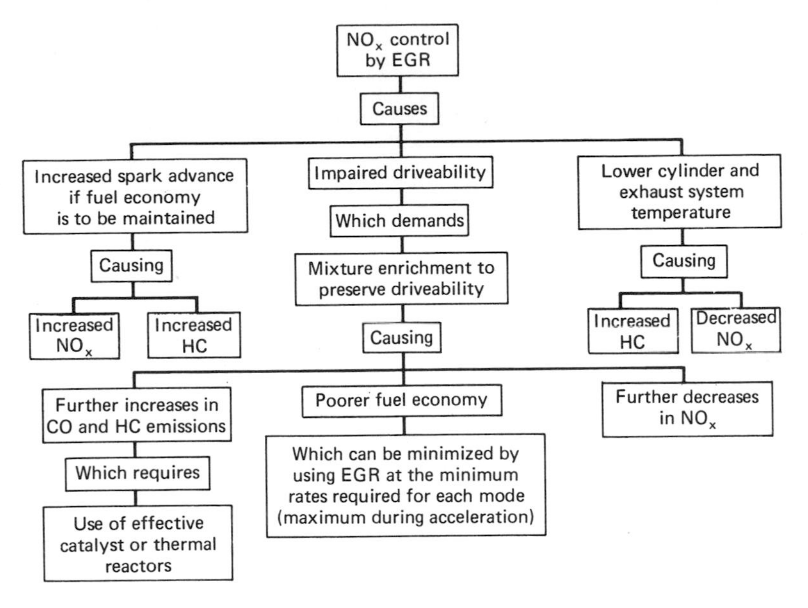

FIGURE 8.6
Effects of exhaust gas recirculation. (*Source: Campau*, 1971.)

Although EGR involves only minor engine modifications, there exist a number of combustion systems which require rather elaborate variations from the conventional spark-ignited, piston/cylinder internal combustion engine.

The first such system is the stratified-charge engine, a spark-ignited engine which employs direct fuel injection into the cylinder in a manner to achieve selective stratification of the air-fuel ratio in the combustion chamber. The stratified-charge engine is illustrated in Fig. 8.7, which shows that the air-fuel ratio must be in the ignitable range only in the immediate vicinity of the spark plug. Thus, the rich air-fuel mixture at the spark plug is ignited, and the flame propagates subsequently throughout the bulk of the chamber, which is at a very lean air-fuel ratio. Even though a local fuel-rich zone exists near the point of ignition, the overall air-fuel ratio supplied to the engine is lean. For this reason, one of the advantages of the stratified-charge engine is excellent fuel economy relative to conventional engines. For emission control the stratified-charge engine is usually outfitted with EGR and exhaust treatment for hydrocarbons. Unfortunately, when EGR is used to reduce NO_x formation, the fuel economy is decreased.

An alternative to the basic stratified-charge engine, shown in Fig. 8.8, achieves the effect of a stratified charge by means of a precombustion chamber. The engine

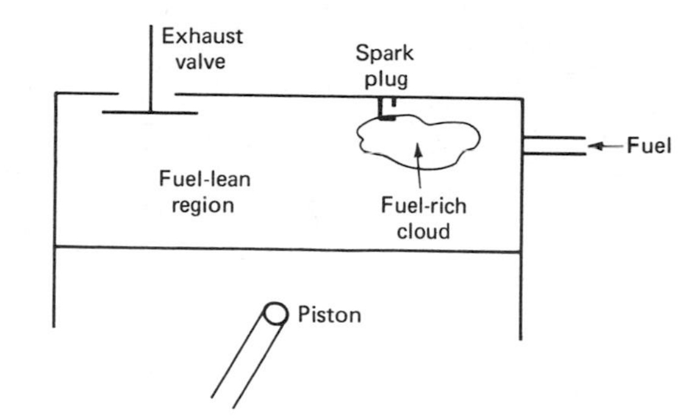

FIGURE 8.7
Stratified charge engine.

uses a conventional engine block, pistons, and spark plugs, with a modified head and an additional valve for each cylinder. The spark plug is located in the smaller combustion chamber, which is supplied with a fuel-rich mixture by a separate carburetor. The main carburetor supplies a lean mixture to the usual intake valve. The rich mixture ignites easily and the flame propagates from the prechamber out into the cylinder. As in the usual stratified-charge engine, the overall mixture is lean. Emissions of HC, CO, and NO_X are all lower than in a conventional engine at the same overall lean air-fuel ratio. Levels of 0.25 g/mi HC, 2.5 g/mi CO, and 0.43 g/mi NO_X have been achieved without EGR or exhaust treatment. In addition, emissions have been found to be relatively insensitive to changes in the air-fuel ratio. Finally, leaded gasoline can be used with this engine, and fuel economy is essentially the same as with a conventional spark-ignition engine.

The third basic engine type is the Wankel or rotary engine, depicted in Fig. 8.9. In this engine the combustion takes place in a chamber formed by the rotating shaft.

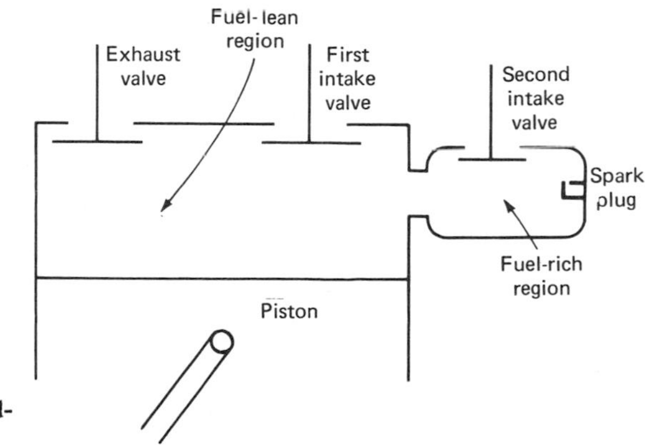

FIGURE 8.8
Three-valve, carbureted, stratified-charge engine.

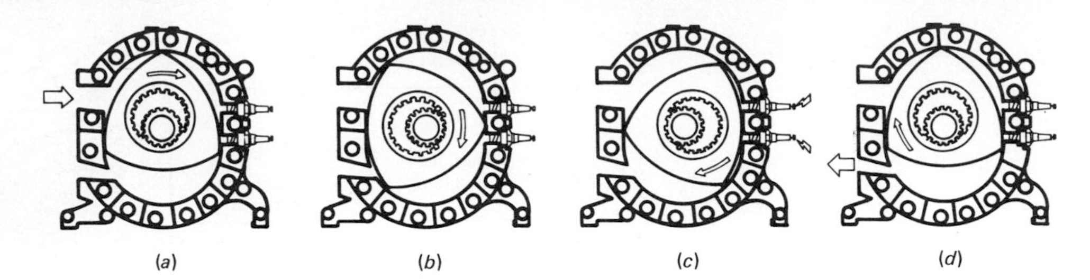

FIGURE 8.9
Wankel engine cycle. (*a*) Fuel-air mixture is drawn into the combustion chamber by a revolving rotor through the intake port (upper left). No valves are necessary. (*b*) Compression. As the rotor continues revolving, it reduces the space in the chamber containing fuel and air. (*c*) Ignition. The fuel-air mixture is now fully compressed. The leading spark plug fires. A fraction of a second later, the following plug fires to ensure complete combustion. The exploding mixture drives the rotor, providing power. (*d*) Exhaust. Combustion products are expelled through the exhaust port.

The Wankel engine offers a cost advantage because of its low weight per horsepower: about 0.68 kg/hp compared with 1.8 to 2.7 kg/hp for a piston/cylinder engine. Emissions from an uncontrolled Wankel engine are two to five times higher in HC, one to three times higher in CO, and 25 to 75 percent lower in NO_X than in an uncontrolled piston engine of equivalent power. Also, fuel economy in the Wankel is somewhat poorer than in a piston engine. An emission control system based on the Wankel engine must employ EGR to lower NO_X formation and exhaust gas treatment to oxidize HC and CO.

8.1.4.3 Modifications in the fuel burned A control technique which may not involve engine design changes is the variation of the composition of gasoline or the substitution of alternative fuels for gasoline. The characteristics of fuel amenable to change are hydrocarbon type and additive content.

In general, fuels containing large amounts of highly reactive hydrocarbons have exhaust emissions with correspondingly high reactivity. For example, fuels containing the largest concentration of reactive olefins produce the most reactive exhaust hydrocarbon emissions. In general, CO and NO emissions are relatively insensitive to the modifications in the hydrocarbon composition of the gasoline burned. With the replacement of gasoline with suitable low-molecular-weight gaseous hydrocarbon fuels, such as natural gas or petroleum gas,[1] hydrocarbon and CO emissions can

[1] Natural gas consists primarily of methane (with small amounts of ethane); petroleum gas contains principally propane and butane. For use in an automobile, both would be stored in liquefied form, called LNG and LPG.

be drastically reduced over those from conventional gasoline-burning engines. A question of some importance, then, is the effect on NO emissions of using such fuels.

Shair and Rupe (1971) studied the formation of NO for fuels consisting of natural gas and mixtures of natural gas and gasoline as a function of air-fuel ratio in a one-cylinder test engine. Fuel compositions were varied from 100 percent gasoline to 100 percent natural gas to (1) evaluate the potential of dual fuel systems and (2) determine whether or not adding small amounts of natural gas to gasoline could yield substantial benefits in terms of NO formation. Thus, the injected fuel could comprise any ratio of natural gas to gasoline at any desired air-fuel ratio. The exhaust NO concentration as a function of fuel-air ratio (the inverse of air-fuel ratio) for 100 percent gasoline and for 100 percent natural gas is shown in Fig. 8.10. The stoichiometric fuel-air ratios for the gasoline and the natural gas (90% CH_4 + 9% C_2H_6 + residual HC + H_2) were 0.0645 and 0.0578, respectively. We see that NO emission levels for natural gas are similar to those observed for gasoline for low fuel-air ratios (lean operation) but substantial reductions in NO formation are realized for fuel-rich operation.

With a substitution of fuels, it is necessary to consider not only exhaust emissions but also engine performance. Use of natural gas was found by Shair and Rupe to result in a relative loss of fuel economy and horsepower as compared with gasoline at the same fuel-air ratio. These results are shown in Fig. 8.11.

If the data are replotted to reflect the emission rate in terms of work done, as shown in Fig. 8.12, the relative advantages associated with substitution of natural gas for gasoline are reduced. In Fig. 8.11, fuel economy is expressed in terms of the specific fuel consumption, that is the grams of fuel expended per kilowatt-second of work delivered by the piston. In Fig. 8.12, NO_X emissions are expressed in terms of grams of NO_X produced per kilowatt-second of work delivered.

Shair and Rupe conclude that:

1 The substitution of natural gas for gasoline yields fractional reductions in peak NO emissions for a given amount of work done, compromises available horsepower by about 20 percent, and imposes a requirement that the engine be run rich in order to minimize NO.

2 The lean stability limit equivalence ratio for gasoline is less than that for natural gas if an efficient atomizer is used.

3 Mixtures of natural gas and gasoline do not produce any appreciable synergistic effect insofar as reduction of NO is concerned.

8.1.4.4 Exhaust gas treatment The fourth mode of reducing HC, CO, and NO_X emissions involves the treatment of exhaust gas in chemical reactors in the exhaust

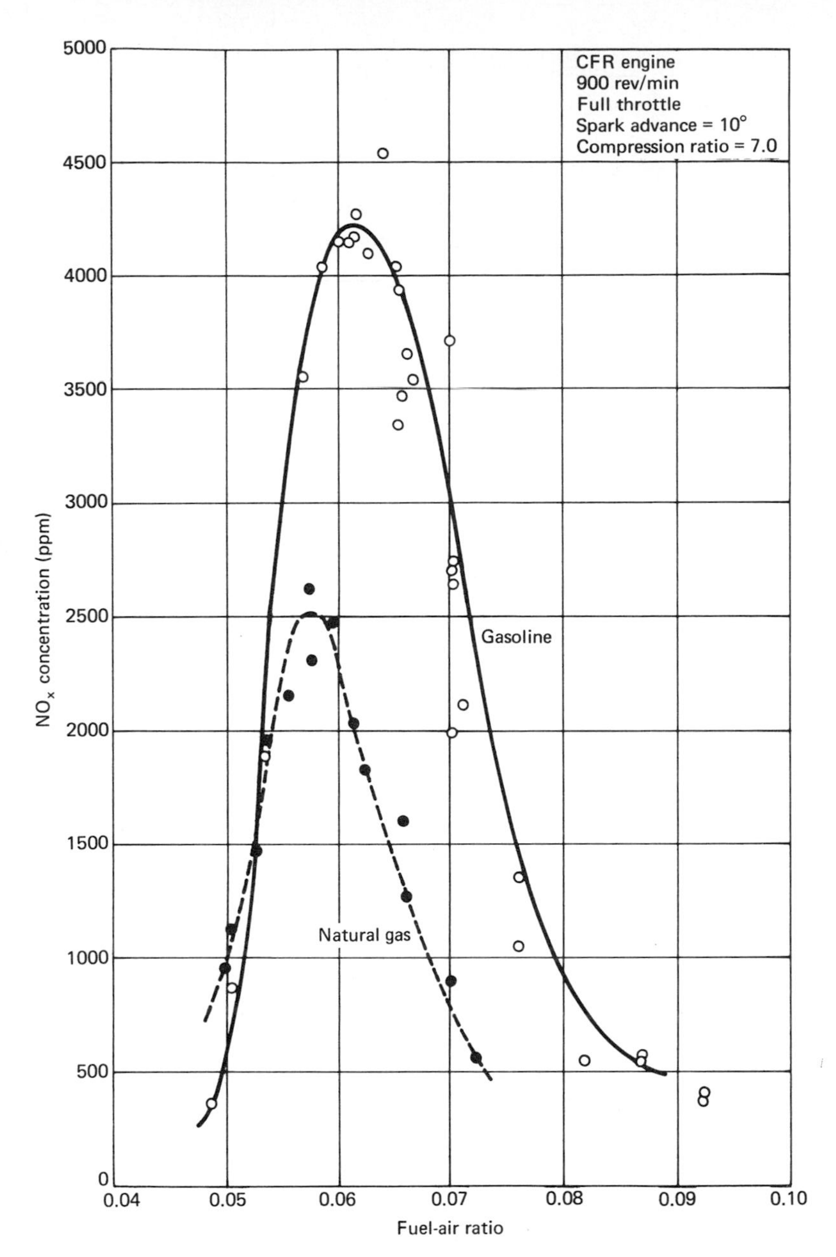

FIGURE 8.10
Concentration of NO_x in the exhaust of an internal combustion engine burning natural gas or gasoline. (*Source: Shair and Rupe*, 1971. *Furnished through the courtesy of the Jet Propulsion Laboratory, California Institute of Technology.*)

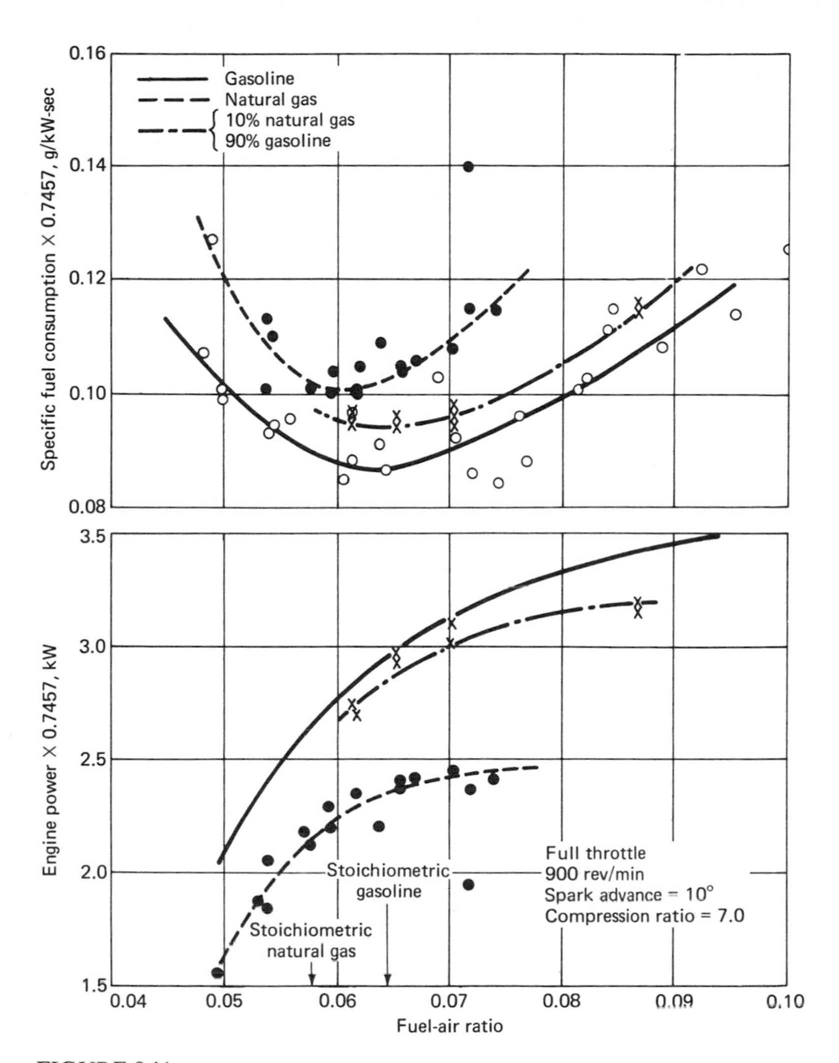

FIGURE 8.11
Engine performance with gasoline and natural gas. (*Source: Shair and Rupe,* 1971. *Furnished through the courtesy of the Jet Propulsion Laboratory, California Institute of Technology.*)

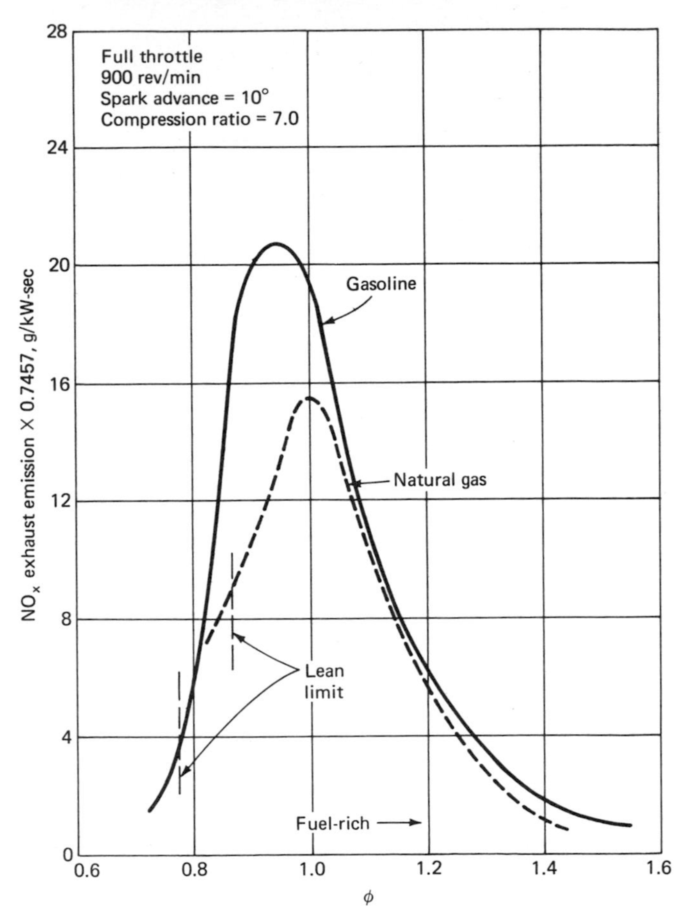

FIGURE 8.12
Emission of NO_X as a function of fuel-air equivalence ratio for natural gas and gasoline. (*Source: Shair and Rupe*, 1971. *Furnished through the courtesy of the Jet Propulsion Laboratory, California Institute of Technology.*)

system. This class can be divided according to catalytic and noncatalytic devices. The basic problem in exhaust gas treatment is that HC and CO must be *oxidized* to CO_2 and H_2O and NO must be *reduced* to N_2 and O_2, if all three pollutants are to be converted in the system. (As we shall see shortly, some proposed systems involve operating at conditions which minimize the formation of one pollutant with subsequent removal of the others in the exhaust system.)

a CATALYTIC TREATMENT The catalytic system which has received the most attention to date is the dual catalytic system depicted in Fig. 8.13. The engine is run rich, with the exhaust first sent to the NO_X-removing bed. This bed provides net

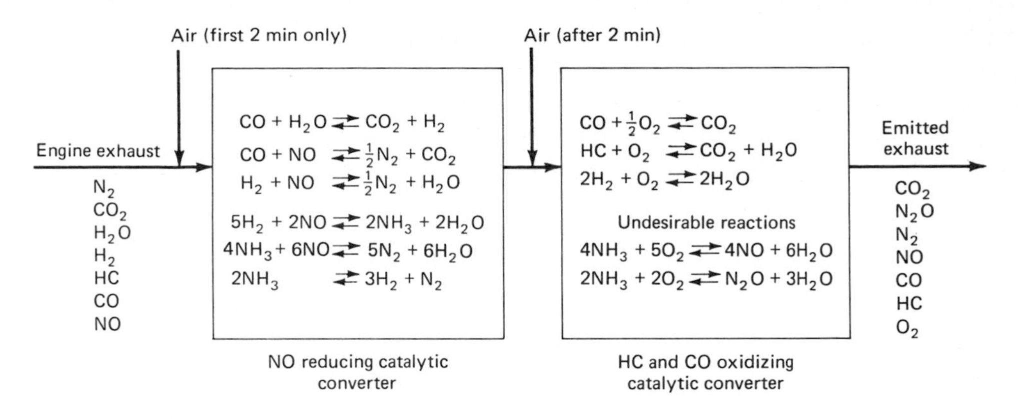

FIGURE 8.13
Dual-catalyst system.

reducing[1] conditions (between 1 and 2 percent CO in the exhaust gas). Air is then added to the exhaust stream between the catalyst beds, and the HC and CO are oxidized in the second bed. The beds may be in separate containers, as shown in Fig. 8.13, or they may be in the same container, as long as provision is made for air introduction between the catalyst beds.

Several metals and alloys have been proposed as catalysts in the dual-catalyst system. Those proposed fall into two categories: (1) the platinum group metals, including platinum, palladium, and ruthenium (also called *noble metals*), and (2) *base metals*, such as copper, nickel, and chromium, and alloys of these (Monel, Inconel, and stainless steel). Certain metals are more effective oxidation catalysts, others perform more favorably as reduction catalysts, and several are equally adept in either role (in general, most catalysts will act to some extent in either capacity).

A *metallic* catalyst consists of a catalyst metal used in its basic metallic form. Examples of systems employing metallic catalysts are those which contain layers of metal screens or are packed with small metal chips. Since only the surface of the catalyst is employed in promoting a chemical reaction, metallic catalysts, by virtue of their ratio of surface area to bulk volume, are an inefficient use of the total mass of metal. Thus, noble-metallic catalysts are quite expensive.

In a *supported* catalyst, a relatively small quantity of noble or base metal is deposited on the surface of an inert support structure, often alumina (Al_2O_3). Alumina, by virtue of its porous structure, has an enormous surface area per unit volume. Two typical forms in which the alumina-support catalyst is used are pellets of about

[1] A reducing atmosphere in this context is one in which the concentration of O_2 is less than that of the sum of CO and H_2. Conversely, in an oxidizing environment, the concentration of O_2 exceeds that of the sum of CO and H_2.

0.25-cm diameter and monolithic supports, thin-walled, honeycombed ceramic structures through which the exhaust gases must pass. Pellet supports are inexpensive and sturdy, but when packed closely in a reactor, they produce large pressure drops across the device, increasing the back pressure in the exhaust system. Monolithic supports allow a freer exhaust gas flow but are expensive and less resistant to mechanical and thermal damage.

The chemical reactions involved in the dual-catalyst system are shown in Fig. 8.13. In using this system the engine is set to run rich so that an abundance of HC, CO, and H_2 is produced during combustion and so that NO levels are reasonably low (see Fig. 7.15). Exhaust gas recirculation can be employed to keep NO levels low. As a result of the reducing atmosphere in the exhaust, NO is reduced to N_2, either by direct reaction with CO or by a two-stage reaction with H_2 that gives NH_3 as an intermediate product. As can be seen from Fig. 8.13, the necessary H_2 comes not only from the combustion gases but also from the reaction of CO and H_2O, the so-called water-gas shift reaction.

In order for the second catalyst bed to oxidize the HC and CO, it is necessary to add air to the exhaust before it enters the second reactor. The basic reactions in the second unit involve oxidation of HC and CO to CO_2 and H_2O. However, some of the ammonia formed in the first reactor may be oxidized back to NO. As we can see, NH_3 formation increases with the H_2 and CO level. Therefore, whereas the engine must be run rich to provide net reducing conditions for the NO_X bed, too much CO and H_2 will lead to significant NH_3 formation and subsequent oxidation of the NH_3 back to NO. The optimum exhaust CO concentration is about 1 to 2 percent, corresponding to air-fuel ratios in the range 13.8 to 14.5.

In order for the oxidizing reactor to function effectively, it should operate at about 400°C. Unfortunately, however, it requires a minute or more of operation before the bed reaches operating temperature. Meanwhile, the engine, during cold operation, is running rich and producing large quantities of HC and CO. One proposed solution to the warm-up problem is to use the reducing reactor as an oxidizing reactor only during warm-up. By temporarily injecting air into the raw engine exhaust, oxidizing conditions can be created in the first reactor. The additional air promotes oxidation in the exhaust manifold and the first reactor, which is able to reach its operating temperature in a few seconds. After a prescribed time, the airstream is switched back and the second converter is then used. The importance of temperature on the conversion of CO is illustrated in Fig. 8.14, in which the percentage of conversion of CO is plotted as a function of inlet gas temperature. The actual catalyst temperature is 100 to 150°C hotter because of the exothermic reactions taking place.

In general, reduction catalysts are less well developed than oxidation catalysts. In addition, the durability of NO_X catalysts has been decidedly inferior to that of the

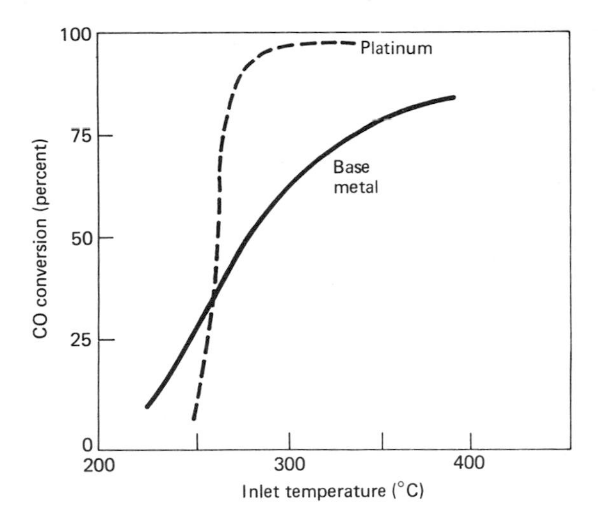

FIGURE 8.14
Effect of temperature on the conversion of CO in an oxidizing exhaust reactor. (*Source: Schlatter et al.*, 1973. *Copyright* 1973 *by the American Association for the Advancement of Science.*)

oxidation catalysts. There are a number of potential reasons for the poor durability of the reducing catalyst. Industrial catalyst use is usually accompanied by steady-state operation at carefully controlled temperatures, pressures, and flow rates, and under these conditions catalyst lives are generally long. In an automobile, the conditions of operation are always changing. The catalyst is cold during start-up and quite hot during a long downhill cruise; the air-fuel ratio is rich on idle and lean at high speed; the exhaust flow rate is low during idle and high during acceleration. In addition, certain compounds in gasoline, notably lead and phosphorus, are extremely efficient catalyst poisons. To alleviate the potential effect of poisons, lead-free gasoline to be available in 1975 and 1976 will have average lead and phosphorus levels of 0.03 and 0.005 g/gal, respectively.

An alternative to the dual catalytic system is the so-called three-way catalyst system, a single catalyst which will simultaneously promote oxidation of HC and CO and reduction of NO. Such a catalyst can operate successfully only in a narrow window of air-fuel ratios, slightly on the rich side of a stoichiometric mixture. The width of this window, shown in Fig. 8.15, is only about ± 0.1 in air-fuel ratio. Deviations on the lean side result in a rapid decrease of NO conversion, and rich-side deviations lead to substantial loss of ability to oxidize HC and CO, coupled with increased NH_3 formation. If the problems of controlling the air-fuel ratio so closely can be overcome in the three-way catalyst system, (1) one bed would be eliminated, (2) difficulties associated with heating up the second bed would be alleviated, and (3) fuel economy would be better since the combustion mixture would be closer to stoichiometric.

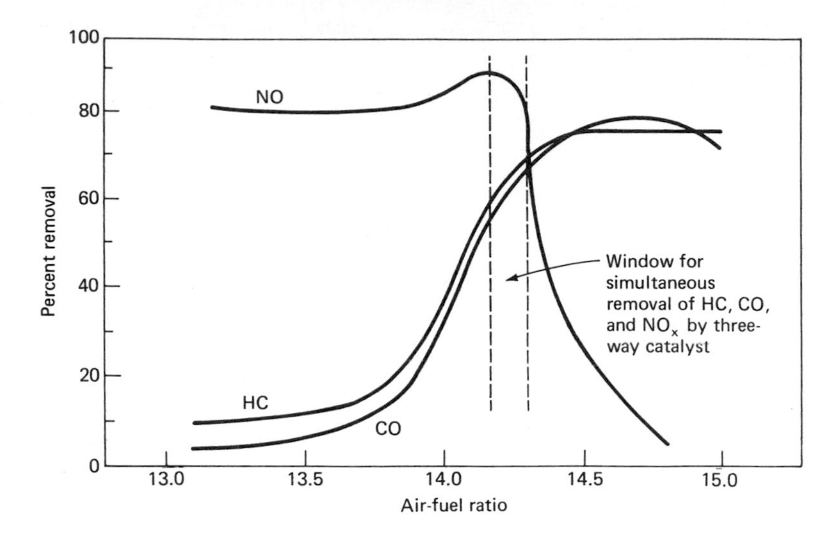

FIGURE 8.15
Effect of air-fuel ratio on the percent removal of HC, CO, and NO_X in a three-way catalyst system.

b THERMAL REACTORS The second type of exhaust treatment system is the so-called thermal reactor, a chamber in which HC and CO are further burned after leaving the cylinders. The thermal reactor can be used under either rich or lean operation. With rich operation, additional air injection is required. The Wankel engine, for example, is well suited to a thermal reactor since exhaust temperatures are quite high and all the exhaust ports are close together. The major problem associated with thermal reactors is achieving high enough temperatures inside the reactor to burn adequately the exhaust HC and CO. Rich-running systems are able to achieve higher temperatures (about 1000°C) but result in a loss in fuel economy and present problems in reactor durability.

8.1.5 Systems Capable of Meeting the 1976 Emissions Standards

We have discussed the four general classes of approaches to the control of exhaust HC, CO, and NO_X emissions. Now we shall consider those systems that offer the most promise for meeting the 1976 emissions standards.[1] In many cases the systems under current study and testing involve combinations of operating and engine modifications together with exhaust treatment. It should be clear that in developing an

[1] In spite of the fact that the original 1976 standards given in Table 8.1 have been postponed, we continue to refer to them as the 1976 standards.

overall system the effect of each modification on all other components in the system must be considered. For example, measures which have a high potential for HC and CO control often complicate the control of NO_X, and vice versa. Also, assuming several systems are capable of meeting the 1976 standards, the question arises as to the basis of comparison in order to choose a "best" system. For example, some systems have good durability characteristics but poor fuel economy, whereas others require frequent maintenance but have good fuel economy. Perhaps the best measure of comparison is the total cost of the system over the life of the vehicle, including initial purchase, maintenance, replacements, and increased fuel use. An up-to-date (as of Feb. 1, 1973) comparison of proposed systems for 1976 automobiles is given in Report by the Committee on Motor Vehicle Emissions (1973).[1] The engines evaluated for 1976 vehicles by the committee were (1) the dual-catalyst system, (2) the diesel, (3) the Wankel, (4) the three-valve stratified charge, and (5) a three-way catalyst system with feedback-controlled electronic fuel injection.

The interest in the dual-catalyst system stems, in part, from the desire of the automobile manufacturers both to protect their investments in the internal combustion engine and to utilize their vast experience with this engine. In spite of considerable study, as of mid-1973, no experimental engine modified to include the dual-catalyst system had demonstrated the durability required to meet the 1976 standards. A particular problem is the development of a catalyst capable of withstanding the strenuous conditions of actual use over a 50,000-mi life.

As of 1973, prototype cars equipped with the carbureted three-valve stratified-charge engine had met the 1975 standards for 50,000 mi and probably can be developed in time for mass production to meet the 1976 standards. The maintenance required for such an engine is apparently no more than that required on a conventional 1973 automobile. The fuel economy of the engine is comparable to that of a conventional 1972 engine and considerably superior to that of a dual-catalyst-equipped 1976 engine.

Diesel engines are capable of achieving emissions of 0.15, 2.5, and 1.65 g/mi for HC, CO, and NO_X, respectively. However, at this time the engine does not appear to be able to meet the 1976 standards. Because of its potential for greatly improved fuel economy, the diesel engine still warrants consideration for future vehicles.

The Wankel engine with a thermal reactor (for HC and CO) can attain the 1976 standards. In doing so, however, a fuel penalty of about 30 percent is incurred. The addition of EGR could probably enable meeting the 1976 NO_X standard, but at a still greater fuel consumption. Certain durability problems related to the thermal reactor still require work.

[1] Since fuel costs are changing so rapidly, and since the ultimate stable price of gasoline cannot yet be predicted, the CMVE conclusions clearly must be reviewed in light of price adjustments and fuel availability.

The three-way catalyst system enables the use of one reactor, as long as the air-fuel ratio is closely controlled. A feedback control system has been proposed in which a sensor measures exhaust oxygen concentration and controls the air-fuel ratio based on this measurement. As of 1973, adequate data on this system were apparently not available.

In summary, the two systems which appear to be most likely candidates for 1975 and 1976 vehicles are the dual-catalyst system and the carbureted stratified-charge engine. The former system appears to have several undesirable characteristics, even assuming the catalyst durability problem can be solved, when compared with the latter, namely:

1 Poorer fuel economy
2 Higher initial cost
3 Lower durability and higher maintenance difficulties
4 More difficulty in starting

8.2 REMOVAL OF GASEOUS POLLUTANTS FROM EFFLUENT STREAMS

As we noted in the introduction to this chapter, there are essentially three ways in which gaseous pollutants may be removed from an effluent stream: (1) absorption of the pollutant(s) in a liquid, (2) adsorption of the pollutant(s) on a solid surface, and (3) chemical alteration of the pollutant(s), as, for example, through combustion or catalytic treatment of the gases. In Subsecs. 8.2.1 and 8.2.2 we will discuss the principles underlying the first two of these techniques, with somewhat more emphasis on gas absorption, the more widely used method of the two. Combustion and catalytic treatment of exhaust gases, at least those containing CO, hydrocarbons, and NO, have already been discussed in Subsec. 8.1.4, and, for this reason, we do not cover these methods further here. In Subsecs. 8.2.3 and 8.2.4 we will briefly summarize current methods employed for the removal of SO_2 and NO_X from gases.

8.2.1 Absorption of Gases by Liquids

In the process of gas absorption the gaseous effluent stream containing the pollutant to be removed is brought into contact with a liquid in which the pollutant will dissolve. The mechanism by which the pollutant is removed from the gas consists of three steps: (1) diffusion of the pollutant molecules through the gas to the surface of the absorbing liquid, (2) dissolution into the liquid at the interface, and (3) diffusion of the dissolved pollutant from the interface into the bulk of the liquid. In order to predict the extent to which a pollutant can be removed by gas absorption, we must be able to describe these diffusion processes. Thus, we begin by reviewing some fundamentals of mass transfer.

Some Fundamentals of Mass Transfer

Let us consider diffusion in a binary mixture of species A and B. We consider a situation in which the total molar concentration of the mixture, c, is a constant, with molar concentrations and mole fractions of A and B, c_A and c_B and x and $1 - x$, respectively. The fundamental relation describing mass transfer by molecular diffusion in a binary mixture is *Fick's law*, which relates the flux of a species at a point to the gradient in its concentration. Considering one-dimensional diffusion in the z direction, Fick's law can be written in the form

$$J_{A_z}^* = -c\mathscr{D}_{AB} \frac{dx}{dz} \tag{8.1}$$

where $J_{A_z}{}^*$ is the molar flux of component A (moles A/area-time) relative to axes moving at the molar average velocity,

$$u^* = xu_A + (1 - x)u_B \tag{8.2}$$

$\mathscr{D}_{AB}$ is the molecular diffusivity of species A in a mixture of A and B, and u_A and u_B are the individual velocities of species A and B in the mixture. The usual convention relative to species fluxes like $J_A{}^*$ is that uppercase letters denote *molar* fluxes, and lowercase letters denote *mass* fluxes. For fluxes referred to *stationary* coordinates, we use n (or N), and for fluxes relative to *moving* coordinates, we employ j (or J). In the case of moving coordinates, the speed at which the axis moves is either the mass average velocity u or the molar average velocity u^*, where the asterisk denotes the latter frame.

An important point concerning (8.1) is that this relation is written relative to moving coordinate axes rather than to stationary ones. To appreciate why we express Fick's law in this way, let us consider a hypothetical diffusion process involving bowling balls and billiard balls. Let us suppose the floor of a room is partially covered with bowling balls, and we release through a door to the room a swarm of billiard balls. The billiard balls will collide with and diffuse among the larger balls, with little effect on the velocities of the latter. The diffusion of billiard balls occurs as a result of collisions with the bowling balls and is dependent on the relative velocities of the two, not on the velocities of the billiard balls relative to the floor. Thus, the proper way to describe the diffusion process is to do so by using some velocity which takes into account both species' velocities. Since the diffusion flux depends on the frequency of collisions, and thus on the number density rather than on the mass density, we choose the molar rather than the mass average velocity as that at which the coordinates move.

If we choose to express Fick's law relative to a fixed coordinate system, the molar flux of A, N_A, is a result of *two* contributions, the molecular diffusion of A and the flux of A due to the bulk motion of the whole system. Thus, since

$$N_A = \underset{\text{(molecular diffusion)}}{J_A{}^*} + \underset{\text{(bulk flow)}}{(N_A + N_B)x}$$

where $N_A = c_A u_A$, $N_B = c_B u_B$, (8.1) becomes

$$N_{A_z} = -c\mathscr{D}_{AB} \frac{dx}{dz} + (N_{A_z} + N_{B_z})x \tag{8.3}$$

Thus, the total flux of the entire fluid is $N_{A_z} + N_{B_z}$. The significance of the two terms in (8.3) can be understood with the aid of the following example. Let us suppose we inject a spot of dye in a flowing fluid. Even if diffusion is so slow that the spot of dye remains intact, as the spot passes a point of observation, there is a nonzero mass flux due solely to the fact that the stream is carrying the dye by its bulk motion. If the dye also spreads out because of diffusion, the *net* flux N_A observed as the stream flows by is that resulting from both diffusion (the spreading) and bulk flow.

Let us consider some simple examples of binary mass transfer. If one mole of A diffuses in a given direction for each mole of B diffusing in the opposite direction, then $N_A = -N_B$ and there is no net molal flow in the system. (Assume that we stand in the center of the room and watch bowling balls going to the right and billiard balls going to the left in equal numbers. There would be no net flow of numbers, or moles, although there would be a net *mass* flow toward the right.) In this case, referred to as equimolal counterdiffusion, (8.3) becomes

$$N_{A_z} = -c\mathscr{D}_{AB} \frac{dx}{dz} \tag{8.4}$$

Figure 8.16*a* shows a situation of equimolal counterdiffusion. A mass balance on component A yields $N_{A_z} = \text{const}$ as long as there are no sources of A between $z = 0$ and $z = l$. Thus, (8.4) may be integrated to give

$$N_{A_z} = \frac{-c\mathscr{D}_{AB}}{l} (x_l - x_0) \tag{8.5}$$

For gases at moderate pressures, partial pressures can be used in place of mole fractions, and

$$N_{A_z} = \frac{-\mathscr{D}_{AB}}{RTl} (p_{A_l} - p_{A_0}) \tag{8.6}$$

Another example of simple mass transfer is diffusion of A through stagnant B, in which case $N_{B_z} = 0$, and

$$N_{A_z} = -c\mathscr{D}_{AB} \frac{dx}{dz} + N_{A_z} x \tag{8.7}$$

Integrating (8.7), in accordance with the situation shown in Fig. 8.16*b*, again with $N_{A_z} = \text{const}$, we obtain

$$N_{A_z} = \frac{-c\mathscr{D}_{AB}}{l} \ln \frac{1 - x_0}{1 - x_l} \tag{8.8}$$

We can rewrite (8.8) as

$$N_{A_z} = \frac{-c\mathscr{D}_{AB}}{l(x_B)_{lm}} (x_l - x_0) \tag{8.9}$$

where $(x_B)_{lm}$, the log mean mole fraction of component B, is defined as

$$(x_B)_{lm} = \frac{(1 - x)_l - (1 - x)_0}{\ln((1 - x)_l/(1 - x)_0)} \tag{8.10}$$

Again, partial pressures may be used instead of mole fractions, in which case (8.9) becomes

$$N_{A_z} = \frac{-p\mathscr{D}_{AB}}{RTl(p_B)_{lm}} (p_{A_l} - p_{A_0}) \tag{8.11}$$

where p is the total pressure.

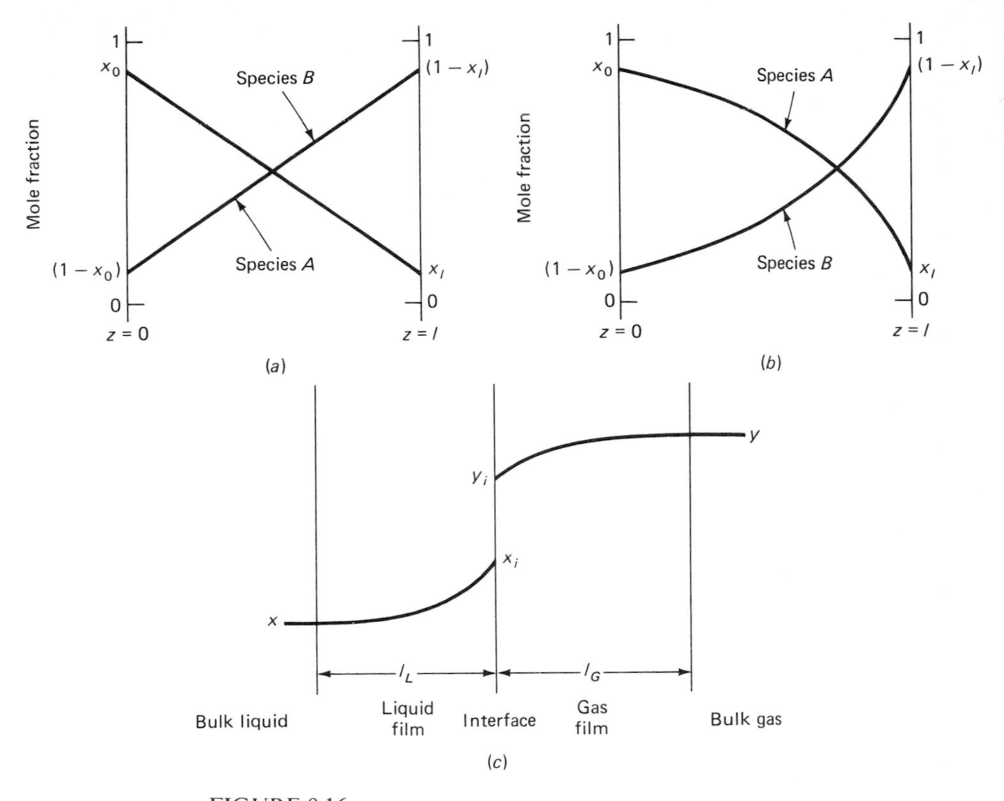

FIGURE 8.16
Mass transfer situations: (*a*) equimolal counterdiffusion; (*b*) diffusion of A through stagnant B; (*c*) the two-film model for interfacial mass transfer.

Up to now we have discussed mass transfer by molecular diffusion and within a single phase. In order to treat gas absorption, we must consider the case in which the gas is in turbulent flow and the transport of species occurs across a gas-liquid interface. Most situations of mass transfer in turbulent flow near an interface are too complicated to allow an exact evaluation of profiles and fluxes. Thus, certain idealized models are postulated for the mass transfer in such a situation, models which enable the solution for the flux of a species in terms of certain empirical coefficients.

Turbulent motion maintains a fairly uniform composition in the bulk gas. Close to the surface of the liquid, a laminar boundary layer exists in the gas across which species in the bulk gas must diffuse to reach the liquid surface. Similarly, on the liquid side, the bulk liquid is at a uniform composition with a thin layer near the surface of the liquid through which species diffuse from the interface into the bulk liquid. At steady state it can be assumed that the flux of species A from the bulk gas to the interface equals the flux of A from the interface to the bulk liquid. The simplest model one can envision for this situation is two stagnant layers on either side of the interface, as shown in Fig. 8.16*c*. Based on the form of (8.11) we assume that the flux of A is given by

$$N_{A_z} = k_G(p_A - p_{A_i}) = k_L(c_{A_i} - c_A) \tag{8.12}$$

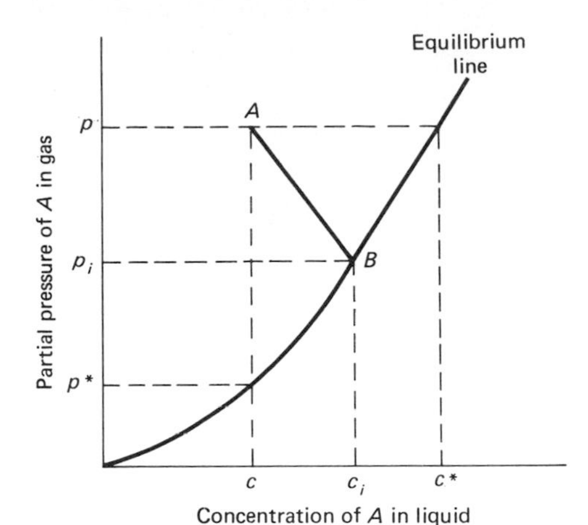

FIGURE 8.17
Driving forces in the two-film model of interfacial mass transfer.

where p_A and p_{A_i} are the partial pressures of A in the bulk gas and at the interface, respectively, c_A and c_{A_i} are the concentrations of A in the bulk liquid and at the interface, respectively, and k_G and k_L are *mass transfer coefficients* for the gas and liquid films, respectively. For the cases considered earlier, we see that k_G and k_L are given by

$$k_G = \frac{p\mathscr{D}_{AB}}{RTl_G(p_B)_{lm}} \tag{8.13}$$

and

$$k_L = \frac{\mathscr{D}_{AB}}{l_L(x_B)_{lm}} \tag{8.14}$$

For dilute mixtures of A in B, $(x_B)_{lm} \cong 1.0$.

Thus, the mass transfer coefficients depend on the molecular diffusivity of A in B and on the thickness of the film over which the diffusion takes place. Unfortunately, in mass transfer between turbulent gas and liquid streams, it is virtually impossible to specify l_G and l_L and, in fact, even to specify the precise location of the interface or the values of p_{A_i} and c_{A_i} at any time. Thus, we usually write (8.12) as

$$N_{A_z} = K_G(p_A - p_A^*) = K_L(c_A^* - c_A) \tag{8.15}$$

where p_A^* is the equilibrium partial pressure of A over a solution of A and B having the bulk concentration c_A, and c_A^* is the concentration of a solution which would be in equilibrium with the partial pressure p_A of the bulk gas stream. We illustrate these points on the equilibrium diagram in Fig. 8.17. The new coefficients K_G and K_L are called *overall* mass transfer coefficients. These must be determined experimentally.

Originally, in (8.12), the driving forces for diffusion were based on the actual interfacial compositions, p_{A_i} and c_{A_i}. Since we do not know these in general, we replaced (8.12) with (8.15), in which the new overall mass transfer coefficients K_G and K_L were defined. The new driving forces for diffusion, $p_A - p_A^*$ and $c_A^* - c_A$ are shown on Fig. 8.17. The point B on the equilibrium curve represents the interfacial composition which we assume to be (p_{A_i}, c_{A_i}). The line AB has a slope $-k_L/k_G$ and is given by

$$\frac{p_A - p_{A_i}}{c_{A_i} - c_A} = \frac{k_L}{k_G} \tag{8.16}$$

When the equilibrium line is given by

$$p_A^* = Hc_A \tag{8.17}$$

with H constant (Henry's law), we can explicitly relate k_G and k_L and K_G and K_L through (8.12) and (8.15) by

$$\frac{1}{K_G} = \frac{1}{k_G} + \frac{H}{k_L} \tag{8.18}$$

and

$$\frac{1}{K_L} = \frac{1}{k_L} + \frac{1}{Hk_G} \tag{8.19}$$

If $H \ll 1$, then $K_G \simeq k_G$, and the overall process is controlled by diffusion through the gas film. On the other hand, if $H \gg 1$, the process is liquid-film-controlled. Note that we have drawn the equilibrium line in Fig. 8.17 as curved, since H is not usually constant (and thus (8.18) and (8.19) are not generally valid).

It is also possible to express (8.12) in terms of mole fractions, in which case

$$N_{A_z} = k_y(y - y_i) = k_x(x_i - x) \tag{8.20}$$

where y and x refer to the gas and liquid phases, respectively.

Let us now consider the design of industrial gas absorption operations. We assume that the gas and liquid phases are immiscible. For example, SO_2 is removed from air by absorption in a liquid amine of low vapor pressure. The low vapor pressure of the amine ensures that virtually no amine evaporates into the gas phase, and operation at atmospheric pressure ensures that no air dissolves in the amine. Thus, even though SO_2 is transferred between phases, the assumption of immiscibility refers to the fact that the two carrier streams, in this case air and amine, do not dissolve in each other to an appreciable extent.

Gas absorption is usually carried out in a column or tower, in which the gas to be cleaned (the rich gas) enters at the bottom and flows countercurrent to the fresh liquid (the lean liquid) which is introduced at the top. The column is often packed with inert solids (e.g. ceramic beads) to promote better contact between the two streams. Such a tower is shown diagrammatically in Fig. 8.18. The basic gas absorption design problem is the following. Given:

1 A rich gas stream entering at a rate G(moles/hr–m^2 of empty tower) containing a known mole fraction of component A, y_0
2 A desired exit gas mole fraction y_1
3 A specified mole fraction of A in the inlet liquid, x_1
4 The equilibrium curve of y^* versus x^* for the system

we wish to compute the height of the tower required to carry out the separation.[1]

[1] One would normally also determine the column diameter based on the liquid flow rate and desired pressure drop characteristics. We do not consider this aspect of the design here; rather, we simply assume that the column diameter is constant and known.

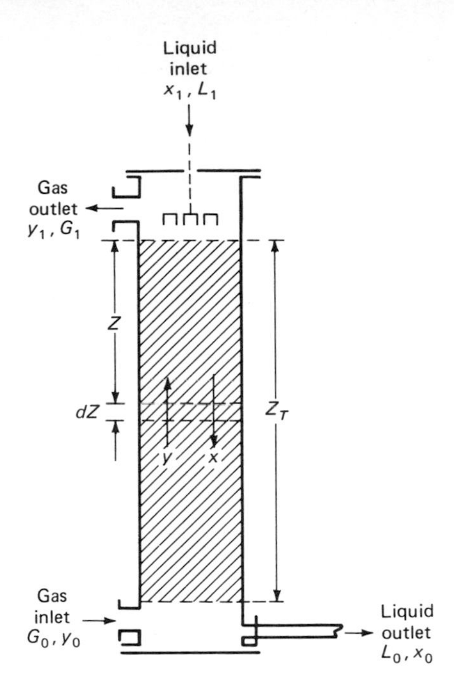

FIGURE 8.18
Diagram of a countercurrent gas absorption tower.

We note that, although the molal gas flow rate G is usually specified, that for the liquid phase is not. A little reflection will show that there is no maximum to the value of L, but, indeed, there is a minimum value of L below which the required A cannot be separated from the gas. Actually the total gas and liquid flows need not be constant through the tower. The gas flow G consists of A + inert gas, and L consists of A + inert liquid. By our assumption of immiscibility, the flow rates of inert gas and inert liquid always remain constant down the tower. We denote these flow rates by G' and L', respectively. Thus, we really want to determine the minimum L'.

In Fig. 8.19 point (x_1, y_1) denotes the top of the tower and point A represents any point in the column. The driving force for mass transfer is proportional to the line AB, as shown previously in Fig. 8.17. Point A must always lie above the equilibrium line; however, as A approaches B, the driving force for mass transfer approaches zero. When A actually coincides with B at any point in the tower, mass transfer ceases, because, of course, the two phases are in equilibrium at that point. Clearly, if a point is reached in an actual column where A and B coincide, no more mass transfer can take place past that point regardless of the height of the column.

In order to determine the minimum value of L', we must perform a material balance on species A for the absorption tower. A balance on species A over the whole tower gives

$$L_1 x_1 + G_0 y_0 = L_0 x_0 + G_1 y_1 \tag{8.21}$$

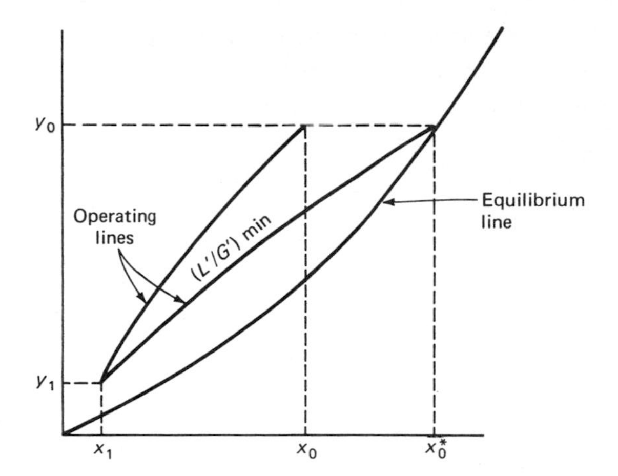

FIGURE 8.19
Equilibrium line and operating lines for a gas absorption tower.

where (G_0, L_0) and (G_1, L_1) represent the flows at the bottom and top of the column, respectively. At any point in the tower, where the flow rates are G and L, a balance around the top of the column gives

$$L_1 x_1 + Gy = Lx + G_1 y_1 \tag{8.22}$$

Rearranging (8.22), we have

$$y = \frac{L}{G} x + \frac{1}{G}(G_1 y_1 - L_1 x_1) \tag{8.23}$$

On a plot of y versus x, (8.23) represents a line, not necessarily straight (unless L and G are constant through the whole column), which relates the compositions of passing streams at any point. Such a line is called an *operating line*. The two ends of the column are represented by points (x_0, y_0) and (x_1, y_1).

In order to draw the operating line, we need to know L and G at each point in the column. In the case of gas absorption, in which only component A is transferred between phases, we know that

$$G = \frac{G'}{1-y} \qquad L = \frac{L'}{1-x} \tag{8.24}$$

where G' and L' are constant. Thus, (8.22) becomes

$$L'\left(\frac{x_1}{1-x_1} - \frac{x}{1-x}\right) = G'\left(\frac{y_1}{1-y_1} - \frac{y}{1-y}\right) \tag{8.25}$$

If the mole fraction of A in each phase is small, then, for all practical purposes, $G \cong G'$, $L \cong L'$, and the operating line is straight, with a slope of L'/G'.

We now consider the two operating lines shown in Fig. 8.19, drawn for the case in which L/G varies over the tower. The average slope of the operating line is L/G, so that as L is decreased, the slope decreases. Point (x_1,y_1) is fixed, and so as L is decreased, the upper end of the operating line, that is, x_0, moves closer to the equilibrium line. The maximum possible value of x_0 and the minimum possible value of L' are reached when the operating line just touches the equilibrium line, as shown in Fig. 8.19. At this point, an infinitely long column would be required to achieve the desired separation. We can find the minimum value of L'/G' by setting $y = y_0$ and $x = x_0^*$ in (8.25), where x_0^* is the abscissa of the point on the equilibrium line corresponding to y_0. Customarily, a value of L'/G' about 1.5 times the minimum is employed. This choice is an economic one. If L'/G' is large, the distance between the operating and equilibrium lines is large, the driving force is large, and a short column is needed. On the other hand, a high liquid flow rate may be costly. Thus, the optimum L'/G' results from a balance between capital equipment costs and operating costs.

Assuming L' has been specified, we wish to determine the required column height. Let us consider a differential height of the column dZ. If the interfacial area per unit volume is a, a material balance for component A in the gas phase yields

$$Gy|_Z = Gy|_{Z+dZ} - k_y a(y - y_i)\, dZ$$

which, upon rearrangement, division by dZ, and letting $dZ \to 0$, gives

$$\frac{d(Gy)}{dZ} = k_y a(y - y_i) \tag{8.26}$$

where y_i is a point on the equilibrium curve. Using (8.24), we see that

$$d(Gy) = G' d\frac{y}{1 - y} = G' \frac{dy}{(1 - y)^2} = G \frac{dy}{1 - y} \tag{8.27}$$

Integrating (8.26) with the aid of (8.27) produces

$$\int_0^{Z_T} dZ = \int_{y_1}^{y_0} \frac{G}{k_y a} \frac{dy}{(1 - y)(y - y_i)} \tag{8.28}$$

To determine the total height Z_T we must evaluate the integral in (8.28). The method of integration depends on the shape of the equilibrium line, the variation in G, and the relative importance of the two mass transfer coefficients $k_x a$ and $k_y a$.

From (8.16) we note that

$$\frac{y - y_i}{x_i - x} = \frac{k_x a}{k_y a} \tag{8.29}$$

Thus, at any point, (8.29) describes a straight line with slope $-k_x a/k_y a$, passing through (x,y) and (x_i,y_i). From a knowledge of $k_x a/k_y a$ we can determine x_i and y_i corresponding to any (x,y) on the operating line. Then (8.28) can be integrated graphically. It is common to express (8.28) as

$$Z_T = \overline{\left(\frac{G}{k_y a}\right)} \int_{y_1}^{y_0} \frac{dy}{(1-y)(y-y_i)} \tag{8.30}$$

where $\overline{(G/k_y a)}$ is the average value of this group over the column. (Since G decreases from bottom to top, and $k_y a$ also decreases from bottom to top, these changes somewhat compensate each other.) The functional dependence of $k_x a$ and $k_y a$ on the molal flow rates must be determined experimentally.

In deriving an expression for Z_T we could have considered a liquid-side balance, in which case the equation corresponding to (8.30) is

$$Z_T = \overline{\frac{L}{k_x a}} \int_{x_1}^{x_0} \frac{dx}{(1-x)(x_i-x)} \tag{8.31}$$

Either (8.30) or (8.31) is suitable for carrying out calculations.

The design method embodied in (8.30) and (8.31) is applicable to an equilibrium line of arbitrary shape. A strongly curved equilibrium line is often due to a significant temperature variation over the height of the tower. Appreciable temperature differences result from the heat of solution of a highly concentrated solute in the rich gas. If the rich gas contains a rather dilute concentration of solute, the temperature gradient in the column is small, and the equilibrium line is approximately straight. When the equilibrium line is straight, overall mass transfer coefficients, which are easier to determine experimentally than $k_x a$ and $k_y a$, can be used. The overall coefficients $K_x a$ and $K_y a$ are defined on the basis of the fictitious driving forces $x - x^*$ and $y - y^*$. From (8.18) and (8.19), if the slope of the equilibrium line is h, then

$$\frac{1}{K_x a} = \frac{1}{k_x a} + \frac{1}{h k_y a} \qquad \frac{1}{K_y a} = \frac{1}{k_y a} + \frac{h}{k_x a} \tag{8.32}$$

The design equations analogous to (8.30) and (8.31) are in this case

$$Z_T = \overline{\frac{G}{K_y a}} \int_{y_1}^{y_0} \frac{dy}{(1-y)(y-y^*)} \tag{8.33}$$

and

$$Z_T = \overline{\frac{L}{K_x a}} \int_{x_1}^{x_0} \frac{dx}{(1-x)(x^*-x)} \tag{8.34}$$

which can be evaluated graphically given the y^* versus x^* equilibrium line.

It has been customary in gas absorption design to express the equations for Z_T, that is (8.30), (8.31), (8.33), and (8.34), as the product of a number of *transfer units* and the depth of packing required by a single of these units (the height of a transfer unit). Then Z_T is written

$$Z_T = NH \qquad (8.35)$$

where N is the number of transfer units and H is the height of a transfer unit (HTU). For example, using (8.30) and (8.33), we define

$$N_y = \int_{y_1}^{y_0} \frac{dy}{(1-y)(y-y_i)} \qquad (8.36)$$

and

$$N_{0y} = \int_{y_1}^{y_0} \frac{dy}{(1-y)(y-y^*)} \qquad (8.37)$$

where N_y and N_{0y} are based on the individual and overall driving forces, respectively. Of course, N_y and N_{0y} are different, and, in order to produce the same Z_T in (8.35), they are compensated for by the corresponding H's. Thus,

$$H_y = \frac{\overline{G}}{k_y a} \qquad (8.38)$$

and

$$H_{0y} = \frac{\overline{G}}{K_y a} \qquad (8.39)$$

Similar relations hold for the liquid-side equations. When the equilibrium line is straight, and G and L are constant throughout the tower, H_{0y} (and H_{0x}) are constant.

Clearly, the method merely represents a different manner of viewing Z_T. Its advantage is that the HTU is usually fairly constant for a particular type of tower (usually with a value in the range of 0.1 to 1.5 m), and data are often reported in terms of the HTU. Correlations for HTUs for packed absorption towers are presented by McCabe and Smith (1956).

Absorption of SO_2 from Air by Water

A packed tower is to be designed for absorption of SO_2 from air by contact with freshwater. The entering gas has a mole fraction of SO_2 of 0.10, and the exit gas must contain a mole fraction of SO_2 no greater than 0.005. The water flow rate used is to be 1.5 times the minimum, and the inlet air flow rate (on an SO_2-free basis) is 500 kg/m^2-hr. The column is to be operated at 1 atm and 30°C. We wish to determine the required depth of the packed section for such a tower.

The following correlations are available for absorption of SO_2 at 30°C in towers packed with 1-in. rings (McCabe and Smith, 1956):

$$k_x a = 0.202\tilde{L}^{0.82} \qquad k_y a = 0.0303\tilde{L}^{0.25}\tilde{G}^{0.7}$$

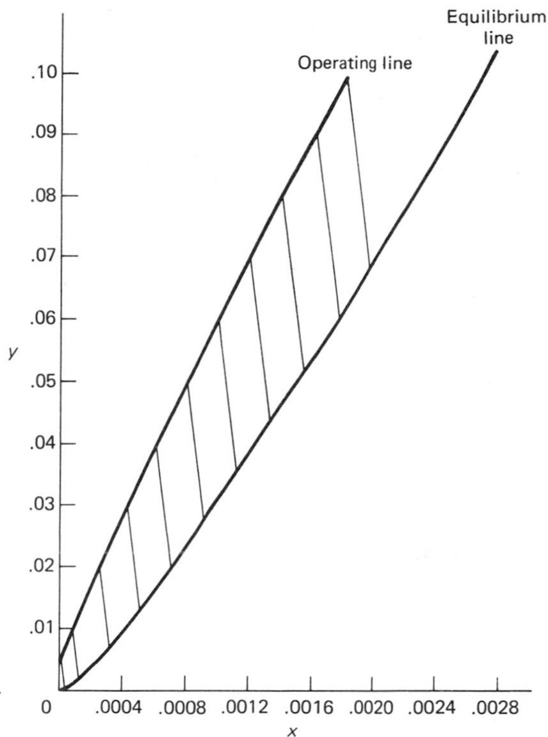

FIGURE 8.20
Equilibrium line and operating line for SO_2 absorption in water.

where $\tilde{L}$ and $\tilde{G}$ are the mass velocities of liquid and gas, respectively, in kg/m²-hr, and $k_x a$ and $k_y a$ are in kg-moles/m³-hr-mole fraction.

Equilibrium data for SO_2 in air and water are available:

$p_{SO_2}{}^*$, mm Hg	0.6	1.7	4.7	8.1	11.8	19.7	36.0	52.0	79.0
c^*, g SO_2/100 g H_2O	0.02	0.05	0.10	0.15	0.20	0.30	0.50	0.70	1.00

From these data, we can calculate the equilibrium curve:

$$y^* = \frac{p_{SO_2}{}^*}{760} \qquad x^* = \frac{c^*/64}{c^*/64 + 100/18}$$

The equilibrium curve is shown in Fig. 8.20.

The first step in the solution is calculation of the minimum water flow rate. Using (8.25) with $y_0 = 0.10$, $x_1 = 0$, $y_1 = 0.005$, and $x_0{}^* = 0.0027$, we obtain $L'_{\min} = 667$ kg-mole/m²-hr. Thus, the actual water rate to be used is $667 \times 1.5 = 1000$ kg-mole/m²-hr.

The equation for the operating line is

$$\frac{x}{1-x} = 0.0172 \frac{y}{1-y} - 0.000086$$

This line is shown in Fig. 8.20.

The SO_2 enters at a rate of 122 kg/m²-hr and leaves at a rate of 5.5 kg/m²-hr. The total exit gas rate is 505.5 kg/m²-hr. The freshwater feed at the top is 18,000 kg/m²-hr, and the rich liquor leaving at the bottom is 18,116.5 kg/m²-hr.

The liquid-side mass transfer coefficient will not change appreciably from the top to the bottom since $\tilde{L}$ is nearly constant. We can calculate $k_x a$ from the average mass velocity of 18,058 kg/m²-hr:

$$k_x a = 625$$

Because of the change of the total gas velocity from the top to bottom, $k_y a$ will change somewhat over the tower. The values at the top and bottom are

$$(k_y a)_0 = 31.74$$
$$(k_y a)_1 = 27.4$$

We shall use the average value of 29.57.

Therefore, from any point (x,y) on the operating line, we can determine $x_i y_i$ by drawing a straight line with slope $-625/29.57 = -21.2$. The integral in (8.30) can be evaluated graphically. Table 8.2 shows the calculation of the quantity $1/(1-y)(y-y_i)$ and the graphical integration of (8.30). The value of the integral in (8.30) is found to be 5.72.

Finally, we evaluate the quantity $k_y a/G$ at the two ends of the tower,

$$\left(\frac{k_y a}{G}\right)_0 = 1.66$$

$$\left(\frac{k_y a}{G}\right)_1 = 1.585$$

and use the average value of 1.62 to calculate Z_T as 3.54 m.

8.2.2 Adsorption of Gases on Solids

An alternative to absorption by liquids is adsorption of pollutants on solids. When a gas molecule is adsorbed on the surface of a solid, it can be retained by either physical or chemical forces. The measurement of hydrocarbon air pollutants by gas chromatography, for example, is achieved by the selective adsorption of certain gases

Table 8.2 EVALUATION OF INTEGRAND IN (8.30)

y	$1-y$	y_i	$y-y_i$	$(1-y)(y-y_i)$	$\frac{1}{(1-y)(y-y_i)}$	ΔI	$\Delta I \Delta Y$
0.005	0.995	0.0005	0.0045	0.00448	223.		
0.01	0.99	0.002	0.0080	0.00792	126.5	164	0.82
0.02	0.98	0.0075	0.0125	0.01225	81.7	102	1.02
0.03	0.97	0.014	0.0160	0.01552	64.5	72	0.72
0.04	0.96	0.0215	0.0185	0.01775	56.4	60	0.60
0.05	0.95	0.0285	0.0215	0.0204	49	52.5	0.525
0.06	0.94	0.036	0.0240	0.0226	44.2	46.5	0.465
0.07	0.93	0.044	0.0260	0.0242	41.4	42.8	0.428
0.08	0.92	0.0520	0.0280	0.0258	38.8	40	0.400
0.09	0.91	0.0605	0.0295	0.0268	37.3	38	0.380
0.10	0.90	0.0685	0.0315	0.0283	35.3	36	0.360
							5.718

on the solid packing in a chromatographic column. The solids most suited for adsorption are those with large surface-to-volume ratios, that is, very porous. Common solids with these properties are activated charcoal (carbon), alumina, and silica gel.

The process of gas adsorption on a solid, like gas absorption in a liquid, proceeds by a series of steps. First, the gas molecules must diffuse from the bulk gas to the outer surface of the solid. Second, the gas molecules must diffuse into the pores of the solid, and, third, the molecule actually adsorbs on the surface of the solid. The first step can be described similarly to the gaseous diffusion step in absorption by a liquid, namely by means of a mass transfer coefficient multiplying a driving force. The description of the rate of internal diffusion is highly complicated and depends, in part, on the relation of the mean pore diameter to the mean free path of the gas molecules.

Potential adsorbents can be classified into three groups (Strauss, 1966):

1 Nonpolar solids, where the adsorption is mainly physical

2 Polar solids, where the adsorption is chemical and no change in the chemical structure of the molecules or the surface occurs

3 Chemical adsorbing surfaces, which adsorb the molecules and then release them after reaction, which may be either catalytic, leaving the surface unchanged, or noncatalytic, requiring replacement of the surface atoms

The only important nonpolar adsorbing solid is carbon, which is very effective in binding nonpolar molecules, such as hydrocarbons. Activated carbon (charcoal, if the source is wood) is made by the decomposition of coals and woods. Activated carbon is used for the removal of hydrocarbons, odors, and trace impurities from gas streams.

The polar adsorbents generally used are oxides, either of silicon or other metals (e.g. aluminum). These materials adsorb both polar and nonpolar molecules, but they exhibit preference for polar molecules. Thus, silicon and aluminum oxides are used to adsorb polar molecules such as water, ammonia, hydrogen sulfide, and sulfur dioxide.

Adsorbents which react chemically with the gas molecules are of great variety, depending on the particular gas involved. Several examples of such materials will be cited in the sections to follow.

The equilibrium characteristics of a gas-solid system are described by a curve of the concentration of adsorbed gas on the solid as a function of the equilibrium partial pressure of the gas at constant temperature. Such a curve is called an *adsorption isotherm*. In the case in which only one component of a binary gas mixture is adsorbed, the adsorption of that species is relatively uninfluenced by the presence of the other gas, and the adsorption isotherm for the pure vapor is applicable as long as the equilibrium pressure is taken as the partial pressure of the adsorbing gas.

Separation of one component from a gaseous mixture by adsorption on a solid may be carried out in a batchwise or continuous manner of operation. Continuous operation can, in turn, be employed in a series of distinct stages or in continuous contact, such as in gas absorption. When one component is being adsorbed, the design of the operation is, from the point of view of the calculational procedure, analogous to gas absorption, in that only one component is transferred between two essentially immiscible phases. A rather thorough treatment of gas adsorption operations is given by Treybal (1968). We consider here only the process of adsorption of a species as the gas is passed through a stationary (fixed) bed of adsorbent.

The key difference between gas absorption with two continuous countercurrent streams and gas adsorption in a fixed bed is that the former is a steady-state process whereas the latter, due to the accumulation of adsorbed gas on the solid, is an unsteady-state process.

We will consider a mixture of two gases, one strongly adsorbed, which is to be passed through a bed initially free of adsorbent. When the mixture first enters the fresh bed, the solid at the entrance to the bed at first adsorbs the gas almost completely. Thus, initially the gas leaving the bed is almost completely free of the solute gas. As the layers of solid near the entrance to the bed become saturated with adsorbed gas, the zone of solid in which the major portion of the adsorption takes place moves slowly through the bed, at a rate generally much slower than the actual gas velocity through the bed. Finally, the so-called adsorption zone reaches the end of the bed. At this point, the exit concentration of solute gas rises sharply and approaches its inlet concentration, since, for all practical purposes, the bed is saturated and at equilibrium with the inlet gas. The curve of effluent concentration as a function of time thus has an S-shaped appearance and is commonly referred to as the break-through curve. The S-shaped break-through curve may be steep or relatively flat, depending on the rate of adsorption, the nature of the adsorption equilibrium, the fluid velocity, the inlet concentration, and the length of the bed. The time at which the break-through curve first begins to rise appreciably is called the breakpoint.

When a bed reaches saturation, the adsorbed material must be removed from the solid. Desorption of an adsorbed solute by passing a solvent through the bed is called *elution*. The process of gas chromatography, mentioned in Chap. 4 as a means of measuring concentrations of gaseous hydrocarbons, is based on the elution of a bed which contains small quantities of several adsorbed gases. As a suitable eluent is passed through such a bed, the adsorbed solutes are desorbed at different rates and pass out of the bed at different times, enabling their identification by comparison with eluent curves previously established for known species.

The design of a fixed-bed adsorption column would normally require that one predict the break-through curve, and thus the length of the adsorption cycle between elutions of the bed, given a bed of certain length and equilibrium data. Alternatively, one could seek the bed depth required for operation over a specified period of time to achieve a desired degree of separation. Because of the different types of equilibrium relationships which can be encountered, and the unsteady nature of the process, prediction of the solute break-through curve is, in general, quite difficult. We present here a design method applicable only when the solute concentration in the feed is small, when the adsorption isotherm is concave to the gas-phase concentration axis, when the adsorption zone is constant in height as it travels through the column, and when the length of the column is large compared with the height of the adsorption zone (Treybal, 1968).

Let us consider the idealized break-through curve shown in Fig. 8.21*a* resulting from flow of an inert gas through a bed with a rate $\tilde{G}'$ kg/m^2-hr containing an inlet solute concentration of Y_0 kg solute/kg inert gas. The total amount of solute-free gas which has passed through the bed up to any time is w kg/m^2 of bed cross section. Values Y_B and Y_E, shown in Fig. 8.21*a*, mark the breakpoint and equilibrium concentrations, respectively; w_B and w_E denote the values of w at Y_B and Y_E, respectively. The adsorption zone, taken to be of constant height Z_a, is that part of the bed in which the concentration profile from Y_B to Y_E exists at any time.

If θ_a and θ_E are the times required for the adsorption zone to move its own length and down the entire bed, respectively, then

$$\theta_a = \frac{w_a}{\tilde{G}'} \tag{8.40}$$

and

$$\theta_E = \frac{w_E}{\tilde{G}'} \tag{8.41}$$

If θ_F is the time required for the adsorption zone to form, and if Z is the length of the bed,

$$Z_a = Z \frac{\theta_a}{\theta_E - \theta_F} \tag{8.42}$$

The solute removed from the gas in the adsorption zone is U kg/m^2 of bed cross section; U is shown in Fig. 8.21*a* by the shaded area, which is

$$U = \int_{w_B}^{w_E} (Y_0 - Y)\, dw \tag{8.43}$$

If all the adsorbent in the zone were saturated, the solid would contain $Y_0 w_a$ kg solute/m^2. Thus, the fractional capacity of the adsorbent in the zone to continue absorbing solute is $f = U/Y_0 w_a$. The shape of the break-through curve is thus characterized by f. If $f = 0$, the time of formation θ_F of the zone should be the same as the

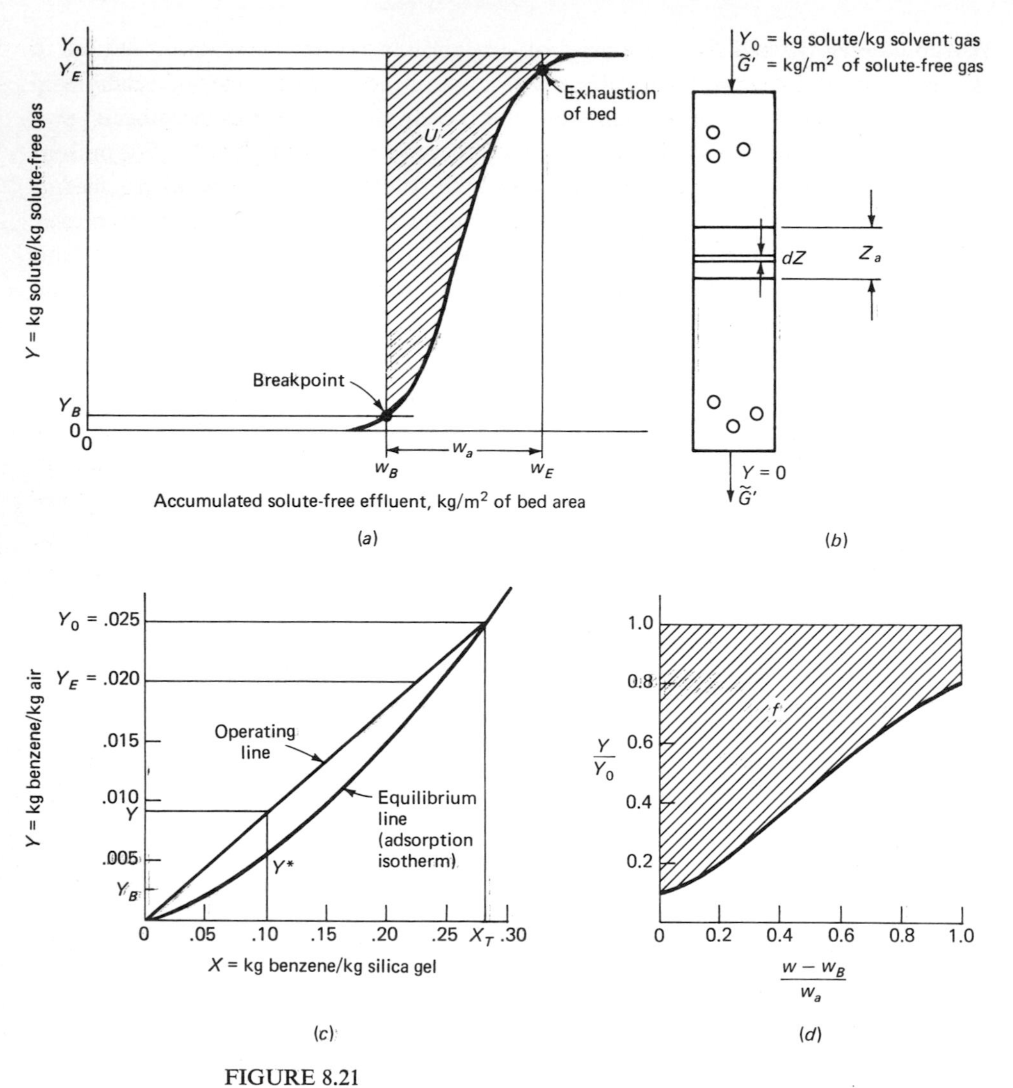

FIGURE 8.21
Adsorption of a gas on a solid. (*a*) A typical break-through curve; (*b*) a fixed bed adsorber with adsorption zone of depth Z_a; (*c*) equilibrium and operating lines for adsorption of benzene on silica gel; (*d*) break-through curve for adsorption of benzene on silica gel.

time required for the zone to travel its own thickness, θ_a, since the break-through curve will be a vertical line. If $f = 1$, the time to establish the zone should be zero. To satisfy these two limiting cases, one sets $\theta_F = (1 - f)\theta_a$. Thus, (8.42) becomes

$$Z_a = Z \frac{\theta_a}{\theta_E - (1 - f)\theta_a} = Z \frac{w_a}{w_E - (1 - f)w_a} \tag{8.44}$$

If the column contains $ZA\rho_s$ kg of adsorbent, where A is the cross-sectional area of the bed and ρ_s is the solid density in the bed, at complete saturation, the bed would contain $ZA\rho_s X_T$ kg of solute, where X_T is the solute concentration on the solid in equilibrium with the feed. At the breakpoint, $Z - Z_a$ of the bed is saturated, and Z_a of the bed is saturated to the extent of $1 - f$. The degree of overall bed saturation at the breakpoint is thus

$$\alpha = \frac{(Z - Z_a)\rho_s X_T A + Z_a \rho_s (1 - f) X_T A}{Z \rho_s X_T A} = \frac{Z - f Z_a}{Z} \tag{8.45}$$

The determination of the break-through curve can be carried out in the following way. Let us consider the adsorption column shown in Fig. 8.21*b*, where the adsorption zone Z_a is in the column, and the solute composition in the gas is Y_0 and 0 at the entrance and exit, respectively. Corresponding to these gas-phase compositions, we assume that those on the solid are X_T (saturation at the entrance to the column) and 0 (no adsorbed solute at the exit). If the column is considered to be infinitely long, the situation depicted in Fig. 8.21*b* is applicable. This point will not really concern us since our only real interest is in the adsorption zone Z_a. The operating line, which relates Y and X at any point in the column, is then a straight line connecting the origin with the point (Y_0, X_T) on the equilibrium curve.

Over a differential depth dZ in Z_a the rate of adsorption is

$$\tilde{G}'\, dY = K_Y a (Y - Y^*)\, dZ \tag{8.46}$$

where $K_Y a$ is the overall mass transfer coefficient for transfer from gas to solid phase. Thus, over the adsorption zone

$$Z_a = \frac{\tilde{G}'}{K_Y a} \int_{Y_B}^{Y_E} \frac{dY}{Y - Y^*} \tag{8.47}$$

and for any value of Z less than Z_a, but within the zone,

$$\frac{Z}{Z_a} = \frac{w - w_B}{w_a} = \frac{\int_{Y_B}^{Y} dY/(Y - Y^*)}{\int_{Y_B}^{Y_E} dY/(Y - Y^*)} \tag{8.48}$$

The break-through curve can be plotted directly from (8.48). The following example illustrates certain aspects of the design of a gas adsorption tower.

Adsorption of Benzene from Air

Benzene vapor present to the extent of 0.025 kg benzene/kg air (benzene-free basis) is to be removed by passing the gas mixture downward through a bed of silica gel at 25°C and 2 atm pressure at a linear velocity of 1 m/sec (based on the total cross-sectional area). It is desired to operate for 90 min. The breakpoint will be considered as that time when the effluent air has a benzene content of 0.0025 kg benzene/kg air, and the bed will be considered exhausted when the effluent air contains 0.020 kg benzene/kg air. Determine the depth of bed required.

Silica gel has a bulk density of 625 kg/m³, an average particle diameter D_p of 0.60 cm, and surface area a of 600 m²/m³. For this temperature, pressure, and concentration range, the adsorption isotherm is

$$Y^* = 0.167X^{1.5}$$

where Y^* = kg benzene/kg air (benzene-free basis) and X = kg benzene/kg gel. We assume that the height of a gas-phase transfer unit is given by

$$H_{OY} = \frac{1.42}{a}\left(\frac{D_p\tilde{G}'}{\mu_{\text{air}}}\right)^{0.51}$$

The cross-sectional area of the bed is 1 m.

First, we can compute H_{OY}. The density of air at 25°C and 2 atm is 2.38 kg/m³, and so $\tilde{G}' = 2.38$ kg/m²-sec. The viscosity of air at 25°C is 0.018 cp $= 1.8 \times 10^{-5}$ kg/m-sec. Thus, $H_{OY} =$ 0.071 m.

The adsorption isotherm is shown in Fig. 8.21*c*. The operating line has been drawn to intersect the equilibrium curve at $Y_0 = 0.025$. From the problem specifications, $Y_B = 0.0025$ and $Y_E = 0.020$. From Fig. 8.21*c* we see that $X_T = 0.284$.

The integral in (8.47) can be evaluated numerically (see Table 8.3) as 5.925. Thus, the height of the adsorption zone Z_a is $0.071 \times 5.925 = 0.42$ m. The extent of saturation

$$f = \frac{\int_{w_B}^{w_E}(Y_0 - Y)\,dw}{Y_0 w_a} = \int_0^1 \left(1 - \frac{Y}{Y_0}\right) d\,\frac{w - w_B}{w_a}$$

By graphical integration, f is found to be 0.55.

Let us suppose the height of the bed is Z meters. The degree of saturation of the bed at the breakpoint is $\alpha = (Z - 0.231)/Z$. The bed area is 1 m²; the apparent density of the packing is 625 kg/m³; thus, the weight of the bed is $625Z$ kg. The weight of benzene adsorbed on the gel is then

$$625Z\,\frac{Z - 0.231}{Z}\,0.284 = 177(Z - 0.231)$$

The weight of benzene that must be removed from the air over a 90-min period is 322 kg. Equating this weight removed with that on the packing at the breakpoint,

$$177(Z - 0.231) = 322$$

we obtain the required bed depth of 2.04 m.

Table 8.3 NUMERICAL EVALUATION OF INTEGRAL IN (8.47)

Y	Y^*	$Y - Y^*$	$\frac{1}{Y - Y^*}$	$\int_B^Y \frac{dY}{Y - Y^*}$	$\frac{w - w_B}{w_a}$	$\frac{Y}{Y_0}$
0.0025	0.0009	0.0016	625		0	0.1
0.0050	0.0022	0.0028	358	1.1375	0.192	0.2
0.0075	0.0042	0.0033	304	1.9000	0.321	0.3
0.0100	0.0063	0.0037	270	2.6125	0.441	0.4
0.0125	0.0089	0.0036	278	3.3000	0.556	0.5
0.0150	0.0116	0.0034	294	4.0125	0.676	0.6
0.0175	0.0148	0.0027	370	4.8375	0.815	0.7
0.0200	0.0180	0.0020	500	5.9250	1.00	0.8

8.2.3 Removal of SO_2 from Gases

The technical and economic feasibility of an SO_2 removal process depends on the type and quantity of effluent gases that must be cleaned. With regard to SO_2 removal, there are essentially two types of effluent gas treatment problems. The first is the problem of removing SO_2 from power plant flue gases. Power plant flue gases generally contain low concentrations of SO_2 (<0.5 percent by volume), but they are emitted at tremendous volumetric flow rates. For example, a coal-fired power plant burning 2 percent sulfur coal (by weight) will produce 40,000 kg of SO_2 for every 10^6 kg of coal burned. Typical concentrations of SO_2 in power plant flue gases (with 15 percent excess air in combustion) are (Strauss, 1966):

Coal (1% sulfur)	0.11% SO_2 by volume
Coal (4% sulfur)	0.35
Fuel oil (2% sulfur)	0.12
Fuel oil (5% sulfur)	0.31

Thus, the key problems in removing SO_2 from power plant flue gases are related to the low concentrations of SO_2 and the enormous gas flow rates involved.

The second class of SO_2 effluent gas treatment problems is that resulting from the need to remove SO_2 from streams containing relatively high concentrations of SO_2 at low flow rates. Streams of this type are typical of those emitted from smelter operations.

In this section we shall consider largely the problem of SO_2 removal from power plant flue gases, since it represents a more prevalent and, in many respects, a more difficult problem than that of SO_2 removal from smelting and other industrial operations. Strauss (1971) has reviewed a number of processes for the cleaning of smelter gases, and the reader may consult this source for details.

The development of flue gas treatment processes for SO_2 is at present a very lively area of industrial concern. Current SO_2 removal processes can be divided into six categories (Davis, 1972):

1 Regenerative alkaline An alkaline agent strips SO_2 from the flue gas stream, combining chemically with SO_2. In a separate regeneration step, the agent is reconstituted and sulfur is recovered, usually as liquid SO_2 or sulfuric acid. Some of the agents used include magnesium oxide, sodium sulfite, metal carbonates, and magnesium dioxide.

2 Nonregenerative alkaline This category is made up of so-called throwaway processes. As in the regenerative routes, an agent combines chemically with SO_2 in the flue gas stream, but in this case it is cheap enough to throw away the product on a once-through basis. Commonly used agents are limestone, lime, dolomite, and solid carbide wastes from acetylene manufacturing.

3 Furnace injection This method, also a throwaway process, differs from the last in that the agent is injected directly into the furnace, and the sulfated product is subsequently scrubbed out of the flue gas with water. Part of the SO_2 is captured chemically within the furnace, the rest in the scrubbing step.

4 Catalysis Using a catalyst to promote oxidation of SO_2 to SO_3, this method recovers sulfuric acid. High inlet-gas temperatures are required.

5 Regenerative solid adsorption This category comprises several activated-char processes, in which SO_2 is adsorbed on char. In most routes, subsequent desorption leads to the production of sulfuric acid. None of these processes has reached true commercial scale.

6 Regenerative organic absorption This differs from alkaline absorption in that an organic absorbing medium is used. A few processes are under investigation, but none is truly commercial.

Several SO_2 removal processes are summarized in Table 8.4. The regenerative alkaline methods in Table 8.4 are those listed as sulfite absorption, magnesium oxide, manganese dioxide, and molten salt. The Chemico process, for example, involves scrubbing the flue gas with a slurry of MgO and recycle $MgSO_3$ and $MgSO_4$ (Shah, 1972). Absorption takes place by the reactions:

$$MgO + SO_2 + 6H_2O \rightarrow MgSO_3 \cdot 6H_2O$$

$$MgO + SO_2 + 3H_2O \rightarrow MgSO_3 \cdot 3H_2O$$

The absorbate enters a centrifuge system where the hydrated crystals of $MgSO_3$ and $MgSO_4$ are separated from the mother liquor. The liquor is returned to the absorber and the centrifuged wet cake is sent to a dryer. Regeneration takes place upon heating:

$$MgSO_3 \rightarrow MgO + SO_2$$

$$MgSO_4 + \underset{\text{(coke)}}{\tfrac{1}{2}C} \rightarrow MgO + SO_2 + \tfrac{1}{2}CO_2$$

The flue gas contains 15 to 16% SO_2 which can then be used for sulfuric acid production.

The nonregenerative processes usually involve the use of cheap reactants such as lime, limestone, or dolomite, which can be injected directly into the furnace or employed in a wet scrubbing operation. The dry dolomite process (Combustion Engineering), for example, is based on the injection of pulverized dry dolomite ($MgCO_3$) and limestone ($CaCO_3$) into a boiler burning a sulfur-containing fossil fuel. The mixture reacts with SO_2 in the combustion gases directly in the boiler forming gypsum ($CaSO_4$). The principal reaction is

$$CaCO_3{\cdot}MgCO_3 + SO_2 + \tfrac{1}{2}O_2 \rightarrow CaSO_4 + 2CO_2$$

Sulfates, unreacted materials, and fly ash are subsequently removed by wet scrubbing. The advantage of the system is that capital costs are low; however, early reported tests have shown rather low removal efficiencies for SO_2.

The Monsanto catalytic process converts SO_2 to H_2SO_4 by passing the flue gases over a vanadium pentoxide (V_2O_5) catalyst, which oxidizes SO_2 to SO_3, and the subsequent contacting of the SO_3 with water to form H_2SO_4. In the process, gas enters the catalyst bed, after particulate removal, at temperatures of 425 to 455°C. After the catalyst bed, the SO_3 is contacted with water, and the H_2SO_4 is condensed. The advantages of the process are that the system is basically simple and catalyst recycle is not necessary. Disadvantages are that expensive, corrosion-resistant materials are needed, the catalyst is easily deactivated by certain particles, and the sulfuric acid produced is usually only 75 to 80% and thus is not too salable.

Since the development of SO_2 removal processes is in such a rapid state of change, it is best not to devote significant coverage to the subject here. For additional information on the removal of SO_2 from stack gases the reader should consult Wiedersum (1970), Slack et al. (1971), and Horlacher et al. (1972).

8.2.4 Removal of NO_X from Gases

For a number of reasons, NO_X removal is more difficult than SO_2 removal, and, as a result, technology for NO_X cleaning of flue gases is not as advanced as for SO_2. The key problem is that NO, the principal NO_X species in flue gas, is relatively stable and unreactive. In addition, flue gases containing NO often also contain H_2O, CO_2, and SO_2 in greater concentrations than NO. These species are more reactive than NO and interfere with its removal.

Thus, the development of technology for removal of NO_X from flue gases is at this time an important industrial problem. A list of possible control methods has been given by Bartok et al. (1971):

1 Catalytic decomposition
2 Catalytic reduction
  a Nonselective in net reducing environment
  b Selective in net oxidizing environment
3 Adsorption-reaction by solids
4 Absorption-reaction by liquids
5 Physical separations

The decomposition of NO is favored thermodynamically at low temperatures; however, a suitable catalyst to promote this decomposition has yet to be found despite the considerable effort that has been expended (Shelef and Kummer, 1969).

Table 8.4 SUMMARY OF SO_2 REMOVAL PROCESSES

Process and developer	Description	Chemistry
Sulfite absorption (Wellman-Power Gas)	A solution method for concentrating dilute SO_2 via bisulfite formation, crystallization, and thermal regeneration. No reduction or oxidation in the solution step	$SO_2(\text{dil.}) + H_2O + Na_2SO_3 \longrightarrow NaHSO_3$; $NaHSO_3 \xrightarrow{\text{Heat}} SO_2(\text{conc.}) + H_2O + Na_2SO_3$ (with Na_2SO_3 returned)
Magnesium oxide (Chemico/Basic)	Essentially a concentration process using MgO as a collector, followed by regeneration and the production of an SO_2 stream	$SO_2 \xrightarrow[\text{MgO}]{200\text{–}300°F} MgSO_3 \xrightarrow{1400°F} SO_2 \xrightarrow{\text{contact process}} H_2SO_4$ (MgO regeneration)
Molten salt (Atomics International)	Dilute SO_2 is concentrated by absorption in molten salt as sulfite, and then reduced to sulfide and hence H_2S. "M" stands for metal	$SO_2 + M_2CO_3 \xrightarrow{800°F} M_2SO_3 + CO_2$; $M_2SO_3 \xrightarrow{H_2,\ CO} M_2S + H_2O + CO_2$; $M_2S + H_2O + CO_2 \longrightarrow H_2S + M_2CO_3$ (M_2CO_3 returned); $H_2S \xrightarrow[SO_2]{\text{Claus}} S + H_2O$
Manganese dioxide (Mitsubishi)	SO_2 is initially concentrated and oxidized to metal sulfate, followed by regeneration of MnO_2 and production of ammonium sulfate	$SO_2 \xrightarrow{MnO_2} MnSO_4 \xrightarrow[\text{air}]{NH_4OH} (NH_4)_2SO_4 + H_2O$ (MnO_2 regeneration)
Limestone (TVA, Combustion Engineering, Chemico, others)	Simultaneous reaction of SO_2 with limestone and air oxidation of resulting sulfite to sulfate results in a slag that requires suitable disposal. Reaction may take place inside furnace or in flue-gas scrubber	$CaCO_3 \xrightarrow{SO_2,\ SO_3,\ \text{air}} CaSO_4 + CO_2$
Catalytic (Monsanto)	Accepts hot dilute SO_2 gas stream rather than high-concentration SO_2 for acid-plant feed	$\text{Air} + SO_2 \xrightarrow[V_2O_5]{900°F} SO_3 \xrightarrow{H_2O} H_2SO_4$
Activated carbon (Westvaco, Hitachi, Chemiebau, others)	All methods depend on adsorptive powers of various forms of active carbon to first concentrate and then catalyze oxidation of SO_2 to SO_3 for acid or sulfate production. Fluidized, fixed, and plugged-flow beds have all been employed.	$SO_2 \xrightarrow[\text{active carbon}]{\text{air},\ H_2O} H_2SO_4$
Ammonia scrubbing (Showa Denko)	Absorption and concentration of SO_2 and air in ammonia solution yields bisulfite and thiosulfate, which then forms sulfate, water, and sulfur	$SO_2 + NH_4OH \xrightarrow{\text{air}} HN_4HSO_3 + (NH_4)_2S_2O_3$; $HN_4HSO_3 + (NH_4)_2S_2O_3 \longrightarrow (NH_4)_2SO_4 + H_2O + S$

SOURCE: Davis (1972).

Catalytic reduction of NO to N_2 by both selective and nonselective means has been applied to nitric acid production "tail gas" emissions (Gerstle and Peterson, 1967). Nonselective reduction under net reducing conditions usually involves either addition of CO, H_2, or CH_4 to the flue gas before a catalytic reactor or operation at airflow rates below stoichiometric. Because this would result in lower plant efficiency and higher CO emissions, nonselective reduction is not an attractive method. Selective reduction involves addition of a species, usually ammonia, which selectively reduces NO in an oxygen-containing environment. This method is potentially attractive because SO_2 emissions could be similarly controlled by the same process.

Bartok et al. (1971) surveyed all common adsorbents as potential candidates for removing NO_X from flue gas, e.g., silica gel, alumina, molecular sieves, char, and ion-exchange resins. Each exhibits some capacity for oxidizing NO to NO_2 and then adsorbing the NO_2. However, the capacities of these adsorbents are quite low at typical NO concentrations.

Physical separations based on molecular properties of NO such as molecular size, condensation temperature, and magnetic susceptibility offer little promise because the physical properties of NO are similar to those of other more abundant components of the flue gas.

Thus, aqueous absorption appears at this time to be the most promising gas treatment method for combined control of NO_X and SO_2 emissions. Normally, alkaline solutions or sulfuric acid is used. These require that there be equimolar concentrations of NO and NO_2 in the gas, since absorption of the combined oxide, N_2O_3, is the most favorable (Koval and Peters, 1960). Recycle of NO_2 in the flue gas appears to be the best way of achieving this balance. The recycle of NO_2 may be supplied by oxidizing the NO recovered from the absorbent.

8.3 ADDITIONAL CONTROL TECHNIQUES FOR GASEOUS POLLUTANTS FROM STATIONARY SOURCES

In this section we discuss those techniques used for control of SO_2 and NO_X emissions not treated in Sec. 8.2.

8.3.1 Sulfur Dioxide Emissions

There are three basic strategies for controlling SO_2 emissions from stationary combustion sources:

1 Fuel substitution
2 Fuel desulfurization
3 Removal of SO_2 from effluent gases

Having already discussed category 3, we will concentrate here on classes 1 and 2.

8.3.1.1 Fuel substitution Use of fuel with little or no sulfur is probably the best way of controlling potential SO_2 emissions. The major drawback to wholesale use of low-sulfur fuels is lack of their availability. Another potential problem associated with fuel substitution is that equipment designed to burn a particular type of fuel may not be readily convertible to burning another fuel. A survey of the convertibility of domestic, commercial, and industrial heating equipment is summarized in Control Techniques for Sulfur Oxide Air Pollutants (1969).

8.3.1.2 Fuel desulfurization. *a* COAL Coal generally contains between 0.2 and 7 percent sulfur by weight on a dry basis, with the high-sulfur coals usually defined as those with more than 3 percent sulfur. The average sulfur content of bituminous coal mined in the United States is about 2 percent.

Sulfur exists in coal in three forms: pyrites (FeS_2), organic compounds, and sulfates. Sulfates are usually present only in very small quantities. Coals having a high sulfur content generally contain FeS_2 as the major constituent. Pyrites occur as fine particles ($> 50\ \mu m$) or as discrete layers. Organic sulfur is bound molecularly in the coal and cannot be removed without chemically altering the coal.

Pyrites can be removed from coal by exploiting the differences in specific gravity between FeS_2 (about 5) and clean coal (1.25). This is done by water-washing after crushing the coal into the millimeter range. Even at this level of crushing, the fine-grained, intimately mixed FeS_2 will remain in the coal. Coal is often crushed to below 100 μm for firing, at which size FeS_2 particles are released and can be separated by suitable means. At this degree of fineness, dry processes (not requiring water), such as air classification and electrostatic precipitation, for separating the pyrites can be used.

Liquefaction of coal involves hydrogenation to produce liquid products, including gasoline. The primary aim of liquefaction is not sulfur removal, although the sulfur does concentrate in the end products, such as a high-sulfur char. Low-sulfur fuel could result as a primary product or by desulfurization of the char. Several liquefaction processes are described in Control Techniques for Sulfur Oxide Air Pollutants (1969).

Gasification of coal is a process whereby coal is reacted with oxygen, steam, hydrogen, and carbon dioxide to produce a gaseous product suitable for use as a fuel. Gasification essentially involves the breaking of the carbon rings and subsequent reaction with hydrogen. As a result, the sulfur is readily removed as H_2S. Coal gasification, although carried out for over 150 yr to obtain coke and town gas, is still an active area of study, particularly because of dwindling supplies of oil and natural gas. A short survey of some industrial coal gasification processes is given by Strauss (1971).

b OIL Heavy fuel oil, or residual fuel oil, contains sulfur from 0.5 to 5 percent, depending on its source and the subsequent refining. Actually, all crude oil contains some sulfur, but the refining operation, which separates the crude into various petroleum fractions, tends to concentrate the sulfur in the high-boiling-temperature products, such as residual fuel oil. Low-sulfur fuel oil is usually classified as having a sulfur content <1 percent.

Processes for removing sulfur from fuel oil are essentially based on reaction of the fuel oil with hydrogen at high pressure in the presence of catalysts. The process is commonly called *hydrodesulfurization*. Surveys of current desulfurization processes are given by Strauss (1971) and in Control Techniques for Sulfur Oxide Air Pollutants (1969).

8.3.2 Nitrogen Oxides Emissions

About one-half of the total NO_X emissions in the United States comes from stationary sources, with the largest contribution from fossil-fuel-fired boilers for electric power generation. Unlike SO_2 emissions, which are directly proportional to the sulfur content of the fuel, NO is, as we know, formed by the high-temperature reaction of atmospheric N_2 and O_2 during combustion. (Depending on the particular fuel used, some NO_X may be formed from nitrogen compounds in the fuel itself.)

As in the internal combustion engine, the amount of NO formed during fuel combustion in a boiler or furnace depends on the time-temperature-composition history of the gases. The major factors affecting the formation of NO in combustion processes are:

1 Combustion temperature. Nitric oxide formation rates are extremely temperature-dependent, with higher temperatures favoring higher NO concentrations.
2 Air-fuel ratio.
3 Degree of mixing of fuel, air, and combustion products. If the fuel and air are mixed so that most of the combustion takes place under rich conditions, NO formation can be reduced. Mixing of combustion products back into the combustion zone lowers the flame temperature and reduces NO formation.
4 Heat transfer rates. High rates of heat removal tend to reduce peak combustion temperature and thus NO formation.
5 Fuel type. On an equivalent heat-generation basis, fuels in order of decreasing NO formation are coal, oil, gas.

There are three major categories of NO_X control methods for stationary sources: modification of operating conditions, modification of design conditions, and removal of NO_X from exhaust gases. Combustion and design modifications techniques appear to be the most economical means of achieving substantial NO_X emission reductions.

8.3.2.1 Modification of operating conditions. *a* LOW-EXCESS-AIR COMBUSTION This method involves supplying as close to stoichiometric requirements of air as possible, thereby limiting the amount of O_2 available to form NO. The effectiveness of low-excess-air firing has been demonstrated for gas and oil combustion by Sensenbaugh and Jonakin (1960). Much less work appears to have been carried out for coal combustion; however, in coal firing with low excess air, imbalances in air-fuel distribution may result, and problems of unburned fuel or CO emissions may limit the utility of this method (Bienstock et al., 1966).

b TWO-STAGE COMBUSTION This technique is based on supplying substoichiometric quantities of primary air to the burners in oil or gas-fired combustion and then injecting secondary air at lower temperatures so that the partially burned fuel and combustion products are allowed to cool before the completion of combustion. Thus, the simultaneous exposure of both N_2 and O_2 to high temperatures is avoided and most of the combustion takes place under fuel-rich conditions (Barnhart and Diehl, 1960). The principal problem in staged combustion is to avoid a significant production of CO, HC, and particles.

c FLUE-GAS RECIRCULATION In this control method, part of the flue gases (about 10 to 20 percent) is recirculated to the flame zone. The recirculated gas must be returned into the combustion zone in order to lower the flame temperature and the available oxygen. Thus, the effect of recirculation depends on where the gases are injected and how much is recycled. The greatest reduction in flame temperature is achieved by mixing the gases directly with the combustion air. Emissions reductions of 80 percent can be achieved by the combination of reducing excess air from 50 to 20 percent and recirculating 50 percent of the flue gas (Andrews et al., 1968).

The best approach to NO_X control by combustion modification appears to be a combination of the above methods. Examples are the combination of low-excess-air firing and staged combustion and low-excess-air firing and flue-gas recirculation. These combined techniques have been used successfully in gas and oil firing; however relatively little work has been done to apply them to coal-fired units.

We summarize the major factors affecting NO formation in stationary combustion processes in Table 8.5. Of the three fuels, gas, oil, and coal, gas allows the most precise control in attaining low NO levels. Emission levels from coal burning vary greatly because of the variety of solids which fall into the "coal" category. Consequently, of the three fuels, the least is known about coal relative to minimizing NO formation.

Table 8.5 MAJOR FACTORS AFFECTING NITRIC OXIDE FORMATION IN STATIONARY COMBUSTION

Factor	Change in factor	Effect in NO_x emissions
Excess air	Decrease	Decrease
Preheat temperature	Decrease	Decrease
Heat release rate	Decrease	Decrease
Heat removal rate	Decrease	Increase
Backmixing of combustion gases	Increase	Decrease
Fuel type	Coal → oil → gas	Coal > oil > gas
Fuel nitrogen content	Decrease	Decrease

SOURCE: Bartok et al. (1971).

8.3.2.2 Modifications of design conditions The configuration of the burners in a boiler can have a substantial influence on NO formation. For example, two basic furnace designs are the tangentially fired boiler and the horizontally fired boiler. In the tangentially fired boiler the flame and combustion products rotate in a spiral upward and around the walls of the furnace. In a horizontally fired furnace, the flame is at right angles to the walls of the firebox. Horizontal firing tends to concentrate the hot gases, producing higher flame temperatures and more NO. [Sensenbaugh and Jonakin (1960) compiled many values of the NO emission rates from both tangentially and horizontally fired units.] Thus, burner configurations which produce more intense combustion and higher temperatures result in more NO than those which permit combustion to occur over a wider area.

Fluidized-bed combustion is a relatively new concept which offers great promise for low NO formation (Proceedings of the First International Conference on Fluidized Bed Combustion, 1968). In this process pulverized coal (or oil or gas) can be completely burned at 750 to 1000°C within a bed of limestone, dolomite, ash, or inert particles suspended by the upward flowing air and combustion products. A fluidized bed offers very high heat transfer rates with resulting low combustion temperatures. In addition, the limestone or dolomite absorbs much of the sulfur in the fuel, and the quantity of particles produced is diminished because of the pulverized state of the coal.

8.4 PRINCIPLES OF PARTICULATE EMISSION CONTROL TECHNIQUES

Particulate emission control techniques for stationary sources fall basically into two categories: (1) gas cleaning and (2) fuel substitution. Gas cleaning is the more common and the one on which we shall concentrate here. Particulate gas cleaning devices

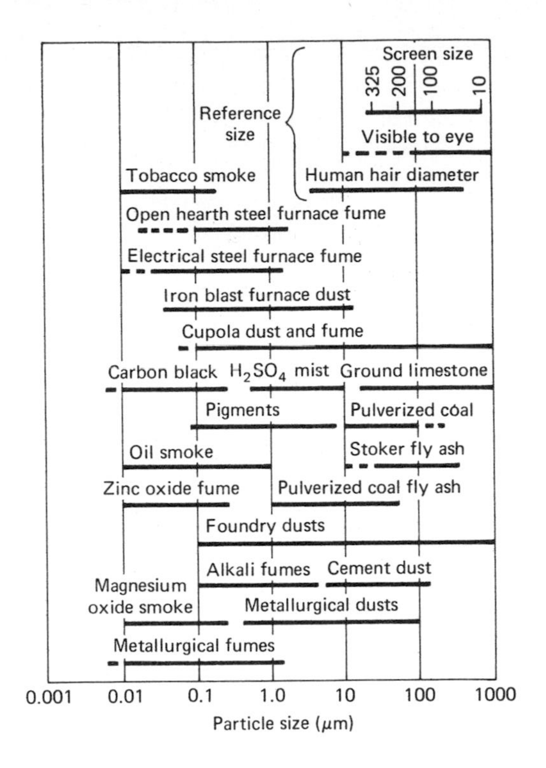

FIGURE 8.22
Sizes of typical stationary-source particulate emissions.

are usually designed to take advantage of certain physical, chemical, or electrical properties of the particles. The performance of particulate collection devices is measured by their efficiency, usually calculated as the percentage by weight of the total particles removed. Ultimate selection of a gas cleaning device will be influenced by the efficiency required; by the type of effluent gas containing the particles; by the characteristics of the particles; and by operating, construction, and economic factors. One of the most important factors in the choice of a technique is the particle size distribution in the gas stream. Figure 8.22 summarizes the size ranges of stationary-source particulate emissions.

As a prelude to the consideration of particulate emission control equipment, we must discuss briefly certain fundamental elements of particle motion.

The removal of particles from a gas stream depends on a number of mechanisms, such as diffusion (both molecular and turbulent); migration under the influence of gravitational, centrifugal, and electrical forces; and interception and inertial impaction. Particulate removal devices are basically designed on the principle that a gas stream containing particles is passed through a region where the particles are acted on by external forces or caused to intercept obstacles, thereby separating them from the gas stream. When acted upon by external forces, the particles acquire a velocity component in a direction different from that of the gas stream. In order to design a separation device based on particulate separation by external forces, one must be able to compute the motion of a particle under such circumstances. It is normally assumed that the fluid resistance that a particle encounters when acted upon by external forces is the same as that which the particle would experience in moving through a stationary fluid.

The subject of the dynamics of the motion of small particles through fluids is a very complicated one. We will consider here only the bare rudiments of this subject as they apply to particulate separation equipment. In particular, we assume in all cases that (1) the motion is steady, (2) the fluid is of infinite extent (no walls), (3) the particles do not influence each other's motion, and (4) all the particles are spherical.

8.4.1 Drag on a Sphere in Steady Motion Through a Fluid

The drag force on a spherical particle of diameter D_p moving at a steady speed u through a fluid of density ρ and viscosity μ can be written

$$F = \tfrac{1}{8} C_D \pi D_p^2 \rho u^2 \qquad (8.49)$$

where C_D is the dimensionless drag coefficient, a function of the particle Reynolds number

$$\mathrm{Re} = \frac{D_p u \rho}{\mu}$$

The experimentally determined curve of C_D versus Re is shown in Fig. 8.23. At very low Reynolds number, that is, $\mathrm{Re} < 0.1$, the drag force can be determined directly as

$$F = 3\pi\mu D_p u \qquad (8.50)$$

by neglecting any inertial effects in the flow field around the particle. Equation (8.50), known as *Stokes' law*, indicates that in this region $C_D = 24/\mathrm{Re}$.

The velocity that a particle attains when in steady motion (such that the external force is exactly balanced by the resistance offered by the fluid) is called its *terminal velocity*. For example, for a sphere falling under the influence of gravity in the Stokes' law region, the terminal velocity is

$$u = \frac{D_p^2(\rho_p - \rho)g}{18\mu}$$

In the general case,

$$u = \left[\frac{4D_p(\rho_p - \rho)g}{3\rho C_D}\right]^{1/2} \qquad (8.51)$$

Evaluation of u by (8.51) is difficult because C_D is itself a function of u through Re. If we consider, however, the group

$$C_D \mathrm{Re}^2 = \frac{4\rho(\rho_p - \rho)D_p^3 g}{3\mu^2} \qquad (8.52)$$

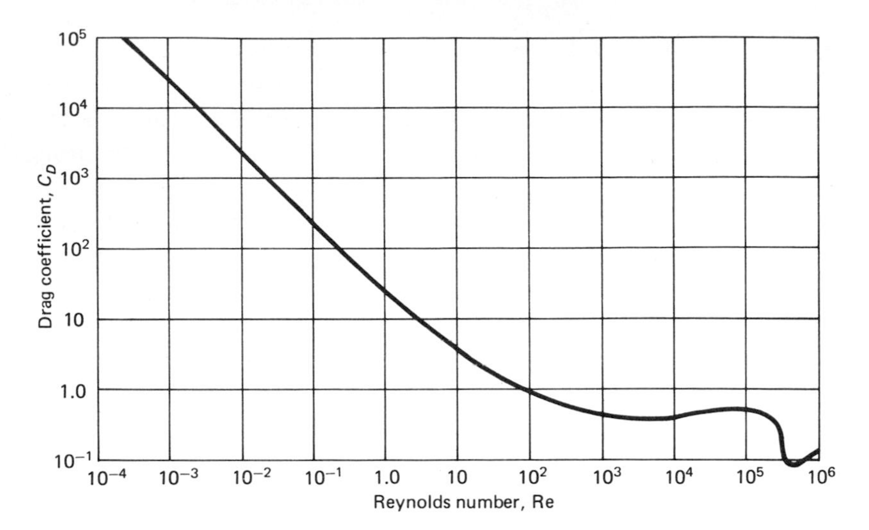

FIGURE 8.23
Drag coefficient C_D as a function of Reynolds number for spheres.

and then replot Fig. 8.23 as $C_D\text{Re}^2$ versus Re, the value of Re, where (8.52) is satisfied, will enable determination of u.

When the particles are very small, i.e., of the order of the mean free path of gas molecules, the assumption inherent in the above developments, that the fluid may be considered a continuous medium, breaks down. In this case, particles tend to move more rapidly than predicted by the classic theories of continuum mechanics. To account for this increased velocity, Cunningham computed a correction to the continuum models based on the kinetic theory of gases. Thus, Stokes' law can be rewritten as

$$F = \frac{3\pi\mu D_p u}{C} \tag{8.53}$$

where C, the Cunningham correction factor, is given by

$$C = 1 + \frac{2\lambda}{D_p}(1.257 + 0.4e^{-1.1D_p/2\lambda}) \tag{8.54}$$

where λ is the mean free path of gas molecules, given by

$$\lambda = \frac{\mu}{0.499\rho\sqrt{8RT/\pi M}}$$

The correction C is about 5 percent for 5-μm diameter particles, 16 percent for 1-μm diameter particles, and 300 percent for 0.1-μm diameter particles (Strauss, 1966).

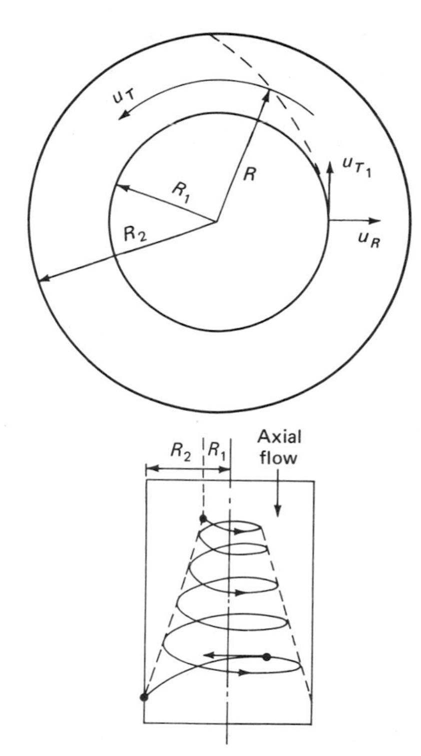

FIGURE 8.24
Particle velocity components in a spiraling gas stream.

8.4.2 Particle Motion in a Rotating Fluid

Centrifugal collectors, or cyclones, are gas cleaning devices that utilize the centrifugal force created by a spinning gas stream to separate particles from a gas. The particles are thrown outward to the wall of the device from which they travel to a receiver. Most industrial cyclones involve rather complicated flow patterns, and thus theoretical evaluation of particle trajectories is not particularly reliable in cyclone design. As we shall see in Subsec. 8.5.2, it is best to use experimentally determined efficiency curves for the unit in question. Nevertheless, we present here a rather brief analysis of particle motion in a rotating fluid to enable the reader to assess the type of analysis necessary to determine cyclone efficiencies. The development which follows has been adapted from Strauss (1966).

Let us consider the motion of a particle in a spiraling gas stream. For a particle moving around a circle with a radius R with tangential velocity u_T and mass m, the centrifugal force, normal to the tangent of the arc, is given by (see Fig. 8.24)

$$F = m\frac{u_T^2}{R} \qquad (8.55)$$

In a centrigual collector, the particles in the spinning gas stream move progressively closer to the outer wall as they flow through the device. The particle velocity thus consists of three components, the tangential velocity u_T, the radial drift velocity u_R, and the axial velocity u_H. The basic question in the design of a simple cyclone separator, such as depicted in Fig. 8.24, is, for a given gas flow rate and inner and outer radii R_1 and R_2: How long must the body of the cyclone be to ensure that particles which enter at the top at R_1 reach R_2 by the exit from the cyclone?

If a particle is at position (R,θ) at time t, the radial and tangential velocity components of the particle are

$$u_R = \frac{dR}{dt} \qquad u_T = R\frac{d\theta}{dt} \tag{8.56}$$

The radial and tangential accelerations are then[1]

$$\frac{du_R}{dt} = \frac{d^2R}{dt^2} - R\left(\frac{d\theta}{dt}\right)^2 \tag{8.57}$$

and

$$\frac{du_T}{dt} = R\frac{d^2\theta}{dt^2} + 2\frac{d\theta}{dt}\frac{dR}{dt} \tag{8.58}$$

The force components on the particles are given by the product of the mass and these accelerations. These forces are balanced by the resistance of the fluid to motion. Thus the radial equation of motion for flow in the Stokes' law region is

$$\frac{m}{3\pi D_p \mu}\left[\frac{d^2R}{dt^2} - R\left(\frac{d\theta}{dt}\right)^2\right] = -u_R = -\frac{dR}{dt} \tag{8.59}$$

[1] These accelerations are determined as follows: From a lagrangian (particle-oriented) point of view, the R- and θ- particle velocity components are functions only of time and are given by (8.56), since the motion of the particle is defined by $R(t)$ and $\theta(t)$. (Note that the tangential velocity component is identical to the u_θ component in a cylindrical, R,θ,Z coordinate system.) From an eulerian point of view, however, the velocity components u_R and u_θ are functions of time and the location of the particle in the field (R,θ,t). If u_R and u_θ are the components of the velocity vector $\mathbf{u}$, the fundamental relation between the lagrangian and eulerian descriptions of the particles' motion is

$$\frac{d\mathbf{u}}{dt}\left(=\frac{D\mathbf{u}}{dt}\right) = \frac{\partial \mathbf{u}}{\partial t} + \mathbf{u}\cdot\nabla\mathbf{u}$$

where the derivative on the left-hand side is that obtained by following the motion of the particle. In cylindrical coordinates

$$\frac{du_R}{dt} = \frac{\partial u_R}{\partial t} + u_R\frac{\partial u_R}{\partial R} + \frac{u_\theta}{R}\frac{\partial u_R}{\partial \theta} - \frac{u_\theta^2}{R}$$

$$\frac{du_\theta}{dt} = \frac{\partial u_\theta}{\partial t} + u_R\frac{\partial u_\theta}{\partial R} + \frac{u_\theta}{R}\frac{\partial u_\theta}{\partial \theta} + \frac{u_R u_\theta}{R}$$

Setting $u_R = dR/dt$ and $u_\theta = R\, d\theta/dt$ in these equations, where R and θ are considered functions of t only, we obtain (8.57) and (8.58).

The tangential forces are zero since the particle is assumed to have the same tangential velocity as the gas, so

$$\frac{m}{3\pi D_p \mu}\left(R\frac{d^2\theta}{dt^2} + 2\frac{dR}{dt}\frac{d\theta}{dt}\right) = 0 \tag{8.60}$$

We nondimensionalize (8.59) and (8.60) by letting $\zeta = R/R_2$, $u = u_R/u_{T_2}$, and $\tau = tu_{T_2}/R_2$, where u_{T_2} is the velocity at R_2. Thus, (8.59) and (8.60) become

$$\xi\left[\frac{d^2\zeta}{d\tau^2} - \zeta\left(\frac{d\theta}{d\tau}\right)^2\right] = -\frac{d\zeta}{d\tau} \tag{8.61}$$

and

$$\xi\left[\zeta\frac{d^2\theta}{d\tau^2} + 2\frac{d\xi}{d\tau}\frac{d\theta}{d\tau}\right] = 0 \tag{8.62}$$

where

$$\xi = \frac{mu_{T_2}}{3\pi\mu D_p R_2}$$

Replacing m by $\pi D_p{}^3(\rho_p - \rho)/6$,

$$\xi = \frac{D_p{}^2(\rho_p - \rho)u_{T_2}}{18\mu R_2}$$

Equation (8.62) can be written

$$\frac{\xi}{\zeta}\frac{d}{d\tau}\left(\zeta^2\frac{d\theta}{d\tau}\right) = 0$$

which implies that $\zeta^2\, d\theta/d\tau$ is constant. Bu multiplying the definition of u_T by R [the second equation in (8.56)],

$$R^2\frac{d\theta}{dt} = u_T R = u_{T_2}R_2$$

or $\zeta^2\, d\theta/d\tau = 1$. Using this relation in (8.61) yields

$$\frac{d^2\zeta}{d\tau^2} + \frac{1}{\xi}\frac{d\zeta}{d\tau} - \frac{1}{\zeta^3} = 0 \tag{8.63}$$

This equation describes the normalized trajectory of a particle in a rotating fluid in the Stokes' law regime. Because of its nonlinearity, (8.63) must be solved numerically. The results of such an integration are given in Table 8.6, in which the minimum length over which the requisite separation of all particles can be accomplished is given as a function of D_p.

Table 8.6 SEPARATING DISTANCES FOR SPHERICAL PARTICLES IN A SPIRALING FLUID

D_p, μm	3	5	10
Cyclone length, cm	90	29.2	8.9

SOURCE: Strauss (1966).
Data: Air at 18.7°C; mean axial velocity, $u_H = 12.2$ cm/sec; $R_1 = 1.11$ cm; $R_2 = 2.54$ cm; $uR = \text{const} = 3225$ cm^2/sec; $\rho_p = 2.7$ g/cm^3.

8.4.3 Aerodynamic Capture of Particles

Filtration through fabric of gases containing particles is one of the oldest gas cleaning methods, but it is still one of the most effective. The particle-laden gas is passed through a fabric so that particles are trapped on the upstream side of the filter. The mechanisms of fabric filtration are more complex than might be initially thought. The fiber filter elements are usually separated by interfiber distances of 100 μm or more. Thus, the filtering process for small particles is not simply sieving, in which particles become trapped in the void spaces between fibers. Rather, the particles are removed when they impinge on the fibers. There are several mechanisms by which a small particle may impinge on an individual fiber in a bed:

1 *Inertial impaction* occurs when a particle is unable to follow the rapidly curving streamlines around an obstacle, and, because of its inertia, continues to move toward the obstacle along a path of less curvature than the flow stream lines. Thus, collision occurs because of the particle's momentum, which is proportional to its mass. Path *A* in Fig. 8.25 depicts inertial impaction.

2 *Direct interception* takes place when a particle, following the streamlines of flow around a fiber, is of a size sufficiently large that its surface and that of the fiber come into contact. Thus, if the streamline on which the particle center lies is within a distance of $D_p/2$ to the fiber, interception occurs (see path *B* in Fig. 8.25). Note that the mechanism of inertial impaction is based on the premise that the particle has mass but no size, whereas direct interception is based on the premise that the particle has size but no mass.

3 For particles ranging in size from less than 0.01 to 0.05 μm in diameter, *diffusion* is the predominant mechanism of deposition. Such small particles do not follow the fluid streamlines because collision with gas molecules occurs, resulting in a random brownian motion that increases the chance of contact between the particles and the collection surfaces. A concentration gradient is established after the collection of a few particles and acts as a driving force to increase the rate of deposition.

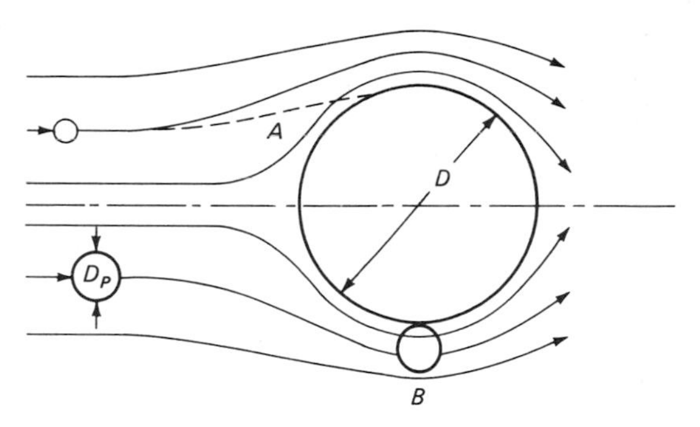

FIGURE 8.25
Inertial impaction and direct interception of particles by a cylindrical fiber placed in a flowing fluid containing particles.

Collection may also result from electrostatic attraction when either particles or fiber or both possess a static charge. These electrostatic forces may be either direct, when both particle and filter are charged, or induced, when only one of them is charged. Such charges are usually not present unless deliberately introduced during the manufacture of the fiber. We will not discuss the mechanisms of electrostatic attraction here. Such a discussion is presented by Strauss (1966).

The size ranges in which the various mechanisms of collection are important are:

Inertial impaction	>1 μm
Direct interception	>1 μm
Diffusion	0.001 to 0.5 μm
Electrostatic attraction	0.01 to 5 μm

We now consider each of the first three mechanisms.

8.4.3.1 Inertial impaction Let us consider the situation in which a gas stream containing particles approaches an infinitely long cylinder of diameter D placed normal to the flow. The appropriate parameter which governs the nature of the flow streamlines around the cylinder is the Reynolds number based on the undisturbed upstream velocity u_0 and D, $\text{Re} = u_0 D\rho/\mu$. When the streamlines begin to diverge as the body is approached, the particles, because of their mass, do not follow the streamlines around the cylinder but rather have a tendency to continue their motion toward the cylinder.

We will consider the trajectory followed by a particle as it approaches a cylinder placed in a flowing stream. The motion of the particle in the absence of external forces is described by

$$m\frac{d\mathbf{u}}{dt} = \mathbf{F} \qquad (8.64)$$

where $\mathbf{u} = (u_x, u_y)$ is the velocity of the particle and $\mathbf{F} = (F_x, F_y)$ is the resistance offered by the fluid. In the Stokes' law region,

$$\mathbf{F} = 3\pi\mu D_p(\mathbf{u} - \mathbf{v}) \tag{8.65}$$

where $\mathbf{v}$ is the fluid velocity. Thus, substituting for m,

$$\frac{(\rho_p - \rho)D_p^2}{18\mu}\frac{d\mathbf{u}}{dt} = -(\mathbf{u} - \mathbf{v}) \tag{8.66}$$

We let $v_x = v_x/u_0$, $v_y = v_y/u_0$, $X = x/D$, $Y = y/D$, and $\tau = u_0 t/D$. The trajectory of the particle is then defined by $X(t)$ and $Y(t)$ and is described by

$$\psi\frac{d^2X}{d\tau^2} + \frac{dX}{d\tau} - v_x = 0 \tag{8.67}$$

$$\psi\frac{d^2Y}{d\tau^2} + \frac{dY}{d\tau} - v_y = 0 \tag{8.68}$$

where

$$\psi = \frac{(\rho_p - \rho)D_p^2 u_0}{18\mu D} \tag{8.69}$$

Thus, the particle trajectory in the Stokes' law region depends only on the fluid streamlines (v_x and v_y) and the parameter ψ. In order to solve (8.67) and (8.68) we need the prior computed fluid streamlines, and so (8.67) and (8.68) cannot be solved directly. Nevertheless, the collection efficiency by inertial impaction can be correlated with the single parameter ψ for a particular obstacle in the Stokes region. In the case of nonvanishingly small Re, (8.67) and (8.68) can be rederived and seen to depend on ψ and Re.

The efficiency of capture by inertial impaction is defined as the fraction of particles (considered to be evenly distributed in the gas stream) which can be collected from the area of the gas stream equal to the frontal area of the collector.

Figure 8.26 shows experimental results of inertial impaction efficiencies determined by Wong and Johnstone (1953). We see that there is a critical value of ψ below which no inertial impaction takes place. This critical value has been determined in theory to be about 0.0625 for cylinders. This prediction is confirmed by the experimental results shown in Fig. 8.26.

8.4.3.2 Direct interception The interception of a particle is governed by the ratio of the diameters of the particle and the intercepting body, $\gamma = D_p/D$. Ranz (1953) obtained the collection efficiency of a cylinder as

$$\eta = \frac{1}{2.002 - \ln \mathrm{Re}}\left[(1 + \gamma)\ln(1 + \gamma) - \frac{\gamma(2 + \gamma)}{2(1 + \gamma)}\right] \tag{8.70}$$

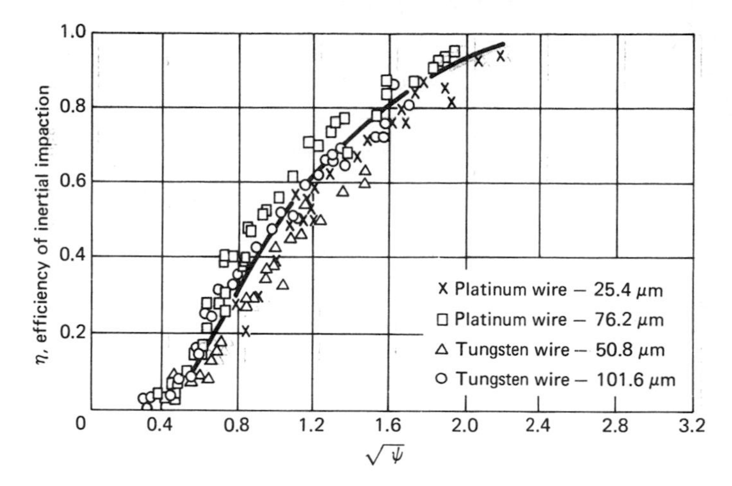

FIGURE 8.26
Experimental efficiencies of inertial impaction of spheres on wires. (*Source: Wong and Johnstone*, 1954.)

by considering the streamlines around the cylinder.

Clearly, the mechanisms of inertial impaction and interception are not independent. A reasonable estimate for the combined efficiency of both mechanisms at any Reynolds number is

$$\eta_{\mathrm{TOT}} = 1 - (1 - \eta_{\mathrm{impaction}})(1 - \eta_{\mathrm{interception}}) \tag{8.71}$$

8.4.3.3 Diffusion The diffusion process of very small ($<0.1\ \mu\mathrm{m}\ D_p$) particles by brownian motion can be described by the classic diffusion equation. Dimensional analysis of this equation reveals that the collection efficiency should depend on the Reynolds number, the Schmidt number, $\mathrm{Sc} = \mu/\rho\mathscr{D}$, and the Peclet number $\mathrm{Pe} = u_0 D/\mathscr{D}$. Ranz (1953) suggested that for cylinders for $0.1 < \mathrm{Re} < 10^4$ and $\mathrm{Sc} < 100$,

$$\eta = \frac{\pi}{\mathrm{Pe}}\left(\frac{1}{\pi} + 0.55\ \mathrm{Re}^{1/2}\mathrm{Sc}^{1/3}\right) \tag{8.72}$$

Friedlander (1958) included both interception and diffusion mechanisms and obtained

$$\eta = 6\mathrm{Sc}^{-2/3}\mathrm{Re}^{-1/2} + 3\gamma^2\mathrm{Re}^{1/2} \tag{8.73}$$

which correlates available data quite well.

8.4.3.4 Combined inertial impaction, interception, and diffusion In considering the combined efficiency from all three mechanisms above, one can write

$$\eta_{\mathrm{TOT}} = 1 - (1 - \eta_{\mathrm{impaction}})(1 - \eta_{\mathrm{interception}})(1 - \eta_{\mathrm{diffusion}}) \tag{8.74}$$

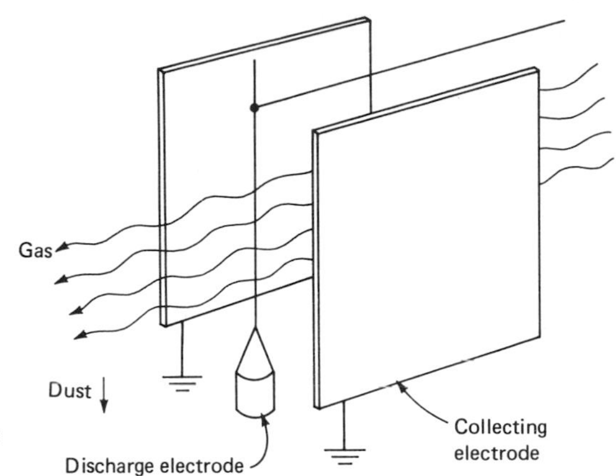

FIGURE 8.27
Principle of operation of the electrostatic precipitator.

8.4.4 Electrostatic Precipitation

In the removal of particulate matter from a gas stream by electrostatic precipitation, three essential steps are involved:

1 Electric charging of the suspended particulate matter
2 Collection of the charged particles on a grounded surface
3 Removal of the particles from the surface

The gas is passed horizontally through parallel rows of electrode plates. On the center line between each set of plates are electrically insulated, high-voltage wires. One typical set of plates and a wire is shown in Fig. 8.27. If a sufficient difference of potential exists between the two electrodes, the discharging electrode (the wire) and the collecting electrode (the plates), a migration of negatively charged gas ions away from the negatively charged wire toward the relatively positively charged plates will occur. This flow is called the *electric wind.* Electrostatic precipitators operate at 40,000 to 50,000 volts below ground potential. At this potential, electrons leave the wires and bombard the gas molecules in the vicinity of the wires, leading to the formation of a visible blue corona. The negatively charged ions then move toward the collecting plates, as shown in Fig. 8.28*a*.

The ions, in migrating from the discharging to the collecting electrode, collide with the particulate matter and charge the particles negatively. Because the gas molecule ions are many orders of magnitude smaller than even the smallest particles and because of their great number, virtually all particles which flow through the plates become charged. The negatively charged particles then migrate toward the grounded collecting plates to which they are held by electrostatic attraction. This migration is shown in Fig. 8.28*b*. The particles build a thickening layer on the collecting plates.

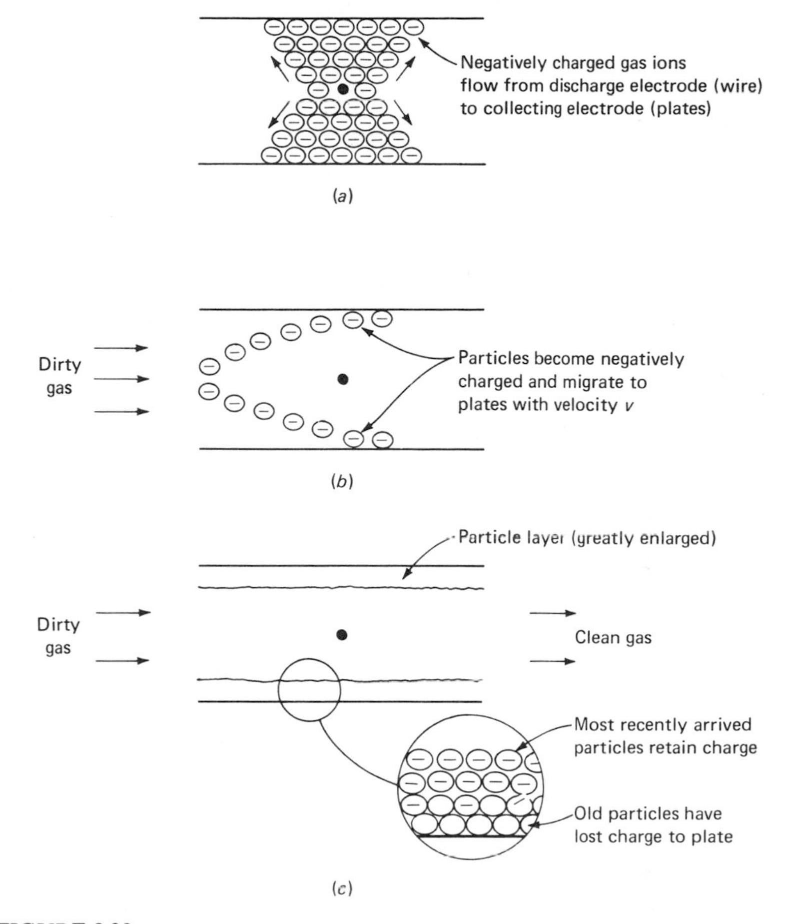

FIGURE 8.28
Processes in continuous electrostatic precipitation. (*a*) Gas ions are formed. (*b*) Particles are charged and migrate to collection plates. (*c*) Layer of particulate matter builds up on plates.

The negative charge slowly bleeds from the particles to the plate. As the layer grows, the charges on the most recently collected particles must be conducted through the layer of previously collected particles. The resistance of the dust layer is called the *dust resistivity*. Resistivities of dust layers in electrostatic precipitators range from 10^7 to 10^{11} ohms/cm.

As the particle layer grows in thickness, the particles closest to the plates lose most of their charge to the plate. As a result, the electrical attraction between the plate and these particles is weakened. However, the newly arrived particles on the

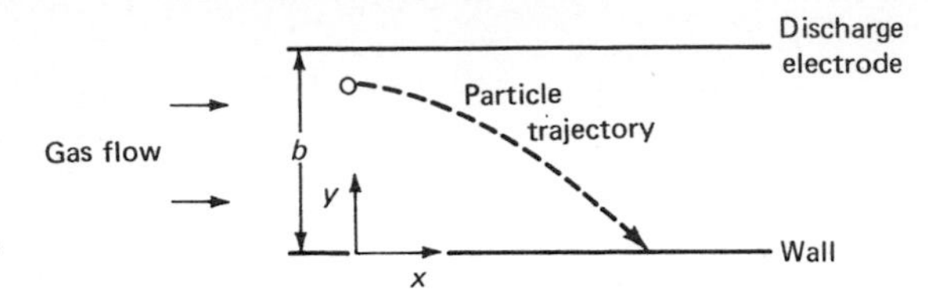

FIGURE 8.29
Simplified geometry of an electrostatic precipitator.

outside of the layer have a full charge. Because of the insulating layer of particles, these new particles do not lose their charge immediately and thus serve to hold the entire layer against the plate. Finally, the layer is removed by rapping, so that the layer breaks up and falls into a collecting hopper.

Of direct interest is the determination of the collection efficiency of a given precipitator as a function of precipitator geometry, gas flow rate, particle size, and gas properties. The particle collection process in an electrostatic precipitator is based on particle migration in a direction transverse to the mean velocity. Three mechanisms can be identified as contributing to the overall mass transfer: (1) electrostatic convection (the process described above), (2) turbulent diffusion, and (3) inertial drift. If the gas flow in the precipitator were laminar, particle movement would result solely from electrostatic convection. Flow in commercial precipitators is, however, turbulent. The prediction of the migration of particles, therefore, requires the consideration of the motion of particles in turbulent flow subject to both electrostatic and inertial forces. As we know, the description of the behavior of particles in turbulence, even in the absence of electric forces, is immensely difficult. There does not now exist, therefore, a rigorous general theory for the design of turbulent-flow electrostatic precipitators. In order to obtain design equations for collection efficiency, we resort, as we have been doing, to rather idealized models for particle motion.

Let us consider the situation depicted in Fig. 8.29 in which a gas is flowing between parallel plates down the center line of which is a discharging electrode. If E_p is the average electric field between $y = 0$ and $y = b$, the electrostatic force on a charged particle is given by $E_p Q$, where the particle charge Q is found from

$$Q = \pi \varepsilon_0 p D_p^2 E_0 \qquad (8.75)$$

where ε_0 is the permittivity of free space (8.35×10^{-12} farad/m),

$$p = 2\left(\frac{\kappa_p - 1}{\kappa_p + 1}\right) + 1$$

κ_p is the relative dielectric constant of the particle, and E_0 is the charging field, i.e. the maximum field near the discharging electrode.

Equating drag and electrostatic forces, the constant migration velocity for Re $\ll 1$ is

$$v = \frac{p\varepsilon_0 E_p E_0 D_p}{3\mu}$$

This expression is highly idealized since it does not include effects of turbulence, nor does it take into account the variation of electric field intensity with position.

Let us assume the precipitator consists of two zones: a laminar boundary layer close to the collecting wall and a turbulent core in which a uniform particle concentration c exists. In the boundary layer, the velocity component of particle motion toward the wall is v. Over a time interval dt, the particulate matter lying within a distance $v\,dt$ of the wall is precipitated on an area dA, thus reducing the content of the turbulent core by dc. Equating rates of particle loss we have

$$V_g\,dc = -vc\,dA \qquad (8.76)$$

where V_g is the volumetric flow rate of gas. Integrating (8.76) between the inlet and outlet of the unit, we have the efficiency in terms of the fraction of particles collected,

$$\eta = 1 - \frac{c(\text{out})}{c(\text{in})} = 1 - e^{-Av/V_g} \qquad (8.77)$$

where A is the total collecting area. Equation (8.77) is called the *Deutsch equation.*

The migration velocity v clearly depends on the size of the particles. Thus, there is really a distribution of efficiencies corresponding to the distribution of particle sizes. Nevertheless, it is common practice to use in (8.77) a single effective migration velocity v_e, which is determined empirically for the specific type of particulate matter.

A design criterion due to Masuda (1966) is the following: Let us assume that the effective migration velocities v_e are known for a sample of nominally identical precipitators and that these velocities are normally distributed with mean value $\bar{v}$ and standard deviation $\sigma \leq \bar{v}/3$. Since the particle migration velocity will not be known exactly, we can ask only for the probability that η is greater than some specified value η_0. The results of this approach are summarized as plots of the probability P of $\eta \geq \eta_0$ as a function of B and K, where

$$B = (1 - \eta_0)e^{A\bar{v}/V_g} \qquad K = \frac{A}{V_g}\sigma$$

Figure 8.30 presents the design curves corresponding to this method.

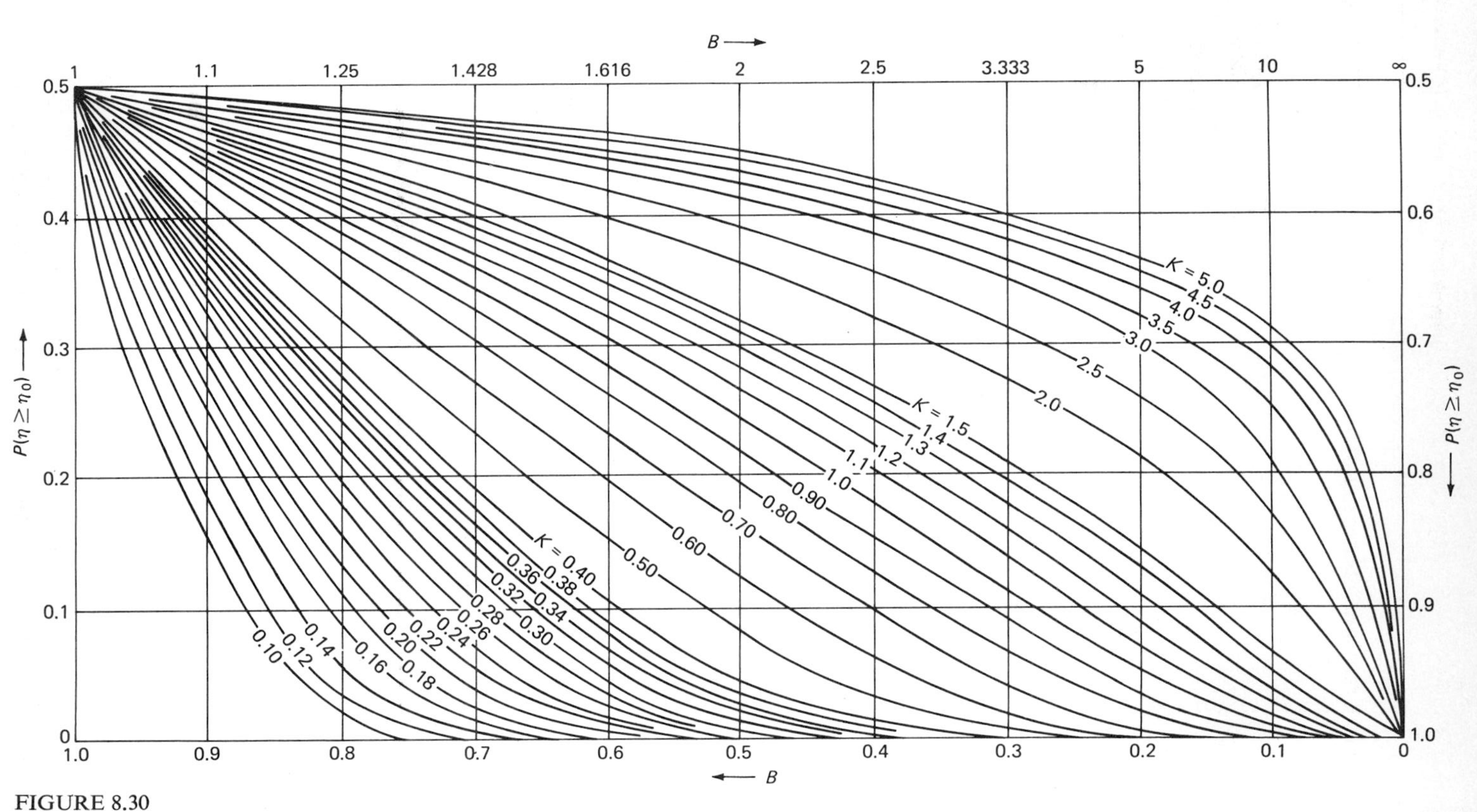

FIGURE 8.30
Design curves for electrostatic precipitation. (*Source: Masuda*, 1966.)

8.5 PARTICULATE EMISSION CONTROL EQUIPMENT

A preliminary selection of suitable particulate emission control systems is generally based on knowledge of four items: particulate concentration in the stream to be cleaned, the size distribution of the particles to be removed, the gas flow rate, and the final allowable particulate emission rate. Once the systems that are capable of providing the required efficiencies at the given flow rates have been chosen, the ultimate selection is generally made on the basis of the total cost of construction and operation. The size of a collector, and therefore its cost, is directly proportional to the volumetric flow rate of gas that must be cleaned. The operating factors which influence the cost of a device are the pressure drop through the unit, the power required, and the quantity of liquid needed (if a wet scrubbing system). In this section we shall concentrate on the design equations which are generally used for calculating efficiencies of various types of particulate emission control equipment. We shall not consider the estimation of capital or operating costs.

8.5.1 Settling Chambers

Gravitational settling is perhaps the most obvious means of separating particles from a flowing gas stream. A settling chamber is, in principle, simply a large box through which the effluent gas stream flows and in which particles in the stream settle to the floor by gravity. Gas velocities through a settling chamber must be kept low enough so that settling particles are not reentrained. The gas velocity is usually reduced by expanding the ducting into a chamber large enough so that sufficiently low velocities result. Although in principle settling chambers could be used to remove even the smallest particles, practical limitations in the length of such chambers restrict their applicability to the removal of particles larger than about 50 μm. Thus, settling chambers are normally used as precleaners to remove large and possibly abrasive particles, prior to passing the gas stream through other collection devices. A typical settling chamber is shown in Fig. 8.31.

Assuming that the gas flows through the chamber with an average velocity u_0, the average residence time θ of the gas in the device is L/u_0, where L is the length of the chamber. Thus, θ is the time available for a particle to settle to the bottom of the chamber. If the settling chamber has a height H, and if a particle of diameter D_p settles a distance h in θ sec, the efficiency of collection of particles of that size is

$$\eta = \frac{h}{H}$$

The settling distance h is determined from $u\theta$, where u is the mean settling velocity of the particle of size D_p. The terminal velocity can be computed from (8.52) and Fig.

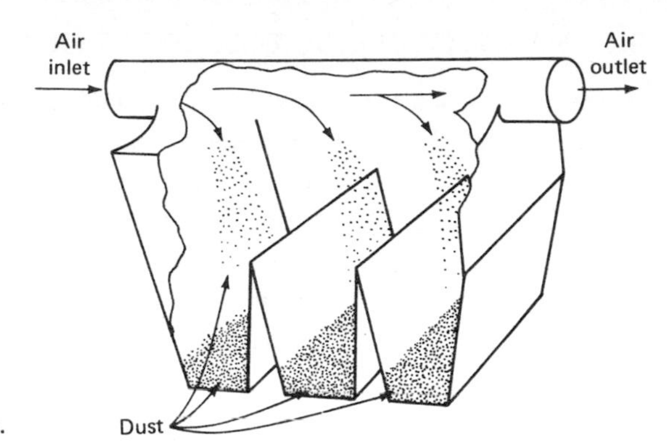

FIGURE 8.31
Typical settling chamber.

8.23. The efficiency η can be determined as a function of particle size to generate an efficiency curve for the particular settling chamber.

Settling chambers offer the advantages of (1) simple construction and low cost, (2) small pressure drops, and (3) collection of particles without need for water. The main disadvantage of settling chambers is the large space that they require.

8.5.2 Cyclone Separators

There are a variety of designs of cyclone separators, differing in the manner in which the rotating motion is imparted to the gas stream. Conventional cyclones can be placed in the following categories:

1 Reverse-flow cyclones (tangential inlet and axial inlet)
2 Straight-through-flow cyclones
3 Impeller collectors

Figure 8.32 shows a conventional reverse-flow cyclone with a tangential inlet. The dirty gas enters at the top of the cyclone and is given a spinning motion because of its tangential entry. Particles are forced to the wall by centrifugal force and then fall down the wall due to gravity. At the bottom of the cyclone the gas flow reverses to form an inner core which leaves at the top of the unit. In a reverse-flow axial-inlet cyclone, the inlet gas is introduced down the axis of the cyclone, with centrifugal motion being imparted by permanent vanes at the top.

In straight-through-flow cyclones the inner vortex of air leaves at the bottom (rather than reversing direction), with initial centrifugal motion being imparted by vanes at the top. This type is used frequently as a precleaner to remove fly ash and large particles. The chief advantages of this unit are low pressure drop and high volumetric flow rates.

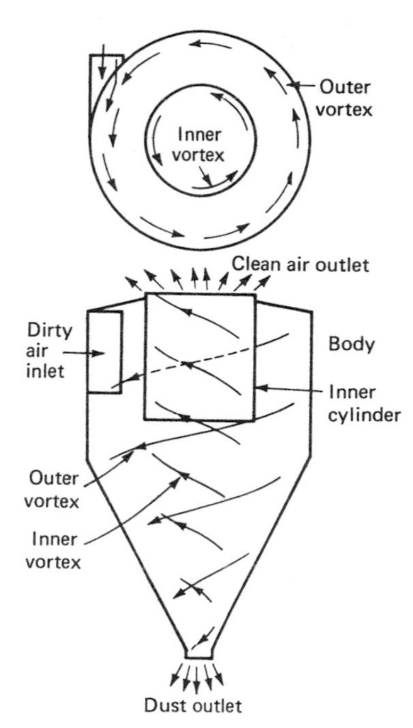

FIGURE 8.32
Principle of operation of the reverse-flow cyclone separator.

In the impeller collector, gases enter normal to a many-bladed impeller and are swept out by the impeller around its circumference while the particles are thrown into an annular slot around the periphery of the device. The principal advantage of this unit is its compactness; its chief disadvantage is a tendency toward plugging from solid buildup in the unit.

Cyclone collection efficiency increases with (1) particle size, (2) particle density, (3) inlet gas velocity, (4) cyclone body length, (5) number of gas revolutions, and (6) smoothness of the cyclone wall. On the other hand, cyclone efficiency decreases with (1) gas viscosity, (2) cyclone diameter, (3) gas outlet duct diameter, and (4) gas inlet area. For any specific cyclone whose ratio of dimensions is fixed, the collection efficiency increases as the diameter is decreased. The design of a cyclone separator represents a compromise among collection efficiency, pressure drop, and size. Higher efficiencies require higher pressure drops (i.e. inlet gas velocities) and larger sizes (i.e. body length).

Each cyclone type can be characterized by efficiency curves as a function of particle size. Although we were able to derive an equation for the trajectories of particles in a spiraling gas stream in Subsec. 8.4.2, prediction of the actual collection efficiency in an industrial cyclone has proved to be a difficult task. At this time the

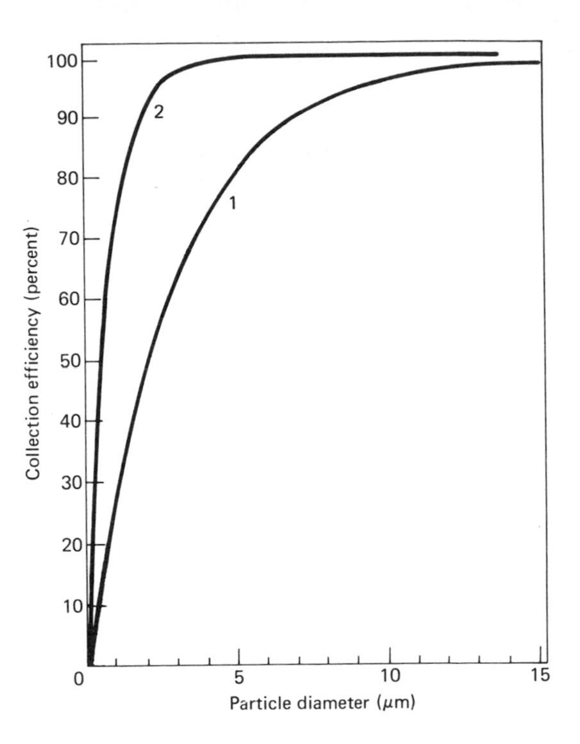

FIGURE 8.33
Efficiency of cyclone separators as a function of particle diameter for two different cyclone types:

Curve	1	2
Type	Axial vane	Reverse flow
Diameter, cm	25.4	150
Capacity, m^3/min	65	325
Pressure drop, cm water	6.1	15

most reliable means of determining the efficiency of a particular cyclone is by experimentally measuring the degree of particle removal for the unit. Figure 8.33 shows two experimentally determined efficiency curves. In addition to the overall collection efficiency, cyclone performance can also be measured by the *cut size*, the particle diameter collected with 50 percent efficiency on a weight basis. The design of cyclones is treated in more detail in the Air Pollution Engineering Manual (1967).

8.5.3 Wet Collectors

Wet collectors employ water washing, usually in the form of a spray, to remove particles directly from a gas stream. Removal of particles results primarily because of collisions between particles and water droplets. Separation can also occur to a lesser extent because of gravitational forces on large particles or electrostatic or thermal forces on small particles Efficiency is improved over dry collectors because small particles grow in size by condensation of water, and also because reentrainment is minimized by trapping of particles in a liquid film and washing them away.

Figure 8.34 illustrates four types of wet collection equipment. The simplest type of wet collector is a spray tower into which water is introduced by means of spray

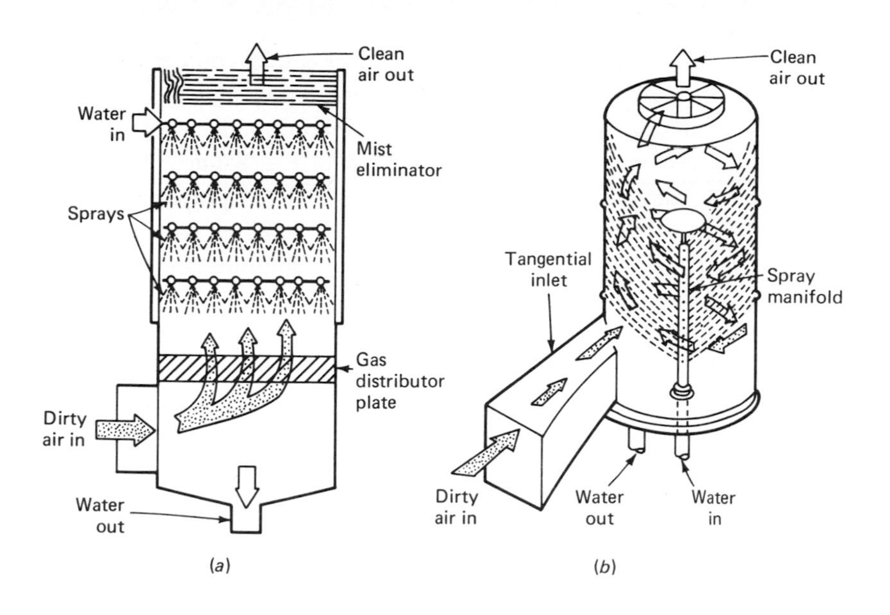

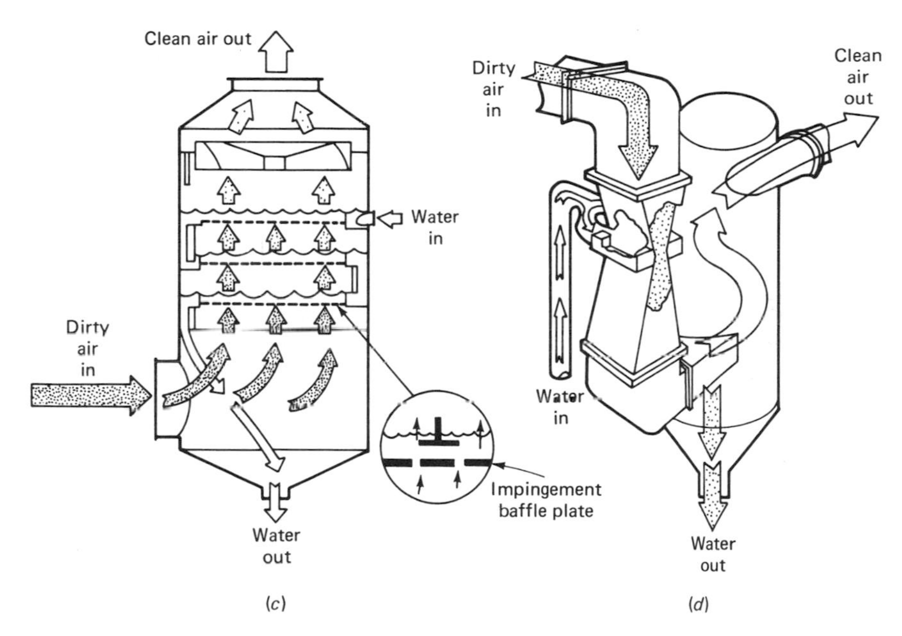

FIGURE 8.34
Wet collectors: (*a*) spray tower; (*b*) cyclone spray tower; (*c*) impingement scrubber; (*d*) venturi scrubber.

nozzles (Fig. 8.34*a*). Gas flow in a spray chamber can be countercurrent, cocurrent, or cross flow to the water flow. Figure 8.34*a* illustrates countercurrent flow, the flow configuration leading to maximum efficiency. Collection efficiency can be improved with the use of a cyclonic spray tower, as shown in Fig. 8.34*b*. The liquid spray is directed outward from nozzles in a central pipe. An unsprayed section above the nozzles is provided so that the liquid drops with the collected particles will have time to reach the walls of the chamber before exit. An impingement plate scrubber, as shown in Fig. 8.34*c*, consists of a tower containing layers of baffled plates with holes (5000 to 50,000/m^2) through which the gas must rise and over which the water must fall. Highest efficiencies are obtained in a venturi scrubber, shown in Fig. 8.34*d*, in which water is introduced at right angles to a high-velocity gas flow in a venturi tube (a contraction), resulting in the formation of very small water droplets by the flow and high relative velocities of water and particles. The high gas velocity is responsible for the breakup of the liquid. Aside from the small droplet size and high impingement velocities, particles grow by condensation.

The collection efficiency of wet collectors can be related to the total energy loss in the equipment. Almost all the energy is introduced in the gas, and thus the energy loss can be measured by the pressure drop of gas through the unit. A comparison of the efficiencies and requirements of the four classes of wet collection devices shown in Fig. 8.34 is given in Table 8.7.

The major advantage of wet collectors is the wide variety of types, allowing the selection of a unit suitable to the particular removal problem. As disadvantages, high pressure drops (and therefore energy requirements) must be maintained, and the handling and disposal of large volumes of scrubbing liquid must be undertaken.

The prediction of collection efficiencies in the removal of airborne particles by water sprays is generally based on the assumption that inertial impaction is the prime mechanism of entrapment, an assumption valid for particles greater than about 0.5μm in diameter. The overall collection efficiency of a water spray is estimated in the following manner (Ekman and Johnstone, 1951). Let η_0 be the efficiency of impaction of particles on the droplets, defined as the ratio of the cross-sectional area of

Table 8.7 EFFICIENCIES AND REQUIREMENTS OF WET COLLECTION UNITS

Unit	Efficiency	Water requirements, gal/100 m^3 gas	Pressure drop, cm water
Spray tower	High for >10 μm	5–20	2
Cyclonic spray tower	96% for 2 to 3 μm	5	2–5
Impingement scrubber	98% for >1 μm (per plate)	3–5	2–20 (per plate)
Venturi scrubber	99% for >1 μm	5	15–150

the hypothetical tube of gas from which the particles are all removed to the frontal area of the droplet. If it is assumed that each water droplet sweeps out an effective path of length L, the effective volume of the tube from which all particles are removed by the passage of drop is $\eta_0 \pi D^2 L/4$, where D is the droplet diameter. The number of droplets in a certain volumetric quantity of water Q_w is equal to $6Q_w/\pi D^3$. Then the total volume swept out by all the drops is $3\eta_0 L Q_w/2D$. If Q_w is the volume of water introduced per unit volume of air, and the rate of collection of particles is proportional to the concentration of the particles and to the volume of gas swept by the droplets per unit volume of air, the rate of change of the concentration of particles per unit length of the device is given by

$$\frac{dc}{dL} = -\left(\frac{3\eta_0 Q_w}{2D}\right)c \qquad (8.78)$$

Therefore, at any point L, the ratio of the concentration of particles to that which entered the device is given by

$$\frac{c}{c_0} = e^{-3\eta_0 L Q_w/2D} \qquad (8.79)$$

The overall efficiency η is then determined by

$$\eta = 1 - \frac{c}{c_0} = 1 - e^{-3\eta_0 L Q_w/2D} \qquad (8.80)$$

If the impaction efficiency η_0 and the effective path of the water droplets through the air are constant, a plot of $\ln(1 - \eta)$ versus Q_w should yield a straight line. However, in general, both the impaction efficiency and the distance the droplets travel depend on drop size and gas velocity, so that a straight line is usually not obtained.

According to (8.80) the total collection efficiency of a spray is directly proportional to the impaction efficiency, the water flow rate per unit flow rate of air, and the total length of the spray, and is inversely proportional to the mean diameter of the drops. (Note that η is also inversely proportional to the air flow rate since Q_w has been defined per unit flow rate of air.) For particles of a certain size, maximum collection efficiency requires as large a D as possible according to (8.80). However, as D decreases the total length of the spray decreases due to increased air resistance on the droplets. Thus, we expect that there should exist for a given particle size a water droplet size at which total collection efficiency is a maximum. Also, since η_0 increases as the particle size increases, we expect that the maximum efficiency of collection should increase as the particle size increases.

Collection efficiencies for a spray tower have been calculated by Stairmand (1950) based on inertial impaction for droplets falling under gravity. Figure 8.35

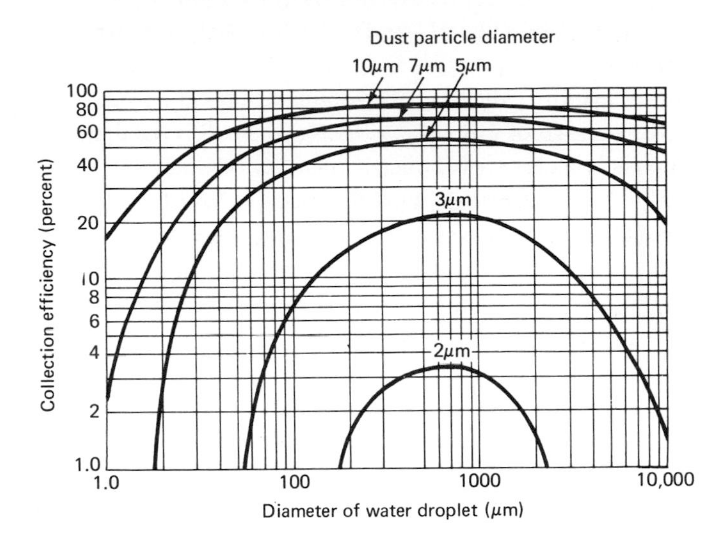

FIGURE 8.35
Collection efficiency by inertial impaction by water droplets falling under gravity. (*Source: Stairmand*, 1950.)

shows the efficiency as a function of the water droplet size for various diameters of the particles to be collected. According to Fig. 8.35, the optimum water-drop size is about 800 μm (0.8 mm), regardless of the size of the particles to be collected. For a cyclonic spray tower, Johnstone and Roberts (1949) determined the collection efficiency

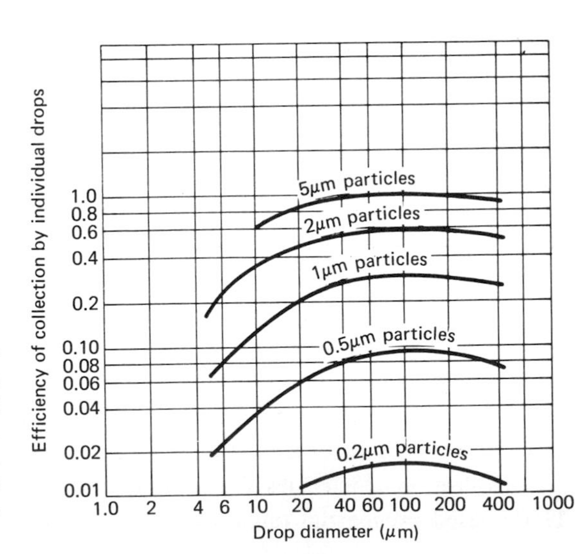

FIGURE 8.36
Collection efficiency by inertial impaction by water droplets moving in a gravitational field of 100g in a centrifugal spray tower. [*Reprinted from H. F. Johnstone and M. H. Roberts* (1949). *Copyright* 1949 *by the American Chemical Society. Reprinted by permission of the copyright owner.*]

by inertial impaction as a function of droplet size and collected-particle size. Figure 8.36 shows the results of that calculation. Again we see that the optimum water-drop size is independent of collected-particle size. Because of the greater relative velocity, the optimum water-drop size in a cyclonic spray-tower is about 100 μm as opposed to 800 μm in the spray tower.

Cheng (1973) has surveyed the prior work in, and presented an analysis of, the collection of airborne particulate matter with spray drops ejected from a high-pressure nozzle, both for unconfined sprays and conical sprays confined in a horizontal duct. In addition, a comprehensive survey of particle collection by liquid scrubbing is presented by Strauss (1966).

8.5.4 Fabric Filters

Industrial fabric filtration is usually accomplished in a so-called baghouse in which are hung a number of filter bags through which the particulate-laden gases are forced. Particles are generally removed from the bags by gravity. Figure 8.37 shows three baghouse designs, in which cleaning is accomplished by vibration (Fig. 8.37*a*), air jet (Fig. 8.37*b*), or traveling ring (Fig. 8.37*c*).

In order to compute the collection efficiency of a whole bed of fibers from that for a single fiber, the following model is generally used (Löffler, 1971). The fibers are assumed to lie in regular layers perpendicular to the gas flow. The particles are assumed to be evenly distributed in the gas phase upstream of the filter, so that the conditions of particle deposition are identical for all fibers. Let l be the total fiber length per unit volume of filter. Then in a unit area of surface, in a layer of depth dL, the fiber has length $l dL$. Assuming all particles have diameter D_p and all fibers have diameter D, the concentration of particles deposited in layer dL is given by

$$u_0\, dc = \eta (l dL) D \left(\frac{u_0 c}{\varepsilon} \right) \tag{8.81}$$

where c is particle concentration, η is the single fiber efficiency, u_0 is the upstream velocity of the flow, and ε is the void fraction of the bed. We compute l from

$$1 - \varepsilon = \frac{\pi D^2}{4} l \tag{8.82}$$

Substituting (8.82) in (8.81) and integrating over the entire bed, we obtain the overall efficiency as

$$\eta_T = 1 - \frac{c(\text{out})}{c(\text{in})} = 1 - \exp\left(-\frac{4}{\pi} \frac{L}{D} \frac{1-\varepsilon}{\varepsilon} \eta \right) \tag{8.83}$$

This result can be extended to a distribution of particle sizes and fiber lengths.

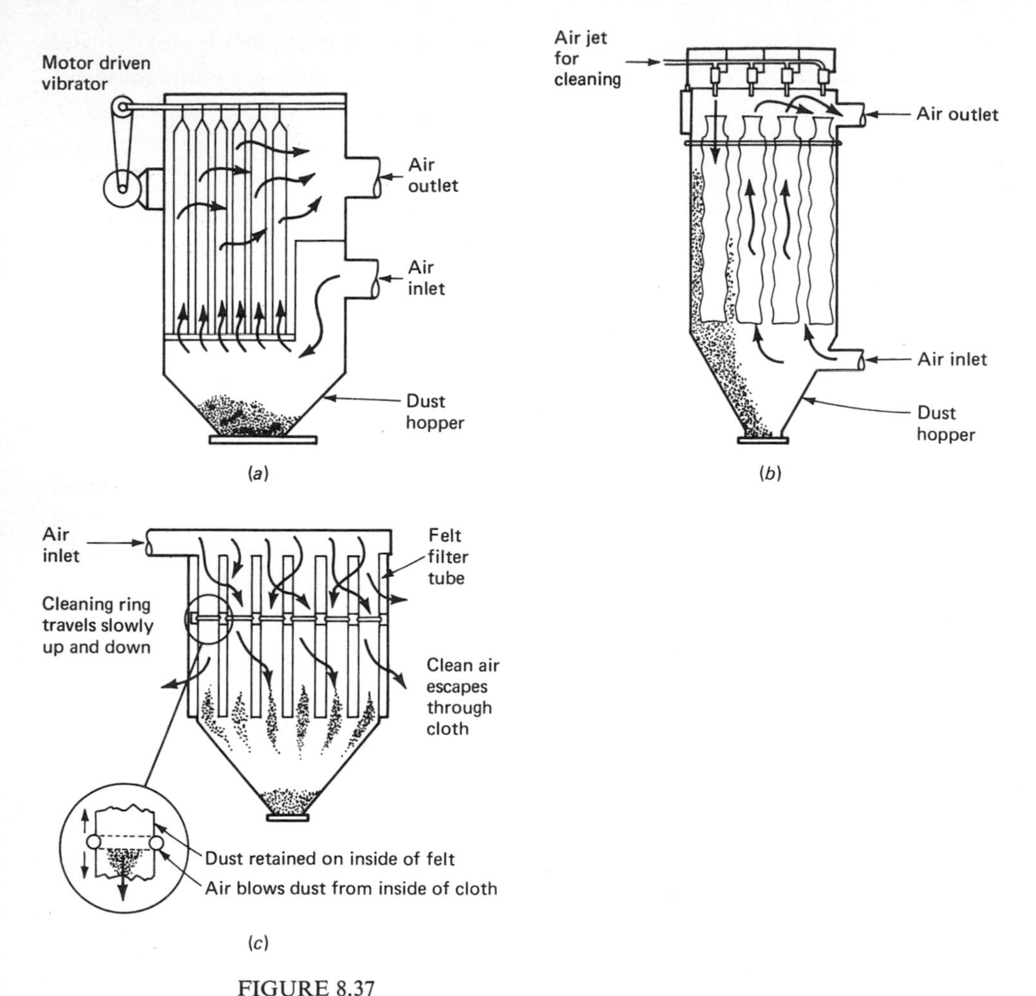

FIGURE 8.37
Three designs for a baghouse: (*a*) motor-drive vibrator; (*b*) air jet; and (*c*) cleaning ring for removing particles from fabric filters.

The above development applies only to a fresh bed of fibers. During the course of the filtration both the efficiency and the pressure drop change depending on the layer of particles existing on the filter. In general, the collection efficiency improves with particulate loading, because the deposited particles provide additional surface for collection.

8.5.5 Electrostatic Precipitators

Electrostatic precipitators are commonly employed for gas cleaning when the volumetric throughput of gas is high. Such units are used routinely for fly ash removal from power plant flue gases. Electrostatic precipitators are also widely employed for the collection of particles and acid mists in the chemical and metallurgical process industries. Design efficiencies for electrostatic precipitation were given in Subsec. 8.4.4. Extensive discussions of actual industrial units are given by Strauss (1966) and Robinson (1971).

8.5.6 Calculation of Total Efficiencies of Particulate Control Devices

In discussing each of the common methods of removing particles from gas streams our main objective has been to determine the efficiency of the technique as a function of the particle size. The total efficiency of a device depends not only on the efficiency as a function of particle size for the unit but also on the size distribution of the incoming particles. If the efficiency of the device is $\eta(D_p)$ and the normalized size density function for the incoming particles is $g(D_p)$, the total efficiency E is given by

$$E = \int_0^\infty \eta(D_p) g(D_p)\, dD_p \qquad (8.84)$$

In general, both $\eta(D_p)$ and $g(D_p)$ are determined experimentally and do not have convenient forms which would enable (8.84) to be integrated analytically. Thus, the integral usually must be evaluated numerically.

8.5.7 Summary of Particulate Emission Control Techniques

Table 8.8 presents a summary of particulate emission control techniques, including minimum particle sizes, ranges of efficiency, and advantages and disadvantages of each type of unit. In selecting a method to meet a particular gas cleaning need, the most important consideration is the total cost (operating and equipment) of the method. The advantages and disadvantages listed in Table 8.8 give an indication of the considerations which enter into a determination of the cost of a particular device.

Table 8.8 SUMMARY OF PARTICULATE EMISSION CONTROL TECHNIQUES

Device	Minimum particle size†, μm	Efficiency, % (mass basis)	Advantages	Disadvantages
Gravitational settler	>50	<50	Low pressure loss, simplicity of design and maintenance	Much space required. Low collection efficiency
Cyclone	5–25	50–90	Simplicity of design and maintenance Little floor space required Dry continuous disposal of collected dusts Low to moderate pressure loss Handles large particles Handles high dust loadings Temperature independent	Much head room required Low collection efficiency of small particles Sensitive to variable dust loadings and flow rates

(continued)

Table 8.8—*continued*

Device	Minimum particle size†, μm	Efficiency, % (mass basis)	Advantages	Disadvantages
Wet collectors			Simultaneous gas absorption and particle removal Ability to cool and clean high-temperature, moisture-laden gases Corrosive gases and mists can be recovered and neutralized Reduced dust explosion risk Efficiency can be varied	Corrosion, erosion problems Added cost of wastewater treatment and reclamation Low efficiency on submicron particles Contamination of effluent stream by liquid entrainment Freezing problems in cold weather Reduction in buoyancy and plume rise Water vapor contributes to visible plume under some atmospheric conditions
Spray tower	>10	<80		
Cyclonic	>2.5	<80		
Impingement	>2.5	<80		
Venturi	>0.5	<99		
Electrostatic precipitator	>1	95–99	99+ percent efficiency obtainable Very small particles can be collected Particles may be collected wet or dry Pressure drops and power requirements are small compared with other high-efficiency collectors Maintenance is nominal unless corrosive or adhesive materials are handled Few moving parts Can be operated at high temperatures (300 to 450°C)	Relatively high initial cost Precipitators are sensitive to variable dust loadings or flow rates Resistivity causes some material to be economically uncollectable Precautions are required to safeguard personnel from high voltage Collection efficiencies can deteriorate gradually and imperceptibly
Fabric filtration	<1	>99	Dry collection possible Decrease of performance is noticeable Collection of small particles possible High efficiencies possible	Sensitivity to filtering velocity High-temperature gases must be cooled to 100 to 450°C Affected by relative humidity (condensation) Susceptibility of fabric to chemical attack

† Collected at 90 percent efficiency

REFERENCES

Air Pollution Engineering Manual, *U.S. Dept. Health, Education, and Welfare, Environ. Protection Agency Rept. AP*–40, 1967.

ANDREWS, R. L., C. W. SIEGMUND, and D. G. LEVINE: Effect of Flue Gas Recirculation on Emissions from Heating Oil Combustion, *Air Pollut. Control Assoc. Paper* 68–21, 1968.

BARNHART, D. H., and E. K. DIEHL: Control of Nitrogen Oxides in Boiler Flue Gases by Two-stage Combustion, *J. Air Pollut. Control Assoc.*, **10**:397 (1960).

BARTOK, W., A. R. CRAWFORD, and A. SKOPP: Control of NO_X Emissions from Stationary Sources, *Chem. Eng. Prog.*, **67**:2, 64 (1971).

BIENSTOCK, D., R. L. AMSLER, and E. R. BAUER: Formation of Oxides of Nitrogen in Pulverized-coal Combustion, *J. Air Pollut. Control Assoc.*, **16**:442 (1966).

BLUMBERG, P., and J. T. KUMMER: Prediction of NO Formation in Spark-ignited Engines—An Analysis of Methods of Control, *Combust. Sci. Technol.*, **4**:73 (1971).

CAMPAU, R. M.: Low Emission Concept Vehicles, *SAE J.*, **80**:1182 (1971).

CHENG, L.: Collection of Airborne Dust by Water Sprays, *Ind. Eng. Chem. Process Des. Develop.*, **12**: 221 (1973).

Control Techniques for CO, NO and Hydrocarbons from Mobile Sources, *U.S. Dept. Health, Education and Welfare, Environ. Protection Agency Rept. AP*–66, 1970.

Control Techniques for Particulate Air Pollutants, *U.S. Dept. Health, Education, and Welfare, Environ. Protection Agency Rept. AP*–51, 1969.

Control Techniques for Sulfur Oxide Air Pollutants, *U.S. Dept. Health, Education, and Welfare, Environ. Protection Agency Rept. AP*–52, 1969.

DAVIS, J. C.: SO_2 Removal Still Prototype, *Chem. Eng.*, **79**: 13, 52 (1972).

EKMAN, F. O., and H. F. JOHNSTONE: Collection of Aerosols in a Venturi Scrubber, *Ind. Eng. Chem.*, **43**: 1358 (1951).

FRIEDLANDER, S. K.: Theory of Aerosol Filtration, *Ind. Eng. Chem.*, **50**:1161 (1958).

GALLAER, C. A., and J. W. SCHINDLER: Mechanical Dust Collectors, *J. Air Pollut. Control Assoc.*, **13**:574 (1963).

GERSTLE, R. W., and R. F. PETERSON: Control of Nitrogen Oxide Emissions from Nitric Acid Manufacturing Processes, *Air Eng.*, **9**:24 (1967).

HORLACHER, W. R., R. E. BARNARD, R. K. TEAGUE, and P. L. HAYDEN: Four SO_2 Removal Systems, *Chem. Eng. Prog.*, **68**:8, 43 (1972).

JOHNSTONE, H. F., and M. H. ROBERTS: Deposition of Aerosol Particles from Moving Gas Streams, *Ind. Eng. Chem.*, **41**:2417 (1949).

KOVAL, E. J., and M. S. PETERS: How Does Nitric Oxide Affect Reactions of Aqueous Nitrogen Dioxide?, *Ind. Enq. Chem.*, **52**:1011 (1960).

LÖFFLER, F.: Collection of Particles by Fiber Fliters, in W. STRAUSS (ed.), "Air Pollution Control," Wiley Interscience, New York, 1971.

MASUDA, S.: Statistical Observations on the Efficiency of Electrostatic Precipitation, Staub (in English), **26**:11, 6 (1966).

MCCABE, W. L., and J. C. SMITH: "Unit Operations of Chemical Engineering," McGraw-Hill, New York, 1956.

Proceedings of the First International Conference on Fluidized Bed Combustion, sponsored by the National Air Pollution Control Administration, Heuston Woods State Park, Oxford, Ohio, 1968.

RANZ, W. E.: *Univ. Illinois Eng. Exp. Stn. Tech. Rept.* 8, 1953.

Report by the Committee on Motor Vehicle Emissions, National Academy of Sciences, Feb. 1, 1973.

ROBINSON, M.: Electrostatic Precipitation, in W. STRAUSS (ed.), "Air Pollution Control," Wiley Interscience, New York, 1971.

SCHLATTER, J. C., R. L. KLIMISCH, and K. C. TAYLOR: Exhaust Catalysts: Appropriate Conditions for Comparing Platinum and Base Metal, *Science*, **179**:798 (1973).

SENSENBAUGH, J. D., and J. JONAKIN: Effect of Combustion Conditions on Nitrogen Oxide Formation in Boiler Furnaces, *Am. Soc. Mech. Eng. Paper* 60–WA–334, 1960.

SHAH, I. W.: MgO Absorbs Stackgas SO_2, *Chem. Eng.*, **79**:14, 80 (1972).

SHAIR, F. H., and J. H. RUPE: Nitric Oxide Emission Studies of Internal Combustion Engines, *JPL Q. Tech. Rev.* **1**:2, 23 (1971).

SHELEF, M., and J. T. KUMMER: The Behavior of Nitric Oxide in Heterogeneous Catalytic Reactions, 62nd Annual Meeting, *Am. Inst. Chem. Eng. Paper* 13f, 1969.

SLACK, A. V., G. G. MCGLAMERY, and H. L. FALKENBERRY: Economic Factors in Recovery of Sulfur Dioxide from Power Plant Stack Gas, *J. Air Pollut. Control Assoc.*, **21**:9 (1971).

SMITH, M., and M. J. MANOS: Determination and Evaluation of Urban Vehicle Operating Patterns 65th Annual Meeting, Miami Beach, June 18-12, 1972, *Air Pollut. Control Assoc. Paper* 72–177.

STAIRMAND, C. J.: Dust Collection by Impingement and Diffusion, *Trans. Inst. Chem. Eng. (London)*, **28**:130 (1950).

STRAUSS, W.: "Industrial Gas Cleaning," Pergamon, New York 1966.

STRAUSS, W.: The Control of Sulfur Emissions from Combustion Processes in W. STRAUSS (ed.), "Air Pollution Control," Wiley Interscience, New York, 1971.

TEMPLE, R. G.: Control of Internal Combustion Engines, in W. STRAUSS (ed.), "Air Pollution Control," Wiley Interscience, New York, 1971.

TREYBAL, R. E.: "Mass Transfer Operations," 2nd ed., McGraw-Hill, New York, 1968.

VATAVUK, W. M.: A Technique for Calculating Overall Efficiencies of Particulate Control Devices, *Environ. Protection Agency Rept. EPA*–450/2–73–002, 1973.

WIEDERSUM, G. C.: Control of Power Plant Emissions, *Chem. Eng. Prog.*, **66**:11, 49 (1970).

WONG, J. B., and H. F. JOHNSTONE: *Univ. Illinois Eng. Exp. Stn. Tech. Rept.* 11, 1953.

PROBLEMS

8.1 Ninety-five percent of the SO_2 in a process effluent stream of SO_2 and air is to be removed by gas absorption with water. The entering gas contains a mole fraction of SO_2 of 0.08; the entering water contains no SO_2. The water flow rate is to be twice the minimum. The entering gas flow rate is 100 moles/min.

(*a*) Assume for the purposes of the calculation that the equilibrium line for SO_2 is straight with a slope of 35 (see Fig. 8.20). Determine the depth of the packing needed.

(*b*) In the case in which both the operating and equilibrium lines are straight, i.e. when the concentration of solute is lean ($y \ll 1, x \ll 1$), the integral in (8.33) can be approximated by

$$\int_{y_1}^{y_0} \frac{dy}{y - y^*}$$

which can be integrated analytically. Show that in this case

$$Z_T = \frac{G}{K_y a} \frac{y_0 - y_1}{(y - y^*)_{lm}}$$

where $$(y - y^*)_{lm} = \frac{(y_0 - y_0^*) - (y_1 - y_1^*)}{\ln[(y_0 - y_0^*)/(y_1 - y_1^*)]}$$

Repeat case (*a*), assuming that the operating and equilibrium lines are both straight.

8.2 An absorber is to be used to remove acetone from an airstream by contact with water. The entering air contains an acetone mole fraction of 0.11, and the entering water is acetone-free. The inlet gas flow rate is 10 m^3/min. The mole fraction of acetone in the air leaving the column is to be 0.02. The equilibrium curve for acetone-water at 1 atm and 26.6°C, the conditions of operation of the tower, is given by (McCabe and Smith, 1956)

$$y^* = 0.33x^* e^{1.95(1-x^*)^2}$$

(*a*) What is the water flow rate if it is to be 1.75 times the minimum?

(*b*) What is the required height of the tower if the gas-phase HTU is given by

$$H_{oy} = 3.3\tilde{G}^{0.33}\tilde{L}^{-0.33} \qquad \text{meters}$$

where $\tilde{G}$ and $\tilde{L}$ are the mass velocities, in kg/m^2-hr?

8.3 Benzene vapor present at a concentration of 0.030 kg benzene/kg air is to be removed by passing the gas mixture downward through a bed of silica gel at 50°C and 2 atm pressure at a linear velocity of 0.5 m/sec (based on the total cross-sectional area of 1 m^2). The bed has a packing depth of 3 m. The breakpoint will be considered that time when the effluent air has a benzene content of 0.0030 kg benzene/kg air, and the bed will be considered exhausted when the effluent air contains 0.024 kg benzene/kg air. Determine the time required to reach the breakpoint.

Under these conditions the adsorption isotherm is

$$Y^* = 0.167X^{1.3}$$

The bulk density of silica gel is 625 kg/m^3, the average particle diameter is 0.60 cm, and $a = 600$ m^2/m^3. Assume that the height of a gas-phase transfer unit is given by

$$H_{OY} = \frac{1.42}{a}\left(\frac{D_p \tilde{G}'}{\mu_{air}}\right)^{0.51}$$

8.4 Air laden with acid fog is led from a process to a square horizontal settling chamber 8 m long and 50 cm high. The fog can be considered to consist of spherical droplets of diameter 0.8 mm. Assume the fog is uniformly distributed at the entrance to the duct and is at a concentration of 50 ml of acid per cubic meter of air. It is desired to remove 90 percent of the fog from the stream. Find the volumetric flow rate, in cubic meters per hour, which will allow 90 percent removal. Find the weight of acid removed, in kilograms per hour.

$$\mu_{\text{air}} = 0.0002 \text{ p}$$

$$\rho_{\text{air}} = 0.001 \text{ g/cm}^3$$

$$\rho_{\text{acid}} = 1 \text{ g/cm}^3$$

8.5 Show that for a rectangular settling chamber of height h, breadth b, and length l, with volumetric gas throughput of Q, in the Stokes' law region the particle diameter that will be wholly retained within the chamber is given by

$$D_{p_{\min}} = \left[\frac{18Q\mu}{(\rho_p - \rho)gbl}\right]^{1/2}$$

8.6 Settling chambers are commonly used in a sinter plant to remove large particles of quartz and iron oxide from effluent gas streams. A settling chamber 3 m high and wide and 6 m long is available. The volumetric flow rate of gas through the chamber is 5000 m^3/hr. The densities of quartz and iron oxide particles are 2.6 and 4.5, respectively. Compute and plot efficiency curves for this unit at the given gas flow rate for both quartz and iron oxide particles as a function of particle diameter.

8.7 A fly ash has the distribution of weight percent by particle sizes

Size range, μm	0–1	1–5	5–10	over 10
% by weight	36	25	24	15

The two cyclones, whose efficiencies, by weight percent, are shown in Fig. 8.33, are available. Making any assumptions which are deemed reasonable, determine the overall efficiencies of the two cyclones for this fly ash.

8.8 The fly ash in Prob. 8.7 is to be removed from its flue gas either by a simple spray tower or by a centrifugal wet collector, the efficiencies of which, by particle number, are shown in Fig. 8.35 and 8.36, respectively. Assuming that the fly ash particles have a density of 1.5 g/cm^3 and are present at a total number density of 10^2 particles/cm^3, determine the overall efficiencies of the two wet collectors for this fly ash.

8.9 Compute the collection efficiency of a cigarette filter which is a fiber layer of thickness 1 cm and void fraction 0.5. Assume that the smoke particles are a monodisperse aerosol of diameter 0.1 μm and that the fiber filaments have a diameter of 50 μm. Smoke is inhaled at a velocity of 3 cm/sec. The diffusion coefficient for the smoke particles can be taken to be 6.63×10^{-6} cm^2/sec.

8.10 The particulate matter in many process streams often has a weight distribution which is nearly log-normal. The form of the log-normal distribution unfortunately does not make possible analytical evaluation of E by (8.84). However, the cumulative distribution function $G_c(D_p)$,

$$G_c(D_p) = \int_{D_p}^{\infty} G(\ln \alpha)\, d \ln \alpha$$

expressing the mass fraction of particles having diameters larger than D_p of a log-normal mass distribution, is closely approximated by the simple expression (Vatavuk, 1973)

$$G_c(D_p) = e^{-\beta D_p}$$

where β can be determined from the slope of the approximate straight-line portion of a plot of $\ln G_c$ versus D_p. Show that the total efficiency of a device with an efficiency curve $\eta(D_p)$ can in this situation be approximated by

$$E = \beta \int_{0}^{\infty} \eta(D_p) e^{-\beta D_p}\, dD_p$$

8.11 Consider the cross-sectional view of an electrostatic precipitator shown in Fig. 8.29. This precipitator consists of parallel plates at a separation of $2b$, with a discharge electrode wire halfway between them. Assume that the gas flow between the plates is laminar and fully developed with a centerline value of u_{max}. The particles can be assumed to travel in the x direction at the prevailing fluid velocity at any value of y. Assume that the particle drift velocity in the direction of the wall, v, is constant. Show that for all particles to be collected the length of this precipitator has to be

$$x = \frac{2u_{max} b}{3v}$$

8.12 A manufacturer has specified that, for its line of electrostatic precipitators, the average migration velocity of fly ash particulate matter is 0.05 m/sec with a standard deviation of 0.01 m/sec. You wish to design a new precipitator of the same type for which there will be 90 percent assurance that an efficiency of 99 percent will be achieved. In the design, you need to determine the ratio of total collection area to volumetric throughput of gas of the new unit. Determine this ratio on the basis of the given information and compare the value of A/V_g obtained with that from the Deutsch equation using $v = \bar{v} = 0.05$ m/sec. Discuss.

8.13 The particle size distribution of fly ash from a coal-burning furnace has the approximately log-normal distribution shown in Fig. 1. The particles are to be collected in an electrostatic precipitator for which the particle migration velocity as a function of particle size is shown in Fig. 2. For the precipitator in question $A/V_g = 100$ sec/m. Calculate the collection efficiency of the precipitator for this effluent. (Assume that the Deutsch equation is valid.)

8.14 (*a*) Show that a rectangular settling chamber operating in the Stokes' law regime of particle motion has an efficiency given by

$$\eta(D_p) = \frac{(\rho_p - \rho) g B L D_p^2}{18 \mu V_g}$$

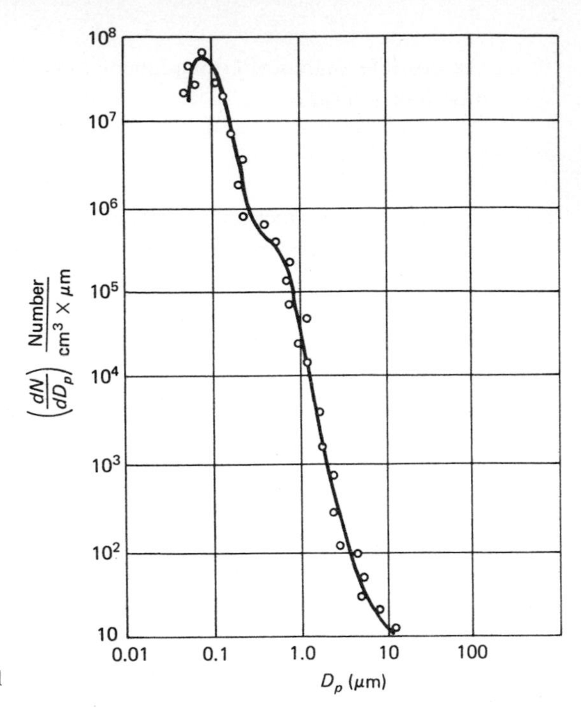

FIGURE 1

where

B = width of chamber
L = length of chamber
V_g = volumetric flow rate of gas

(*b*) In Prob. 8.10 it was stated that the total efficiency of a device collecting log-normally distributed particulate matter could be approximated:

$$E = \beta \int_0^\infty \eta(D_p) e^{-\beta D_p} \, dD_p$$

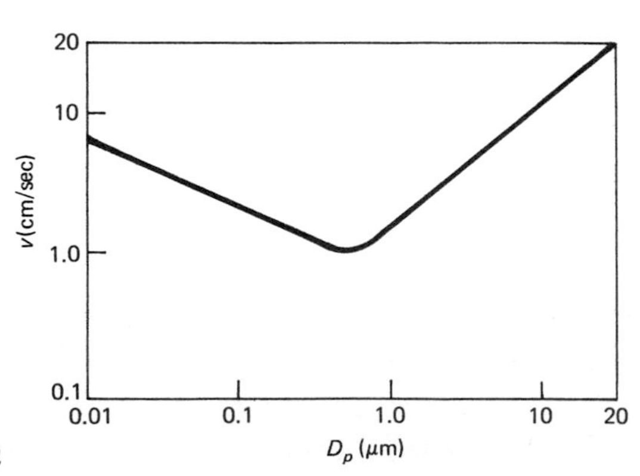

FIGURE 2

Show that the total efficiency of a settling chamber operating in the Stokes' law regime is given by

$$E = \frac{2K}{\beta^2}\left[1 - \exp\left(\frac{-\beta}{\sqrt{K}}\right)\left(1 + \frac{\beta}{\sqrt{K}}\right)\right]$$

where $K = (\rho_p - \rho)gBL/18\mu V_g$.

(c) Determine the total efficiency of a settling chamber under the following circumstances:

$$\rho_p = 1.5 \text{ g/cm}^3 \qquad B = 1 \text{ m}$$
$$\rho = 0.001 \text{ g/cm}^3 \qquad L = 10 \text{ m}$$
$$\mu = 0.0002 \text{ p} \qquad V_g = 3 \text{ m}^3\text{/sec}$$
$$\beta = 0.03 \mu m^{-1}$$

8.15 Gallaer and Schindler (1963) suggested that a useful empirical equation for the efficiency of a cyclone is

$$\eta(D_p) = 1 - e^{-\alpha D_p}$$

where α is a constant characteristic of the particular cyclone.

(a) Using the result of Prob. 8.10, show that the total efficiency of a cyclone collecting log-normally distributed particulate matter can be approximated by

$$E = \frac{\alpha}{\alpha + \beta}$$

(b) The cut size D_{pc} of a cyclone has been defined as that particle diameter collected with 50 percent efficiency. In addition, the mean particle diameter D_{pm} of a cyclone is defined as the size above which 50 percent of the particles lie. Show that the total efficiency may be expressed on the basis of D_{pc} and D_{pm} as

$$E = \frac{1}{1 + (D_{pc}/D_{pm})}$$

8.16 The collection efficiency of an electrostatic precipitator is often expressed by the Deutsch equation. Compute the total collection efficiency of an electrostatic precipitator for an inlet gas stream which has an approximate log-normal size distribution and for which the result of Prob. 8.10 is valid. The parameters of the gas stream and the device are:

$$E_p = 1.0 \text{ kV/cm} \qquad A = 200 \text{ m}^2$$
$$E_0 = 2.0 \text{ kV/cm} \qquad V_g = 5 \text{ m}^3\text{/sec}$$
$$\kappa_p = 7 \qquad \beta = 0.025 \text{ }\mu\text{m}^{-1}$$
$$\mu = 0.0002 \text{ p}$$

8.17 Several particulate collection devices are often operated in series, with each succeeding device used to collect smaller and smaller particles. Consider n particulate removal devices connected in series, such that the outlet stream from unit 1 is the inlet stream to unit 2, etc. If the efficiencies of the n devices are $\eta_1(D_p)$, $\eta_2(D_p)$, ... , $\eta_n(D_p)$, show that the total efficiency of the n units is

$$E_n = \int_0^\infty \{\eta_1(D_p) + \eta_2(D_p)[1 - \eta_1(D_p)] + \cdots + \eta_n(D_p)[1 - \eta_{n-1}(D_p)] [1 - \eta_{n-2}(D_p)] \cdots [1 - \eta_1(D_p)]\} g(D_p)\, dD_p$$

APPENDIX A
STATISTICAL DESCRIPTION OF TURBULENCE

The description of the diffusion of marked particles in turbulent fluid requires a knowledge of the statistical properties of velocity fluctuations. We not only need to know the average values of variables, such as $\bar{u}_i$ and $\overline{u'_i u'_j}$, but also to have some information on the probability distributions of these variables. Also, we want to know the degree of correlation that exists between fluctuations at different locations at any one time and at different times at any one location, as well as the correlation between the velocities of a marked particle at different times. Our objective is to be able to predict the behavior of particles released into a turbulent flow, specifically with regard to the rate of their dispersion.

A.1 SOME DEFINITIONS

The velocity component in direction i at location $\mathbf{x} = (x_1, x_2, x_3)$, $u_i(\mathbf{x}, t)$, is a random function of time t, as shown in Fig. 5.1. The statistical theory of turbulence is concerned with description of the properties of the random velocities, such as the variance and auto- and cross-correlation functions (these will be defined shortly). The random functions, $u_i(\mathbf{x}, t)$, $i = 1, 2, 3$, and their various correlations can be characterized in several ways:

1 Turbulence is *stationary* if all statistical properties are independent of time. [The variables themselves, such as $u_i(\mathbf{x},t)$, are random functions of time, but their statistical properties, such as the mean, are independent of time.]

2 Turbulence is *homogeneous* if all statistical properties are independent of location in the field.

3 Turbulence is *isotropic* if all statistical properties are independent of the orientation of the coordinate axes. This is the most "idealized" state of turbulence.

The mean of a random function $u_i(\mathbf{x},t)$ should be computed in theory by an *ensemble average*. The velocity record in Fig. 5.1 represents only one of an infinite number of similar records which would be obtained if the experiment from which the record is observed were repeated an infinite number of times. The ensemble average of $u_i(\mathbf{x},t)$, $\langle u_i(\mathbf{x},t)\rangle$, is the average value of $u_i(\mathbf{x},t)$ over the infinite number of possible velocity records at time t. Unfortunately, in the real world we cannot repeat an experiment indefinitely to obtain an ensemble average. Consequently, we would like to replace the ensemble average with a time average over a single velocity record:

$$\bar{u}_i(\mathbf{x}) = \lim_{T\to\infty} \frac{1}{T}\int_{t_0}^{t_0+T} u_i(\mathbf{x},t)\,dt \qquad \text{(A.1)}$$

A direct correspondence between $\langle u_i(\mathbf{x},t)\rangle$ and $\bar{u}_i(\mathbf{x})$ exists only when $u_i(\mathbf{x},t)$ is a stationary function, because then $\langle u_i(\mathbf{x},t)\rangle$ is independent of t, and $\bar{u}_i(\mathbf{x})$ is independent of t_0. This correspondence cannot be proved and is called the *ergodic hypothesis*. Subsequently we will consider only stationary turbulence.

The *mean kinetic energy per unit mass* is $\frac{1}{2}\overline{u_\beta u_\beta} = \frac{1}{2}(\overline{u_1^2} + \overline{u_2^2} + \overline{u_3^2})$ which can be written $\frac{1}{2}\bar{u}_\beta\,\bar{u}_\beta + \frac{1}{2}\overline{u'_\beta u'_\beta}$.[1] Because of stationarity, the mean kinetic energy is a function of position but not of time. The first term is the kinetic energy of the mean motion; the second is the mean kinetic energy of the fluctuations.

The *intensity* of turbulence is defined by $(\overline{u'_i u'_i})^{1/2}/\bar{u}_i$, a measure of the size of the fluctuations relative to the mean velocity at the same location.

A random function, such as $u_i(\mathbf{x},t)$ can be characterized by a probability density $p_i(u)$, where $p_i(u)\,du$ is the probability that $u \le u_i(\mathbf{x},t) < u + du$. By definition,

$$\int_{-\infty}^{\infty} p_i(u)\,du = 1 \qquad p_i(u) \ge 0 \qquad \text{(A.2)}$$

The ensemble mean of u_i is therefore computed by

$$\langle u_i\rangle = \int_{-\infty}^{\infty} u p_i(u)\,du \qquad \text{(A.3)}$$

which, by the ergodic hypothesis, is equal to $\bar{u}_i$. Decomposing u_i into $\bar{u}_i + u'_i$, the variance of $u_i(\mathbf{x},t)$ is seen to be the mean-square value of the fluctuations:

$$\langle (u_i - \bar{u}_i)^2\rangle \equiv \overline{u_i'^2} = \int_{-\infty}^{\infty} (u - \bar{u}_i)^2 p_i(u)\,du \qquad \text{(A.4)}$$

[1] In this appendix only those terms containing a repeated Greek subscript are to be summed over the three components of the term.

The intensity is just the square root of the mean-square fluctuation normalized by the mean velocity. Since only in the most idealized cases would $p_i(u)$ be known, the statistical properties of u_i are normally computed by time averages of actual experimental records.

The properties $\bar{u}_i$ and $\overline{u_i'^2}$ are based on a single velocity component at a single location and time. Although these are very useful characterizations of the velocity field, other statistical properties measuring the amount of correlation between different velocity components at different times and locations are equally useful. We define the eulerian space-time correlation function by

$$R_{ij}(\mathbf{x},\mathbf{r},t,\tau) = \overline{u_i(\mathbf{x},t)u_j(\mathbf{x}+\mathbf{r},\, t+\tau)} \tag{A.5}$$

For stationary turbulence, R_{ij} does not depend on the time origin t, only on the time separation τ, and so

$$R_{ij}(\mathbf{x},\mathbf{r},t,\tau) = R_{ij}(\mathbf{x},\mathbf{r},\tau)$$

If the turbulence is also homogeneous, R_{ij} does not depend on the reference location $\mathbf{x}$, only on the spatial separation $\mathbf{r}$, and so the space-time correlation reduces to

$$R_{ij}(\mathbf{x},\mathbf{r},\tau) = R_{ij}(\mathbf{r},\tau)$$

Then, $R_{ij}(\mathbf{r}) = \overline{u_i(\mathbf{x},t)u_j(\mathbf{x}+\mathbf{r},\, t)}$ is the space correlation, and $R_{ij}(\tau) = \overline{u_i(\mathbf{x},t)u_j(\mathbf{x},\, t+\tau)}$ is the time correlation between different velocity components in stationary, homogeneous turbulence. It might be noted that

$$\frac{1}{2}\sum_{i=1}^{3} R_{ii}(0,0) = \text{kinetic energy per unit mass}$$

We have defined the correlation functions on the basis of the total velocities u_i and u_j. These can be decomposed into two terms, the first being the product of the average velocities $\bar{u}_i\,\bar{u}_j$ and the second being the average of the product of the fluctuating values, $\overline{u_i' u_j'}$. This second term is of primary importance in characterizing the turbulence. Let us denote this correlation by

$$R'_{ij}(\mathbf{r},\tau) = \overline{u_i'(\mathbf{x},t)u_j'(\mathbf{x}+\mathbf{r},\, t+\tau)} \tag{A.6}$$

Let us consider the autocorrelation function $R'_{ii}(\tau) = \overline{u_i'(\mathbf{x},t)u_i'(\mathbf{x},\, t+\tau)}$, expressing the degree of dependence between the values of the same velocity component at times separated by τ. Based in $R'_{ii}(\tau)$ we can define the eulerian integral time scale $\mathscr{T}_E$ by

$$\mathscr{T}_E = \max_i \left[\frac{1}{\overline{u_i'(\mathbf{x},t)u_i'(\mathbf{x},t)}} \int_0^\infty R'_{ii}(\tau)\, d\tau \right] \tag{A.7}$$

where $\overline{u_i'^2}$ is constant in stationary turbulence. The term $\mathscr{T}_E$ is a measure of the length of time over which u_i' is correlated with itself. A typical form of $R'_{ii}(\tau)$ is shown in Fig. A.1.

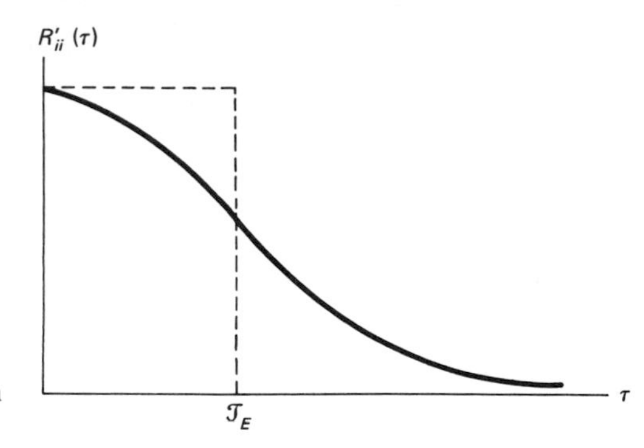

FIGURE A.1
Typical form of the autocorrelation function $R'_{ii}(\tau)$.

A.2 FOURIER TRANSFORMS

It is quite useful to deal with the Fourier transforms of the probability density functions $p_i(u)$ and of the various moments and correlation functions of the velocity field. In this section we will summarize some of the properties of the Fourier transform that are useful in the analysis of turbulent velocities.

The representation of a periodic function of time by a Fourier series is based on an expansion in its frequency components, $\omega_i = 2\pi i/T$, where T is the period of the function. A nonperiodic function of time, $f(t)$, can be represented as a continuous sum (integral) over an infinite number of frequencies by

$$f(t) = \frac{1}{2\pi} \int_{-\infty}^{\infty} \phi(\omega) e^{i\omega t} \, d\omega \tag{A.8}$$

where

$$\phi(\omega) = \int_{-\infty}^{\infty} e^{-i\omega t} f(t) \, dt \tag{A.9}$$

The pair of equations (A.8) and (A.9) is called the Fourier transform pair. We will not be concerned with conditions for the existence of a Fourier transform, since all functions with which we will deal are relatively well behaved.

The Fourier transform $\phi(\omega)$ is essentially the frequency domain representation of $f(t)$. Whereas $f(t)$ specifies a function of time, $\phi(\omega)$ specifies the relative amplitudes of the frequency components of the function. However, in general $\phi(\omega)$ is complex, although if $f(t)$ is a real function (as it will be for all our applications), the complex conjugate of $\phi(\omega)$, $\phi^*(\omega)$, equals $\phi(-\omega)$.

Let us show the Fourier transforms of some simple functions. We consider first the double-sided exponential $e^{-a|t|}$, for which

$$\phi(\omega) = \int_{-\infty}^{\infty} e^{-a|t|} e^{-i\omega t} \, dt$$

$$= \frac{2a}{a^2 + \omega^2}$$

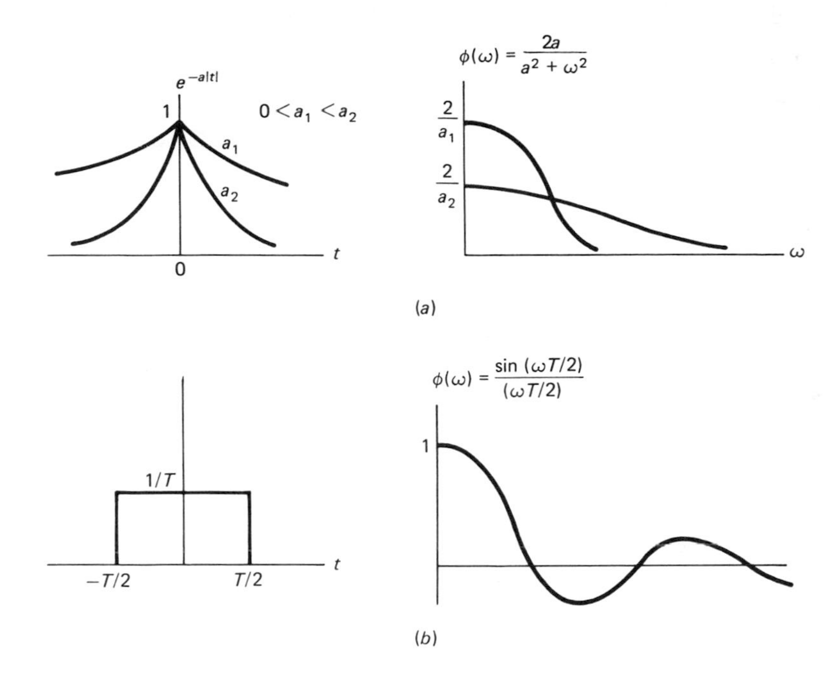

FIGURE A.2
Time domain and frequency domain representations. (*a*) $e^{-a|t|}$. (*b*) Rectangular pulse of width T and height $1/T$.

Figure A.2*a* shows $e^{-a|t|}$ and $\phi(\omega)$ for two values of a. We note that as a increases, the width of $f(t)$ decreases and the width of $\phi(\omega)$ increases.

We next consider the gate function shown in Fig. A.2*b*, a rectangular pulse of width T and height $1/T$. For this function

$$\phi(\omega) = \int_{-T/2}^{T/2} \frac{1}{T} e^{-i\omega t} \, dt$$

$$= \frac{\sin(\omega T/2)}{\omega T/2}$$

which is shown in Fig. A.2*b*.

The reader may easily verify that the Fourier transform of the gaussian density

$$f(t) = \frac{1}{\sqrt{2\pi}\,\sigma} e^{-t^2/2\sigma^2}$$

is

$$\phi(\omega) = e^{-\sigma^2\omega^2/2}$$

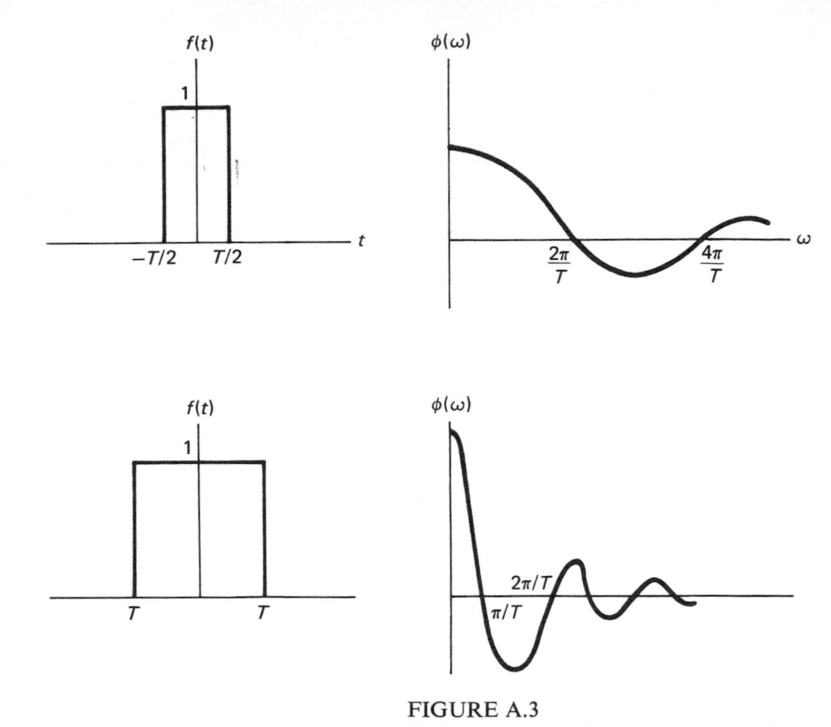

FIGURE A.3
Scaling property of the Fourier transform.

Let us now consider some properties of the Fourier transform. The Fourier transform is simply another way of specifying a function, specifically in terms of its exponential components of various frequencies. It is therefore important to know how behavior in the time domain corresponds to behavior in the frequency domain. First, the Fourier transform pair has the symmetry property that if $f(t)$, $\phi(\omega)$ is a pair, then $\phi(t)$, $2\pi f(-\omega)$ is also a pair. The second property is quite important for our purposes and is called the *scaling* property. Let us first show that if $f(t)$, $\phi(\omega)$ is a pair then $f(at)$, $(1/|a|)\phi(\omega/a)$ is also a pair and then discuss its significance. By definition the Fourier transform of $f(at)$ is

$$F[f(at)] = \int_{-\infty}^{\infty} f(at)e^{-i\omega t}\,dt$$

Let $x = at$. Then for $a > 0$,

$$F[f(at)] = \frac{1}{a}\int_{-\infty}^{\infty} f(x)e^{-i\omega x/a}\,dx$$

$$= \frac{1}{a}\,\phi\left(\frac{\omega}{a}\right)$$

For $a < 0$, it can be shown similarly that

$$F[f(at)] = -\frac{1}{a}\phi\left(\frac{\omega}{a}\right)$$

The function $f(at)$ represents $f(t)$ compressed in the time domain by a factor of a. Also, $\phi(\omega/a)$ represents $\phi(\omega)$ expanded in the frequency domain by a factor of a. Thus, compression in the time domain is equivalent to expansion in the frequency domain and vice versa. Intuitively this makes sense, since compression in the time domain by a factor a means that the function is varying rapidly by the same factor and the frequencies of its components will be increased by the factor a. The effect is illustrated in Fig. A.3.

The third property relates to the effect of time shifts on the Fourier transform. If $f(t)$, $\phi(\omega)$ is a pair, then $f(t-\tau)$, $\phi(\omega)e^{-i\omega\tau}$ is a pair. This result indicates that if a function is shifted in the time domain by τ sec the magnitude of its transform $\phi(\omega)$ remains unchanged, but the phase is shifted by $-\omega\tau$.

The final important property is the *convolution theorem.* The so-called convolution integral

$$f(t) = \int_{-\infty}^{\infty} f_1(\tau)f_2(t-\tau)\,d\tau \qquad \text{(A.10)}$$

arises frequently in applied studies. If $f_1(t)$, $\phi_1(\omega)$ and $f_2(t)$, $\phi_2(\omega)$ are pairs, then $f(t)$, $\phi_1(\omega)\phi_2(\omega)$ are also pairs. This is shown as follows:

$$\begin{aligned}
F[f(t)] &= \int_{-\infty}^{\infty} e^{-i\omega t}\left[\int_{-\infty}^{\infty} f_1(\tau)f_2(t-\tau)\,d\tau\right]dt \\
&= \int_{-\infty}^{\infty} f_1(\tau)\left[\int_{-\infty}^{\infty} e^{-i\omega t}f_2(t-\tau)\,dt\right]d\tau \\
&= \int_{-\infty}^{\infty} f_1(\tau)\phi_2(\omega)e^{-i\omega\tau}\,d\tau \\
&= \phi_1(\omega)\phi_2(\omega) \qquad \text{(A.11)}
\end{aligned}$$

Similarly, we can show that

$$F[f_1(t)f_2(t)] = \frac{1}{2\pi}\int_{-\infty}^{\infty} \phi_1(\nu)\phi_2(\omega-\nu)\,d\nu \qquad \text{(A.12)}$$

Property (A.11) is one of the prime reasons for the attractiveness of the Fourier transform representation.

Everything we have developed up to this point has been concerned with the Fourier transform of $f(t)$, a function of a single independent variable t. However, the Fourier transform of a function of several variables can also be defined and is quite useful in the study of turbulence. Let us consider, for example, a function $f(\mathbf{x})$, where $\mathbf{x} = (x_1, x_2, x_3)$. We define its Fourier transform by

$$\phi(\boldsymbol{\kappa}) = \int_{-\infty}^{\infty}\int_{-\infty}^{\infty}\int_{-\infty}^{\infty} f(\mathbf{x})e^{-i\boldsymbol{\kappa}\cdot\mathbf{x}}\,d\mathbf{x} \qquad \text{(A.13)}$$

and the inverse by

$$f(\mathbf{x}) = \frac{1}{(2\pi)^3} \int_{-\infty}^{\infty} \int_{-\infty}^{\infty} \int_{-\infty}^{\infty} \phi(\boldsymbol{\varkappa}) e^{i\boldsymbol{\kappa} \cdot \mathbf{x}} \, d\boldsymbol{\varkappa} \qquad \text{(A.14)}$$

The interpretation of $\boldsymbol{\varkappa} = (\kappa_1, \kappa_2, \kappa_3)$ depends on the nature of the original variables x_1, x_2, and x_3. In our applications these will be spatial variables. Thus, $\boldsymbol{\varkappa}$ is not a frequency, as was ω, but rather is now interpreted as a *wave number*, waves being the spatial counterpart of frequencies when the original independent variable is time. Therefore, (A.13) can be interpreted as a decomposition of the spatially varying function $f(\mathbf{x})$ into waves of different periods or wavelengths. Whereas the units of ω are sec^{-1}, those of $\boldsymbol{\varkappa}$ are cm^{-1}.

A.3 ENERGY SPECTRUM

The energy spectrum enables us to decompose the total energy of the turbulence into its various wave number contributions. Since wave numbers are inversely related to spatial lengths, this decomposition indicates how the energy of turbulence is distributed over the various length scales (or eddy sizes) of the motion. Therefore, we will be interested in the eulerian spatial correlation function $R_{ij}(\mathbf{r})$, which depends in general on the length and orientation of the separation vector between two points.

The energy spectrum will be shortly defined as the Fourier transform of the eulerian spatial autocorrelation. Since the correlation depends on the vector $\mathbf{r}$, measurements must be made in many different directions at a given separation. The Fourier transform of such a correlation is, as we have seen, a function of a wave number vector $\boldsymbol{\varkappa}$. In order to remove the overwhelming amount of information due to measurements in many directions, the transform, or spectrum, is integrated over spherical shells at a certain magnitude of the wave number vector. The result is the total amount of energy of the turbulence contained in all eddies of that wave number magnitude.

The eulerian spatial correlation function, in terms of the fluctuating velocities, is defined by (homogeneous turbulence)

$$R'_{ij}(\mathbf{r}) \equiv \overline{u'_i(\mathbf{x},t) u'_j(\mathbf{x} + \mathbf{r}, t)} \qquad \text{(A.15)}$$

The Fourier transform of $R'_{ij}(\mathbf{r})$ is given by

$$\phi_{ij}(\boldsymbol{\varkappa}) = \frac{1}{(2\pi)^3} \int_{-\infty}^{\infty} \int_{-\infty}^{\infty} \int_{-\infty}^{\infty} R'_{ij}(\mathbf{r}) e^{-i\boldsymbol{\kappa} \cdot \mathbf{r}} \, d\mathbf{r} \qquad \text{(A.16)}$$

where, by convention, we use the $(2\pi)^3$ factor in the transform partner of the pair. Thus,

$$R'_{ij}(\mathbf{r}) = \int_{-\infty}^{\infty} \int_{-\infty}^{\infty} \int_{-\infty}^{\infty} \phi_{ij}(\boldsymbol{\varkappa}) e^{i\boldsymbol{\kappa} \cdot \mathbf{r}} \, d\boldsymbol{\varkappa} \qquad \text{(A.17)}$$

The turbulent kinetic energy (of the fluctuations) is given by

$$\tfrac{1}{2}\overline{u'_\beta u'_\beta} = \tfrac{1}{2}(\overline{u'_1 u'_1} + \overline{u'_2 u'_2} + \overline{u'_3 u'_3})$$

From its definition (A.15), $R'_{11}(0) + R'_{22}(0) + R'_{33}(0)$ equals twice the turbulent kinetic energy. But

$$R'_{ii}(0) = \int_{-\infty}^{\infty} \int_{-\infty}^{\infty} \int_{-\infty}^{\infty} \phi_{ii}(\boldsymbol{\kappa})\, d\boldsymbol{\kappa} \qquad \text{(A.18)}$$

Thus, the kinetic energy contained by all wave numbers with magnitude κ, $E(\kappa)$, is obtained by integrating (A.18) over a sphere in κ space at a radius of κ:

$$E(\kappa) = \frac{1}{2} \sum_{i=1}^{3} \int_{|\boldsymbol{\kappa}|=\kappa} \phi_{ii}(\boldsymbol{\kappa})\, dA(\boldsymbol{\kappa}) \qquad \text{(A.19)}$$

Then, the total kinetic energy of the fluctuations per unit mass is the integral of $E(\kappa)$ over all κ from 0 to ∞. Therefore,

$$\begin{aligned} \int_0^{\infty} E(\kappa)\, d\kappa &= \frac{1}{2} \int_0^{\infty} \left[\sum_{i=1}^{3} \int_{|\boldsymbol{\kappa}|=\kappa} \phi_{ii}(\boldsymbol{\kappa})\, dA(\boldsymbol{\kappa}) \right] d\kappa \\ &= \frac{1}{2} \sum_{i=1}^{3} \int_{-\infty}^{\infty} \int_{-\infty}^{\infty} \int_{-\infty}^{\infty} \phi_{ii}(\boldsymbol{\kappa})\, d\boldsymbol{\kappa} \\ &= \frac{1}{2} \sum_{i=1}^{3} R_{ii}(0) \end{aligned} \qquad \text{(A.20)}$$

Thus, $E(\kappa)\, d\kappa$ is the fraction of the turbulent kinetic energy contained in wave numbers κ to $\kappa + d\kappa$.

Since wave numbers are the inverse of lengths, large wave numbers correspond to small eddies and vice versa. However, the idea of an "eddy" may be misleading, and so let us discuss this concept (Tennekes and Lumley, 1972). An eddy of size l can be associated with a wave number of size κ. An eddy of wave number κ can be envisioned as a fluctuation in the field that contains energy near κ. If would be tempting, as Tennekes and Lumley point out, to think of an eddy as a fluctuation contributing a narrow spike to the spectrum $E(\kappa)$ at κ. However, a narrow spike in the spectrum corresponds to slowly damped oscillations (of wavelength $2\pi/\kappa$) in the correlation. But on physical grounds we expect the correlation to go to zero within one or two wavelengths, since eddies lose their identity rather quickly. Thus, the contribution of an eddy of wave number κ to the spectrum should be a fairly broad peak, wide enough to avoid oscillatory behavior of the correlation. For example, we could define an eddy of wave number κ as a fluctuation containing energy between 0.62κ and 1.62κ, which centers the energy around κ on a logarithmic scale. The eddy size l would then be roughly equal to $2\pi/\kappa$. It should be pointed out that an eddy cannot really be associated with a single wavelength, and so a decomposition of a turbulent velocity field into waves is only an artifact of the convenience of using Fourier transforms.

The energy of turbulence is transferred from larger eddies to smaller ones and ultimately dissipated. Most of the energy that crosses a given wave number comes from the next larger eddies and goes to the next smaller eddies. This picture leads to the description of an energy cascade, like a series of waterfalls and pools.

This cascade picture of the turbulent spectrum was put forth by Kolmogorov. In brief, Kolmogorov proposed that (1) the large eddies contain most of the energy and (2) the dissipation of energy occurs in the small eddies. Thus, there is a net transfer of energy from the large to the small eddies. In addition, it is supposed that there is a decoupling between the large and small eddies, so that, regardless of the mechanism of generation of the large eddies, the small-scale eddies are isotropic. If the properties of the small-scale eddies are independent of those of the large scale, these properties can be dependent only on viscosity (since dissipation depends on viscosity) and the rate of energy transfer ε ($\text{cm}^2\ \text{sec}^{-3}$) into the small eddies. This is Kolmogorov's first hypothesis. Between the large-energy-containing eddies and the small dissipation eddies there is a region of eddies containing little energy and dissipating little energy whose properties must depend solely on ε. This region is called the *inertial subrange* and constitutes Kolmogorov's second hypothesis.

In the equilibrium range, time scales are very short and length scales are small so that the details of the large-scale turbulence are unimportant. The parameters important in governing the behavior of $E(\kappa)$ are the total rate of energy transfer (equal to the dissipation rate) ε, the viscosity ν ($\text{cm}^2\ \text{sec}^{-1}$), and κ (cm^{-1}), that is, $E = E(\kappa,\varepsilon,\nu)$. The only group with the dimensions of E is $\nu^{5/4}\varepsilon^{1/4}$, and so

$$\frac{E(\kappa)}{\nu^{5/4}\varepsilon^{1/4}} = \frac{E(\kappa)}{v^2\eta} = f(\kappa\eta) \qquad \text{(A.21)}$$

where $\eta = (\nu^3/\varepsilon)^{1/4}$, called the *Kolmogorov microscale*, and $v = (\nu\varepsilon)^{1/4}$, called the *Kolmogorov velocity*. Since the left-hand side of (A.21) is dimensionless, the right-hand side can only be a function of the dimensionless wave number $\kappa\eta$. Most of the viscous dissipation occurs at length scales of the order of η. The equilibrium range includes the dissipation range.

At the small-wave-number end of the spectrum we expect, on one hand, the viscosity to be unimportant but, on the other hand, the rate of supply of turbulent energy from the mean flow and the energy transfer rate to smaller scales to be important in determining the behavior of $E(\kappa)$. In the inertial subrange, the energy spectrum should depend only on the rate of energy transfer from smaller wave numbers ε and on κ itself, that is, $E = E(\kappa,\varepsilon)$. It can be shown in several different ways that in the inertial subrange the functional form of E should be (Hinze, 1958; Tennekes and Lumely, 1972)

$$E(\kappa,\varepsilon) = \alpha\varepsilon^{2/3}\kappa^{-5/3} \qquad \text{(A.22)}$$

An inertial subrange will exist if there is a sufficient range of eddy sizes which is free from external influences, so that the only source of energy for this range is the transfer of energy from larger eddies. Thus, the eddies to which we apply the similarity hypothesis of Kolmogorov must be small compared with those subject to direct external influences. Let us consider the range of eddy sizes in atmospheric turbulence. Because of the extremely large size of the energy-containing eddies in the atmosphere and the extremely small value of the Kolmogorov microscale $\eta = (\nu^3/\varepsilon)^{1/4}$, which is of the order of the eddy sizes influenced by viscosity, the inertial subrange can be expected to be very large. For example, the size of the energy-containing eddies in the atmosphere might be estimated to be of the order of 100 m.

On the other hand, using estimates of ν of 0.15 cm^2 sec^{-1} and ε of 5 cm^2 sec^{-3} [estimated by Brunt and reported by Batchelor (1950)], we find the microscale to be 0.2 cm. Thus, the range of eddy sizes over which the inertial subrange hypothesis might be expected to hold (in the absence of external influences) is quite extensive.

Up to this point we have considered only spatial spectra, i.e. Fourier transforms of eulerian spatial correlation functions. We can also consider temporal spectra, obtained from either eulerian or lagrangian time correlations. The eulerian time correlation is defined by

$$R'_{ij}(\tau) = \overline{u'_i(\mathbf{x},t)u'_j(\mathbf{x}, t+\tau)} \qquad \text{(A.23)}$$

Corresponding to $R'_{ij}(\tau)$ we can define the eulerian time spectrum

$$\psi_{ij}(\omega) = \frac{1}{2\pi}\int_{-\infty}^{\infty} R'_{ij}(\tau)e^{-i\omega\tau}\,d\tau \qquad \text{(A.24)}$$

so that

$$R'_{ij}(\tau) = \int_{-\infty}^{\infty} \psi_{ij}(\omega)e^{i\omega\tau}\,d\omega \qquad \text{(A.25)}$$

Since time is a scalar, the spectra are one-dimensional, depending on the frequency ω.

All the correlation functions introduced so far in this appendix have been eulerian correlations, taken with respect to fixed points in space. Alternatively, a lagrangian correlation based on the velocity of a marked particle as it meanders through the fluid, can be defined, e.g.

$$\mathscr{R}_{ij}(\tau) = \overline{v_i(\mathbf{x}_0,t)v_j(\mathbf{x}_0, t+\tau)} \qquad \text{(A.26)}$$

where $v_i(\mathbf{x}_0,t)$ is the ith velocity component of a marked particle at time t that was at $\mathbf{x}_0$ at $t=0$, and $v_j(\mathbf{x}_0, t+\tau)$ is the jth velocity component of the same marked particle at time $t+\tau$. Corresponding to $\mathscr{R}_{ij}(\tau)$ we can define the lagrangian time spectrum

$$\chi_{ij}(\omega) = \frac{1}{2\pi}\int_{-\infty}^{\infty} \mathscr{R}_{ij}(\tau)e^{-i\omega\tau}\,d\tau \qquad \text{(A.27)}$$

The lagrangian integral time scale $\mathscr{T}_L$ is defined by

$$\mathscr{T}_L = \max_i \left[\frac{1}{\overline{u'_i(\mathbf{x},t)u'_i(\mathbf{x},t)}}\int_0^{\infty} \mathscr{R}_{ii}(\tau)\,d\tau\right] \qquad \text{(A.28)}$$

Thus, $\mathscr{T}_L$ is a measure of how long it takes a wandering particle to "forget" its former velocity, and the eulerian integral time scale, defined by (A.7), is a measure of the time over which the velocity at a fixed point remains correlated with a prior value. The relation of eulerian and lagrangian correlation data is basically an unsolved problem in turbulence theory. We expect $\mathscr{T}_L > \mathscr{T}_E$, since a wandering particle is expected to maintain a "memory" of a past velocity somewhat longer than does the velocity at a fixed location.

atmosphere. Although conditions of no wind are rare at the heights of actual stacks, the study of plume behavior during this condition serves as a stepping stone to the more common windy case. When a buoyant plume is released from a stack into a calm atmosphere, it rises because of the difference in density between the plume and the ambient air. As the plume rises, it will entrain outside air as a result of turbulence induced by its motion through the air. The entrainment velocity is usually assumed to be proportional to the vertical velocity of the plume. As the plume grows with height due to entrainment of ambient air, the density deficit depends on the temperature of the entrained air. In an unstable atmosphere, the plume is always warmer than the surrounding air and should, in theory, continue to rise. In a stable environment, a hot plume entrains air from below and carries it upward into regions of warmer air. Hence, the plume's buoyancy decays, and if the ambient air is stable throughout, the plume eventually stops rising. The final plume height attained in a stable environment can be expected to depend on the initial flux of buoyancy at the source (to be defined below) and the degree of stability of the atmosphere. The stability of the atmosphere can be characterized by the difference between the actual lapse rate dT/dz and the adiabatic lapse rate Γ. This difference is called the potential lapse rate and is given by (3.15)[1]:

$$\frac{d\theta}{dz} = \frac{dT}{dz} + \Gamma$$

When a wind is present, the entrained air gives the plume horizontal momentum and the plume bends over. Having bent over, the plume moves horizontally at the speed of the entrained air and continues to rise. As entrainment subsides, atmospheric turbulence becomes the dominant mixing mechanism. The height that a buoyant or forced plume attains in a windy stratified environment is a function not only of the initial buoyancy flux and the atmospheric stability but also of the wind speed.

The description of the behavior of a turbulent plume in a stratified environment involves, in principle, simultaneous solution of the equations of continuity, motion, and energy. As in all turbulence problems, such a solution is based on the premise that, despite the seemingly chaotic motion of a plume, its statistically averaged behavior can be predicted by solution of the time-averaged conservation equations. Nevertheless, such a solution can only be carried out numerically (see, for example, Shir, 1970) and is impractical for routine plume rise estimates. Thus, most current formulas for plume rise are based on correlations arising from dimensional analysis and rather idealized plume models.

B.1 BUOYANT PLUMES IN A CALM ATMOSPHERE

The behavior of a plume in a thermally stratified atmosphere is governed by the equations of continuity, motion, and energy for an ideal gas, subject to the Boussinesq approximation (see Sec. 5.1):

[1] Recall that Γ is defined as positive.

$$\frac{\partial u}{\partial x}+\frac{\partial v}{\partial y}+\frac{\partial w}{\partial z}=0 \tag{B.1}$$

$$\frac{Du}{Dt}=-\frac{1}{\rho_0}\frac{\partial p}{\partial x}+\nu\left(\frac{\partial^2 u}{\partial x^2}+\frac{\partial^2 u}{\partial y^2}+\frac{\partial^2 u}{\partial z^2}\right)$$

$$\frac{Dv}{Dt}=-\frac{1}{\rho_0}\frac{\partial p}{\partial y}+\nu\left(\frac{\partial^2 v}{\partial x^2}+\frac{\partial^2 v}{\partial y^2}+\frac{\partial^2 v}{\partial z^2}\right) \tag{B.2}$$

$$\frac{Dw}{Dt}=-\frac{1}{\rho_0}\frac{\partial p}{\partial z}+\nu\left(\frac{\partial^2 w}{\partial x^2}+\frac{\partial^2 w}{\partial y^2}+\frac{\partial^2 w}{\partial z^2}\right)+\frac{T-T_0}{T_0}g$$

$$\frac{DT}{Dt}=\kappa\left(\frac{\partial^2 T}{\partial x^2}+\frac{\partial^2 T}{\partial y^2}+\frac{\partial^2 T}{\partial z^2}\right) \tag{B.3}$$

where z is the vertical coordinate directed upward, w is the vertical velocity, ρ_0 and T_0 are the (reference) density and temperature at some given point in the fluid, $\nu=\mu/\rho_0$, and $\kappa=k/\rho_0 C_p$. As we noted, (B.1) to (B.3) can be solved, in principle, subject to appropriate boundary conditions, to predict plume behavior. However, other approaches of less complexity have been used to analyze plume rise, and we shall concentrate on these other approaches here.

B.1.1 Dimensional Analysis (Batchelor, 1954)

B.1.1.1 Laminar plume in a neutrally stratified environment Laminar flow rarely occurs in the atmosphere; however, we shall consider a laminar plume initially to illustrate the method of dimensional analysis. An example of a laminar plume is the smoke rising from a cigarette. The plume widens as it rises due only to molecular diffusion of heat and momentum. (A few centimeters above the source, instability sets in, and turbulent motion develops.)

When there is no characteristic length in one or more of the coordinate directions of a problem (such as in this case where the flow develops from a point source), it is prudent to look for a *similarity solution* of the governing equations. Such a solution assumes:

1 Similar distributions for the velocity and temperature at different distances from the source; i.e. the forms of the radial variations of w and $g(T-T_0)/T_0$ are the same for all z (e.g. gaussian).

2 The magnitudes of the velocity and temperature distributions vary as some power of z; that is the magnitudes of w and $g(T-T_0)/T_0$ are scaled by some power of the distance from the source.

Let us consider the plume shown in Fig. B.1. For such a plume we seek a similarity solution of the form

$$R \sim z^l$$

$$w \sim z^m \times \text{function}\left(\frac{r}{R}\right)$$

$$\frac{g(T-T_0)}{T_0} \sim z^n \times \text{function}\left(\frac{r}{R}\right) \tag{B.4}$$

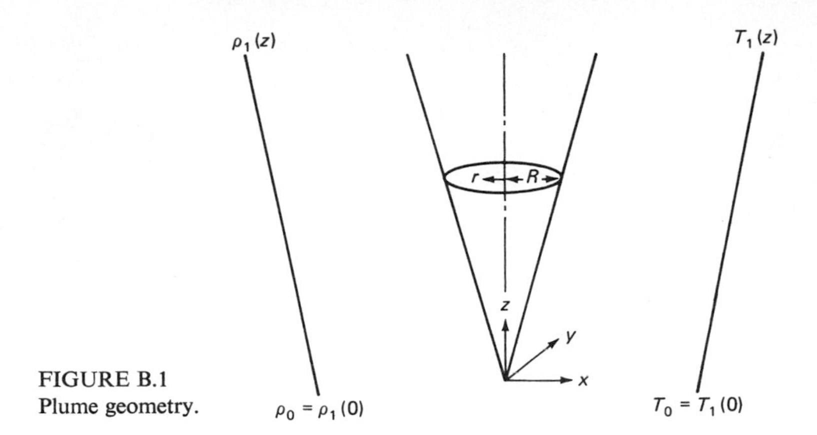

FIGURE B.1
Plume geometry.

Let us determine the exponents l, m, and n.

We define the excess heat due to buoyancy as the quantity $g(T - T_0)/T_0$. The initial flux of excess heat at the source is $F = w_0 R_0 g[(T - T_0)/T_0]$, where w_0 is the initial exit velocity from the source, R_0 is the radius of the source, and T_0 is the ambient temperature at the stack. Since there are no other sources of heat, the flux of excess heat due to buoyancy across all horizontal planes is the same, i.e.

$$\int_0^\infty wg\left(\frac{T - T_0}{T_0}\right) 2\pi r\, dr = F \qquad \text{(B.5)}$$

where $(\rho_0 C_p T_0/g)F$ is the rate at which heat is being released at the source. The dimensions of F are $(\text{length})^4(\text{time})^{-3}$. Substituting (B.4) into (B.5) and setting the exponent of z equal to zero, since (B.5) is valid for all z, yield

$$2l + m + n = 0$$

In the z component of the equation of motion we assume that all terms are of equal magnitude for all z. Thus, they must have the same dependence on z. An order of magnitude analysis of the terms of importance gives

$$w\frac{\partial w}{\partial z} \sim \frac{w^2}{z}$$

$$\nu\left(\frac{\partial^2 w}{\partial x^2} + \frac{\partial^2 w}{\partial y^2}\right) \sim \frac{\nu w}{R^2}$$

$$g\left(\frac{T - T_0}{T_0}\right) \sim g\left(\frac{T - T_0}{T_0}\right)$$

Assuming that the z dependence of each of these three terms is the same, we obtain

$$2m - 1 = n = m - 2l$$

Thus, $l = \frac{1}{2}$, $m = 0$, and $n = -1$.

Since $R \sim z^{1/2}$, we need to multiply $z^{1/2}$ by a group whose dimensions are (length)$^{1/2}$ to make the expression dimensionally consistent. The independent variables of the problem are z, r, F, κ, and ν. Thus, R must depend on z, F, κ, and ν. The choice of variable combinations is somewhat arbitrary, as long as the dimensions are consistent. First, we choose to replace κ with the Prandtl number, $\mathrm{Pr} = \kappa/\nu$. Then, we select $(\nu^3/F)^{1/4}$ as the group having the dimensions of (length)$^{1/2}$. Thus, we write

$$R = \frac{z^{1/2}\nu^{3/4}}{F^{1/4}} \times \text{function (Pr)} \tag{B.6}$$

indicating that the laminar plume is a paraboloid.

Next, from (B.4), and since $m = 0$,

$$w \sim \text{function}\left(\frac{r}{R}\right)$$

Making this expression dimensionally consistent and including (B.6), we have

$$w = \left(\frac{F}{\nu}\right)^{1/2} \times \text{function}\left(\frac{rF^{1/4}}{z^{1/2}\nu^{3/4}}, \mathrm{Pr}\right) \tag{B.7}$$

Likewise, we find that

$$g\left(\frac{T - T_0}{T_0}\right) = \frac{F}{\nu z} \times \text{function}\left(\frac{rF^{1/4}}{z^{1/2}\nu^{3/4}}, \mathrm{Pr}\right) \tag{B.8}$$

B.1.1.2 Turbulent plume in a neutrally stratified environment In a turbulent plume we would expect molecular viscosity and thermal conductivity to have little effect on the mean profiles of velocity and temperature. In such a case, $\bar{w}$ and $\bar{T}$ should depend only on z, r, and F. Dimensional analysis then leads immediately to

$$R = c_1 z \qquad c_1 = \text{const} \tag{B.9}$$

$$\bar{w} = \left(\frac{F}{z}\right)^{1/3} \times \text{function}\left(\frac{r}{R}\right) \tag{B.10}$$

$$g\left(\frac{\bar{T} - T_0}{T_0}\right) = \frac{F^{2/3}}{z^{5/3}} \times \text{function}\left(\frac{r}{R}\right) \tag{B.11}$$

The mass flow across a plane at any z, $\int_0^\infty \rho\bar{w}2\pi r\, dr$, is proportional to $\bar{w}R$, which, by virtue of (B.9) and (B.10), is proportional to $z^{5/3}$. Thus, the mass flow increases with z, for which entrainment of air from the edge of the plume is responsible. From the continuity equation, $\bar{v} \sim \bar{w}R/z \sim z^{-1/3}$. Thus, the mean horizontal velocity $\bar{v}$ is proportional to the mean vertical velocity $\bar{w}$, and the ratio of the two determines the angle of the conical plume. Empirically, the form of the function of r/R is often chosen to be gaussian, that is, $e^{-c_2 r^2/R^2}$.

B.1.1.3 Turbulent plume in a nonneutrally stratified environment Let $T_1(z)$ be the temperature of the ambient atmosphere, and let $T_0 = T_1(0)$. We shall seek similarity solutions of the form (B.4) for R, $\bar{w}$, and $g(\bar{T}-T_1)/T_0$. We assume that the horizontal inflow velocity at the edge of the plume due to entrainment is proportional to the mean vertical velocity at the same level. If

$$\text{Inflow velocity} \sim \frac{1}{2\pi R}\frac{d}{dz}\int_0^\infty \bar{w}2\pi r\,dr \sim \bar{w} \tag{B.12}$$

then $l=1$, indicating that the plume is again conical. Since the increase of flux of vertical momentum with z is due to the buoyancy force, i.e.

$$\frac{d}{dz}\int_0^\infty \overline{w^2}2\pi r\,dr = \int_0^\infty g\left(\overline{\frac{T-T_1}{T_0}}\right)2\pi r\,dr$$

then $2m+1=n+2$. The increase of heat with z is due to the entrainment, i.e.

$$\frac{d}{dz}\int_0^\infty \bar{w}\bar{T}2\pi r\,dr = T_1\frac{d}{dz}\int_0^\infty \bar{w}2\pi r\,dr$$

which can be rewritten

$$\frac{d}{dz}\int_0^\infty \bar{w}g\left(\overline{\frac{T-T_1}{T_0}}\right)2\pi r\,dr = -\frac{g}{T_0}\frac{dT_1}{dz}\int_0^\infty \bar{w}2\pi r\,dr \tag{B.13}$$

Relation (B.13) shows that the environment must have the lapse rate:

$$\frac{g}{T_0}\frac{dT_1}{dz} = -c_3 z^p$$

where c_3 is a positive constant. This indicates that the dimensional-analysis, similarity solution approach is restricted to unstably stratified environments. Relation (B.13) also indicates that $m+n+1=p+m+2$; hence, $m=(p+2)/2$ and $n=p+1$, and furthermore that

$$R = c_4 z \qquad c_4 = \text{const}$$

$$\bar{w} = (c_3 z^p)^{1/2} z \times \text{function}\left(\frac{r}{R}\right)$$

$$g\left(\overline{\frac{T-T_1}{T_0}}\right) = (c_3 z^p) z \times \text{function}\left(\frac{r}{R}\right)$$

The vertical flux of heat, $\int_0^\infty \bar{w}\bar{T}2\pi r\,dr \sim z^{(3p+8)/2}$, and the value at $z=0$ is zero for $p>-\frac{8}{3}$, finite for $p=-\frac{8}{3}$, and infinite for $p<-\frac{8}{3}$. For $p=-\frac{8}{3}$, $c_3=0$ for the above relations to be satisfied, and the analysis reduces to that for a neutral atmosphere. Since the case of a source of infinite strength is not practical, we are left with $p>-\frac{8}{3}$ and a zero value of the heat flux at $z=0$. As intuitively expected, the development of a turbulent heat plume in an unstable environment does not need a heat source; i.e. any small disturbance allows the plume to develop.

In summary, dimensional analysis makes it possible to predict the functional dependence of the growth of both laminar and turbulent plumes in a calm, neutral, or unstable atmosphere.

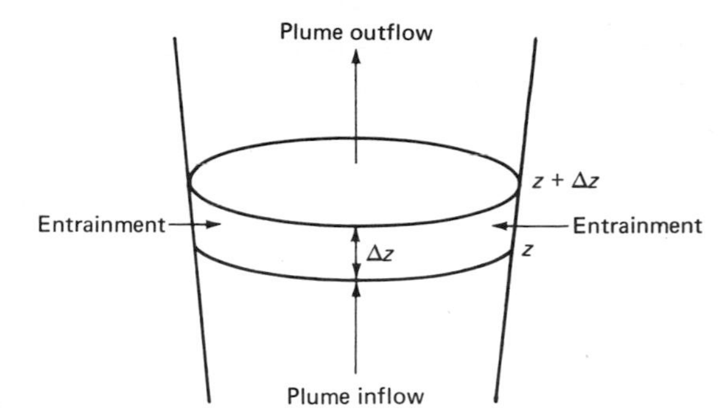

FIGURE B.2
Small element of plume.

Of most importance from a practical standpoint, however, is plume behavior in a stably stratified atmosphere, since in this case the plume will attain only a finite height above the source. Thus, we now turn to analyses based on physical models of the plume, the most important of which is that of Morton et al. (1956).

B.1.2 Morton, Taylor, and Turner's Approach (1956)

The three main assumptions in the quantitative analysis of plume rise by Morton et al. (1956) are:

1 The entrainment rate at the edge of the plume is proportional to the vertical velocity of the plume at the height.

2 The profiles of mean vertical velocity and buoyancy force in horizontal sections are of similar form at all heights.

3 The largest local variations of density are small in comparison with the reference density, the density of the ambient fluid at the level of the source.

Let us consider a cross-sectional element of the plume as shown in Fig. B.2. Assuming that the vertical velocity and temperature are uniform across the plume at any height, we can perform a balance on plume volume as follows:

$$\pi R^2 \overline{w}|_{z+\Delta z} - \pi R^2 \overline{w}|_z = 2\pi R_{av}\, \Delta z \alpha \overline{w}_{av}$$

where R_{av} and $\overline{w}_{av}$ are the average values of the radius and the vertical velocity over the element of thickness Δz. The term α is the *entrainment coefficient*, such that the inflow velocity at the edge of the plume at any z is $\alpha\overline{w}(z)$. Dividing by Δz and taking the limit as $\Delta z \to 0$ gives

$$\frac{d(\pi R^2 \overline{w})}{dz} = 2\pi R \alpha \overline{w} \tag{B.14}$$

A momentum balance on the element shown in Fig. B.2 gives

$$\pi R^2 \overline{w}^2 \rho|_{z+\Delta z} - \pi R^2 \overline{w}^2 \rho|_z = \pi R_{av}^2\, \Delta z g(\rho_{1,av} - \rho_{av})$$

where ρ and ρ_1 are the densities inside and outside the plume, respectively. Dividing by Δz and taking the limit as $\Delta z \to 0$ gives

$$\frac{d(\pi R^2 \bar{w}^2 \rho)}{dz} = \pi R^2 g(\rho_1 - \rho) \tag{B.15}$$

A balance on the density deficiency in the plume element can be expressed as

$$\pi R^2 \bar{w}(\rho_0 - \rho)|_{z+\Delta z} - \pi R^2 \bar{w}(\rho_0 - \rho)|_z = 2\pi R_{av}\, \Delta z \alpha \bar{w}_{av}(\rho_{0,av} - \rho_{1,av})$$

where $\rho_0 = \rho_1\ (z = 0)$. Dividing by Δz and taking the limit as $\Delta z \to 0$ give

$$\frac{d[\pi R^2 \bar{w}(\rho_0 - \rho)]}{dz} = 2\pi R \alpha \bar{w}(\rho_0 - \rho_1) \tag{B.16}$$

Equations (B.14) and (B.16) can be combined to yield

$$\frac{d}{dz}[\pi R^2 \bar{w}(\rho_0 - \rho)] = (\rho_0 - \rho_1)\frac{d}{dz}(\pi R^2 \bar{w})$$

$$= \frac{d}{dz}[\pi R^2 \bar{w}(\rho_0 - \rho_1)] - \pi R^2 \bar{w}\frac{d}{dz}(\rho_0 - \rho_1)$$

and hence

$$\frac{d}{dz}[\pi R^2 \bar{w}(\rho_1 - \rho)] = \pi R^2 \bar{w}\frac{d\rho_1}{dz}$$

Since the density variations are assumed to be small with respect to ρ_0, the density ρ on the left-hand side of (B.15) can be replaced by ρ_0, yielding the conservation relations

$$\frac{d}{dz}(R^2 \bar{w}) = 2\alpha R \bar{w} \tag{B.17}$$

$$\frac{d}{dz}(R^2 \bar{w}^2) = R^2 g\left(\frac{\rho_1 - \rho}{\rho_0}\right) \tag{B.18}$$

$$\frac{d}{dz}\left(R^2 \bar{w} g \frac{\rho_1 - \rho}{\rho_0}\right) = R^2 \bar{w}\frac{g}{\rho_0}\left(\frac{d\rho_1}{dz}\right) \tag{B.19}$$

For boundary conditions, the radius and momentum flux at the source are zero, and the buoyancy flux is known. Equation (B.19) yields, under neutral conditions,

$$R^2 \bar{w} g\left(\frac{\rho_1 - \rho}{\rho_0}\right) = Q = \text{const}$$

and (B.17) and (B.18) can be solved to yield

$$R^2 \bar{w} = \frac{6\alpha}{5}\left(\frac{9}{10}\alpha Q\right)^{1/3} z^{5/3}$$

$$R\bar{w} = \left(\frac{9}{10}\alpha Q\right)^{1/3} z^{2/3}$$

Therefore, under neutrally stable conditions,

$$R = \frac{6\alpha}{5} z \tag{B.20}$$

$$\bar{w} = \frac{5}{6\alpha}\left(\frac{9}{10}\alpha Q\right)^{1/3} z^{-1/3} \tag{B.21}$$

$$g\left(\frac{\rho_1 - \rho}{\rho_0}\right) = \frac{5Q}{6\alpha}\left(\frac{9}{10}\alpha Q\right)^{-1/3} z^{-5/3} \tag{B.22}$$

Comparing (B.20) to (B.22) with (B.9) to (B.11), we now have explicit relationships for the constants and functions in the latter three equations.

In the above analysis we assumed that the velocity and temperature are constant across the plume (so-called top-hat profiles). Since it has been observed that velocity and temperature profiles, if anything, are closer to gaussian distributions than uniform distributions, it seems reasonable to generalize the above analysis by assuming that

$$\bar{w}(z,r) = \bar{w}(z)e^{-r^2/b^2} \tag{B.23}$$

$$g\left(\frac{\rho_1 - \rho}{\rho_0}\right) = g\phi(z)e^{-r^2/b^2} \tag{B.24}$$

Note that b is proportional to but not necessarily equal to the plume radius R. The entrainment constant α is chosen such that the rate of volume entrainment at any z is $2\pi b\alpha\bar{w}$. After writing the equations for the conservation of volume, momentum, and density deficiency, for an annular plume element between r and $r + dr$, and z and $z + dz$, substituting (B.23) and (B.24) into these equations, and then integrating in the horizontal plane from $r = 0$ to $r = \infty$, we arrive at the governing equations

$$\frac{d}{dz}(\pi b^2 \bar{w}) = 2\pi b\alpha\bar{w} \tag{B.25}$$

$$\frac{d}{dz}\left(\frac{1}{2}\pi b^2 \bar{w}^2 \rho\right) = \pi b^2 g(\rho_1 - \rho) \tag{B.26}$$

$$\frac{d}{dz}\left[\frac{1}{2}\pi b^2 \bar{w}(\rho_1 - \rho)\right] = \pi b^2 \bar{w}\frac{d\rho_1}{dz} \tag{B.27}$$

As above, (B.25) to (B.27) can be rearranged, retaining the density only to the necessary order of accuracy, to yield the reduced set of governing equations:

$$\frac{d}{dz}(b^2\bar{w}) = 2\alpha b\bar{w} \tag{B.28}$$

$$\frac{d}{dz}(b^2\bar{w}^2) = 2b^2 g\left(\frac{\rho_1 - \rho}{\rho_0}\right) \tag{B.29}$$

$$\frac{d}{dz}\left(b^2\bar{w}g\frac{\rho_1 - \rho}{\rho_0}\right) = 2b^2\bar{w}\frac{g}{\rho_0}\left(\frac{d\rho_1}{dz}\right) \tag{B.30}$$

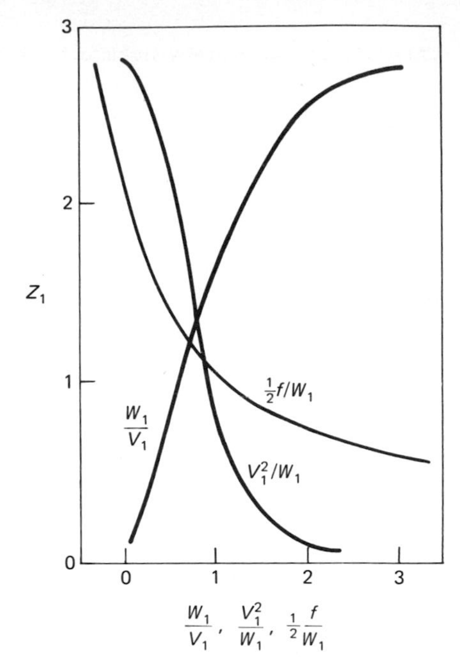

FIGURE B.3
Variation with height of the horizontal extent ($\sim W_1/V_1$), the vertical velocity ($\sim V_1/W_1$), and the buoyancy ($\sim f/W_1$), calculated for a turbulent buoyant plume from a maintained point source in a calm, stable environment of constant density gradient. (*Source: Morton et al.* 1956.)

Defining new variables $V = b\bar{w}$, $W = b^2\bar{w}$, $F^* = b^2\bar{w}g[(\rho_1 - \rho)/\rho_0]$, and $G = -g/\rho_0(d\rho_1/dz)$ [= positive constant for a constant negative (stable) density gradient], the equations take on the form

$$\frac{dW}{dz} = 2\alpha V \tag{B.31}$$

$$\frac{dV^4}{dz} = 4F^*W \tag{B.32}$$

$$\frac{dF^*}{dz} = -2WG \tag{B.33}$$

The boundary conditions associated with (B.31) to (B.33) are

$$W(0) = 0 \qquad V(0) = 0 \qquad F^*(0) = F_0^* = \frac{2}{\pi}F_0 \tag{B.34}$$

The two physical parameters governing the problem are F_0 and G.

The following transformations, which can be derived from dimensional arguments,

$$z = 2^{-7/8}\pi^{-1/4}\alpha^{-1/2}F_0^{1/4}G^{-3/8}z_1$$

$$V = 2^{3/4}\pi^{-1/2}F_0^{1/2}G^{-1/4}V_1$$

$$W = 2^{7/8}\pi^{-3/4}\alpha^{1/2}F_0^{3/4}G^{-5/8}W_1$$

$$F^* = 2\pi^{-1}F_0 f$$

are chosen to reduce (B.31) to (B.34) to their simplest nondimensional form given by

$$\frac{dW_1}{dz_1} = V_1 \qquad \frac{dV_1^4}{dz_1} = fW_1 \qquad \frac{df}{dz_1} = -W_1 \tag{B.35}$$

$$W_1(0) = 0 \qquad V_1(0) = 0 \qquad f(0) = 1 \tag{B.36}$$

Equations (B.35) together with boundary conditions (B.36) can be integrated numerically, producing the solution shown in Fig. B.3. Note that the buoyancy first vanishes at $z_1 = 2.125$, and then the vertical velocity vanishes at $z_1 = 2.8$. The bulk of the fluid in the plume will rise to $z_1 = 2.8$, and then as it spreads out sideways (i.e. "mushrooms"), it will fall back again with the mean height of the sideways flow between $z_1 = 2.125$ and $z_1 = 2.8$. Note that the large values of velocity and buoyancy near the source predicted by the above equations are due to the mathematical singularity at the origin.

B.2 FORCED PLUMES IN A CALM ATMOSPHERE

Morton (1959) extended the earlier analysis of Morton et al. (1956) to forced plumes, those in which the initial fluxes of both buoyancy and momentum are important. The strength of the source is defined by specifying the rates of discharge of buoyancy ($\pi \rho F_0$), momentum ($\pi \rho V_0^2$), and mass ($\pi \rho W_0$).

If there is a finite flux of buoyancy, $F_0 \sim b^2 \bar{w} g(\rho_1 - \rho)/\rho_0 |_{z=0}$, and of momentum, $V_0^2 \sim b^2 \bar{w}^2 |_{z=0}$, from a point source, then $\bar{w} \sim 1/b$ and $g[(\rho_1 - \rho)/\rho_0] \sim 1/b$, and hence the corresponding mass flux, $W_0 \sim b^2 \bar{w} \sim b$, must be zero. Therefore, to consider $W_0 \neq 0$, we must consider sources of finite area. However, as pointed out by Morton, such sources can be considered as virtual point sources with $W_0 = 0$ and $F_0 \neq 0$, $V_0 \neq 0$, situated at a different level than $z = 0$. Let us consider the case in which the mean vertical velocity is constant across a section of width $2b(z)$ and zero outside it, and the mean buoyancy is constant across a section of width $2\lambda b(z)$ and zero outside it. If the case has a constant density gradient, there are three relevant parameters, F_0, V_0, and $G = -(g/\rho_0(d\rho_1/dz))$. With the transformations $V = b\bar{w}$, $W = b^2\bar{w}$, $F = b^2 \bar{w} g[(\rho_1 - \rho)/\rho_0]$, the conservation equations (B.31) to (B.33) can be written

$$\frac{dW}{dz} = 2\alpha V \tag{B.37}$$

$$\frac{dV^4}{dz} = 2\lambda^2 FW \tag{B.38}$$

$$\frac{dF}{dz} = -GW \tag{B.39}$$

With the transformations $F = |F_0| f$, $V = 2^{1/4}\lambda^{1/2}|F_0|^{1/2}G^{-1/4}V_1$, $W = 2^{5/8}\alpha^{1/2}\lambda^{1/4}|F_0|^{3/4} \times G^{-5/8}W_1$, and $z = 2^{-5/8}\alpha^{-1/2}\lambda^{-1/4}|F_0|^{1/4}G^{-3/8}z_1$, (B.37) to (B.39) reduce to

$$\frac{dW_1}{dz_1} = V_1 \qquad \text{(B.40)}$$

$$\frac{dV_1^4}{dz_1} = fW_1 \qquad \text{(B.41)}$$

$$\frac{df}{dz_1} = -W_1 \qquad \text{(B.42)}$$

with the corresponding boundary conditions at $z_1 = 0$ of

$$W_1 = 0 \qquad V_1 = 2^{-1/4}\lambda^{-1/2}|F_0|^{-1/2}G^{1/4}V_0 \qquad f = \operatorname{sgn} F_0 \qquad \text{(B.43)}$$

where

$$\operatorname{sgn} F_0 = \begin{cases} +1 & \text{for } F_0 > 0 \\ -1 & \text{for } F_0 < 0 \end{cases}$$

B.3 PLUME RISE IN A WINDY ATMOSPHERE

Usually the environment into which a plume is released is windy rather than calm. In the case of a plume released into a calm atmosphere, the growth of the plume is due solely to entrainment of ambient air as a result of turbulence induced by the plume's motion. When a wind is present there is, in addition to the self-generated turbulence due to the relative motion of plume and air, the turbulence naturally present in the wind. In the early stages of plume growth the self-generated turbulence dominates the entrainment process, whereas far downwind of the stack the natural turbulence of the atmosphere is responsible for plume dispersion. Csanady (1965) noted that the latter stages of plume growth can really be subdivided into two regions: (1) the first in which the important diffusing eddies belong to the inertial subrange (see Appendix A) and (2) the second in which the plume diameter is of the order of the size of the energy containing eddies or larger. Although it is certainly an idealization to speak of three such distinct phases, as Slawson and Csanady (1967) point out, such a classification is useful in constructing an analytical approach to predicting plume behavior.

There are basically two avenues of approach to describing the growth and dispersion of a turbulent plume in a stratified atmosphere: (1) the turbulent entrainment model of Morton et al. (1956), presented in Subsec. B.1.2 (see also Morton, 1971), and (2) the use of eddy viscosity and thermal conductivity in the time-averaged equations of motion and energy for the plume. Slawson and Csanady (1967) considered each approach for plume behavior in a windy atmosphere, showing how the same results can be obtained through each.

Following Slawson and Csanady, we assume that the plume axis is nearly horizontal. Let U be the uniform wind speed in the direction x, and let $\Delta\rho = \rho_1 - \rho$ be the density difference between the outside and inside of the plume. Then the analogs of (B.14) to (B.17) for a neutral atmosphere are

$$U\frac{dR^2}{dx} = 2\alpha\bar{w}R \tag{B.44}$$

$$U\frac{d(R^2\bar{w})}{dx} = gR^2\frac{\Delta\rho}{\rho} = \frac{F}{U} \tag{B.45}$$

$$gR^2U\left(\frac{\Delta\rho}{\rho}\right) = F = \text{const} \tag{B.46}$$

where $\Delta\rho$ and $\bar{w}$ are considered constant across the plume at any x and zero outside it. Equation (B.44) holds only for the initial phase since the influx velocity is assumed to be a result solely of the plume velocity $\bar{w}$. Since we are interested in the plume's behavior far from the source, let us assume that the initial plume radius and velocity are negligible. Hence, integrating (B.45) yields

$$\bar{w} = \frac{Fx}{U^2R^2} \tag{B.47}$$

Writing $dR^2/dx = 2R(dR/dx)$, substituting (B.47) into (B.44), and integrating give

$$R = \left(\frac{3}{2}\alpha\frac{F}{U^3}\right)^{1/3} x^{2/3} \tag{B.48}$$

Noting that

$$\frac{dz}{dx} = \frac{\bar{w}}{U} \tag{B.49}$$

using (B.47) to (B.49), and integrating produce

$$\frac{z}{l} = \left(\frac{3}{2\alpha^2}\right)^{1/3}\left(\frac{x}{l}\right)^{2/3} \tag{B.50}$$

where $l = F/U^3$ is a length scale characteristic of buoyant movements.

We obtained (B.50) by neglecting the size of the source and the initial efflux velocity of the plume, and so we expect (B.50) to apply beyond a certain distance from the source (but still in the initial phase of plume growth). Thus, we may expect that when $x \gg R_0$ and

$$g\left(\frac{\Delta\rho_0}{\rho}\right)\frac{x}{U} \gg w_0$$

or, equivalently, that $x \gg w_0 U\rho/(g\,\Delta\rho_0)$, (B.50) is valid. Slawson and Csanady pointed out that normally these inequalities are satisfied for power station chimneys at distances of $x =$ 100 m or larger.

In the intermediate phase, the rate of plume growth is determined by the atmospheric eddies of the inertial subrange and is known to be given by

$$\frac{1}{2}\frac{dR^2}{dt} = a_1\varepsilon^{1/3}R^{4/3} \tag{B.51}$$

where ε is the turbulent energy dissipation per unit mass (constant) and a_1 is a constant. For a plume traveling with uniform speed U, (B.51) becomes

$$U\frac{dR^2}{dx} = 2a_1\varepsilon^{1/3}R^{4/3} \qquad \text{(B.52)}$$

Equation (B.52) now replaces (B.44) with the influx velocity $\alpha\bar{w}$ of the initial phase replaced by $a_1\varepsilon^{1/3}R^{1/3}$, which is proportional to the velocity scale of eddies of size R in the inertial subrange. Integrating (B.45) and (B.52) yields

$$\bar{w} = U\frac{dz}{dx} = \frac{R_1{}^2\bar{w}_1 + F(x-x_1)/U^2}{[R_1{}^{2/3} + 2a_1\varepsilon^{1/3}(x-x_1)/3U]^3} \qquad \text{(B.53)}$$

where R_1 and $\bar{w}_1$ are the radius and vertical velocity of the plume at the beginning of the intermediate phase. At large $x - x_1$, if the intermediate phase still holds,

$$\frac{dz}{dx} = \text{const} \times (x - x_1)^{-2}$$

In the final phase, the large-energy-containing eddies dominate the mixing and the plume growth is given by

$$\frac{1}{2}\frac{dR^2}{dt} = a_2 v_R L_R \qquad \text{(B.54)}$$

where v_R and L_R are the root-mean-square turbulent-velocity and (diffusion) length scales (constants), respectively. Integration of (B.45) and (B.54) gives

$$\bar{w} = U\frac{dz}{dx} = \frac{R_2{}^2\bar{w}_2 + f(x-x_2)/U^2}{R_2{}^2 + 2a_2 v_R L_R(x-x_2)/U} \qquad \text{(B.55)}$$

where R_2 and $\bar{w}_2$ are the radius and vertical velocity of the plume at the beginning of the final phase. At large $x - x_2$,

$$\frac{dz}{dx} = \frac{F}{2a_2 v_R L_R U^2} = \frac{1}{2a_2}\frac{l}{L_R}\frac{U}{v_R} = \text{const} \qquad \text{(B.56)}$$

In summary, the theory of Slawson and Csanady predicts that:

1 In the initial phase of plume behavior, where the self-generated turbulence is mainly responsible for plume growth, the height of the plume center line, z, varies as $x^{2/3}$.
2 In the intermediate phase, dominated by eddies in the inertial subrange, the plume has a tendency to level off.
3 In the final phase, the height of the plume is predicted to vary linearly with distance x.

Data reported by the authors confirmed a plume rise given by

$$\frac{z}{l} = 2.3\left(\frac{x}{l}\right)^{2/3}$$

in the initial phase which ended approximately at $x/l = 1200$, $z/l = 280$. It was also found that the intermediate phase was very short and that the asymptotic slope of the final phase was almost the same as the slope of the curve for the initial phase at the transition point.

It is interesting to note that the dimensionless plume height z/l is similar to the dimensionless height z/L used in describing the stability properties of the atmospheric surface layer. The Monin-Obukhov length L (see Chap. 5) relates the production rate of turbulent energy by buoyancy to the production rate by mechanical forces and is proportional to the surface heat flux divided by the cube of the mean wind speed.

In the later phases of growth, the plume behavior depends on the intensity and scale of the atmospheric turbulence. In this phase the plume height varies as

$$\frac{z}{l} \sim \frac{x}{l}$$

with a proportionality constant depending on the turbulence parameters. The distance at which the plume height begins to deviate from the two-thirds law (B.50) and where the region of linear rise begins can be used as the definition of the plume rise Δh. At this point, $x = \lambda$, $z = \Delta h$, and

$$\Delta h = \lambda l = \lambda \left(\frac{F}{U^3}\right) \tag{B.57}$$

At heights above 100 m the variation of wind speed with altitude is not strong. Thus, U can be taken as the wind speed at the top of the stack.

The development presented above is concerned with conditions of neutral stability, although under unstable conditions plume growth can be described in a manner similar to that above. However, under unstable conditions, the parameter governing the turbulence level is G/U^2, where

$$G = \frac{g}{\rho_0} \frac{d\rho_1}{dz} = \frac{g}{\theta_a} \frac{d\theta}{dz}$$

As $d\theta/dz$ approaches zero and U increases, the plume rise approaches that of the neutral case. When $d\theta/dz < 0$, the plume may be dispersed more quickly than in the neutral case because of the presence of convective as well as mechanical turbulence in the wind.

Under stable conditions the rise of the plume is retarded both by the entrainment of outside air by the buoyant motion and by the influx of stable air by wind-generated turbulence. The plume will eventually bend over and reach an equilibrium height. Briggs (1965) considered this case and found that the rise of a plume in windy, stable conditions can be given by

$$\Delta h = B \left(\frac{F}{UG}\right)^{1/3} \tag{B.58}$$

where B is a constant containing an entrainment coefficient.

In Secs. B.1 to B.3 we have presented a few of the most widely known approaches to the prediction of plumes. This treatment does not represent, by any means, an exhaustive survey of plume rise theories. Perhaps the most comprehensive source in this regard is the monograph

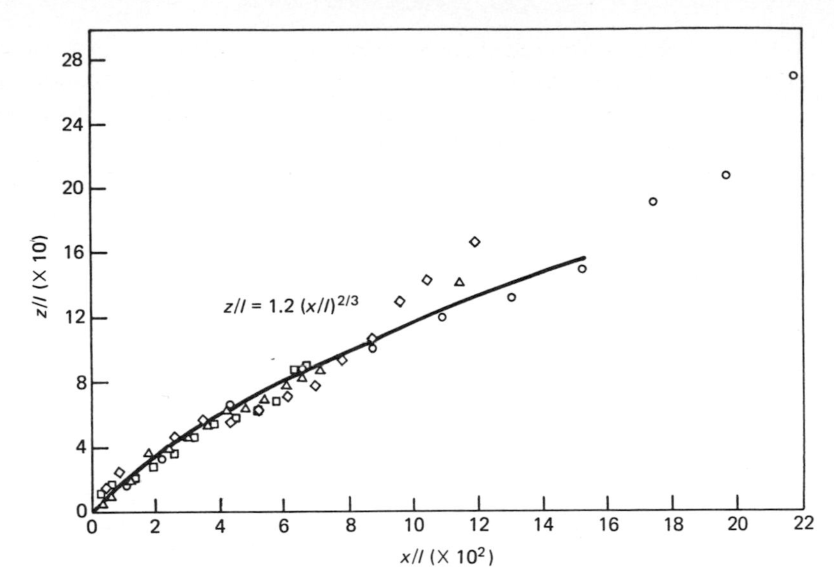

FIGURE B.4
Plume rise during neutrally stratified, windy conditions. A comparison of the two-thirds law of plume growth with the data of Carpenter et al. (1968). (*Source: Frizzola*, 1969.)

of Briggs (1969). In addition, the reader may wish to consult the recent review by Morton (1971). The intent here has been to illustrate how the problem of plume rise has been approached and to obtain some formulas of potential practical value. The final question is: How well do these (and other) formulas perform in practice? The final section is devoted to this issue.

B.4 PLUME RISE FORMULAS COMPARED WITH OBSERVATIONS

There have been published a large number of studies in which various plume rise formulas have been compared with observed plume rise data. Briggs (1969) presented an extensive comparison of observed plume behavior and that calculated from various empirical and theoretical formulas. In addition, comparative studies have been carried out by Frizzola (1969), Hoult et al. (1969), Bringfelt (1969), Thomas et al. (1970), and Shwartz and Tulin (1972), among others. We shall summarize here the results of the studies of Frizzola (1969) and Thomas et al. (1970).

In order to apply the formulas developed earlier, we must know the exit velocity and temperature of the effluent, the radius and height of the stack, and the variation of wind speed and temperature with height in the atmosphere. The required source conditions are usually specified, whereas those associated with the atmosphere are often not known. Most obser-

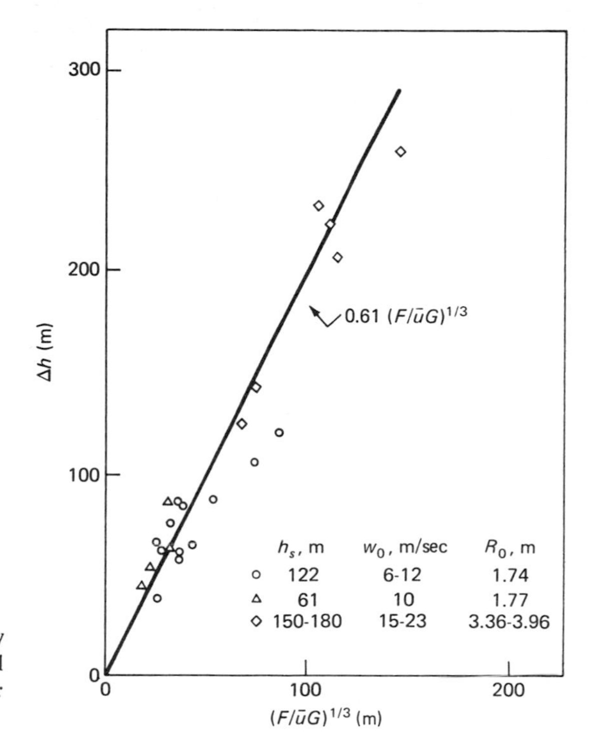

FIGURE B.5
Plume rise during stably stratified, windy conditions. A comparison of several experiments with (B.58). (*Source:* *Frizzola*, 1969.)

vations of plume rise are obtained from photographs, and thus measurements of the mean wind speed and temperature at stack level, as well as of the gradients in these variables, must be obtained by elevated sensors. Measurements of turbulent intensity and scale are almost never available at the height corresponding to the release, so that the terms in plume rise formulas containing these quantities must be treated as empirical parameters chosen to fit the given plume rise data.

Frizzola analyzed plume rise data reported by Carpenter (1968). For neutral stability, cases were selected in which the wind speed was in excess of 7 m sec^{-1} at stack height (122 and 184 m) and the potential temperature gradient near zero. Figure B.4 shows the dimensionless centerline plume height (z/l) versus (x/l). The change of slope from the two-thirds law occurs at z/l of 150, so that the plume rise is given by (B.57) with $\lambda = 150$. A summary of several measurements under stable conditions is presented in Fig. B.5, in which the predicted value using (B.58) and the final height of rise Δh are compared with a value of 0.61 for the constant B.

Thomas et al. considered plume rise data from six power plants with unit ratings from 173 to 704 MW and stack heights from 76.2 to 182.9 m. A number of plume rise formulas, all of the general form $\Delta h = A/U^a$, where A is some function of the kinetic and thermal energy of the plume, were tested. The formulas considered are summarized in Table B.1. Plume

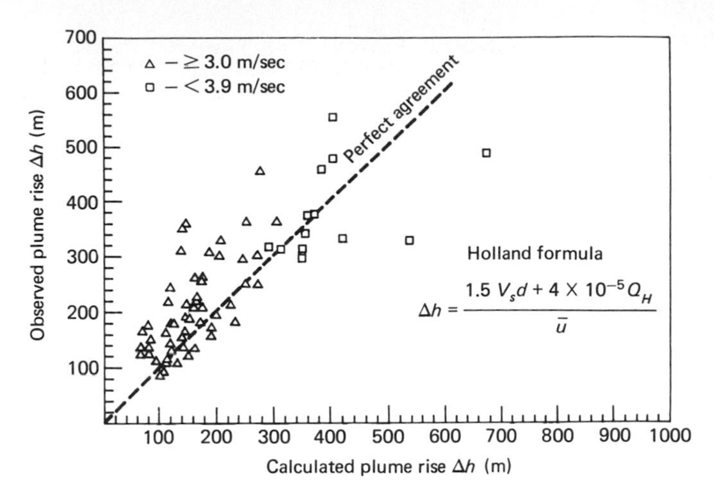

FIGURE B.6
Observed and calculated plume rise with Holland's formula (Table B.1). (*Source: Thomas et al.*, 1970.)

rise was calculated with these relations and plotted against observed values. Figures B.6 and B.7 show the performance of Holland's formula and the CONCAWE formula, both of which showed good agreement with the data.

It is difficult to specify that one particular plume rise formula is superior to all others. Some perform better than others, depending on the situation to which the formulas are applied.

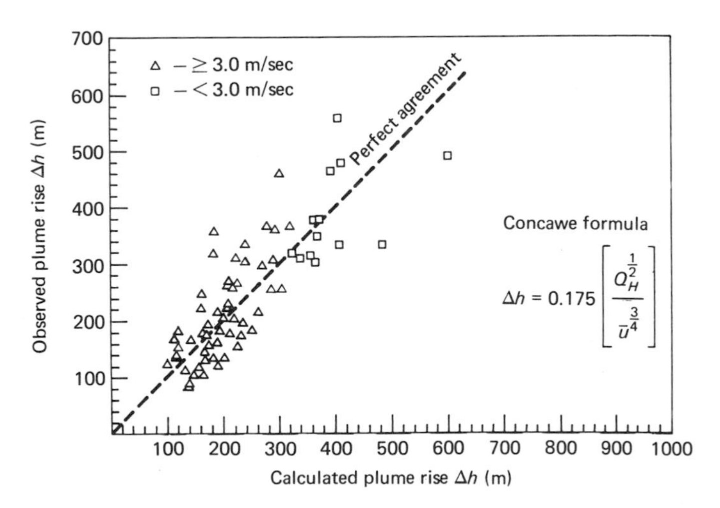

FIGURE B.7
Observed and calculated plume rise with the CONCAWE formulas (Table B.1). (*Source: Thomas et al.*, 1970.)

Table B.1 PLUME RISE FORMULAS TESTED BY CARPENTER ET AL. (1969)

Δh, m	Name
$\dfrac{1.5V_s d + 4 \times 10^{-5} Q_H}{U}$	Holland
$d\left(\dfrac{V_s}{U}\right)^{1.4}\left(1 + \dfrac{\Delta T}{T_s}\right)$	Davidson-Bryant
$250 \dfrac{F}{U^3}$	Slawson and Csanady
$0.175 \dfrac{Q_H^{1/2}}{U^{3/4}}$	CONCAWE
(1) $\dfrac{0.7\alpha}{U}\left(\dfrac{Q_N}{G_N}\right)^{1/4}$	Lucas, Moore, and Spurr
(2) $\alpha \dfrac{Q_N^{1/4}}{U}$	
$\alpha = 475 + 2(h_s - 100)$ m^2 sec^{-1} MW$^{-1/4}$	

$d =$ stack exit diameter, m
$F = gV_s R^2(\Delta T/T)$, m^4/sec^3
$g =$ acceleration due to gravity, m/sec^2
$G_N = \dfrac{108}{U^3} \dfrac{\Delta\theta}{\Delta z}$
$Q_H =$ heat emission, cal/sec
$Q_N =$ heat emission, MW
$\Delta T =$ temperature difference between exit stack gas and ambient air, °K
$U =$ mean horizontal wind speed, m/sec
$V_s =$ stack gas exit velocity, m/sec
$\Delta\theta/\Delta z =$ change of potential temperature with height, °C/100 m

The formulas given in Table B.1 may be used to obtain estimates of plume rise in neutral and stable conditions. For a thorough review and comparison of proposed formulas the reader is urged to consult Briggs (1969).

REFERENCES

BATCHELOR, G. K.: Heat Convection and Buoyancy Effects in Fluids, *Q. J. Roy. Meteorol. Soc.*, **80**: 339 (1954).

BRIGGS, G. A.: A Plume Rise Model Compared with Observations, *J. Air Pollut. Control Assoc.*, **15**: 433 (1965).

BRIGGS, G. A.: "Plume Rise," U.S. Atomic Energy Commission, Washington, D.C., 1969.

BRINGFELT, B.: A Study of Buoyant Chimney Plumes in Neutral and Stable Atmospheres, *Atmos. Environ.*, **3**:609 (1969).

CARPENTER, S. B., J. M. LEAVITT, F. W. THOMAS, J. A. FRIZZOLA, and M. E. SMITH: Full-scale Study of Plume Rise at Large Coal-fired Electric Generating Stations, *J. Air Pollut. Control Assoc.*, **18**: 458 (1968).

CSANADY, G. T.: The Buoyant Motion within a Hot Gas Plume in a Horizontal Wind, *J. Fluid Mech.*, **22**: 225 (1965).

FRIZZOLA, J. A.: The Ascent of Power Plant Plumes During Various Meteorological Conditions, Proceedings of the American Power Conference, 1969.

HOULT, D. P., J. A. FAY, and L. J. FORNEY: A Theory of Plume Rise Compared with Field Observations *J. Air Pollut. Control Assoc.*, **19**: 585 (1969).

MORTON, B. R.: Forced Plumes, *J. Fluid Mech.*, **5**: 151 (1959).

MORTON, B. R.: The Choice of Conservation Equations for Plume Models, *J. Geophys. Res.*, **76**: 7409 (1971).

MORTON, B. R., G. I. TAYLOR, and J. S. TURNER: Turbulent Gravitational Convection from Maintained and Instantaneous Sources, *Proc. Roy. Soc.* (London), *Ser. A*, **234**: 1 (1956).

SHIR, C.: A Pilot Study in Numerical Techniques for Predicting Air Pollutant Distribution Downwind from a Line Stack, *Atmos. Environ.*, **4**: 387 (1970).

SHWARTZ, J., and M. P. TULIN: Chimney Plumes in Neutral and Stable Surroundings, *Atmos. Environ.*, **6**: 19 (1972).

SLAWSON, P. R., and G. T. CSANADY: On the Mean Path of Buoyant, Bent-over Chimney Plumes, *J. Fluid Mech.*, **28**: 311 (1967).

THOMAS, F. W., S. B. CARPENTER, and W. C. COLBAUGH: Plume Rise Estimates for Electric Generating Stations, *J. Air Pollut. Control Assoc.*, **20**: 170 (1970).

INDEX